STRUCTURAL DYNAMICS

An Introduction to Computer Methods

STRUCTURAL DYNAMICS

An Introduction to Computer Methods

ROY R. CRAIG, Jr.

*Department of Aerospace Engineering
and Engineering Mechanics
The University of Texas at Austin*

John Wiley & Sons

New York Chichester Brisbane Toronto Singapore

Library of Congress Cataloging in Publication Data:

Craig, Roy R 1934–
Structural dynamics.

Includes index.
1. Structural dynamics—Data processing. 2. Structural dynamics—Mathematical models. I. Title
TA654.C72 624.1′71′02854 80-39798
ISBN 0-471-04499-7

Printed in the United States of America

10 9 8 7 6 5 4

To Jane, Carole, and Karen

PREFACE

The topic of structural dynamics has undergone profound changes over the past two decades. The reason is the availability of digital computers to carry out numerical aspects of structural dynamics problem solving. Recently, the extensive use of the fast Fourier transform has brought about even more extensive changes in structural dynamics analysis, and has begun to make feasible the correlation of analysis with structural dynamics testing. Although this book contains much of the material that characterizes standard textbooks on mechanical vibrations, or structural dynamics, its goal is to present the background needed by an engineer who will be using structural dynamics computer programs or doing structural dynamics testing, or who will be taking advanced courses in finite element analysis or structural dynamics.

Although the applications of structural dynamics in aerospace engineering, civil engineering, engineering mechanics, and mechanical engineering are different, the principles and solution techniques are basically the same. Therefore, this book places emphasis on these principles and solution techniques, and illustrates them with numerous examples and homework exercises from the various engineering disciplines.

Special features of this book include: an emphasis on mathematical modeling of structures and experimental verification of mathematical models; an extensive introduction to numerical techniques for computing natural frequencies and mode shapes and for computing transient response; a systematic introduction to the use of finite elements in structural dynamics analysis; an application of complex frequency response representations for the response of single and multiple degree-of-freedom systems; a thorough exposition of both the mode-displacement and mode-acceleration versions of mode-superposition for computing dynamic response; an introduction to practical methods of component mode synthesis for dynamic analysis; and the introduction of an instructional matrix manipulation and finite element computer code, ISMIS (*I*nteractive *S*tructures and *M*atrix *I*nterpretive *S*ystem), for solving structural dynamics problems.

Although the emphasis of this book is on linear problems in structural dynamics, techniques for solving a limited class of nonlinear structural dynamics problems are also introduced. On the other hand, the topic of random vibrations is not discussed, since a thorough treatment of the subject is definitely beyond the scope of the book, and a cursory introduction would merely dilute the emphasis on numerical techniques for structural dynamics analysis. However, instructors wishing to supplement the text with material on random vibrations will find the information on complex frequency response to be valuable as background for the study of random vibrations.

A primary aim of the book is to give students a thorough introduction to the numerical techniques underlying finite element computer codes. This is done primarily through "hand" solutions and the coding of several subroutines in FORTRAN (or BASIC). Use of the ISMIS computer program extends the problem-solving capability of the student, while avoiding the "black box" nature of production-type finite element codes. Although the ISMIS computer program is employed in Chapters 14 and 17, its use is by no means mandatory. The FORTRAN source code and a complete *User's Manual* for ISMIS are available for a very nominal fee and can be obtained by contacting the author directly at the University of Texas (Austin, TX 78712).

Computer graphics is beginning to play an important role in structural dynamics, for example, in computer simulations of vehicle collisions and animated displays of structural mode shapes. One feature of this book is that all figures that portray functional representations are direct computer-generated plots.

The text of this book has been used for a one-semester senior level course in structural dynamics and a one-semester graduate level course in computational methods in structural dynamics. The undergraduate course typically covers the following material: Chapters 1 through 6, Sections 9.1, 9.2, 10.1, 10.2, 11.1 through 11.4, and Chapter 12. The graduate course reviews the topics above (i.e., it assumes that the students have had a prior course in mechanical vibrations or structural dynamics) and then covers the remaining topics in the book as time permits. Both undergraduate and graduate courses make use of the ISMIS computer program, while the graduate course also includes several FORTRAN coding exercises.

Portions of this text have been used in a self-paced undergraduate course in structural dynamics. This led to the statements of objectives at the beginning of each chapter and to the extensive use of example problems. Thus, the text should be especially valuable to engineers pursuing a study of structural dynamics on a self-study basis.

I express appreciation to my students who used the notes that led to the present text. Special thanks are due to Arne Berg, Mike Himes, and Rick McKenzie, who generated most of the computer plots, and to Butch Miller and Rodney Rocha, who served as proctors for the self-paced classes. Much of the content and "flavor" of the book is a result of my industrial experience at the Boeing Company's Commercial Airplane Division, at Lockheed Palo Alto Research Laboratory, and at NASA Johnson Space Center. I am indebted to the colleagues with whom I worked at these places.

I am grateful to Dr. Pol D. Spanos for reading Chapter 20 and making helpful comments. Dean Richard Gallagher reviewed the manuscript and offered many suggestions for changes, which have been incorporated into the text. This valuable service is greatly appreciated.

This book might never have been completed had it not been for the patience and accuracy of its typist, Mrs. Bettye Lofton, and to her I am most deeply indebted.

Finally, many of the hours spent in the writing of this book were hours that would otherwise have been spent with Jane, Carole, and Karen—my family. My gratitude for their sacrifices cannot be measured.

Roy R. Craig, Jr.

CONTENTS

1 THE SCIENCE AND ART OF STRUCTURAL DYNAMICS 1

1.1 Introduction to Structural Dynamics *1*
1.2 Analysis of the Dynamical Behavior of Structures *2*
1.3 Dynamical Testing of Structures *8*
1.4 Scope of the Text *10*

PART I SINGLE-DEGREE-OF-FREEDOM SYSTEMS **13**

2 MATHEMATICAL MODELS OF SDOF SYSTEMS 15

2.1 Elements of Lumped-Parameter Models *15*
2.2 Application of Newton's Laws to Lumped-Parameter Models *17*
2.3 Application of the Principle of Virtual Displacements to Lumped-Parameter Models *25*
2.4 Application of the Principle of Virtual Displacements to Continuous Models; the Assumed-Modes Method *32*

3 FREE VIBRATION OF SDOF SYSTEMS 49

3.1 Free Vibration of Undamped SDOF Systems *51*
3.2 Free Vibration of Viscous-Damped SDOF Systems *54*
3.3 Experimental Determination of Fundamental Natural Frequency and Damping Factor of a SDOF System *59*
3.4 Free Vibration of a SDOF System with Coulomb Damping *65*

4 RESPONSE OF SDOF SYSTEMS TO HARMONIC EXCITATION 71

4.1 Response of Undamped SDOF Systems to Harmonic Excitation *72*
4.2 Response of Viscous-Damped SDOF Systems to Harmonic Excitation *76*
4.3 Complex Frequency Response *83*
4.4 Vibration Isolation—Force Transmissibility and Base Motion *87*
4.5 Vibration Measuring Instruments *92*

4.6 Use of Frequency Response Data to Determine Natural Frequency and Damping Factor of Lightly Damped SDOF System *95*
4.7 Equivalent Viscous Damping *97*
4.8 Structural Damping *101*

5 RESPONSE OF SDOF SYSTEMS TO SPECIAL FORMS OF EXCITATION 111
5.1 Response of a Viscous-Damped SDOF System to an Ideal Step Input *111*
5.2 Response of an Undamped SDOF System to Rectangular Pulse and Ramp Loadings *113*
5.3 Response of an Undamped SDOF System to a Short-Duration Impulse; Unit Impulse Response *117*

6 RESPONSE OF SDOF SYSTEMS TO GENERAL DYNAMIC EXCITATION 123
6.1 Response of a SDOF System to General Dynamic Excitation—Duhamel Integral Method *123*
6.2 Response Spectra *127*

7 NUMERICAL EVALUATION OF DYNAMIC RESPONSE OF SDOF SYSTEMS 139
7.1 Numerical Solution Based on Interpolation of the Excitation Function *139*
7.2 Numerical Solution Based on Approximating Derivatives; Step-by-Step Numerical Integration *146*
7.3 Nonlinear SDOF Systems *151*
7.4 Step-by-Step Numerical Solution for Response of Nonlinear SDOF Systems *154*

8 RESPONSE OF SDOF SYSTEMS TO PERIODIC EXCITATION; FREQUENCY DOMAIN ANALYSIS 163
8.1 Response to Periodic Excitation—Real Fourier Series *163*
8.2 Response to Periodic Excitation—Complex Fourier Series *169*
8.3 Response to Nonperiodic Excitation—Fourier Integral *175*
8.4 Relationship Between Complex Frequency Response and Unit Impulse Response *179*
8.5 Discrete Fourier Transforms (DFT) and Fast Fourier Transforms (FFT) *180*

PART II CONTINUOUS SYSTEMS **187**

9 MATHEMATICAL MODELS OF CONTINUOUS SYSTEMS 189

9.1 Application of Newton's Laws—Axial Deformation *189*
9.2 Application of Newton's Laws—Transverse Vibration of Linearly Elastic Beams (Bernoulli-Euler Theory) *192*
9.3 Application of Hamilton's Principle *198*
9.4 Application of Hamilton's Principle—Beam Flexure Including Shear Deformation and Rotatory Inertia (Timoshenko Beam Theory) *202*

10 FREE VIBRATION OF CONTINUOUS SYSTEMS 207

10.1 Free Axial Vibration *207*
10.2 Free Transverse Vibration of Bernoulli-Euler Beams *210*
10.3 Rayleigh's Method of Approximating the Fundamental Frequency of a Continuous System *217*
10.4 Free Vibration of Beams Including Shear Deformation and Rotatory Inertia *219*
10.5 Some Properties of Natural Modes *221*
10.6 Vibration of Thin Flat Plates *226*

PART III MULTIPLE-DEGREE-OF-FREEDOM SYSTEMS **235**

11 MATHEMATICAL MODELS OF MDOF SYSTEMS 237

11.1 Application of Newton's Laws to Lumped-Parameter Models *237*
11.2 Lagrange's Equations *243*
11.3 Application of Lagrange's Equations to Lumped-Parameter Models *247*
11.4 Application of Lagrange's Equations to Continuous Models: The Assumed-Modes Method *251*
11.5 Constrained Coordinates and Lagrange Multipliers *261*

12 VIBRATION OF UNDAMPED 2-DOF SYSTEMS 273

12.1 Free Vibration of 2-DOF Systems *273*
12.2 Further Examples of Modes and Frequencies *278*
12.3 Systems with Rigid-Body Modes *283*
12.4 Response of an Undamped 2-DOF System to Harmonic Excitation: Mode-Superposition *286*

13 FREE VIBRATION OF MDOF SYSTEMS 295

13.1 Some Properties of Natural Frequencies and Natural Modes *295*
13.2 Rayleigh Method; Rayleigh-Ritz Method *313*

14 NUMERICAL EVALUATION OF MODES AND FREQUENCIES OF MDOF SYSTEMS 321
14.1 Introduction to Methods for Solving Algebraic Eigenproblems *321*
14.2 Vector Iteration Methods *323*
14.3 Use of ISMIS to Solve for Modes and Frequencies of MDOF Systems *331*

15 DYNAMIC RESPONSE OF MDOF SYSTEMS: MODE-SUPERPOSITION METHOD 341
15.1 Introduction: Principal Coordinates *341*
15.2 Mode-Displacement Solution for Response of Undamped MDOF Systems *344*
15.3 Mode-Acceleration Solution for Response of Undamped MDOF Systems *350*
15.4 Mode-Superposition Solutions for Response of Certain Viscous-Damped Systems *353*
15.5 Dynamic Stresses by Mode-Superposition *366*
15.6 Mode-Superposition for Undamped Systems with Rigid-Body Modes *368*

16 FINITE ELEMENT MODELING OF STRUCTURES 381
16.1 Introduction to the Finite Element Method *381*
16.2 Element Stiffness and Mass Matrices and Element Force Vector *383*
16.3 Transformation of Element Matrices *393*
16.4 Assembly of System Matrices: The "Direct Stiffness" Method *399*
16.5 Boundary Conditions *406*
16.6 Constraints: Reduction of Degrees of Freedom *409*
16.7 Systems with Rigid-Body Modes *413*

17 VIBRATION ANALYSIS EMPLOYING FINITE ELEMENT MODELS 423
17.1 Finite Element Solutions for Natural Frequencies and Modes *423*
17.2 Finite Element Solution for Dynamic Response by the Mode-Displacement Method *433*

18 DIRECT INTEGRATION METHODS FOR DYNAMIC RESPONSE 447
18.1 Damping in MDOF Systems *447*
18.2 Nonlinear MDOF Systems *452*
18.3 Properties of Step-by-Step Numerical Integration Algorithms *455*

19 COMPONENT MODE SYNTHESIS 467
19.1 Introduction to Component Mode Synthesis *467*
19.2 Component Modes for Constrained Components *469*
19.3 System Synthesis for Undamped Free Vibration *470*
19.4 Component Modes for Unconstrained Components *478*
19.5 Residual Flexibility; Residual Component Modes *483*

20 INTRODUCTION TO EARTHQUAKE RESPONSE OF STRUCTURES 497
20.1 Introduction *497*
20.2 Response of a SDOF System to Earthquake Excitation: Response Spectra *498*
20.3 Response of MDOF Systems to Earthquake Excitation *508*
20.4 Further Considerations *513*

APPENDIX A UNITS **517**

AUTHOR INDEX **521**

SUBJECT INDEX **523**

1 THE SCIENCE AND ART OF STRUCTURAL DYNAMICS

This introductory chapter is entitled *The Science and Art of Structural Dynamics* to emphasize at the outset that the mathematical formulas and principles discussed in this book only make it possible for you to "open the door" on the exciting topic of structural dynamics. Structural dynamicists must also master the "art" of mathematical modeling and, in many cases, structural dynamicists must perform dynamic tests. This demands that skill and judgment be exercised in order that useful results may be obtained.

Upon completion of this chapter you should be able to:

- Indicate when dynamics must be considered in a structural analysis.
- Name three phases of a dynamical investigation.
- Write a paragraph on mathematical modeling of structures for dynamic analysis.
- Write a paragraph on dynamical testing of structures.

1.1 Introduction to Structural Dynamics

What do a high-speed ground transportation vehicle moving along its guideway, an offshore drilling platform located in the North Sea, and a jet airplane flying through a thunderstorm have in common? One answer is that all are structures and all are subjected to dynamic loading. The emphasis placed on safety, performance, and reliability of structures such as these has led to the need for extensive analysis and testing to determine their response to dynamic loading.

In the above context, *dynamic* means time-varying. A dynamic load is one whose magnitude, direction, or point of application varies with time; while the resulting time-varying deflections and stresses constitute the dynamic response. If the loading is a known function of time, the loading is said to be *prescribed,* and the analysis of a specified structural system to a prescribed loading is called a *deterministic analysis.* If the time history of the loading is not completely

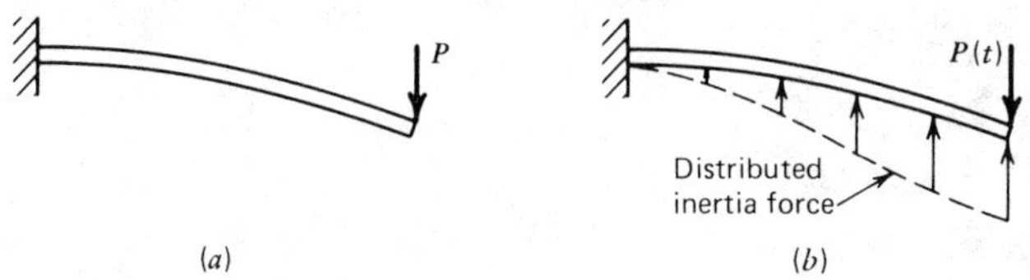

Figure 1.1. A cantilever beam under (a) *static loading,* (b) *dynamic loading.*

known, but is known only in a statistical sense, the loading is said to be *random*. This text treats only prescribed dynamic loading.

A structural dynamics problem differs from the corresponding static problem in two important respects. The first has been noted above, namely, the time-varying nature of the excitation. Of greater importance, however, is the role played by accelerations in a structural dynamics problem. Figure 1.1*a* shows a cantilever beam under static loading. The deflection and internal stresses depend directly on the load P. On the other hand, Fig. 1.1*b* shows a similar cantilever beam subjected to a time-varying load $P(t)$. The acceleration of the beam gives rise to a distributed inertia force. If the inertia force contributes significantly to the deflections and internal stresses in the structure, then a dynamical investigation is required.

Figure 1.2 is a diagram of the typical steps in a dynamical investigation. The three major steps are: *design, analysis,* and *testing*. Frequently only one or two of these steps are required. For example, a civil engineer might be asked to perform a dynamical analysis of an existing dam and confirm the analysis by performing dynamic testing of the dam[1.1]. The results of this analysis and testing might lead to criteria for maximum water depth to ensure safety against failure due to specified earthquake excitation. Automotive engineers perform extensive analysis and testing to determine the dynamical behavior of new car designs[1.2]. Results of this analysis and testing frequently lead to design changes that will improve the ride quality and economy of the vehicle.

This text treats primarily the analysis phase of a dynamical investigation, but techniques of structural dynamics testing will also be discussed in order to emphasize the close relationship between theory and experiment. Also, significant design parameter studies can be carried out using the computer program described later in this text or using other structural dynamics computer programs.

1.2 Analysis of the Dynamical Behavior of Structures

Perhaps the most demanding step in any dynamical analysis is the creation of a mathematical model of the structure. This process is illustrated by steps 2a

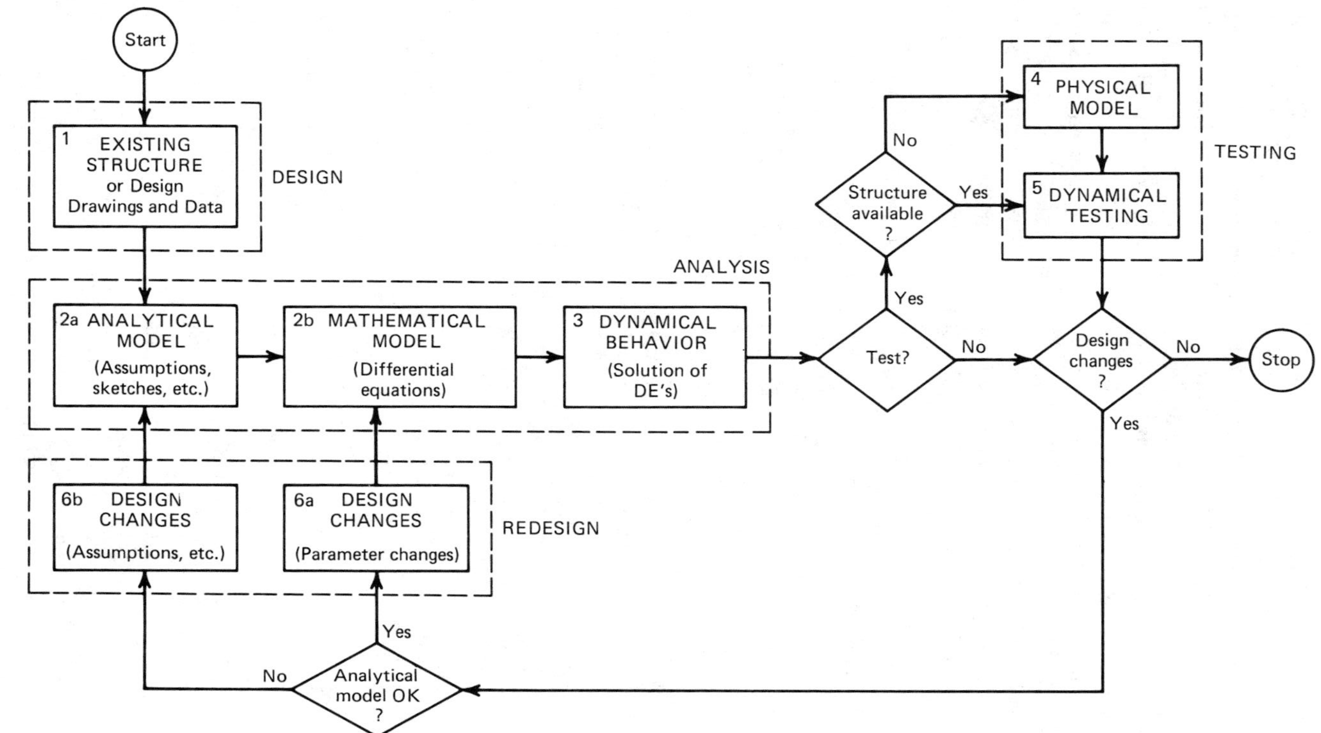

Figure 1.2. Steps in a dynamical investigation.

and 2b of Fig. 1.2. In step 2a you must contrive an idealized model of the structural system to be studied, a model essentially like the real system (which may already exist or may merely be in the planning stages) but easier to analyze mathematically. This *analytical model* consists of: (1) a set of the simplifying assumptions made in reducing the real system to the analytical model, (2) a set of drawings depicting the analytical model, and (3) a list of the design parameters (sizes, materials, etc.).

Analytical models fall into two basic categories: *continuous models* and *discrete-parameter models.* Figure 1.3*a* shows a continuous model of a cantilever beam. The number of displacement quantities which must be considered in order to represent the effects of all significant inertia forces is called the *number of degrees of freedom* (DOF) of the system. Thus, a continuous model

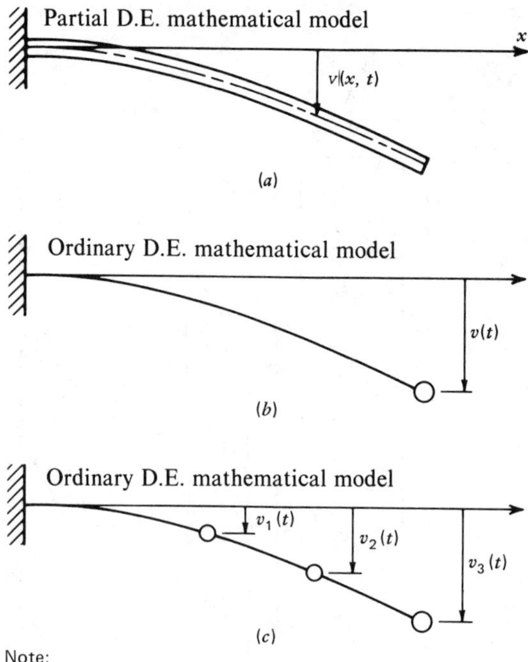

Note:

Throughout this text a special italic typeface will be used to distinguish functions of x and t from functions of t alone, as seen on this figure.

Figure 1.3. *Continuous and discrete-parameter analytical models of a cantilever beam.* (a) *Distributed mass cantilever beam—a continuous model.* (b) *One degree-of-freedom model—a discrete-parameter model.* (c) *Three degree-of-freedom model—a more refined discrete-parameter model.*

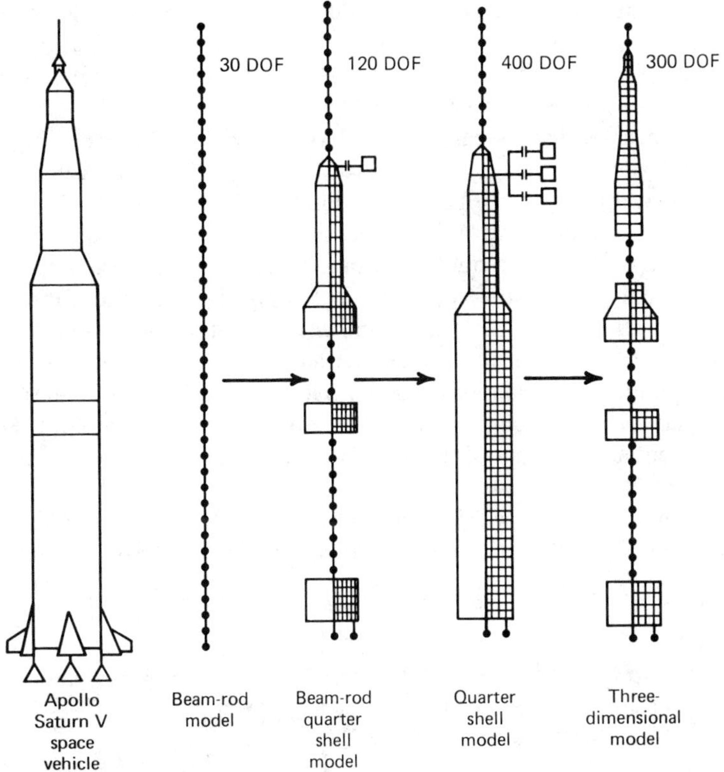

Figure 1.4. *Analytical models of varying complexity used in studying Apollo Saturn V space vehicle dynamics. (C. E. Green et al.,* Dynamic Testing for Shuttle Design Verification, *NASA (1972).)*

represents an infinite DOF system. On the other hand, Figs. 1.3*b* and 1.3*c* depict finite DOF systems. The discrete-parameter models shown here are called *lumped-mass models* because the mass of the system is assumed to be represented by a small number of point masses, or particles. The lumped-mass method and other techniques for creating discrete-parameter models will be discussed further in Chapters 2 and 11.

In order to create a useful analytical model you must have clearly in mind the intended use of the analytical model, that is, the types of behavior of the real system that the model is supposed to represent faithfully. The complexity of the analytical model is determined by the types of behavior it must represent. For example, Fig. 1.4 shows four different analytical models used in studying the dynamical behavior of the Apollo Saturn V space vehicle. The 30-DOF beam-rod model was used for preliminary studies and to determine full-scale testing requirements. The 300-DOF model on the right, on the other hand, was required to give an accurate description of motion at the flight sensor locations. Simplicity of the analytical model is most desirable, so long as the model is adequate to represent the necessary behavior.

Once you have created an analytical model of the structure you wish to study, you can then apply physical laws (e.g., Newton's laws, stress-strain relationship, etc.) to obtain the differential equation(s) of motion which describe, in mathematical language, the analytical model. A continuous model leads to partial differential equations, while a discrete-parameter model leads to ordinary differential equations. The set of differential equations of motion so derived is called the *mathematical model* of the structure. To obtain the mathematical model you will use methods studied in dynamics (e.g., Newton's laws, Lagrange's equation) and in mechanics of deformable bodies (e.g., strain-displacement relations, stress-strain relations), and will combine these to obtain differential equations describing the dynamical behavior of a deformable structure.

In practice you will find that the entire process of creating first an analytical model and then a mathematical model may be referred to simply as *math-*

Figure 1.5. (a) *Bus body frame (D. Radaj et al.,* Finite Element Analysis, an Automobile Engineer's Tool, *Society of Automotive Engineers (1974). Used with permission of the Society of Automotive Engineers, Inc. Copyright © 1974 SAE.)*

ematical modeling. In using a large computer program such as NASTRAN[1.3] or SAP[1.4] to carry out a dynamical analysis, your major modeling task will be to simplify the system and provide input data on dimensions, material properties, loads, and so forth. This is where the "art" of structural dynamics comes into play. On the other hand, actual creation and solution of the differential equations is done by the computer program. Figure 1.5 shows a picture of an actual bus body and a computer-generated plot of the idealized structure, that is, analytical model, which was input to the computer. Computer graphics has become an invaluable tool for use in creating mathematical models of structures.

Once a mathematical model has been formulated, the next step in a dynamical analysis is to solve the differential equation(s) to obtain the dynamical response. The two types of dynamical behavior which are of primary importance in structural applications are *free vibration* and *forced response,* the former being the motion resulting from specified initial conditions, and the latter being the motion resulting from specified inputs to the system from

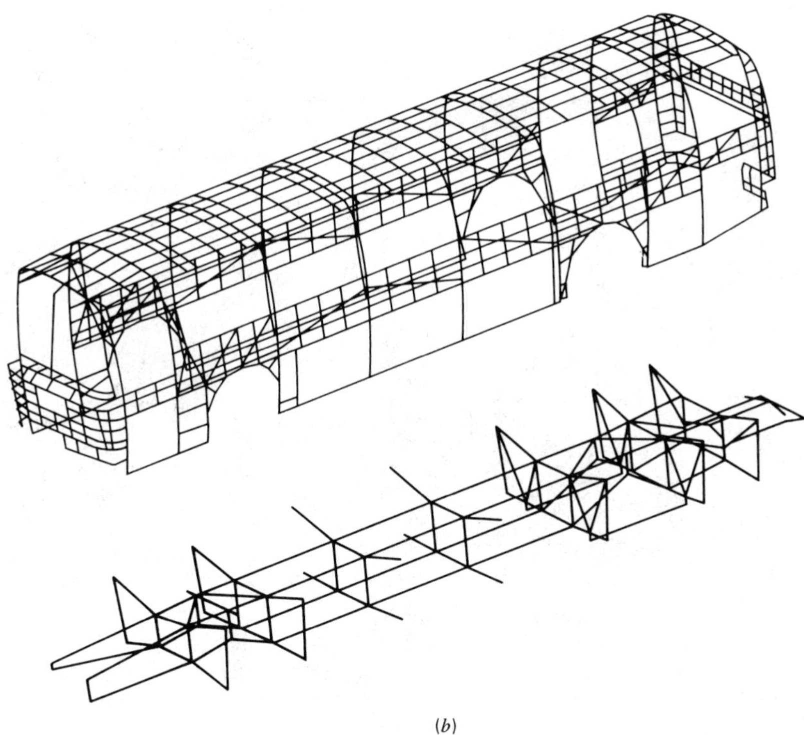

(*b*)

Figure 1.5 (b) *finite element model of bus body/frame.*

external sources. Thus, you solve the differential equations of motion subject to specified initial conditions and inputs from external sources and obtain the resulting time history of the motion of the structure. This constitutes the predicted behavior of the (real) structure, or the *response.*

The *analysis phase* of a dynamical investigation consists of the three steps just described: defining the *analytical model,* deriving the *mathematical model,* and solving for the *dynamical response.* Frequently this is all that is required in a specific dynamical investigation. The majority of this text deals with the second and third steps in the analysis phase of a dynamical investigation.

1.3 Dynamical Testing of Structures

A primary purpose of dynamical testing is to confirm a mathematical model and, in many instances, to obtain important information on loads and other quantities that may be required in the dynamical analysis. In some instances these tests are conducted on reduced-scale physical models, for example, wind

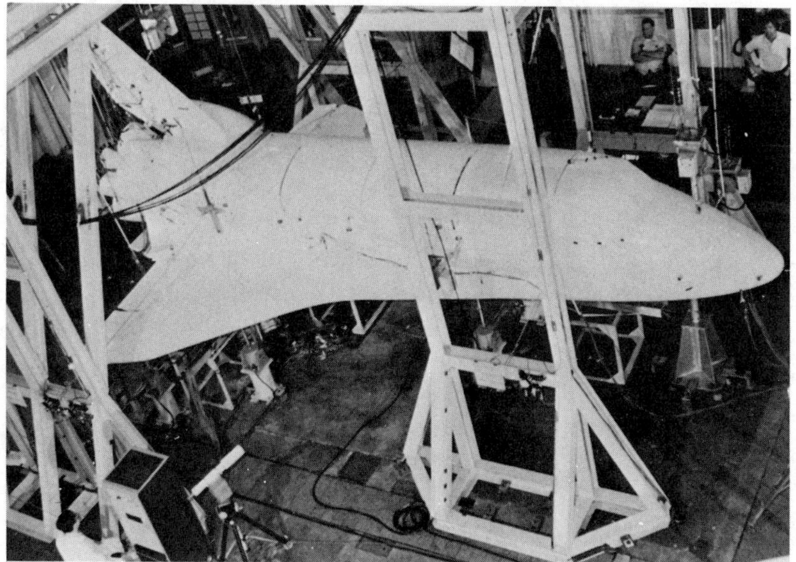

Figure 1.6. *Ground vibration testing of the one-quarter scale model of the space shuttle orbiter vehicle. (D. H. Emero,* The Quarter-Scale Space Shuttle Design, Fabrication and Test, *Rockwell International Corporation (1977). Copyright © 1977, Rockwell International Corp., Quarter Scale Space Shuttle Vibration Test Article.)*

tunnel tests of airplane models. In other cases, when a full-scale structure is available, the tests may be conducted on it.

When a new structure (off-shore platform, airplane, etc.) is being designed, it is frequently necessary early in the design program to confirm the mathematical model experimentally and also to obtain significant data on loads the structure might experience in service. A *physical model* may be constructed for this purpose. Figure 1.6 shows a ground vibration test of a one-quarter-scale model of the Space Shuttle Orbiter vehicle[1.5]. Most scale-model tests are performed on one-tenth-scale models or smaller.

Once the actual structure is available, further dynamical testing may be performed. Aerospace vehicles, for example, must be subjected to extensive static and dynamic testing on the ground prior to actual flight of the vehicle. Figure 1.7 shows a ground vibration test in progress on the Boeing YC-14 airplane. Note the electrodynamic shakers in place under the wing tip and the horizontal tail.

Dynamical testing of physical models may also be employed for determining qualitatively and quantitatively the dynamical behavior characteristics of a class of structures. For example, Fig. 1.8 shows a picture of a cylindrical tank structure containing fluid in place on the shake table[1.6].

Figure 1.7. *Ground vibration test of Boeing YC-14. (H. E. Carr,* YC-14 Ground Vibration Testing, *Boeing Commercial Airplane Co., 1976.)*

Figure 1.8. *Fluid-filled tank subjected to simulated earthquake excitation. (R. W. Clough,* Storage Tank Vibration Test, *R. W. Clough, 1977. Used with permission.)*

1.4 Scope of the Text

The first major section of this text, encompassing Chapters 2 through 8, treats single degree-of-freedom (SDOF) systems. In Chapter 2 procedures are described for developing mathematical models of SDOF systems. Both Newton's laws and the principle of virtual displacements are employed. The free vibration of undamped and viscous-damped systems is the topic of Chapter 3, while in Chapter 4 you will learn about the response of SDOF systems to harmonic (i.e., sinusoidal) excitation. This is, perhaps, the most important topic in the entire book, because it describes the fundamental characteristics of dynamic response. Chapters 5, 6, and 7 continue the discussion of dynamic response of SDOF systems. Chapter 5 treats the response of SDOF systems to particular input time histories, while Chapter 6 describes both the Duhamel integral method for determining response to general prescribed input time histories, and the use of response spectra to present the results of response calculations. In Chapter 7 two approaches to numerical solution of the equation of motion of a SDOF system are presented. Part I is concluded in Chapter 8, where the response of SDOF systems to periodic excitation is first considered, and frequency domain techniques, which are currently widely used in dynamic testing, are introduced as a natural extension of the discussion of periodic excitation.

Part II treats structures that are modeled as continuous systems. While important topics like the determination of partial differential equation mathematical models (Chapter 9) and determination of modes and frequencies (Chapter 10) are treated, the primary purpose of Part II is to provide "exact solutions" that can be used for evaluating the accuracy of approximate multi-degree-of-freedom (MDOF) models as developed in Part III.

The analysis of the dynamical behavior of structures of even moderate complexity requires the use of MDOF models, which are the subject of Part III. Mathematical modeling of MDOF systems is first treated in Chapter 11. While Newton's laws can be used to derive mathematical models of some MDOF systems, the principal tool for deriving such models is the use of Lagrange's equations. Continuous systems are approximated by MDOF models through the use of the assumed-modes method, which is introduced in Sec. 11.4. Chapter 12, which treats the vibration of undamped 2-DOF systems, introduces many of the concepts important to the study of the response of MDOF systems, for example, natural frequencies (eigenvalues), normal modes (eigenvectors), and mode-superposition.

Chapter 13, which discusses many of the mathematical properties related to modes and frequencies and introduces the important Rayleigh-Ritz method for determining approximate natural frequencies of complex systems, is followed in Chapter 14 by a brief introduction to numerical techniques for solving for modes and frequencies.

The most widely-used procedure for determining the response of MDOF systems is mode-superposition. Two versions of mode-superposition, the mode-displacement method and the mode-acceleration method, are described in Chapter 15.

From Chapters 11 through 15 you should have a good understanding of some methods for obtaining MDOF mathematical models, determining their modes and frequencies, and solving for their response to specified dynamical excitation. In Chapter 16, a continuation of Chapter 11, you will be introduced to the important finite element method for creating MDOF mathematical models. Prior study of the finite element method, or of matrix structural analysis, is not a prerequisite for Chapter 16.

In Chapter 17 some example problems are solved using the ISMIS computer program to implement the finite element method. Even readers who do not study the details of the ISMIS programs will want to note the comparison of various finite-DOF models with "exact" continuous models as determined in Chapter 10.

Chapter 18 provides an introduction to numerical techniques for direct integration to obtain the response of MDOF systems including nonlinear systems. In Chapter 19 you will find an introduction to methods of component mode synthesis for modeling very complex structures. Finally, Chapter 20 introduces the topic of earthquake response of structures. While the order of presentation of topics in Part III may appear to be random, that is, modeling (Chapter 11), calculation of modes and frequencies (Chapters 12 through 14), calculation of response (Chapter 15), modeling (Chapter 16), calculation of modes and frequencies (Chapter 17), calculation of response (Chapter 18), modeling (Chapter 19), and calculation of response (Chapter 20), the discussion moves from topics of lesser complexity to those of greater complexity. A reader who merely wishes to attain an introduction to the dynamics of MDOF systems need only study Chapters 11, 12, and 15, where fundamental concepts are presented.

References

1.1 R. M. Stephen and J. G. Bouwkamp, "Dynamic Response of Morrow Point Dam," *Advances in Civil Engineering Through Engineering Mechanics,* 302–305 (1977).

1.2 A. L. Klosterman, "A Combined Experimental and Analytical Procedure for Improving Automotive System Dynamics," *Society of Automotive Engineers,* Paper No. 720093 (1972).

1.3 *NASTRAN Theoretical Manual (Level 16.0),* National Aeronautics and Space Administration, Washington, DC (1976).

1.4 K. J. Bathé, E. L. Wilson, and F. E. Peterson, *SAP-IV—A Structural Analysis Program for Static and Dynamic Response of Linear Systems,* Report EERC 73-11, College of Engineering, University of California, Berkeley (June 1973, rev. Apr. 1974).

1.5 D. H. Emero, "The Quarter-Scale Space Shuttle Design, Fabrication, and Test," *AIAA/ASME/ASCE/AHS 20th Structures, Structural Dynamics, and Materials Conference,* St. Louis, MO, 395–402 (1979).

1.6 D. Clough and R. W. Clough, "Earthquake Simulator Study of Cylindrical Liquid Storage Tanks," *ASCE Fall Convention,* San Francisco, CA (1979).

PART I
SINGLE DEGREE-OF-FREEDOM SYSTEMS

2 MATHEMATICAL MODELS OF SDOF SYSTEMS

In Chapter 1 you were introduced to the concept of mathematical modeling. This chapter deals with single degree-of-freedom (SDOF) models, that is, models which are described by a single, second-order ordinary differential equation. Both Newton's laws and the principle of virtual displacements will be employed to derive the equation of motion of lumped-parameter systems. Then, the principle of virtual displacements will be extended to approximate a continuous system by a SDOF generalized-parameter model.

Upon completion of Chapter 2, you should be able to:

- Use Newton's laws to derive the equation of motion of a single particle or rigid body having 1 DOF.
- Use the principle of virtual displacements, in conjuction with d'Alembert forces, to derive the equation of motion of a particle, rigid body, or assemblage of bodies having 1 DOF.
- Use the principle of virtual displacements, together with an assumed mode, to derive the equation of motion which represents a generalized-parameter SDOF model of a structure.

2.1 Elements of Lumped-Parameter Models

The components that constitute a lumped-parameter model of a structure are those that relate force to displacement, velocity, and acceleration, respectively.

The component that relates force to displacement is usually called a *spring*. Figure 2.1 shows an idealized massless spring and a plot of spring force versus elongation. The spring force always acts along the line joining the two ends of the spring. For most structural materials, for small values of the elongation, $e = u_2 - u_1$, there is a linear relationship between force and elongation (contraction). This relationship is given by

$$f_S = ke \tag{2.1}$$

where k is called the *spring constant*. The units of k are pounds per inch (lb/in.) or Newtons per meter (N/m).*

*The units of various quantities will be given in the fundamental English and SI units. Problems, however, will also be solved using related units, for example, kip (= 1000 lb), foot, kiloNewton, millimeter, and so forth.

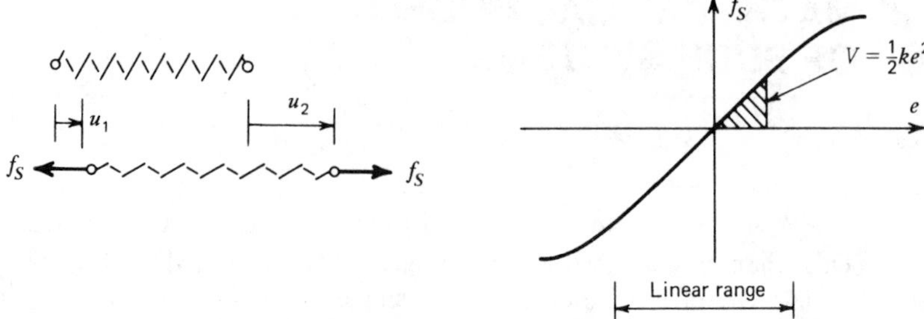

Figure 2.1. Force-deformation behavior of a spring.

Within its elastic range a spring serves as an energy storage device. The energy stored in a spring when it is elongated (contracted) an amount e within its linearly-elastic range is called *strain energy* and is given by

$$V = \tfrac{1}{2}(ke^2) \tag{2.2}$$

The strain energy is indicated as area under the f_S versus e curve, Fig. 2.1.

Just as a spring serves as an energy storage device, there are also means by which energy is dissipated from a deforming structure. These are called *damping mechanisms.* Although research studies have proposed numerous ways of mathematically describing damping[2.1], the exact nature of damping in a structure is usually impossible to determine. The most common analytical model of damping employed in structural dynamics analyses is the *linear viscous dashpot* model, which is illustrated in Fig. 2.2. The damping force f_D is given by

$$f_D = c(\dot{u}_2 - \dot{u}_1) \tag{2.3}$$

and is thus a linear function of the relative velocity between the two ends of the dashpot. The constant c is called the *coefficient of viscous damping* and its units are pounds per inch per second (lb-sec/in.) or Newtons per meter per second (N·s/m).

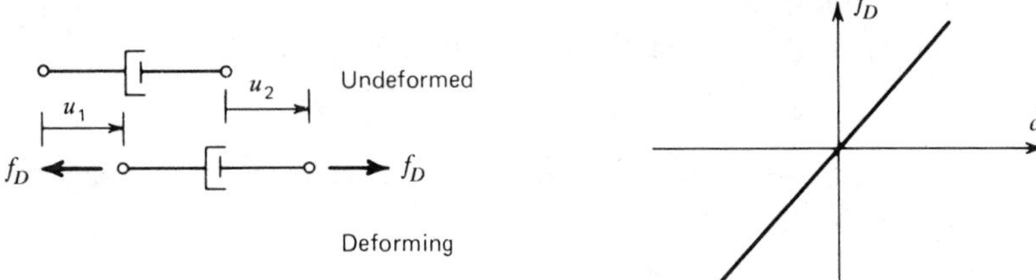

Figure 2.2. A linear viscous dashpot.

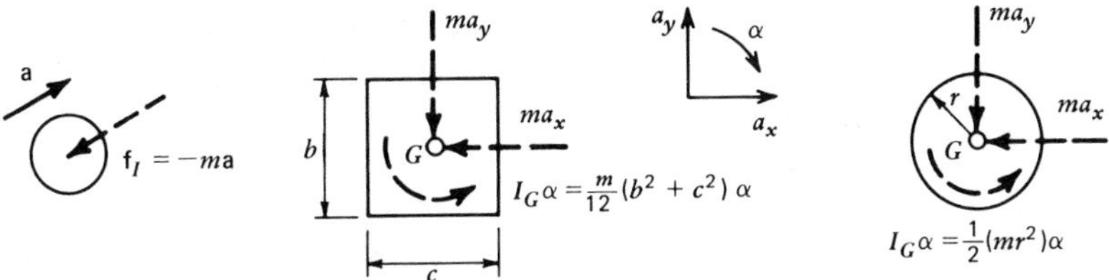

Figure 2.3. *Inertia forces.*

In writing the equation of motion of a particle, *Newton's second law*

$$\sum \mathbf{F} = m\mathbf{a} \tag{2.4}$$

is used, where m is the mass of the particle and $\mathbf{a}$ is its acceleration relative to an inertial reference frame.* The units of mass are (lb-sec²/in.) or (N·s²/m).

For structural dynamics problems it is frequently useful to introduce the d'Alembert force, or *inertia force*

$$\mathbf{f}_I = -m\mathbf{a} \tag{2.5}$$

Then Eq. 2.4 can be written as an equation of dynamic equilibrium

$$\sum \mathbf{F}' \equiv \mathbf{f}_I + \sum \mathbf{F} = 0 \tag{2.6}$$

with the resultant inertia force added to the resultant of the other forces acting on the particle. Figure 2.3 shows the inertia forces for a particle and for two rigid bodies in general plane motion.

The kinetic energy of particles and rigid bodies will be needed in Chapter 11. For the particle and rigid bodies of Fig. 2.3, the kinetic energy is given by

$$\begin{aligned} T &= \tfrac{1}{2} mv^2 \qquad \text{(point mass)} \\ T &= \tfrac{1}{2} m(v_{Gx}^2 + v_{Gy}^2) + \tfrac{1}{2} I_G\omega^2 \qquad \text{(rigid body)} \end{aligned} \tag{2.7}$$

where v is the linear velocity and ω the angular velocity of the respective bodies.

2.2 Application of Newton's Laws to Lumped-Parameter Models

In this section, the equation of motion of several lumped-parameter models will be derived by using Newton's laws or, equivalently, the d'Alembert force

*An inertial reference frame is an imaginary set of rectangular axes assumed to have no translation or rotation in space.

method. This will serve as a review of your previous studies of dynamics and will also introduce you to procedures employed in determining the mathematical model of a SDOF system.

Example 2.1

Use Newton's laws to derive the equation of motion of the simple spring-mass-dashpot system below. Assume only vertical motion, and assume that the spring is linear with spring constant k. Neglect air resistance, the mass of the spring, and any internal damping in the spring. $p(t)$ is the force applied to the mass by an external source.

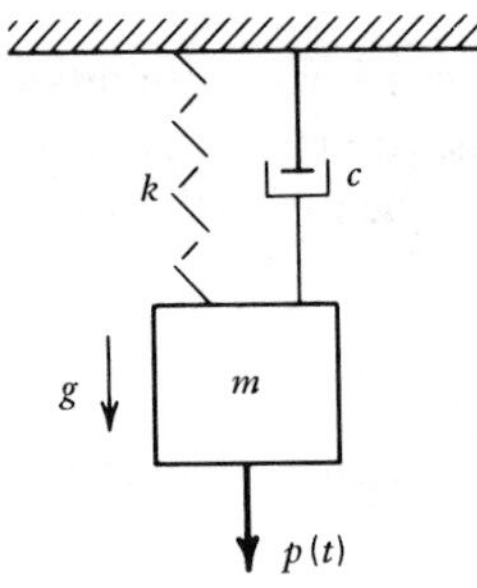

Solution

a. Establish a reference frame and a displacement coordinate:

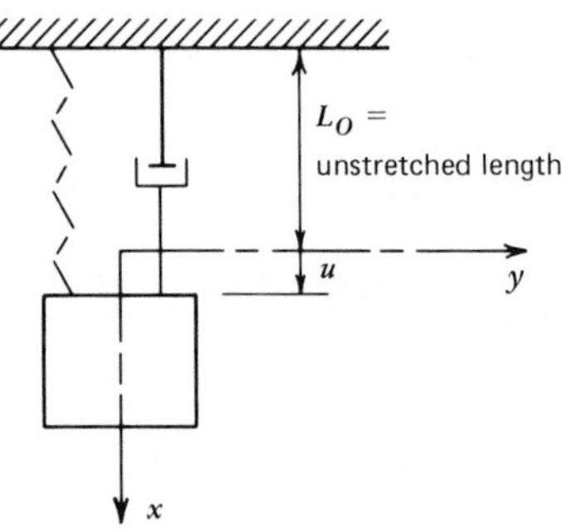

Choose the x-axis along the line of motion and choose the origin (i.e., $x = 0$) at the location where the spring is unstretched. Let u be the displacement in the x-direction.

b. Draw a freebody diagram of the particle in an arbitrary displaced position:

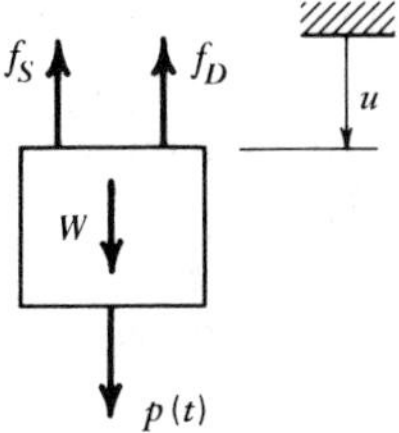

c. Apply Newton's second law:

$$+\downarrow \sum F_x = m\ddot{u} \tag{1}$$

(*Note:* + sense downward since u is positive downward.)
From the freebody diagram, determine the forces for the left-hand side of Eq. 1.

$$p - f_S - f_D + W = m\ddot{u} \tag{2}$$

d. Relate forces to motion variables:

$$f_S = ke = ku \tag{3}$$

$$f_D = c\dot{e} = c\dot{u} \tag{4}$$

($u = 0$ corresponds to the unstretched spring.)

e. Combine and simplify, arranging unknowns on the left-hand side of the equation:

$$\boxed{m\ddot{u} + c\dot{u} + ku = W + p(t)} \tag{5}$$

Note that this is a second-order, linear, nonhomogeneous ordinary differential equation with constant coefficients. This equation can be simplified by the following considerations.

The static displacement of the weight W on the linear spring is given by

$$u_{st} = \frac{W}{k} \tag{6}$$

Let the displacement of the mass measured relative to the static equilibrium position be u_r. That is,

$$u = u_r + u_{st} \tag{7}$$

Since u_{st} is a constant, Eq. 5 can be rewritten as

$$\boxed{m\ddot{u}_r + c\dot{u}_r + ku_r = p(t)} \tag{8}$$

Thus, the weight force can be eliminated if the displacement of the system is measured relative to the static equilibrium configuration.

Equation 8 in Example 2.1 may be considered to be the *fundamental equation in structural dynamics* and linear vibration theory. You will be spending a great deal of time in determining its solution and in applying it to structural dynamics problems both for SDOF systems and MDOF systems.

In Example 2.1, Newton's second law was used directly, so no inertia forces were shown on the freebody diagram. Example 2.2 demonstrates the use of inertia forces and also illustrates the important feature of *support excitation,* or *base motion,* such as a building structure would experience during an earthquake.

Example 2.2

Use the d'Alembert force method to determine the equation of motion of the mass m. Assume that the damping forces in the system can be represented by the linear viscous dashpot shown, and assume that the support excitation, $z(t)$, is known. When $u = z = 0$, the spring is unstretched.

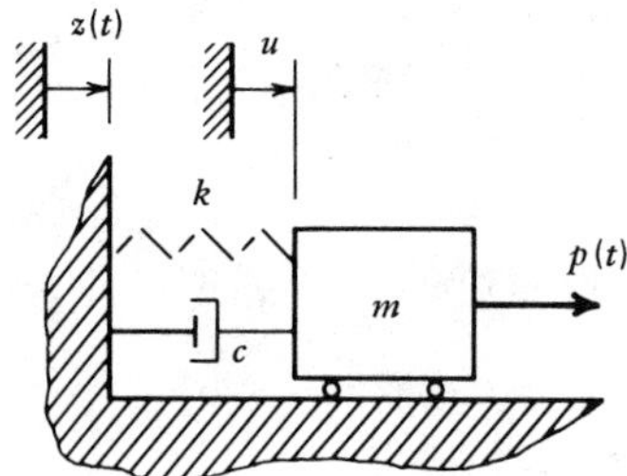

Solution

a. Draw a freebody diagram of the mass including the inertia force along with the real forces:

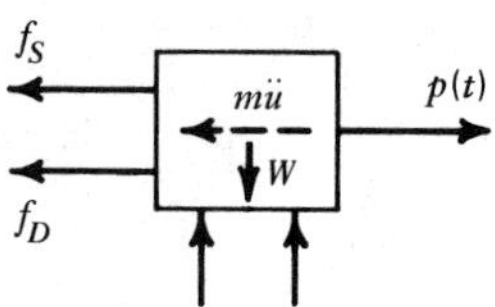

b. Write the dynamic equilibrium equation:

$$\overset{+}{\rightarrow} \sum F'_x = 0 \tag{1}$$

From the freebody diagram

$$p - f_S - f_D - m\ddot{u} = 0 \tag{2}$$

c. Relate forces to motion variables and simplify:

$$m\ddot{u} + c(\dot{u} - \dot{z}) + k(u - z) = p \tag{3}$$

Note that the damping force and spring force are related to the motion of the mass relative to the moving support. Equation 3 can be written with all known quantities on the right-hand side. Thus,

$$\boxed{m\ddot{u} + c\dot{u} + ku = c\dot{z} + kz + p} \tag{4}$$

Equation 4 is the equation of motion for the actual displacement of the mass relative to an inertial reference frame, that is, for $u(t)$. It is frequently desirable to formulate the equation of motion in terms of the relative displacement

$$w = u - z \tag{5}$$

since the forces applied to the mass are directly related to this relative displacement. By subtracting $m\ddot{z}$ from both sides of Eq. 3 and using Eq. 5, you can obtain the following equation

$$\boxed{m\ddot{w} + c\dot{w} + kw = p - m\ddot{z}} \tag{6}$$

Thus, base motion has the effect of adding a reversed inertia force $p_{\text{eff}}(t) = -m\ddot{z}$ to the other applied forces. Equation 6 is much more useful than Eq. 4, since relative motion is usually more important than absolute motion and also since the base acceleration $\ddot{z}$ is much easier to measure than the base velocity and displacement entering into Eq. 4.

Newton's laws for a particle may be extended to give the following equations for a rigid body in plane motion. The force equation is

$$\sum \mathbf{F} = m\mathbf{a}_G \tag{2.8}$$

where $\mathbf{a}_G$ is the acceleration vector of the mass center of the body.

One of the following moment equations is also needed.

$$\sum M_G = I_G\alpha \tag{2.9a}$$

where moments are summed about the mass center, α is the angular acceleration* of the body, and I_G is the mass moment of inertia about G; or

$$\sum M_O = I_O\alpha \tag{2.9b}$$

*α is used later as a phase angle. However, this duplication should not cause confusion.

where O is a fixed axis about which the body is rotating; or

$$\sum M_C = I_C\alpha \tag{2.9c}$$

where C is a point whose acceleration vector passes through G. Equation 2.9a is a general equation and should be used in most cases. Equation 2.9b is valid only for fixed-axis rotation, and Eq. 2.9c is of limited applicability (e.g., a uniform disk rolling without slipping).

Example 2.3

An airplane engine and the pylon that attaches it to the wing are modeled, for a particular study of lateral motion, as a rigid body attached to a rigid, weightless beam, which is elastically restrained as shown below. Derive the equation of motion for small displacements. Neglect damping. The rotational spring shown exerts a restoring moment on the beam which is proportional to the angle the beam makes with the vertical. The engine has a mass moment of inertia I_G about an axis through its mass center, which is located as shown.

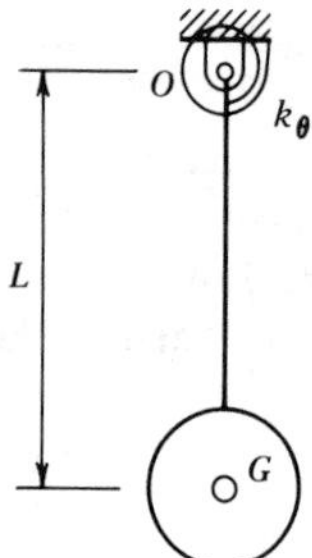

Solution

a. Draw a freebody diagram of the body in an arbitrary displaced position and identify the displacement coordinate:

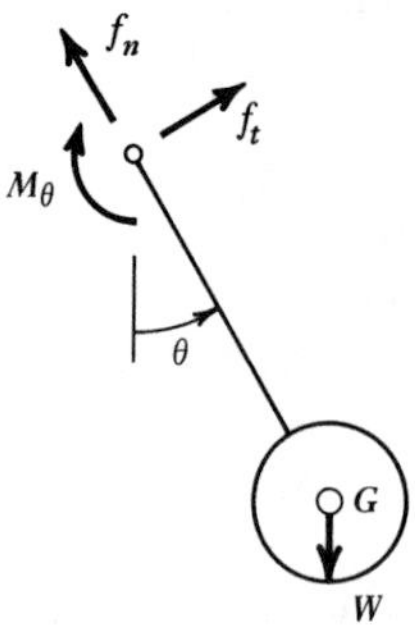

b. Write the equation of motion:

The force equation, Eq. 2.8, will not be needed for this problem unless the pin forces f_n and f_t are desired. Since this is a fixed-axis rotation problem, Eq. 2.9b will be used for the required moment equation.

$$\circlearrowleft + \sum M_O = I_O \ddot{\theta}$$

(+ sense in direction of $+\theta$.)

From the freebody diagram

$$-M_\theta - WL \sin \theta = I_O \ddot{\theta} \tag{1}$$

c. Write the force-displacement equation:

$$M_\theta = k_\theta \theta \tag{2}$$

d. Combine and simplify the equations:

From the parallel-axis theorem,

$$I_O = I_G + mL^2 \tag{3}$$

From Eqs. 1, 2, and 3 the final equation of motion is obtained.

$$\boxed{(I_G + mL^2)\ddot{\theta} + k_\theta \theta + WL \sin \theta = 0} \tag{4}$$

As a final example of using Newton's laws directly, or using inertia forces in a dynamic equilibrium equation, to derive the equation of motion of a SDOF system, consider Example 2.4.

Example 2.4

A heavy uniform disk is mounted at the end of a shaft. Neglecting the weight of the shaft and neglecting damping, derive the equation of motion for torsional oscillation of the disk. The shaft is made of material with a shear modulus of elasticity G.

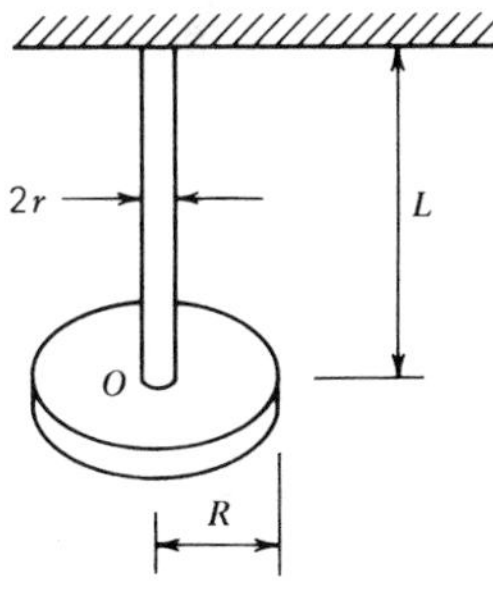

Solution

a. Draw a freebody diagram of the disk in an arbitrary displaced position and identify the displacement coordinate:

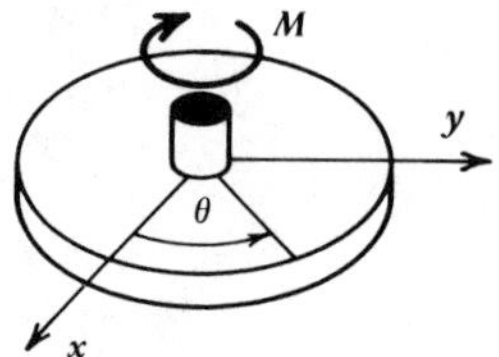

b. Write the equation of motion:
From Eq. 2.9b,

$$\circlearrowleft + \sum M_O = I_O \alpha \tag{1}$$

From the freebody diagram

$$-M = I_O \alpha \tag{2}$$

c. Write the "force-displacement" equation:
From strength of materials, the relationship between the twisting torque, M, and the angle of rotation, θ, at the end of a shaft is

$$M = \left(\frac{GJ}{L}\right)\theta \tag{3}$$

where

$$J = \frac{\pi r^4}{2} \tag{4}$$

d. Combine and simplify:

$$I_O \ddot{\theta} + \left(\frac{GJ}{L}\right)\theta = 0 \tag{5}$$

or

$$\boxed{\frac{1}{2}(mR^2)\ddot{\theta} + \frac{1}{2}\left(\frac{\pi r^4 G}{L}\right)\theta = 0} \tag{6}$$

Notice that this has the form

$$m_\theta \ddot{\theta} + k_\theta \theta = 0$$

which is similar to that obtained in previous examples.

2.3 Application of the Principle of Virtual Displacements to Lumped-Parameter Models

In the previous section you briefly reviewed elementary dynamics and saw that Newton's laws could be used directly or with the introduction of inertia forces to obtain the equation of motion of a SDOF system. In this section you will be introduced to the principle of virtual displacements. Although it can be applied to problems like those in Sec. 2.2, it is particularly powerful when applied to connected rigid bodies or when used to approximate continuous bodies by finite-DOF models (Sec. 2.4). Several definitions are needed at the outset.

- *A displacement coordinate* is a quantity used in specifying the change of configuration of a system.
- *A constraint* is a kinematical restriction on the possible configurations a system may assume.
- *A virtual displacement* is an infinitesimal imaginary change of configuration of a system consistent with its constraints.

Figure 2.4 shows a rigid rod of length L with a mass m attached at its tip. u and v can be used as displacement coordinates. However, the mass m is constrained to move in a circle by the *equation of constraint*

$$u^2 + v^2 = L^2 \tag{2.10}$$

u and v are thus not independent. In fact, they can both be related to the single displacement coordinate θ by

$$u = L \cos \theta, \qquad v = L \sin \theta \tag{2.11}$$

A small change in configuration is also shown in Fig. 2.4. $\delta\theta$ represents a virtual displacement of the system. δu and δv can be obtained graphically or by using Eq. 2.11. For example,

$$\begin{aligned} v + \delta v &= L \sin(\theta + \delta\theta) \\ &= L(\sin \theta \cos \delta\theta + \sin \delta\theta \cos \theta) \end{aligned} \tag{2.12}$$

Since $\delta\theta$ is infinitesimal $\cos \delta\theta \doteq 1$ and $\sin \delta\theta \doteq \delta\theta$. Thus,

$$v + \delta v = L \sin \theta + L \cos \theta \, \delta\theta$$

or

$$\delta v = (L \cos \theta) \, \delta\theta \tag{2.13a}$$

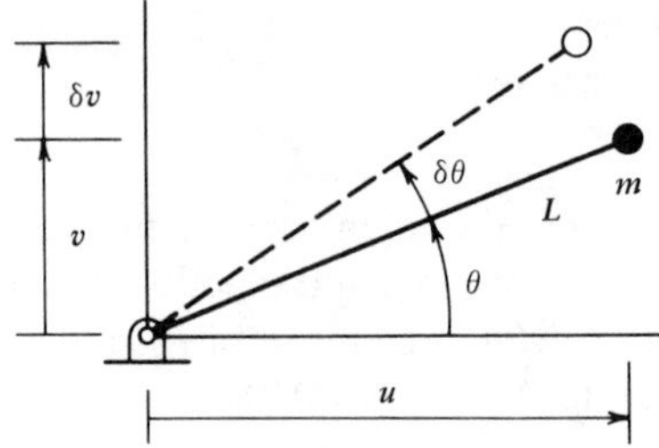

Figure 2.4. *Constraint; virtual displacement.*

Similarly,

$$\delta u = -(L \sin \theta)\, \delta\theta \tag{2.13b}$$

Note that δu and δv behave like differentials.

Continuing with the definitions:

- A *set of generalized coordinates* is a set of linearly independent displacement coordinates which are consistent with the constraints and which are sufficient to describe an arbitrary configuration of the system. For example, θ is a generalized coordinate for the SDOF system in Fig. 2.4. The symbols q_i $(i = 1, 2, \ldots, N)$ are frequently employed as the labels for generalized coordinates of an N-DOF system.
- The *virtual work,* δW, is the work of the forces acting on a system as it undergoes a virtual displacement. The virtual work can be written

$$\delta W = \sum_{i=1}^{N} Q_i\, \delta q_i \tag{2.14}$$

- The *generalized force,* Q_i, is the quantity which multiplies δq_i in forming the virtual work due to δq_i. That is, Q_i is the virtual work done when $\delta q_i = 1$ and $\delta q_j = 0$ for $j \neq i$. The concept of a generalized force is very important in structural dynamics, and you should become very familiar with this way of identifying generalized forces by noting the virtual work they do.

Example 2.5

Determine the virtual work done by the distributed force acting on a rigid beam as shown. Assume that the force remains perpendicular to the beam.

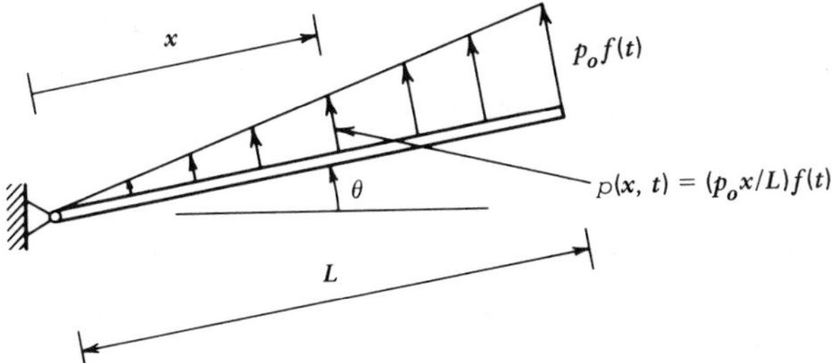

Solution

From the basic definition of work (force times distance moved)

$$\delta W = \int_0^L p(x, t)\, \delta v(x, t)\, dx \tag{1}$$

where δv is the virtual displacement of the bar due to a virtual angular displacement $\delta\theta$, that is,

$$\delta v = x\, \delta\theta \tag{2}$$

Then,

$$\delta W = \int_0^L \left(\frac{p_o x f}{L}\right)(x\, \delta\theta)\, dx \tag{3}$$

or

$$\boxed{\delta W = \left(\frac{p_o L f}{2}\right)\left(\frac{2L\, \delta\theta}{3}\right) = \left(\frac{p_o f L^2}{3}\right)\delta\theta} \tag{4}$$

Two things should be noted about this answer: (1) the result is the same as would be obtained by placing the statically equivalent load $(p_o f L/2)$ at its centroid $(2L/3)$, and (2) the generalized force Q_θ is given by $Q_\theta = p_o f L^2/3$.

• The *principle of virtual displacements* as applied to dynamics problems can now be stated as follows: For any arbitrary virtual displacement of a system, the combined virtual work of real forces and inertia forces must vanish. That is,

$$\delta W' \equiv \delta W_{\substack{\text{real}\\ \text{forces}}} + \delta W_{\substack{\text{inertia}\\ \text{forces}}} = 0 \tag{2.15}$$

Consider the following examples. Example 2.6 could easily be treated by using Newton's laws as in Sec. 2.2. However, Example 2.7 shows you the advantage of the virtual work formulation:

Example 2.6

Use the principle of virtual displacements to derive the equation of motion of the idealized system shown below. Neglect gravity and assume small rotation of the beam.

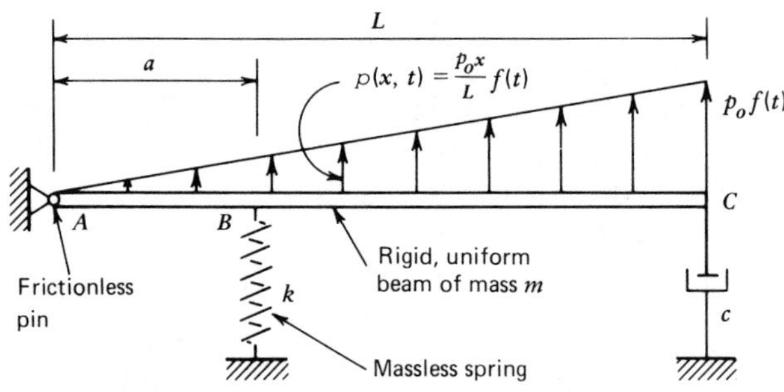

Solution

a. Sketch the body in an arbitrary displaced configuration and also with an additional virtual displacement. Write the necessary kinematical equations:

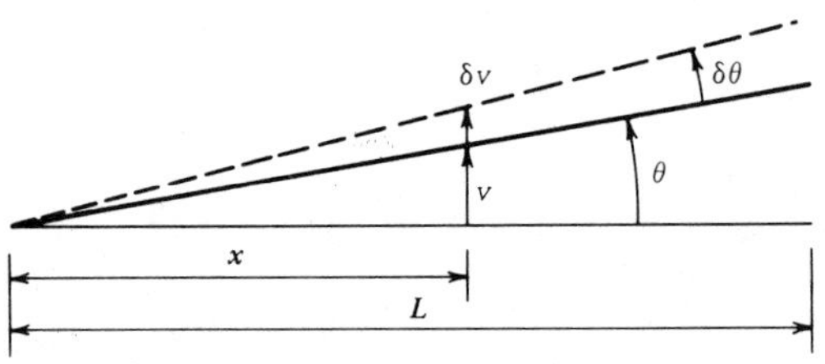

$$v(x, t) = x \tan \theta(t) \tag{1}$$

For small θ,

$$v(x, t) = x\theta(t) \tag{2a}$$
$$\delta v(x, t) = x\, \delta\theta \tag{2b}$$

b. On a sketch of the displaced body show all the forces that can do work, including inertia forces.

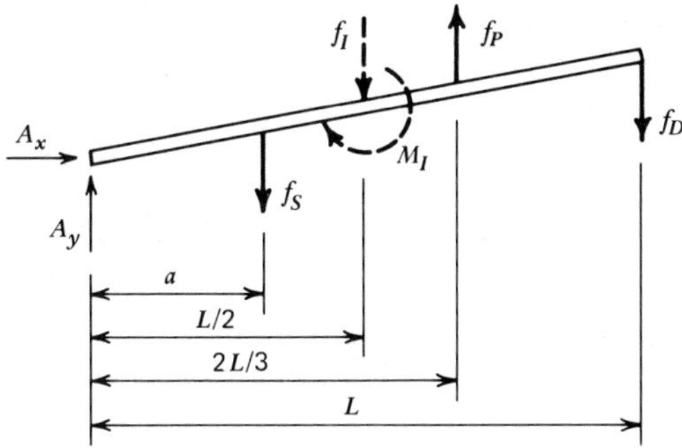

c. Write the virtual work equation:

$$\delta W' = 0 \tag{3}$$

From Eq. 2b and the above figures,

$$\begin{aligned}\delta W' = -f_S(a\,\delta\theta) - f_I\left(\frac{L\,\delta\theta}{2}\right) \\ - M_I(\delta\theta) + f_P\left(\frac{2L\,\delta\theta}{3}\right) - f_D(L\,\delta\theta) = 0\end{aligned} \tag{4}$$

d. Relate forces to motion variables.

$$\begin{aligned} f_S &= ka\theta \\ f_I &= \left(\frac{mL}{2}\right)\ddot{\theta} \\ M_I &= \left(\frac{mL^2}{12}\right)\ddot{\theta} \\ f_P &= \left(\frac{p_0 L}{2}\right)f(t) \\ f_D &= cL\dot{\theta} \end{aligned} \tag{5}$$

e. Combine Eqs. 4 and 5 and simplify:

$$\left[\left(\frac{mL^2}{3}\right)\ddot{\theta} + (cL^2)\dot{\theta} + (ka^2)\theta - \frac{p_0 L^2 f}{3}\right]\delta\theta = 0 \tag{6}$$

Since $\delta\theta \neq 0$, the equation of motion is

$$\boxed{\left(\frac{mL^2}{3}\right)\ddot{\theta} + (cL^2)\dot{\theta} + (ka^2)\theta = \left(\frac{p_0 L^2}{3}\right)f(t)} \tag{7}$$

Notice in Example 2.6 that the solution procedure for applying the principle of virtual displacements is slightly different than the solution procedure for applying Newton's laws. Since the key equation is the virtual work equation, $\delta W' = 0$, the needed quantities are (1) forces that do work, including inertia forces, and (2) kinematics of deformation, or displacement, including virtual displacements. This is the reason that a sketch showing forces that do work and a sketch showing deformation replace the freebody diagram, which is so essential when Newton's laws are being used directly.

Of course, in Example 2.6 it would have been easier just to draw a freebody diagram and use Newton's laws. When there are connected rigid bodies, however, several freebody diagrams and many equations would be required in a solution using Newton's laws. The virtual work approach greatly simplifies the solution, as you will see in Example 2.7.

Example 2.7

An instrument package of mass M is attached to the wall of a moving vehicle by two identical thin rigid beams of mass m as shown. Use the principle of virtual displacements to derive the equation of motion of the instrument package. Neglect gravity and damping forces and assume small angular motions of the support beams.

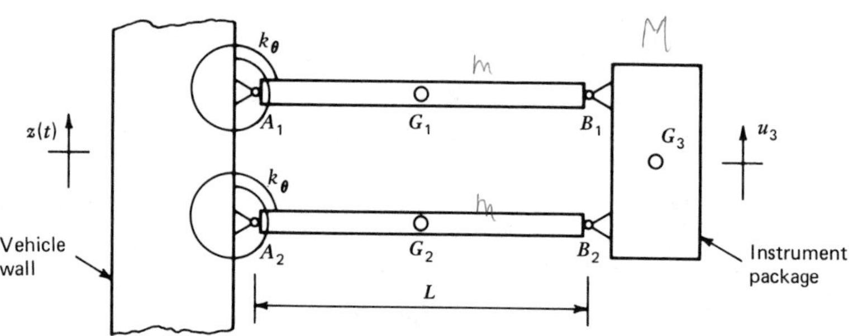

Solution

a. Sketch the system in an arbitrary displaced configuration and also with an additional virtual displacement. Write the necessary kinematical equations:

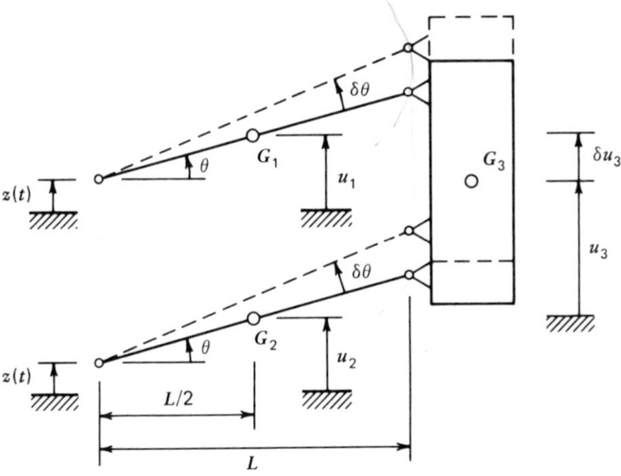

For small θ and small $\delta\theta$,

$$u_1 = u_2 = z + \left(\frac{L}{2}\right)\theta, \qquad \delta u_1 = \delta u_2 = \left(\frac{L}{2}\right)\delta\theta$$
$$u_3 = z + L\theta, \qquad \delta u_3 = L\delta\theta \tag{1}$$

(Notice that since z is a specified function of time, $z(t)$, no virtual displacement δz appears above. Also notice that the motion of G_1, G_2, and G_3 can all be expressed in terms of the known displacement z and the single unknown displacement θ. Therefore, this is a SDOF system even though it involves motion of several rigid bodies.)

b. On a sketch of the displaced body show all of the forces that do work, including all inertia forces:

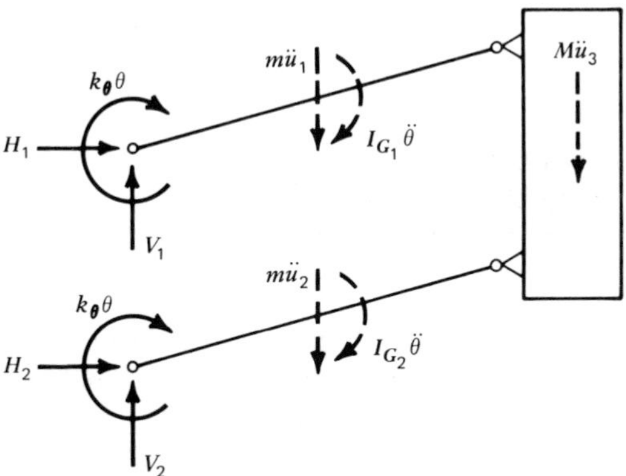

c. Use the virtual work equation, Eq. 2.15:

$$\delta W' = 0 \tag{2}$$

$$-(2k_\theta\theta)\,\delta\theta - m\ddot{u}_1\,\delta u_1 - I_{G1}\ddot{\theta}\,\delta\theta - m\ddot{u}_2\,\delta u_2 - I_{G2}\ddot{\theta}\,\delta\theta - M\ddot{u}_3\,\delta u_3 = 0 \tag{3}$$

d. Combine Eqs. 1 and 3 and simplify:

$$2k_\theta\theta\,\delta\theta + 2m\left[\ddot{z} + \left(\frac{L}{2}\right)\ddot{\theta}\right]\left[\left(\frac{L}{2}\right)\delta\theta\right] + 2\left[\left(\frac{1}{12}\right)mL^2\right]\ddot{\theta}\,\delta\theta + M(\ddot{z} + L\ddot{\theta})(L\,\delta\theta) = 0 \tag{4}$$

or

$$\left[\left(M + \frac{2m}{3}\right)L^2\ddot{\theta} + 2k_\theta\theta + (m + M)L\,\ddot{z}\right]\delta\theta = 0 \tag{5}$$

Finally, since $\delta\theta \neq 0$, Eq. 5 requires that

$$\boxed{\left(M + \frac{2m}{3}\right)L^2\ddot{\theta} + 2k_\theta\theta = -(m + M)L\ddot{z}} \tag{6}$$

This is the equation of motion of the system.

From Examples 2.6 and 2.7 you can see that applying the principle of virtual displacements to a SDOF system will lead to a single virtual work equation of the form

$$[F(\ddot{q}, \dot{q}, q, t)]\,\delta q = 0 \tag{2.16}$$

Since δq is an arbitrary small displacement which is not necessarily zero, the expression in brackets must be zero. This, then, is the differential equation of motion for the system, as you have seen in Examples 2.6 and 2.7.

2.4 Application of the Principle of Virtual Displacements to Continuous Models; The Assumed-Modes Method

In Sec. 2.3 you observed that using the principle of virtual displacements greatly simplifies the task of deriving the equation of motion of connected rigid bodies. But the idea of *rigid body,* in itself, is merely an idealization of a system. For example, if the so-called "rigid" beam in Example 2.6 were to be

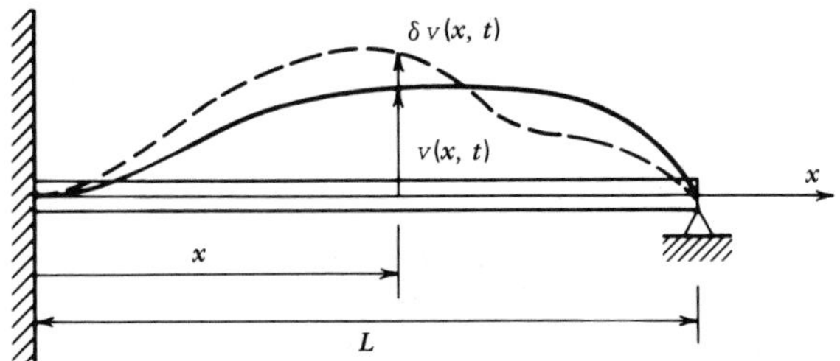

Figure 2.5. An illustration of geometric boundary conditions and virtual displacements.

excited by certain time-dependent functions, $f(t)$, the beam would not remain rigid at all but would deform substantially, that is, its neutral axis would assume a time-varying curved shape. Hence, to assume that it is a rigid body would not be a suitable analytical model at all. The cantilever beam of Fig. 1.3*a* would not be free to deform at all if we had to model it as a rigid body.

Fortunately, the principle of virtual displacements can be extended to produce a *generalized parameter model* of a continuous system in a manner that approximates the flexible behavior of the system. The procedure is referred to as the *assumed-modes method.* It will be employed here to create a SDOF generalized parameter model, but will be extended to MDOF systems in Chapter 11.

A few definitions are needed at the outset.

- A *continuous system* is a system whose deformation is described by one or more functions of one, two, or three spatial variables and time. For example, the deformation of the cantilever beam in Fig. 1.3*a* is specified in terms of the deflection curve, $v(x, t)$ of the neutral axis.
- A *geometric boundary condition* is a specified kinematical constraint placed on displacement and/or slope on portions of the boundary of a body.
- A *virtual displacement* of a continuous system is an infinitesimal, imaginary change in the displacement function(s) consistent with all geometric boundary conditions.

Figure 2.5 illustrates the above definitions. The deformation of this propped cantilever beam is given by $v(x, t)$. The geometric boundary conditions are

$$v(0, t) = v'(0, t) = v(L, t) = 0 \tag{2.17}$$

The dotted curve shows a possible virtual displacement, $\delta v(x, t)$, of the beam. The only condition on $\delta v(x, t)$ is that it satisfy the same geometric boundary conditions as $v(x, t)$, that is,

$$\delta v(0, t) = \delta v'(0, t) = \delta v(L, t) = 0 \tag{2.18}$$

Note that $\delta v(x, t)$ is not a function of time in the same sense as $v(x, t)$. Rather, the notation $\delta v(x, t)$ just means an arbitrary small change of configuration relative to the configuration of the beam at time t.

Continuing now with the definitions:

- An *admissible function* is a function that satisfies the geometric boundary conditions of the system under consideration and possesses derivatives of an order at least equal to that appearing in the strain energy expression for the system (e.g., Eq. 2.23).
- An *assumed-mode,* or *shape function,* is an admissible function that is used to approximate the deformation of a continuous system.

To create a generalized-parameter SDOF model of a continuous system, a single assumed-mode is used. For example, the deflection curve of a beam may be approximated by

$$v(x, t) = \psi(x)v(t) \tag{2.19}$$

Any admissible function may be employed as $\psi(x)$, but a shape that can be expected to be similar to the shape of the deforming structure should be chosen. $v(t)$ in Eq. 2.19 is called a *generalized displacement* for this SDOF model. It will be determined as the solution to an ordinary differential equation, just as in the case of lumped-parameter models.

The principle of virtual displacements may now be employed to create a SDOF generalized-parameter model based on Eq. 2.19. Equation 2.15 is used, as before. For continuous systems, however, it is convenient to introduce potential energy, that is, the work done by conservative forces. Let

$$\delta W_{\substack{\text{real}\\ \text{forces}}} = \delta W_{\text{cons}} + \delta W_{\text{nc}} \tag{2.20}$$

where δW_{cons} is the virtual work of conservative forces and δW_{nc} is the virtual work of nonconservative forces. From the definition of potential energy

$$\delta W_{\text{cons}} = -\delta V \tag{2.21}$$

where δV is the change in potential energy as the conservative forces acting on the system move through a virtual change of configuration, for example, through $\delta v(x, t)$ for a beam. Then, Eq. 2.15 becomes

$$\delta W' = \delta W_{\text{nc}} - \delta V + \delta W_{\text{inertia}} = 0 \tag{2.22}$$

Expressions for strain energy in a bar undergoing axial deformation $u(x, t)$ and a beam undergoing transverse deflection $v(x, t)$ are

$$V_{\text{axial}} = \frac{1}{2}\int_0^L AE(u')^2\,dx \tag{2.23a}$$

and

$$V_{\text{bending}} = \frac{1}{2}\int_0^L EI(v'')^2\,dx \tag{2.23b}$$

Consequently, δV is given by

$$\delta V_{\text{axial}} = \int_0^L (AEu')\,\delta u'\,dx \tag{2.24a}$$

and

$$\delta V_{\text{bending}} = \int_0^L (EIv'')\,\delta v''\,dx \tag{2.24b}$$

respectively.

Example 2.8

An axial force $P(t)$ is applied to the end of a uniform, linearly elastic bar. Choose a simple function for $\psi(x)$, and use the assumed-modes method to derive the equation of motion of the resulting SDOF model of the bar.

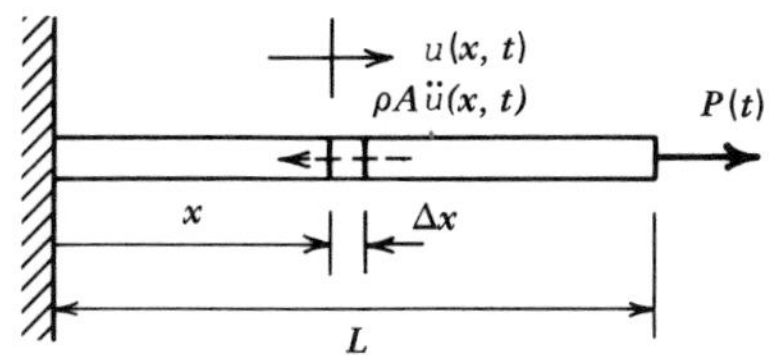

Solution

a. Choose a shape function $\psi(x)$ consistent with the geometric boundary condition:

The only geometric boundary condition here is

$$u(0, t) = 0 \tag{1}$$

Thus, the shape function $\psi(x)$ must satisfy

$$\psi(0) = 0 \tag{2}$$

The simplest function satisfying this condition is $\psi(x) = x$. However, it is convenient to normalize $\psi(x)$ so that it has a value of one at some convenient place, say $x = L$. Therefore, choose

$$\boxed{\psi(x) = \frac{x}{L}} \tag{3}$$

Thus,

$$u(x, t) = \left(\frac{x}{L}\right) u(t) \tag{4}$$

where $u(t)$, the generalized displacement, is actually the time-dependent tip displacement.

b. Set up the virtual work equation, Eq. 2.22:

$$\delta W' = \delta W_{nc} - \delta V + \delta W_{inertia} = 0 \tag{5}$$

$$\delta W_{nc} = P\,\delta u(L, t) = P\,\delta u \tag{6}$$

From Eq. (2.24a)

$$\begin{aligned} \delta V &= \int_0^L (AEu')\,\delta u'\,dx \\ &= AE \int_0^L \left(\frac{u}{L}\right)\left(\frac{\delta u}{L}\right) dx \\ &= \left(\frac{AEu}{L}\right) \delta u \end{aligned} \tag{7}$$

The inertia force per unit length $(-\rho A\ddot{u})$ is distributed along the rod. Hence, the virtual work of inertia forces is

$$\begin{aligned} \delta W_{inertia} &= \int_0^L (-\rho A\ddot{u})\,\delta u\,dx \\ &= -\left(\frac{\rho A\ddot{u}\,\delta u}{L^2}\right) \int_0^L x^2\,dx \\ &= -\left(\frac{\rho AL}{3}\right) \ddot{u}\,\delta u \end{aligned} \tag{8}$$

c. Combine and simplify:

$$\left[P - \left(\frac{AE}{L}\right) u - \left(\frac{\rho AL}{3}\right) \ddot{u}\right] \delta u = 0 \tag{9}$$

Equation 9 has the form of Eq. 2.16. The condition that δu be arbitrary leads to the following differential equation for the generalized-parameter SDOF model.

$$\boxed{\left(\frac{\rho AL}{3}\right) \ddot{u} + \left(\frac{AE}{L}\right) u = P(t)} \tag{10}$$

Note that this equation has the familiar form

$$m\ddot{u} + ku = p(t)$$

seen in previous lumped-parameter examples.

It is very important that you remember that Eq. 10 is based on Eq. 4, which is only an approximation to the real time-dependent motion $u(x, t)$ which the bar would actually experience as a result of the applied force $P(t)$.

The example above shows how a particular choice for the shape function $\psi(x)$ leads to an ordinary differential equation representing a SDOF model of a continuous system. Next, a very important general example of the assumed-modes method will be presented. You should study this example carefully.

Example 2.9

For the system shown below determine expressions for m, c, k, k_G, and p in the following generalized-parameter equation of motion

$$m\ddot{v} + c\dot{v} + (k - k_G)v = p(t)$$

Use the general assumed-modes form

$$v(x, t) = \psi(x)v(t)$$

with $\psi(L) = 1$. The compressive axial force, N, is constant and remains horizontal. Neglect gravitational forces.

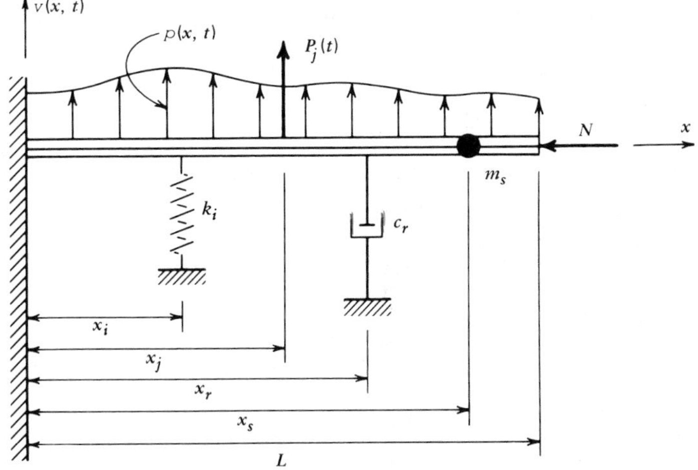

Solution

a. Use the virtual work equation, Eq. 2.22:

$$\delta W' = \delta W_{nc} - \delta V + \delta W_{\text{inertia}} = 0 \tag{1}$$

b. From the inertia forces determine m:

$$\delta W_{\text{inertia}} = \int_0^L (-\rho A \ddot{v})\, \delta v\, dx - m_s \ddot{v}(x_s, t)\, \delta v(x_s, t) \tag{2}$$

Substituting $\ddot{v}$ and $\delta \bar{v}$ into the above equation, we obtain

$$\delta W_{\text{inertia}} = -\left[\int_0^L \rho A \psi^2(x)\, dx + m_s \psi^2(x_s) \right] \ddot{v}\, \delta v \tag{3}$$

In anticipation of the final equation to be derived, let

$$\delta W_{\text{inertia}} = -m \ddot{v}\, \delta v \tag{4}$$

where

$$\boxed{m = \int_0^L \rho A \psi^2\, dx + m_s \psi_s^2} \tag{5}$$

is the *generalized mass*.

c. From the viscous dashpot determine c:

$$\delta W_{\text{damping}} = -c_r \dot{v}(x_r, t)\, \delta v(x_r, t) \tag{6}$$

Let

$$\delta W_{\text{damping}} = -c \dot{v}\, \delta v \tag{7}$$

where

$$\boxed{c = c_r \psi_r^2} \tag{8}$$

is the *generalized viscous damping coefficient*.

d. From the linear spring and the elastic beam determine k:

From Eq. 2.2

$$\delta V_{\text{spring}} = k_i e_i\, \delta e_i = k_i v(x_i, t)\, \delta v(x_i, t) \tag{9}$$

So,

$$\delta V_{\text{spring}} = k_i \psi_i^2 v\, \delta v \tag{10}$$

From Eq. 2.24b

$$\delta V_{\text{bending}} = \int_0^L EI v''\, \delta v''\, dx \tag{11}$$

Thus,

$$\delta V_{\text{bending}} = \left[\int_0^L EI(\psi'')^2 \, dx \right] v \, \delta v \tag{12}$$

Let

$$\delta V_{\text{bending}} + \delta V_{\text{spring}} = kv \, \delta v \tag{13}$$

Then, the *generalized stiffness coefficient* is

$$\boxed{k = \int_0^L EI(\psi'')^2 \, dx + k_i \psi_i^2} \tag{14}$$

e. From the compressive axial force, N, determine k_G:

Due to transverse deflection, the tip of the beam where N is applied will move in the axial direction as well as in the transverse direction. This causes N to do work.

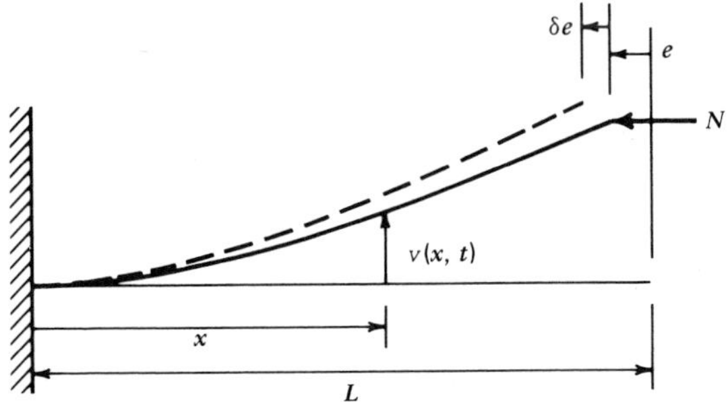

To determine the shortening effect due to transverse deflection, consider an element dx. Assume that the element AB of length dx remains dx in length but

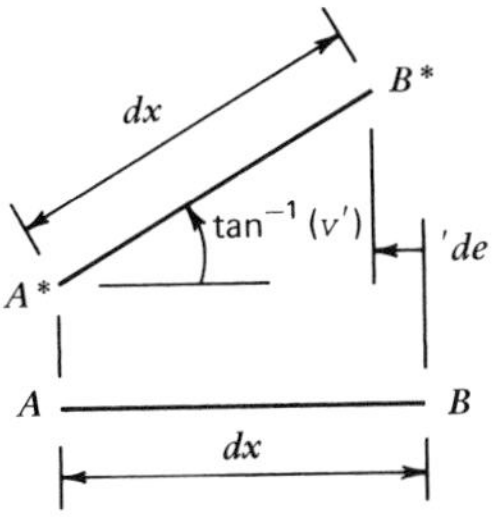

rotates to the position A^*B^* due to the transverse deflection, $v(x, t)$. From the above figure

$$\overline{AB} = \overline{A^*B^*} \cos[\tan^{-1}(v')] + de \tag{15}$$

For $v' \ll 1$,

$$\cos[\tan^{-1}(v')] \doteq \cos(v') \doteq 1 - \tfrac{1}{2}(v')^2 \tag{16}$$

Therefore, from Eqs. 15 and 16,

$$dx = dx[1 - \tfrac{1}{2}(v')^2] + de$$

or

$$de = \tfrac{1}{2}(v')^2\, dx \tag{17}$$

Hence, the axial shortening due to transverse deflection is

$$e = \frac{1}{2}\int_0^L (v')^2\, dx \tag{18}$$

$$\delta W_N = N\,\delta e = N\int_0^L v'\,\delta v'\, dx \tag{19}$$

Substituting the assumed mode into this, we get

$$\delta W_N = k_G v\,\delta v \tag{20}$$

where

$$\boxed{k_G = N\int_0^L (\psi')^2\, dx} \tag{21}$$

is the *generalized geometric stiffness coefficient.*

f. Finally, from the applied transverse force determine $p(t)$:

$$\delta W_p = \int_0^L p(x, t)\,\delta v(x, t)\, dx + P_j\,\delta v(x_j, t) \tag{22}$$

Substituting δv into the above, we get

$$\delta W_p = p(t)\,\delta v \tag{23}$$

where the *generalized external force* is given by

$$\boxed{p(t) = \int_0^L p(x, t)\psi(x)\, dx + P_j(t)\psi_j} \tag{24}$$

g. Combine the above expressions for δV and δW with Eq. 1, and simplify:

$$[-c\dot{v} + p(t) + k_G v - kv - m\ddot{v}]\,\delta v = 0 \tag{25}$$

Since $\delta v \neq 0$, we can write the equation of motion of this system as

$$\boxed{m\ddot{v} + c\dot{v} + (k - k_G)v = p(t)} \tag{26}$$

as required, where the coefficients are defined above.

Notice that, with the exception of the k_G term, the differential equation determined above by the assumed-modes method is identical in form to Eq. 8 of Example 2.1. This latter equation was derived for a simple lumped-parameter spring-mass-dashpot system. The assumed-modes method leads to a *generalized-parameter model* whose displacement function is the *generalized (displacement) coordinate, $v(t)$*. Correspondingly, m is called the generalized mass coefficient, or simply *generalized mass, c* the *generalized viscous damping, k* the *generalized stiffness, k_G* the *generalized geometric stiffness,* and $p(t)$ the *generalized force.* In more general form these would be given by the following expressions*:

$$m = \int_0^L \rho A \psi^2 \, dx + \sum_s m_s \psi_s^2 \tag{2.25}$$

$$c = \int_0^L c(x) \psi^2 \, dx + \sum_r c_r \psi_r^2 \tag{2.26}$$

$$k = \int_0^L EI(\psi'')^2 \, dx + \int_0^L k(x) \psi^2 \, dx + \sum_i k_i \psi_i^2 \tag{2.27}$$

$$k_G = \int_0^L N(x)(\psi')^2 \, dx \tag{2.28}$$

$$\text{and } p(t) = \int_0^L p(x, t) \psi \, dx + \sum_j P_j \psi_j \tag{2.29}$$

where Σ indicates that several discrete elements or forces may be present, where distributed damping $c(x)$ and distributed elastic foundation $k(x)$ are permitted, and where the internal axial load $N(x)$ may vary with position.

Reference

2.1 C. W. Bert, "Material Damping: An Introductory Review of Mathematical Models, Measures and Experimental Techniques," *J. Sound and Vibration,* v. 29, 129–153 (1973).

Problem Set 2.2†

Use Newton's laws to determine the equations of motion of the SDOF systems in problems 2.1 through 2.6. Show necessary freebody diagrams.

*The same symbols m, c, k, p have been used here for generalized parameters as were previously introduced for lumped-parameter models, since the symbols serve the same function in the final differential equation of motion.

†Problem headings refer to the text section to which the problem set pertains.

2.1 A frame structure consists of a rigid horizontal member BC of mass m supported by two vertical members each of whose bending stiffness is EI. Neglect the effect of the weight of BC on the stiffness of the vertical members. Determine the equation of motion.

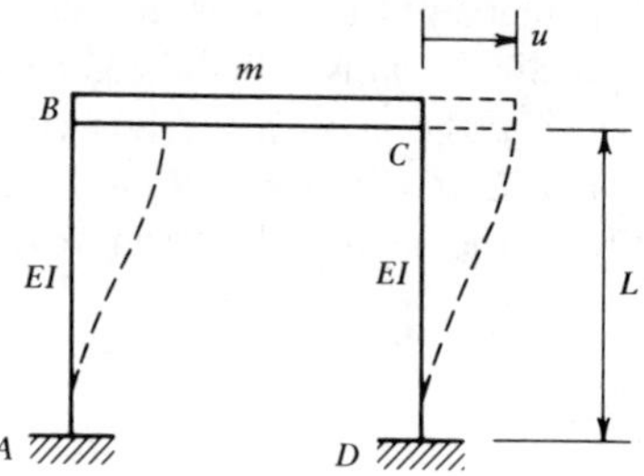

Figure P2.1

2.2 A heavy disk of radius R and mass m is attached to the middle of a shaft of length L, which is rigidly clamped at its ends. The radius of the shaft is r. Determine the equation of motion of the disk. Let θ be the displacement coordinate.

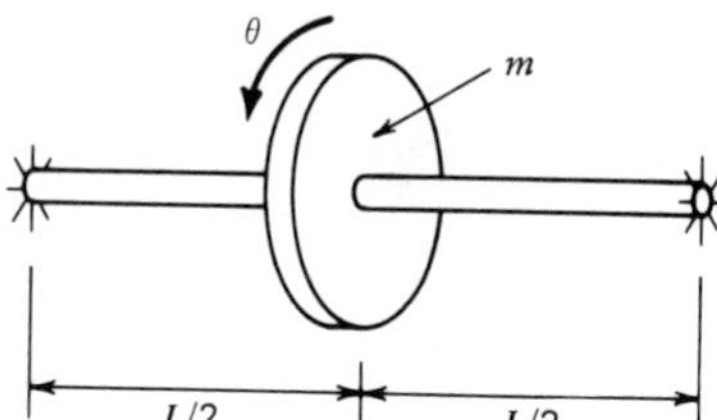

Figure P2.2

2.3 A rigid beam AB is to be excited by the force in spring BC, where the motion of C is specified to be $z(t)$. Determine the equation of motion of the system in terms of the vertical motion u at B. Assume small motions.

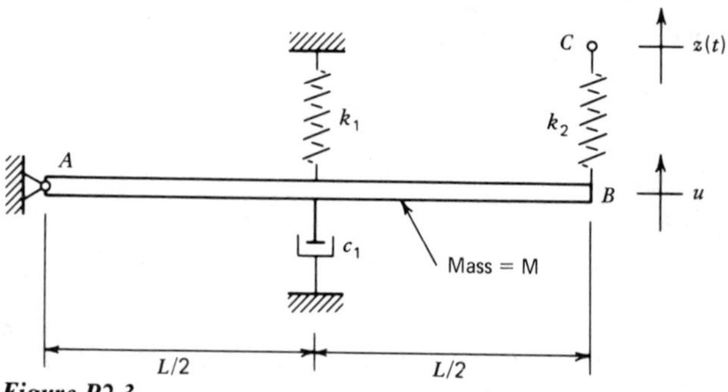

Figure P2.3

2.4 An automobile is modeled as a lumped mass m with input from a spring and dashpot as shown. Assume that the automobile travels at constant speed V over a road whose roughness is known as a function of position along the roadway.

(a) Determine the equation of motion for the system.

(b) Discuss some of the possible limitations of this model.

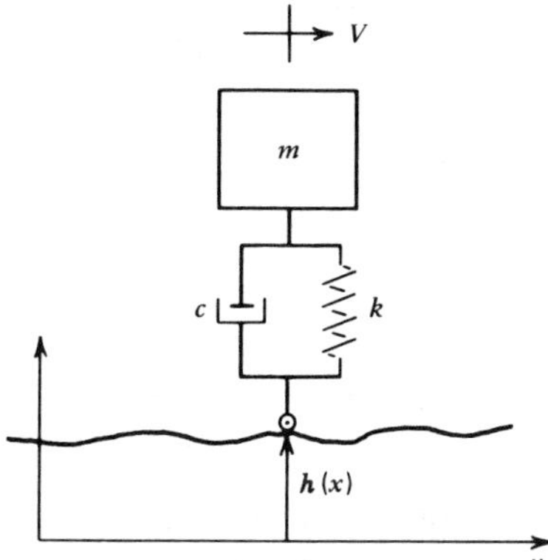

Figure P2.4

2.5 For the system shown below, determine the equation of motion in the form

$$m_u\ddot{u} + c_u\dot{u} + k_u u = p_u(t)$$

where u is the vertical motion of point E. Assume small rotations of the thin, rigid bar AE whose mass is M.

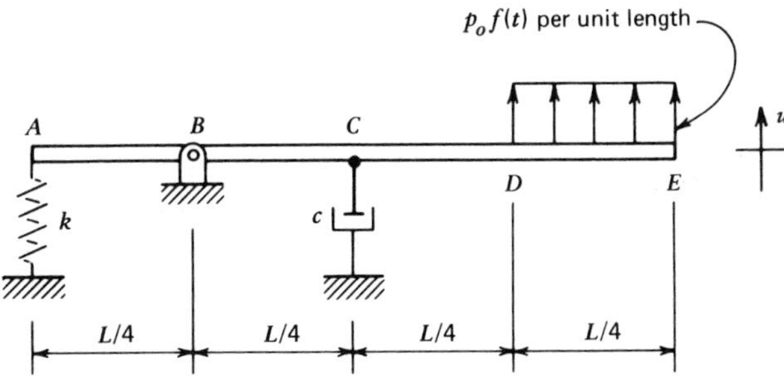

Figure P2.5

2.6 The inertia properties of a control tab of an airplane elevator may be determined by weighing the control tab, determining its center of gravity, and performing a free vibration test from which the moment of inertia I_A can be determined. The figure below shows the setup for the vibration test. Using

Newton's laws, determine the equation of motion for small rotations about axis A.

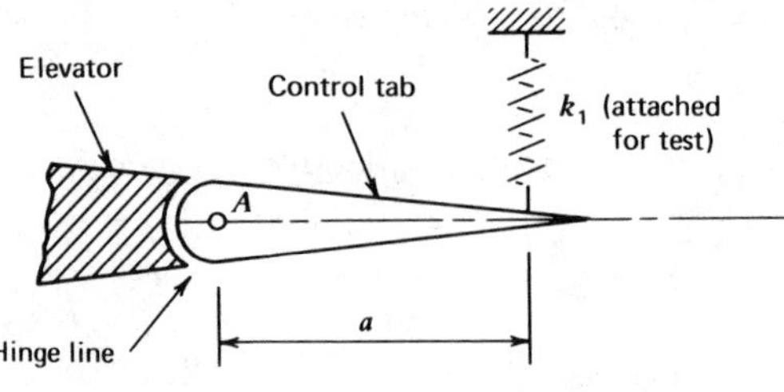

Figure P2.6

Problem Set 2.3

2.7 Use the principle of virtual displacements to derive the equation of motion of the system in Fig. P2.3.

2.8 Use the principle of virtual displacements to derive the equation of motion of the system in Fig. P2.5. Use u as the displacement coordinate.

2.9 Use the principle of virtual displacements to derive the equation of motion of the system shown below. Let the angular displacement of the bar be the displacement coordinate. Assume small angles of rotation. The triangularly distributed load varies with time, but N remains constant and horizontal.

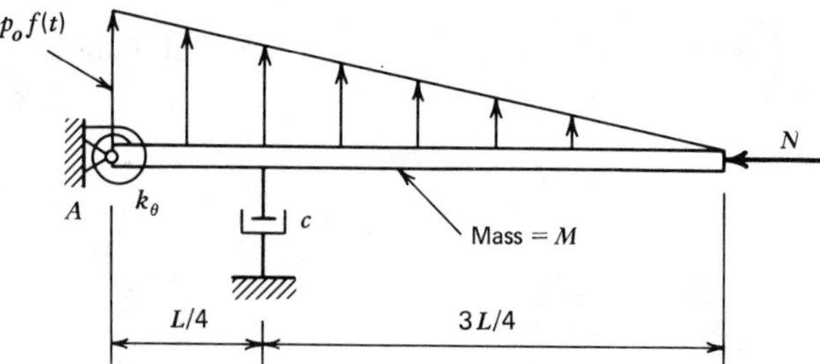

Figure P2.9

2.10 An airplane landing gear system is modeled as a lumped mass m attached to an airplane M by a spring and dashpot. The tire forces to the mass are modeled as resulting from the motion of the bottom of the spring k_2. Assume a constant speed V over a sinusoidally rough runway. Use the principle of virtual displacements to determine the equation of motion of the mass. Assume horizontal motion of the "airplane" mass M.

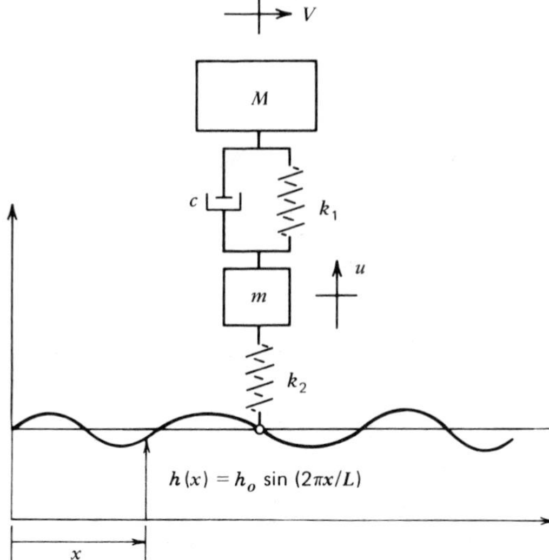

Figure P2.10

2.11 Inertia properties of masses are frequently measured by performing a pendulum test. In the figure below a uniform bar of weight W is suspended by two identical weightless vertical wires. The bar performs small rotational oscillations about the vertical axis. Determine the equation of motion by using the method of virtual displacements. Let the angle of rotation be the displacement coordinate.

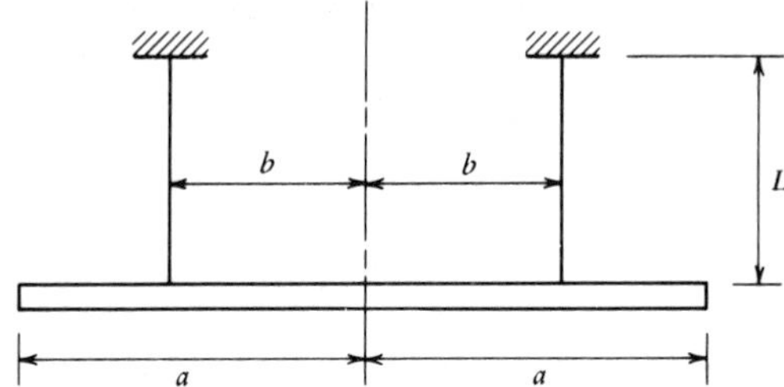

Figure P2.11

2.12 Use the principle of virtual displacements to obtain the equation of motion of the system in Problem 2.6.

Problem Set 2.4

2.13 A uniform bar of mass density ρ per unit volume has a tip mass M. Using the assumed-modes method with $\psi(x) = x/L$, derive the equation of motion for free axial vibration of this system. AE = constant.

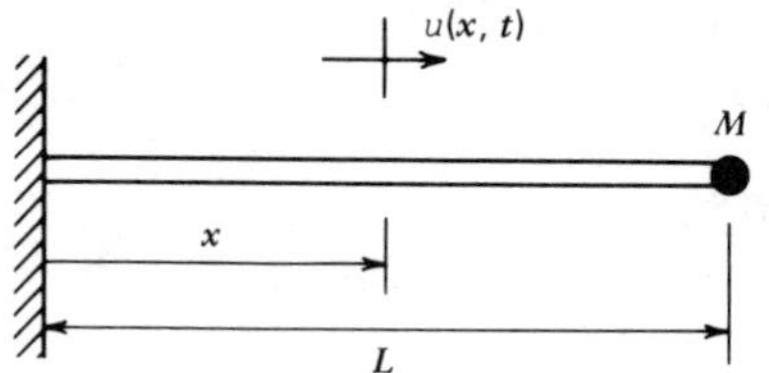

Figure P2.13

2.14 Repeat Problem 2.13 using $\psi(x) = \sin(\pi x/2L)$.

2.15 A tapered bar has a mass density ρ per unit volume and a circular cross section. Using the assumed-modes method with $\psi(x) = x/L$, derive the equation of motion for free axial vibration of the bar.

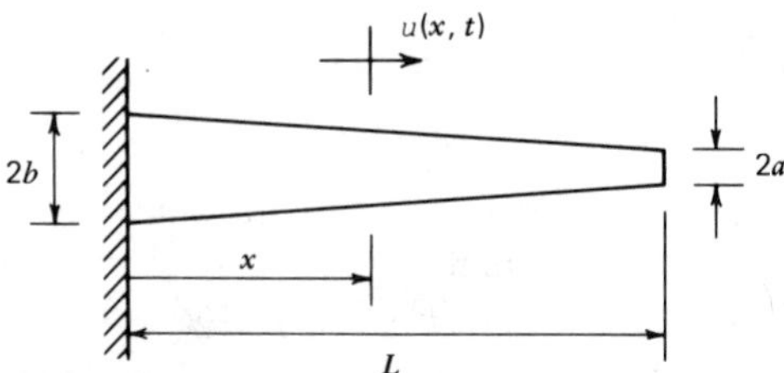

Figure P2.15

2.16 A bar is attached to rigid supports at both ends. Determine appropriate shape functions to be used in representing the axial motion if

(a) $\psi(x)$ is a polynomial,

(b) $\psi(x)$ is sinusoidal.

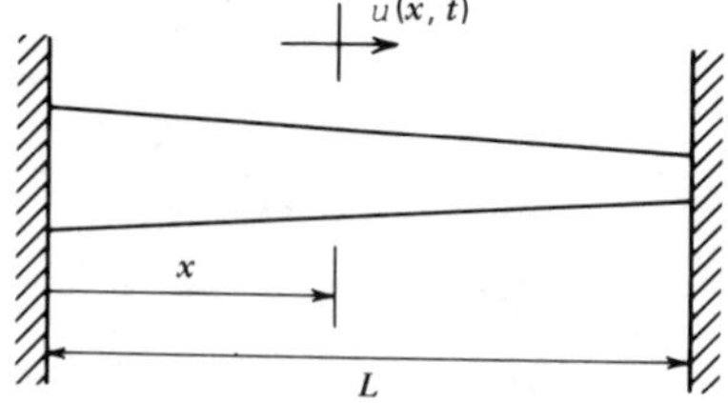

Figure P2.16

2.17 A uniform cantilever beam has a constant horizontal force N and a time-varying distributed transverse load $p(x, t)$ as shown. Using a simple polynominal for $\psi(x)$, derive the equation of motion for transverse vibration of the beam.

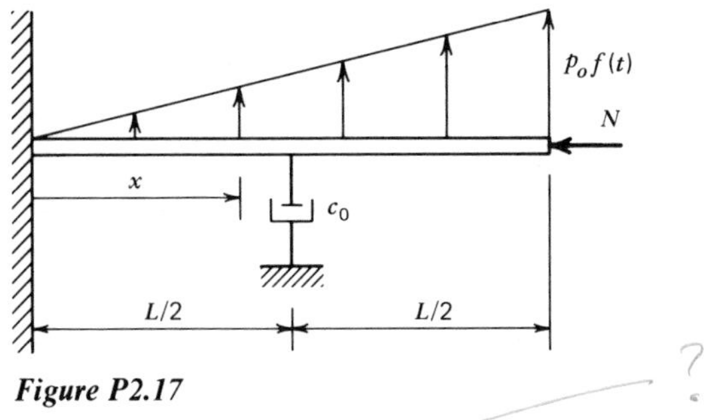

Figure P2.17

2.18 An observation tower in an amusement park can be modeled as a rigid uniform cylinder of mass M_C and radius R atop a slender uniform column whose total mass is M_T and whose stiffness is EI. Assume $H \ll L$, that is, the cylinder can be considered to be a thin "disk" whose center of gravity is at the top of the column. However, translational and rotational inertia of this "disk" must be included.

Use the assumed-modes method to derive the equation of motion for lateral vibration of the system. For $\psi(x)$ use the static deflection curve of a uniform beam with concentrated tip force. Include the geometric stiffness contributions from both M_C and M_T.

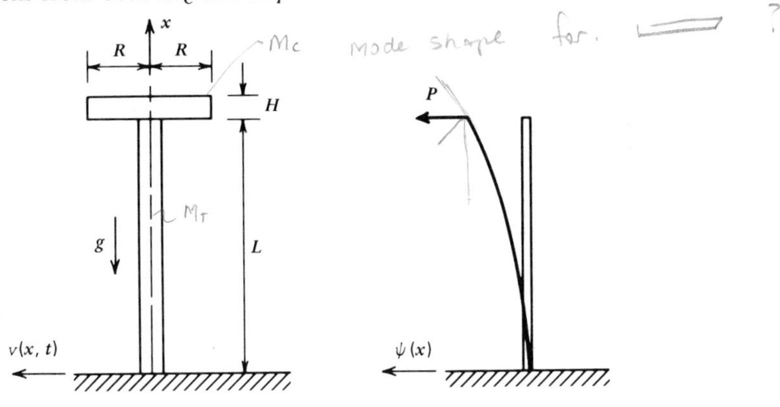

Figure P2.18

3 FREE VIBRATION OF SDOF SYSTEMS

Chapter 2 presented several techniques for deriving the equation of motion of a SDOF system. You observed that in all cases the equation of motion for a linear SDOF system has the form

$$m\ddot{u} + c\dot{u} + ku = p(t) \tag{3.1}$$

where m, u, and so forth, might refer to generalized quantities. Equation 3.1 is the equation of motion for the system of Fig. 3.1. Thus, the system in Fig. 3.1. can be considered to be the prototype SDOF system. We will now begin our study of the response of such a system to an excitation $p(t)$ and to initial conditions, for example, velocity and displacement at $t = 0$, given by

$$u(0) = u_o, \qquad \dot{u}(0) = \dot{u}_o \tag{3.2}$$

where u_o and $\dot{u}_o$ are the specified initial displacement and initial velocity, respectively.

It is convenient to divide Eq. 3.1 by m and rewrite it as

$$\ddot{u} + 2\zeta\omega_n\dot{u} + \omega_n^2 u = \left(\frac{\omega_n^2}{k}\right) p(\text{t}) \tag{3.3}$$

where ω_n is defined by

$$\omega_n^2 = \frac{k}{m} \tag{3.4a}$$

and where ζ is defined by

$$\zeta = \frac{c}{c_{cr}} \tag{3.4b}$$

where

$$c_{cr} = 2m\omega_n = \frac{2k}{\omega_n} = 2\sqrt{km} \tag{3.4c}$$

ω_n is called the *undamped circular natural frequency,* and its units are radians per second (rad/s). ζ is a dimensionless quantity called the *viscous damping factor,* and c_{cr} is called the *critical damping coefficient.* ω_n and ζ are very important parameters determining the response of a SDOF system.

Equation 3.3 is a linear ordinary differential equation with constant coef-

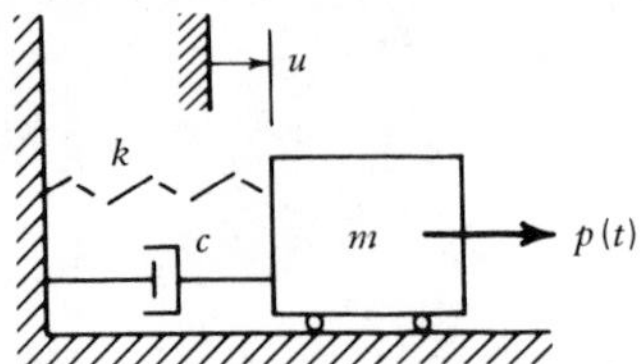

Figure 3.1. Prototype SDOF system.

ficients. Its solution, the *total response,* consists of the linear sum of two distinct parts, a *forced motion,* $u_p(t)$, related directly to $p(t)$, and a *natural motion,* $u_c(t)$, which makes it possible to satisfy arbitrary initial conditions. Thus,

$$u(t) = u_p(t) + u_c(t) \tag{3.5}$$

In mathematical terminology, the *general solution* of the differential equation consists of a *particular solution,* $u_p(t)$, and a *complementary solution,* $u_c(t)$.

In this chapter we will consider only *free vibration,* that is, the natural motion which is the solution of Eq. 3.3 when $p(t) = 0$. Then,

$$\ddot{u} + 2\zeta\omega_n\dot{u} + \omega_n^2 u = 0 \tag{3.6}$$

The general technique for solving Eq. 3.6 is to assume a solution of the form*

$$u = \overline{C}e^{\bar{s}t} \tag{3.7}$$

When Eq. 3.7 is substituted into Eq. 3.6, we obtain

$$(\bar{s}^2 + 2\zeta\omega_n\bar{s} + \omega_n^2)\overline{C}e^{\bar{s}t} = 0 \tag{3.8}$$

For Eq. 3.8 to be valid for all values of t, we must set

$$\bar{s}^2 + 2\zeta\omega_n\bar{s} + \omega_n^2 = 0 \tag{3.9}$$

Equation 3.9 is called the *characteristic equation.*

In Sec. 3.1 we will consider undamped free vibration, that is, the solution of Eq. 3.9 when $\zeta = 0$. Then, in Sec. 3.2 we will consider the solution when $\zeta \neq 0$, that is, when viscous damping is present. Upon completion of this chapter you should be able to:

- Determine, for an undamped SDOF system, the following: undamped natural frequency, period, amplitude, phase, and motion resulting from specified initial conditions.
- Determine, for a viscously damped SDOF system, the following: damped natural frequency, damping factor, logarithmic

*As will be seen later in the chapter, it is possible for $\overline{C}$ and $\bar{s}$ to be complex numbers. In this text complex numbers will be designated by the bar.

decrement, time constant, and motion resulting from specified initial conditions.

- Determine experimentally, for a simple lightly damped SDOF system, the approximate values of natural frequency and damping factor.

3.1 Free Vibration of Undamped SDOF Systems

The equation of motion for an undamped SDOF system is

$$\ddot{u} + \omega_n^2 u = 0 \tag{3.10}$$

and the corresponding characteristic equation is

$$\bar{s}^2 + \omega_n^2 = 0 \tag{3.11}$$

The roots of Eq. 3.11 are

$$\bar{s}_{1,2} = \pm i\omega_n \quad \text{where} \quad i = \sqrt{-1} \tag{3.12}$$

When these roots are substituted into Eq. 3.7, we get the *general solution*

$$u = \bar{C}_1 e^{i\omega_n t} + \bar{C}_2 e^{-i\omega_n t} \tag{3.13}$$

By introducing Euler's equation

$$e^{\pm i\theta} = \cos\theta \pm i\sin\theta \tag{3.14}$$

we can rewrite Eq. 3.13 in terms of trigonometric functions as

$$u = A_1 \cos\omega_n t + A_2 \sin\omega_n t \tag{3.15}$$

where A_1 and A_2 are real constants to be determined from the initial conditions, Eqs. 3.2. Equations 3.2 and 3.15 lead to

$$\begin{aligned} u(0) &= u_o = A_1 \\ \dot{u}(0) &= \dot{u}_o = A_2\omega_n \end{aligned} \tag{3.16}$$

Thus,

$$u = u_o \cos\omega_n t + \left(\frac{\dot{u}_o}{\omega_n}\right) \sin\omega_n t \tag{3.17}$$

is the free vibration response of an undamped SDOF system.

Consider first the case of a system that is displaced from its equilibrium position by an amount u_o and released. Then $\dot{u}(0) = 0$, so

$$u = u_o \cos\omega_n t \tag{3.18}$$

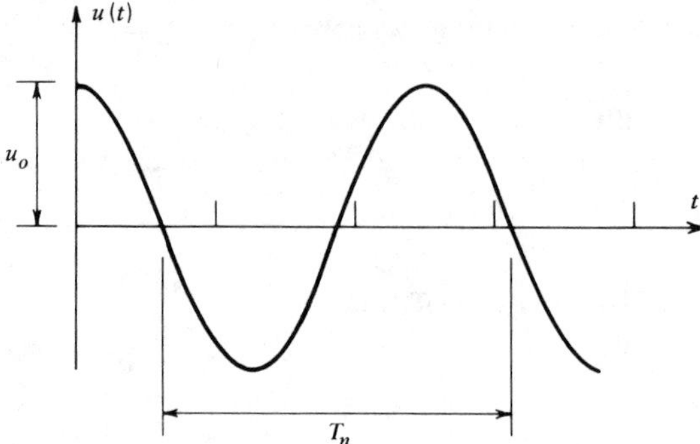

Figure 3.2. Free vibration of an undamped SDOF system with $\dot{u}(0) = 0$.

which is plotted in Fig. 3.2. It can be seen that the resulting motion is *simple harmonic motion* with an amplitude of u_o, an *undamped natural period* of

$$T_n = \frac{2\pi}{\omega_n} \quad \text{(s)} \tag{3.19}$$

and an *undamped natural frequency* of

$$f_n = \frac{1}{T_n} = \frac{\omega_n}{2\pi} \quad \text{(Hz)} \tag{3.20}$$

The symbol Hz stands for hertz. 1 Hz = 1 cycle/s.

Figure 3.3. shows a plot of Eq. 3.17 when neither u_o nor $\dot{u}_o$ is zero. This is still simple harmonic motion with period T_n. $u(t)$ can be expressed by Eq. 3.17 or by the equation

$$u(t) = U\cos(\omega_n t - \alpha) = U\cos\omega_n\left(t - \frac{\alpha}{\omega_n}\right) \tag{3.21}$$

where the *amplitude* is U, and the *phase angle*, α, determines the amount by which $u(t)$ lags the function $\cos\omega_n t$. To relate these quantities to u_o and $\dot{u}_o$ it is convenient to employ the rotating-vector, or complex-plane, representation of simple harmonic motion.

Figure 3.4*a* shows how a complex number $\overline{C}$ can be represented in the complex plane by its polar form

$$\overline{C} = Ce^{i\theta} \tag{3.22a}$$

or by its rectangular form

$$\overline{C} = R(\overline{C}) + iI(\overline{C}) = C_R + iC_I \tag{3.22b}$$

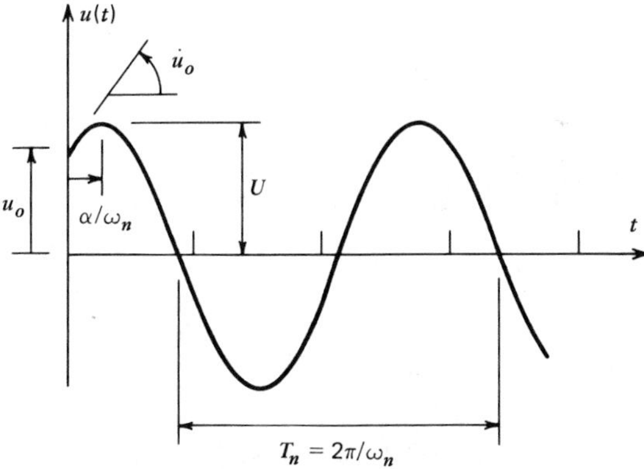

Figure 3.3. General free vibration response of an undamped SDOF system.

where $R(\ )$ stands for "real part of" and $I(\ )$ for "imaginary part of." The two forms are related by Euler's formula, Eq. 3.15. Thus,

$$C_R = C\cos\theta, \qquad C_I = C\sin\theta \tag{3.23}$$

Figure 3.4*b* shows that when the angle θ is taken to be $\omega_n t$, then $\overline{C}$ becomes a *rotating vector*

$$\overline{C} = Ce^{i\omega_n t} = C\cos\omega_n t + iC\sin\omega_n t \tag{3.24}$$

The horizontal projection, or real component, of $\overline{C}$ represents harmonic motion $C\cos\omega_n t$, and the vertical projection, or imaginary component, represents harmonic motion $C\sin\omega_n t$.

Figure 3.5. shows how two rotating vectors can be arranged so that the sum of their projections onto the real axis corresponds to the expression for

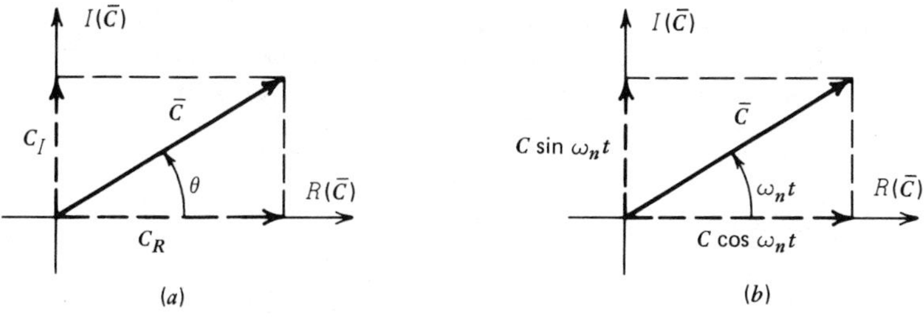

Figure 3.4. Representation of vectors in the complex (Argand) plane. (a) *Complex number representation.* (b) *Rotating vector representation of SHM.*

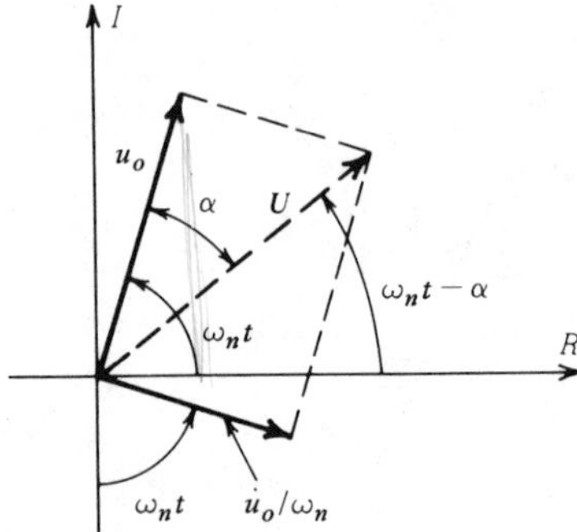

Figure 3.5. Rotating vector representation of undamped free vibration.

$u(t)$ on the right-hand side of Eq. 3.17, and so that the projection of their resultant corresponds to the expression for $u(t)$ on the right-hand side of Eq. 3.21. Thus, from Fig. 3.5. it can be seen that the amplitude, U, and phase angle, α, in Eq. 3.21 are given by

$$U^2 = u_o^2 + \left(\frac{\dot{u}_o}{\omega_n}\right)^2$$

and

$$\tan \alpha = \frac{\dot{u}_o/\omega_n}{u_o} \tag{3.25}$$

3.2 Free Vibration of Viscous-Damped SDOF Systems

From Eq. 3.21 and Fig. 3.3, you can see that once an undamped SDOF system is set into motion with initial conditions u_o and $\dot{u}_o$, that motion will continue (theoretically) indefinitely. In actuality, all systems have some damping that dissipates energy and causes the motion to die out eventually.

Consider now the free vibration of a SDOF system with linear viscous damping. Its equation of motion is Eq. 3.6, repeated here.

$$\ddot{u} + 2\zeta\omega_n\dot{u} + \omega_n^2 u = 0 \tag{3.26}$$

Assuming a solution of the form

$$u = \bar{C}e^{\bar{s}t} \tag{3.27}$$

we obtain the characteristic equation

$$\bar{s}^2 + 2\zeta\omega_n\bar{s} + \omega_n^2 = 0 \tag{3.28}$$

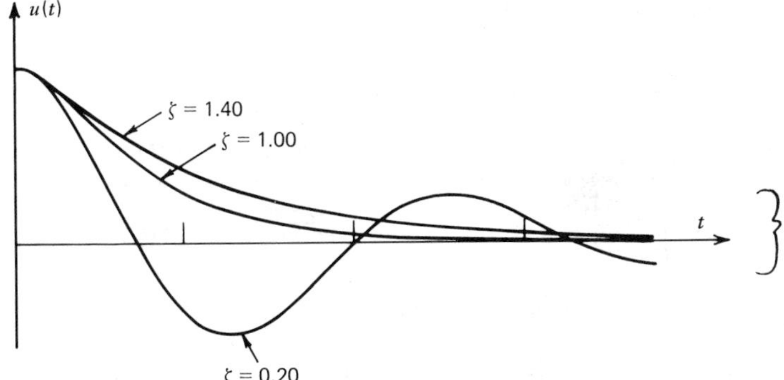

Figure 3.6. *Response of a viscous-damped SDOF system with various damping levels.*

whose roots, $\bar{s}_1$ and $\bar{s}_2$ are given by

$$\bar{s}_{1,2} = -\zeta\omega_n \pm \omega_n\sqrt{\zeta^2 - 1} \tag{3.29}$$

The magnitude of the damping factor, ζ, can be used to distinguish three cases: *underdamped* ($0 < \zeta < 1$), *critically damped* ($\zeta = 1$), and *overdamped* ($\zeta > 1$). Figure 3.6 illustrates these three cases. For the underdamped case the motion is oscillatory with a decaying amplitude. For the overdamped case there is no oscillation, and the amplitude slowly decays. For the critically damped system there is no oscillation, and the amplitude decays more rapidly than in either the underdamped or overdamped cases. Since the underdamped case is the most important case for structural dynamics applications, it will be treated first.

Underdamped Case ($\zeta < 1$)

For $\zeta < 1$ it is convenient to write Eq. 3.29 in the form

$$\bar{s}_{1,2} = -\zeta\omega_n \pm i\omega_d \tag{3.30}$$

where ω_d is the *damped circular natural frequency* given by

$$\omega_d = \omega_n\sqrt{1 - \zeta^2} \tag{3.31a}$$

with corresponding *damped period,* T_d, given by

$$T_d = \frac{2\pi}{\omega_d} \tag{3.31b}$$

With aid of Euler's formula, the general solution, $u(t)$, can be written in the form

$$u(t) = e^{-\zeta\omega_n t}(A_1 \cos \omega_d t + A_2 \sin \omega_d t) \tag{3.32}$$

Again, u_o and $\dot{u}_o$ are used to evaluate A_1 and A_2, with the result that

$$u(t) = e^{-\zeta\omega_n t}\left[u_o \cos \omega_d t + \left(\frac{\dot{u}_o + \zeta\omega_n u_o}{\omega_d}\right) \sin \omega_d t\right] \tag{3.33}$$

Equation 3.33 can be written in the form

$$u(t) = Ue^{-\zeta\omega_n t} \cos(\omega_d t - \alpha) \tag{3.34}$$

and the rotating vector technique can be employed to show that

$$U^2 = u_o^2 + \left(\frac{\dot{u}_o + \zeta\omega_n u_o}{\omega_d}\right)^2 \tag{3.35}$$

and

$$\tan \alpha = \frac{\dot{u}_o + \zeta\omega_n u_o}{\omega_d u_o}$$

Figure 3.7 shows a comparison of the responses of SDOF systems having different levels of subcritical damping. In each case, since $u_o = 0$, the response is given by

$$u(t) = \left(\frac{\dot{u}_o}{\omega_d}\right) e^{-\zeta\omega_n t} \sin \omega_d t \tag{3.36}$$

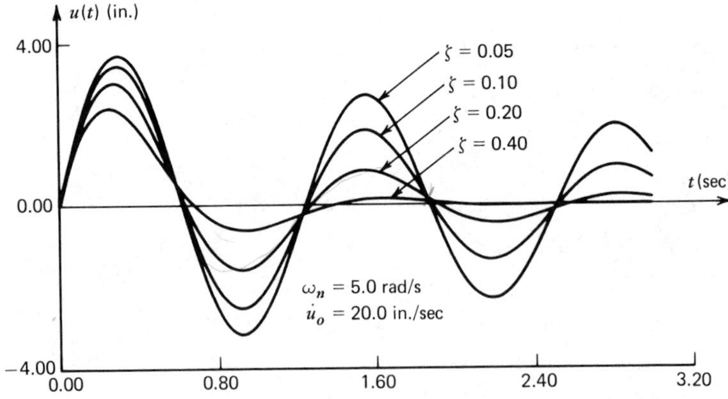

***Figure 3.7.** Effect of damping level on free vibration.*

Although the value of ζ has an effect on the frequency, ω_d, the most pronounced effect of damping is on the rate at which the motion dies out, that is, on the $e^{-\zeta\omega_d t}$ term. This effect will be considered further in Sec. 3.3, which treats measurement of damping.

Example 3.1

A SDOF system has an undamped natural frequency of 5 rad/s and a damping factor of 20%. It is given the initial conditions $u_o = 0$ and $\dot{u}_o = 20$ in./sec. Determine the damped natural frequency, and determine the expression for the motion of the system for $t > 0$.

Solution

$$\omega_n = 5 \text{ rad/s}, \qquad \zeta = 0.2$$
$$u_o = 0, \qquad \dot{u}_o = 20 \text{ in./sec}$$

From Eq. 3.31a

$$\omega_d = \omega_n \sqrt{1 - \zeta^2} \tag{1}$$
$$= 5\sqrt{1 - (0.2)^2} = 4.90 \text{ rad/s}$$
$$f_d = \frac{\omega_d}{2\pi} = \frac{4.90}{6.28} = 0.78 \tag{2}$$

$$\boxed{f_d = 0.78 \text{ Hz}} \tag{3}$$

From Eq. 3.33 or 3.36

$$u(t) = \left(\frac{\dot{u}_o}{\omega_d}\right) e^{-\zeta\omega_n t} \sin \omega_d t \tag{4}$$

Thus,

$$u(t) = \left(\frac{20}{4.90}\right) e^{-(0.2)(5.0)t} \sin(4.90)t \tag{5}$$

or

$$\boxed{u(t) = 4.08e^{-t} \sin(4.90)t} \tag{6}$$

A plot of this result is one of the curves of Fig. 3.7.

Unless significant amounts of damping are intentionally incorporated into a structural system, the damping will fall into the underdamped category, generally in the range of 0.5% to 5%. However, the critical damping case and the overdamped case will be treated briefly for completeness.

Critically-Damped Case ($\zeta = 1$)

When $\zeta = 1$, Eq. 3.29 gives only one solution

$$\bar{s} = -\zeta\omega_n \tag{3.37}$$

Then, the response takes the form

$$u(t) = (C_1 + C_2 t)e^{-\zeta\omega_n t} \tag{3.38}$$

rather than that given in Eq. 3.27. When the initial conditions are taken into account, the response of a critically damped system is found to be

$$u(t) = [u_o + (\dot{u}_o + \zeta\omega_n u_o)t]e^{-\zeta\omega_n t} \tag{3.39}$$

Examples of this type of nonoscillatory response are seen in Fig. 3.6 and in Fig. 3.8.

Overdamped Case ($\zeta > 1$)

When $\zeta > 1$, there are two distinct, negative real roots given by Eq. 3.29. Let

$$\omega^* = \omega_n\sqrt{\zeta^2 - 1} \tag{3.40}$$

Then the response of an overdamped system can be written in the form

$$u(t) = e^{-\zeta\omega_n t}(C_1 \cosh \omega^* t + C_2 \sinh \omega^* t) \tag{3.41}$$

where C_1 and C_2 depend on the initial conditions. Figure 3.8 shows the effect of damping level on the response, indicating that the initial overshoot is greater

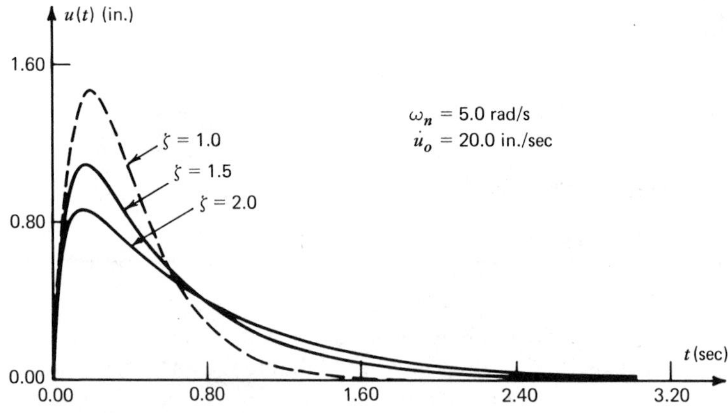

Figure 3.8. Response of overdamped systems.

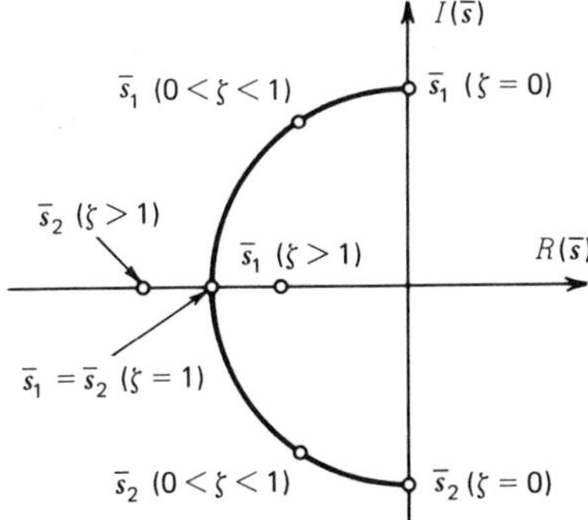

Figure 3.9. *$\bar{s}$-plane plot of roots of characteristic equation.*

for the smaller damping levels, but the final decay is more rapid for damping levels approaching $\zeta = 1$.

From the previous discussion you can see that the form, or character, of the response depends solely on the roots, $\bar{s}_i$, of the characteristic equation. Furthermore, these roots depend solely on the two system parameters ω_n and ζ, and not on the initial conditions. The dependence of the roots on the damping factor ζ is summarized in the $\bar{s}$-plane plot of Fig. 3.9. By comparing this figure with Fig. 3.6 you can see how the position of the roots in the $\bar{s}$-plane is related to the character of the response.

3.3 Experimental Determination of Fundamental Natural Frequency and Damping Factor of a SDOF System

It is frequently necessary to determine the dynamical properties (e.g., natural frequency and damping factor) of a given system by experimental methods. It may be possible to measure the spring constant k and the mass m of a simple SDOF system, but the damping seldom arises in a manner which allows the damping coefficient c to be measured directly. The damping of a real system usually results from looseness of joints, internal damping in the material, and so forth, and does not result from some viscous dashpot. However, as long as the amplitude of vibration decays exponentially (or approximately so) as shown in Fig. 3.7, it may be assumed that viscous type damping can be used in the mathematical model of the system. A damping factor, ζ, is usually measured and, if desired, an effective value of c can be computed from Eq. 3.4b.

The undamped natural frequency of a simple SDOF system may be determined from static measurement, as illustrated by Example 3.2.

Example 3.2

Determine the natural frequency of a simple spring-mass system by using static deflection measurements.

Solution

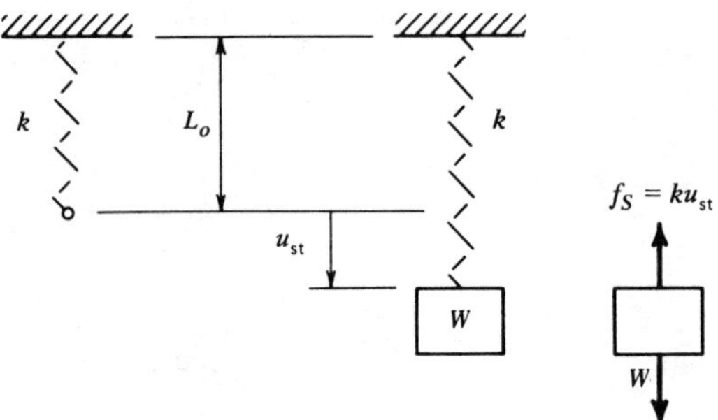

From Eq. 3.4a

$$\omega_n^2 = k/m \tag{1}$$

The equilibrium of the mass as it hangs on the spring is expressed by

$$+\downarrow \sum F = 0 \tag{2}$$

or

$$W - f_s = 0 \tag{3}$$

From the force-elongation equation for the spring

$$f_s = ku_{st} \tag{4}$$

Combining Eqs. 3 and 4 gives

$$f_s = mg = ku_{st} \tag{5}$$

Thus, from Eqs. 1 and 5,

$$\boxed{\omega_n^2 = \frac{g}{u_{st}}} \tag{6}$$

If the damping in the system is small ($\zeta < 0.2$), Eq. 3.31 shows that ω_d is approximately equal to ω_n. Example 3.3. shows how a free vibration experiment could be used to determine the natural frequency of a SDOF system.

Example 3.3

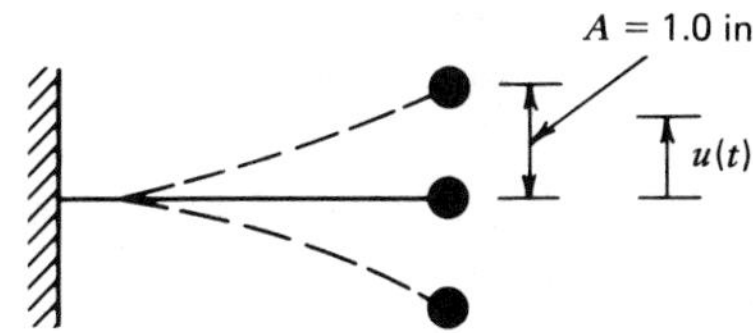

The natural frequency of a cantilever beam with a lumped mass at its tip is to be determined dynamically. The mass is deflected by an amount $A = 1$ in. and released. The ensuing motion is shown below and indicates that the damping in the system is very small. Compute the natural frequency in radians per second and in hertz. What is the period?

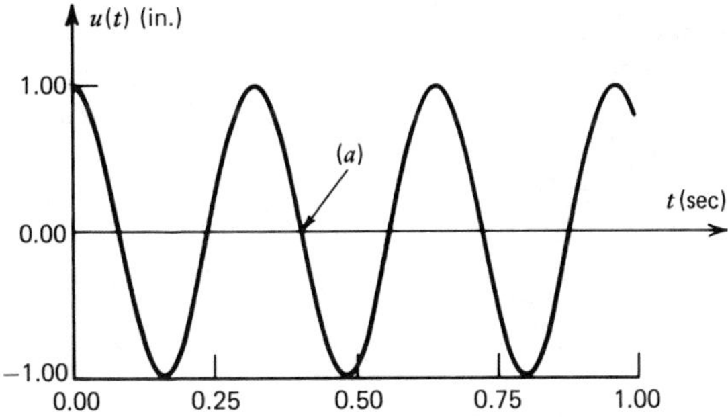

Solution

At point (a) the mass has executed $1\frac{1}{4}$ cycles.

$$f_n \doteq \frac{1.25 \text{ cycles}}{0.4 \text{ s}} = 3.125 \text{ Hz} \tag{1}$$

$$\omega_n = 2\pi f_n \doteq (6.28)(3.125) = 19.6 \text{ rad/s} \tag{2}$$

$$T_n = \frac{1}{f_n} \doteq \frac{1}{3.125} = 0.32 \text{ s} \tag{3}$$

There are two similar methods for determining the damping factor, ζ, using the decay record of free vibration of a SDOF system: the *logarithmic decrement method* and the *half-amplitude method*. Both are based on Eq. 3.34.

(3.34) $u(t) = Ue^{-\zeta\omega_n t}\cos(\omega_d t - \alpha)$

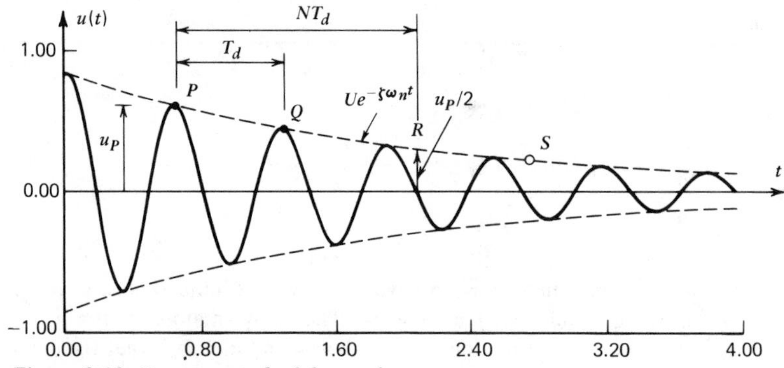

Figure 3.10. Decay record of damped system.

In the logarithmic decrement method the amplitude of motion, u_P, at the beginning of a cycle and the amplitude, u_Q, at the end of the cycle are measured. At the end of a period (i.e., one cycle) the value of $\cos(\omega_d t - \alpha)$ returns to the value it had at the beginning of the cycle. Hence, from Eq. 3.34 is obtained the expression:

$$\frac{u_P}{u_Q} = e^{\zeta\omega_n T_d} \tag{3.42}$$

The *logarithmic decrement* δ is defined by

$$\delta = \ln\left(\frac{u_P}{u_Q}\right) = \zeta\omega_n T_d \tag{3.43}$$

where T_d is the *damped natural period,* given by

$$T_d = \frac{2\pi}{\omega_d} = \frac{2\pi}{\omega_n\sqrt{1-\zeta^2}} \tag{3.44}$$

Thus, from Eqs. 3.42 and 3.43 we obtain

$$\delta = \zeta\omega_n T_d = \frac{2\pi\zeta}{\sqrt{1-\zeta^2}} \tag{3.45}$$

For small damping ($\zeta < 0.2$) the approximation

$$\delta \doteq 2\pi\zeta \tag{3.46}$$

is acceptable, enabling the damping factor to be obtained from the equation

$$\zeta \doteq \left(\frac{1}{2\pi}\right)\ln\left(\frac{u_P}{u_Q}\right) \tag{3.47}$$

A similar procedure leads to the *half-amplitude method,* which results in a much simpler calculation for the damping factor. The half-amplitude method is based on the amplitude of the *envelope curve*

$$\hat{u}(t) = Ue^{-\zeta\omega_n t} \tag{3.48}$$

at two points P and R, where

$$\hat{u}_R = \frac{\hat{u}_P}{2} \tag{3.49}$$

Those points are N damped periods apart, where N is not necessarily an integer. Then,

$$\frac{\hat{u}_P}{\hat{u}_R} = e^{\zeta\omega_n N T_d} = 2 \tag{3.50}$$

From Eqs. 3.44 and 3.50,

$$\frac{2\pi N\zeta}{\sqrt{1-\zeta^2}} = \ln(2) \tag{3.51}$$

Figure 3.11 shows the relationship between ζ and N. However, for small values of damping, $\zeta^2 \ll 1$, so Eq. 3.51 gives

$$2\pi N\zeta \doteq \ln(2) \tag{3.52}$$

or

$$\zeta \doteq \frac{0.11}{N} \tag{3.53}$$

Equation 3.53 provides a very convenient means of estimating the damping in a system that is lightly damped ($\zeta < 0.1$, i.e., $N > 1$).

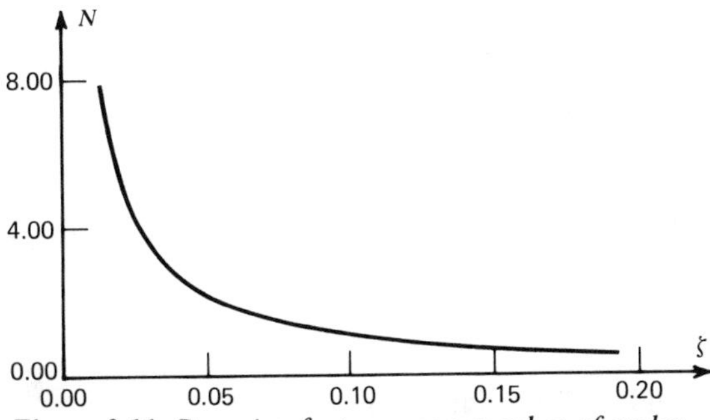

Figure 3.11. *Damping factor versus number of cycles to reduce amplitude by 50%.*

Example 3.4

Use the half-amplitude method to estimate the damping of the system whose motion is recorded in Fig. 3.10.

Solution

a. Sketch the envelope curve (already on the figure).

b. Pick point P at one peak and measure u_P:

$u_P = 0.44$ in.

c. Locate point R, where the amplitude of the envelope curve is $u_P/2 = 0.22$ in.

d. Estimate the number of cycles between P and R:

$N = 2.25$ cycles

e. Use Eq. 3.53 to estimate ζ:

$$\boxed{\zeta = \frac{0.11}{2.25} = 0.049}$$

The level of damping in a system is also reflected in a quantity called the *time constant*, τ. It is defined as the time required for the amplitude to be reduced by a factor of $1/e$. In a manner similar to the way in which the half-amplitude formula was derived, an expression for the time constant can be obtained. Use the envelope curve of Fig. 3.10 again, and let S be the point such that

$$\frac{u_P}{u_S} = \frac{u_P}{u_P(1/e)} = e \tag{3.54}$$

Thus,

$$\frac{u_P}{u_S} = \frac{U\exp(-\zeta\omega_n t_P)}{U\exp[-\zeta\omega_n(t_P + \tau)]} = e \tag{3.55}$$

or

$$e^{\zeta\omega_n\tau} = e \tag{3.56}$$

By taking the logarithm of both sides we get

$$\zeta\omega_n\tau = 1 \tag{3.57}$$

Then, the time constant, τ, is given by

$$\tau = \frac{1}{\zeta\omega_n} = \frac{T_n}{2\pi\zeta} \tag{3.58}$$

Recall that $1/e = 1/2.718 = 0.368$. Hence, the time constant τ, is the time required for the amplitude of motion to be reduced by 63%.

3.4 Free Vibration of a SDOF System with Coulomb Damping

Since viscous damping leads to a linear differential equation of motion, Eq. 3.1, which is relatively easily solved for free or forced response, this model of damping is the one most frequently used in analytical studies. However, the actual damping in a structure may result from looseness of joints, dry friction between components, material damping, and many other complex causes, any of which would lead to nonlinear behavior of the structure. It is not possible to treat all of these forms of damping, but it is instructive to compare and contrast viscous damping with *Coulomb,* or dry-friction, *damping.*

Figure 3.12 shows a mass sliding on a rough surface which produces a force of sliding friction

$$f_D = \mu_k N = \mu_k mg \tag{3.59}$$

where μ_k is the *coefficient of kinetic friction,* or coefficient of sliding friction. The friction force always opposes the motion, that is, its sense is opposite to that of $\dot{u}$. Using Newton's second law, we get

$$-f_S - f_D = m\ddot{u} \tag{3.60}$$

But,

$$f_S = ku$$

and

$$f_D = \mu_k mg \operatorname{sgn}(\dot{u}) \tag{3.61}$$

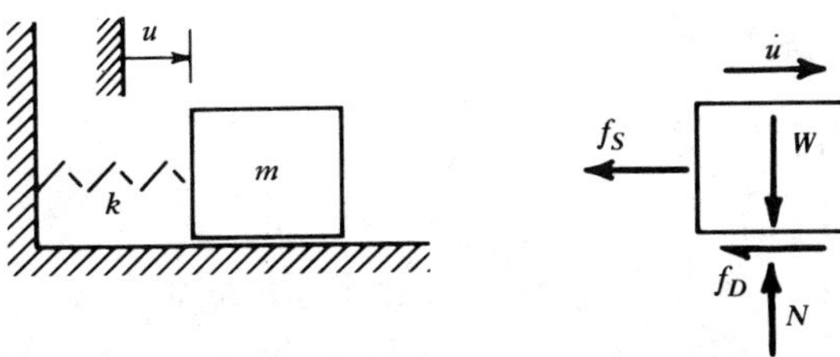

Figure 3.12. *A SDOF system with Coulomb friction.*

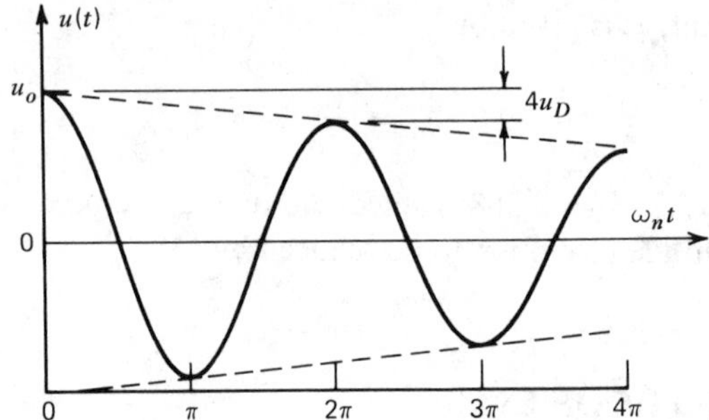

Figure 3.13. *Free vibration response of a system with Coulomb friction.*

Then,

$$
\begin{aligned}
m\ddot{u} + ku &= -\mu_k mg, \qquad \dot{u} > 0 \\
m\ddot{u} + ku &= +\mu_k mg, \qquad \dot{u} < 0
\end{aligned}
\tag{3.62}
$$

Let

$$u_D = |f_D| \left(\frac{1}{k}\right) = \frac{\mu_k g}{\omega_n^2} \tag{3.63}$$

Equations 3.62 and 3.63 may be combined to give

$$
\begin{aligned}
\ddot{u} + \omega_n^2 u &= -\omega_n^2 u_D, \qquad \dot{u} > 0 \\
\ddot{u} + \omega_n^2 u &= +\omega_n^2 u_D, \qquad \dot{u} < 0
\end{aligned}
\tag{3.64}
$$

The resulting motion is plotted in Fig. 3.13. Note in Fig. 3.13 that the Coulomb-damped system behaves like an undamped SDOF system whose equilibrium position is shifted at the end of each half-cycle. A distinguishing feature of the response, as seen in Fig. 3.13, is that the amplitude decays linearly with time, not exponentially as in the case of viscous damping.

Problem Set 3.1

3.1 The structure in Fig. P2.1 has the following properties: $m = 0.10$ k-sec^2/ft, $L = 12$ ft, $EI = 1200$ k-ft^2. Determine its undamped natural period.

3.2 The control surface of Fig. P2.6 has the following properties: $k_1 = 600$ N/m, $a = 0.10$ m. When the surface is displaced slightly and released from rest, it is observed to have very low damping and to have an "undamped" natural frequency of 5.0 Hz. Determine the mass moment of inertia I_A.

3.3 A slender column has a total mass M, and its bending stiffness is EI. A lumped mass μM is located at the top of the column. Determine an expression for the approximate change in ω_n^2 due to the inertia and geometric stiffness effects of the tip mass. Use the assumed-modes expressions of Example 2.9, and for $\psi(x)$ use the static displacement function for a uniform beam with concentrated transverse tip force (See Fig. P2.18).

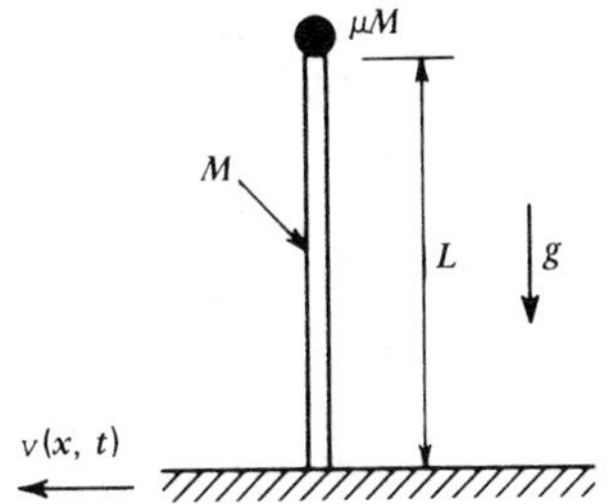

Figure P3.3

3.4 A 22-kg mass m_1 hangs from a spring whose spring constant is k = 17kN/m. A second mass m_2 = 10 kg drops from a height h = 0.2 m and sticks to mass m_1.

(a) Determine an expression for the motion $u(t)$ of the two masses after the moment of impact.

(b) Determine the maximum displacement of the two masses.

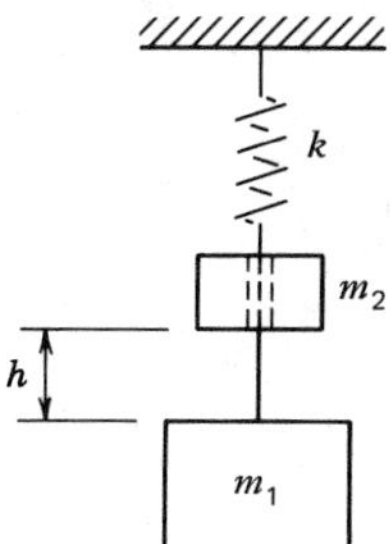

Figure P3.4

Problem Set 3.2

3.5 The system in Fig. P2.3 has the following properties: M = 0.01 k-sec^2/ft, k_1 = 15.0 k/ft, k_2 = 0, and L = 10.0 ft. The damping coefficient c_1 is such that ζ = 0.10.

(a) Determine the numerical value of c_1.

(b) If end B is displaced vertically upward 0.10 ft and released from rest at t = 0, how many complete cycles of vibration will have to occur before the maximum spring force falls below 0.2 k?

3.6 Plot the tip motion $u(t)$ resulting from the "pluck test" described in Problem 3.5(b) above.

3.7 A machine whose mass is 70 kg is mounted on springs whose total stiffness is 50 kN/m, and the total damping is 1.2 kN·s/m. Find the motion $u(t)$ for the following initial conditions:

(a) $u(0) = 10$ mm, $\dot{u}(0) = 0$

(b) $u(0) = 0$, $\dot{u}(0) = 100$ mm/s

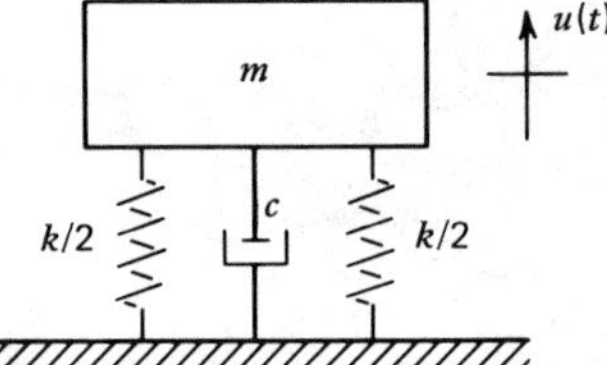

Figure P3.7

3.8 A wind turbine is modeled as a concentrated mass (the turbine) atop a weightless tower of height L. To determine the dynamic properties of the system, a large crane is brought alongside the tower and a lateral force P = 200 lb is exerted along the turbine axis as shown. This causes a horizontal displacement of 1.0 in.

The cable attaching the turbine to the crane is instantaneously severed, and the resulting free vibration of the turbine is recorded. At the end of two complete cycles, the time is 1.25 sec and the amplitude is 0.64 in.

From the above data determine the following:

(a) The undamped natural frequency ω_n.

(b) The effective stiffness k (lb/in.).

(c) The effective mass m (lb-sec^2/in.).

(d) The effective damping factor ζ.

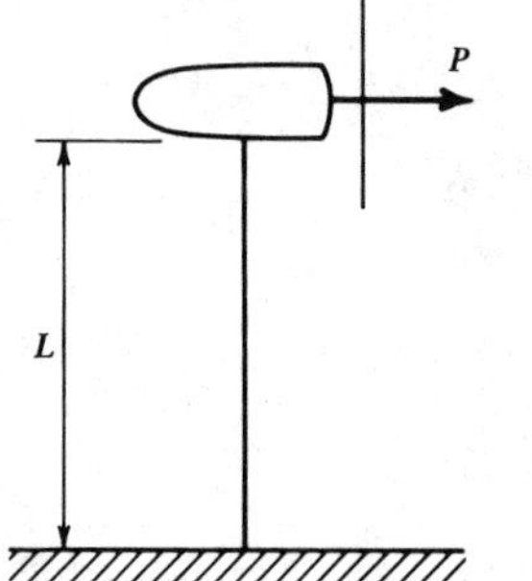

Figure P3.8

3.9 Starting with Eq. 3.41 determine an expression for the displacement, $u(t)$, of an overdamped system with initial conditions $u(0) = u_o$, $\dot{u}(0) = \dot{u}_o$.

3.10 Determine an expression for the critical damping coefficient, c_{cr}, for the system in Fig. P2.5.

Problem Set 3.3

3.11 For the time history plotted below:

(a) Estimate the damped natural frequency in hertz.

(b) Use the half-amplitude method to estimate the damping factor ζ.

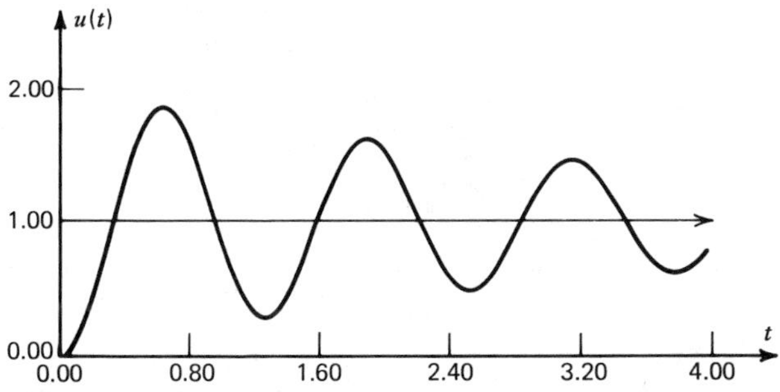

Figure P3.11

3.12 Figure P3.12 is a photograph of an oscilloscope trace of the displacement of a SDOF system.

(a) Estimate the damped natural frequency in hertz.

(b) Use the half-amplitude method to estimate the damping factor ζ.

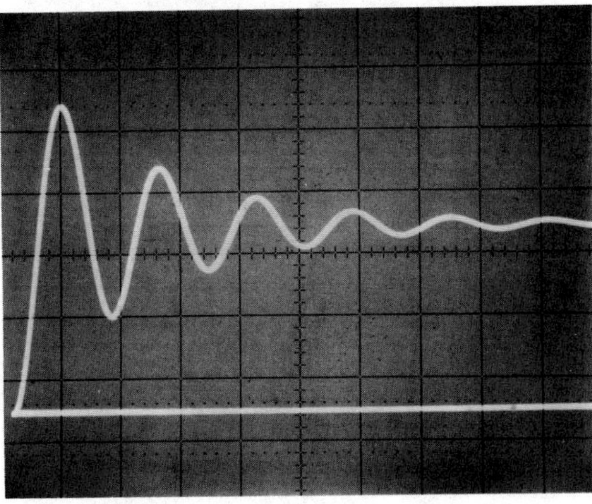

Figure P3.12

1.0 m sec/div

Problem Set 3.4

3.13 (a) Starting with Eqs. 3.64, determine expressions for $u(t)$ for the two cases $\dot{u} > 0$ and $\dot{u} < 0$.

(b) If $u(0) = u_o$ and $\dot{u}(0) = 0$, what is the expression for $u(t)$: (1) for $0 \le t \le \pi/\omega_n$, (2) for $\pi/\omega_n \le t \le 2\pi/\omega_n$?

(c) Determine an expression for the amount of amplitude decay per cycle.

4 RESPONSE OF SDOF SYSTEMS TO HARMONIC EXCITATION

In Chapter 2 several methods were presented for deriving the differential equation of motion of a SDOF system. In Chapter 3 you began to consider the solution of this differential equation by studying the free vibration response of a SDOF system. In this chapter you will begin your study of forced motion by studying the response of undamped and viscous-damped SDOF systems to harmonic excitation. This topic is extremely important, not only because many SDOF structures are subjected to harmonic excitation, but also because the results of this chapter can also be extended to treat MDOF structures and structures subjected to more complex types of excitation.

Upon completion of this chapter you should be able to:

- Determine the steady-state response and total response of an undamped system subjected to harmonic excitation.
- Determine the steady-state response and the total response of a viscous-damped system subjected to harmonic excitation.
- Describe the effect of damping factor and frequency ratio on the amplitude and phase of the steady-state response of a viscous-damped SDOF system subjected to an harmonic excitation force.
- Sketch, for various frequency ratios, force vector polygons that show the phase and amplitude relationships of the steady-state forces acting on a SDOF system.
- Use the complex frequency response method to obtain expressions for the amplitude and phase of the steady-state response of a viscous-damped SDOF system.
- Set up the equation of motion for a "base excitation" problem and obtain the steady-state solutions (frequency response functions) for absolute motion and motion of the mass relative to the base.
- Obtain numerical values for the magnification factor and phase angle of a system with given base excitation.
- Discuss the principle of operation of seismic transducers, namely the vibrometer and accelerometer.
- Describe how frequency response data can be used to determine

the undamped natural frequency and damping factor of a viscous-damped SDOF system.

- Calculate the work done per cycle of harmonic motion by a specified dissipative force.
- Calculate the equivalent viscous damping coefficient for a specified dissipative force.
- Discuss how structural damping differs from viscous damping, and list the principal features of vector response plots for systems with structural damping.

4.1 Response of Undamped SDOF Systems to Harmonic Excitation

As noted in Chapter 3, the total response of a linear system consists of the superposition of a forced motion and a natural motion. In the case of harmonic excitation the forced motion is referred to as the steady-state response. This section treats the response of undamped SDOF systems to harmonic excitation; Secs. 4.2 through 4.8 treat the response of damped systems.

Consider the undamped SDOF system shown in Fig. 4.1. It is assumed that the system is linear and that the excitation amplitude p_o and excitation frequency Ω are constants.

The equation of motion is

$$m\ddot{u} + ku = p_o \cos \Omega t \tag{4.1}$$

From the fact that only even-order derivatives appear on the left-hand side of Eq. 4.1, it is seen that the forced motion, or *steady-state response*, will have the form

$$u_p = U \cos \Omega t \tag{4.2}$$

To determine the amplitude, U, of the steady-state response, Eq. 4.2 is substituted into Eq. 4.1 giving

$$U = \frac{p_o}{k - m\Omega^2} \tag{4.3}$$

provided that $(k - m\Omega^2) \neq 0$. Let

$$U_o = \frac{p_o}{k} \tag{4.4}$$

which is the static deflection. Then, Eq. 4.3 may be written

$$H(\Omega) = \frac{1}{1 - r^2}, \qquad r \neq 1 \tag{4.5}$$

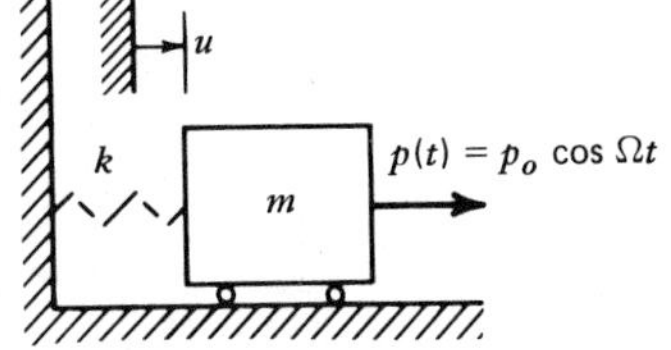

Figure 4.1. Harmonic excitation of an undamped SDOF system.

where

$$r = \frac{\Omega}{\omega_n} \tag{4.6}$$

is called the *frequency ratio,* and

$$H(\Omega) = \frac{U}{U_o} \tag{4.7}$$

is called the *frequency response function.* The frequency response function, $H(\Omega)$, gives the magnitude and sign of the steady-state motion as a function of the frequency ratio r. The magnitude

$$D_s = |H(\Omega)| \tag{4.8}$$

is called the *steady-state magnification factor,* or gain. It is plotted in Fig. 4.2.

Equations 4.2 through 4.5 may be combined to give the steady-state response

$$u_p = \left(\frac{U_o}{1 - r^2}\right) \cos \Omega t, \qquad r \neq 1 \tag{4.9}$$

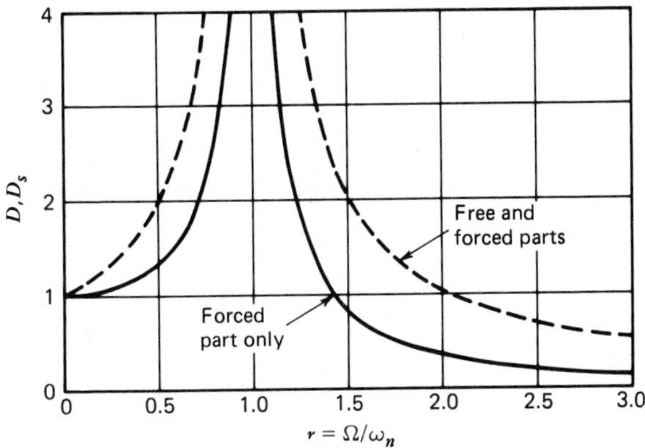

Figure 4.2. Dynamic magnification factors for an undamped SDOF system with p(t) = p$_o$ *sin* Ωt.

If $r < 1$, the response is in-phase with the excitation, since $(1 - r^2)$ is positive. If $r > 1$, the response is 180° out-of-phase with the excitation, that is, u_p can be written

$$u_p = \left(\frac{U_o}{r^2 - 1}\right)(-\cos \Omega t) \tag{4.10}$$

The general form of the natural motion of an undamped SDOF system is given by Eq. 3.15. (Note that Eq. 3.17 is valid only for free vibration!) Hence, Eqs. 3.15 and 4.9 may be combined to give the total response

$$u = \left(\frac{U_o}{1 - r^2}\right) \cos \Omega t + A_1 \cos \omega_n t + A_2 \sin \omega_n t \tag{4.11}$$

The example below will illustrate the use of Eq. 4.11.

Example 4.1

The system shown in Fig. 4.1 has $k = 40$ lb/in., and the mass weighs 38.6 lb. If the system is at rest, that is, $u_o = \dot{u}_o = 0$, when an excitation $p(t) = 10 \cos(10t)$ commences, determine an expression for the resulting motion. Sketch the resulting motion.

Solution

From Eq. 4.11 the total response is given by

$$u = \frac{U_o}{1 - r^2} \cos \Omega t + A_1 \cos \omega_n t + A_2 \sin \omega_n t \tag{1}$$

This may be differentiated to give the velocity

$$\dot{u} = \frac{-U_o\Omega}{1 - r^2} \sin \Omega t - A_1\omega_n \sin \omega_n t + A_2\omega_n \cos \omega_n t \tag{2}$$

From Eq. 3.4a

$$\omega_n = \left(\frac{k}{m}\right)^{1/2} = \left(\frac{kg}{W}\right)^{1/2} = \sqrt{\frac{40(386)}{(38.6)}} = 20 \text{ rad/s} \tag{3}$$

From Eq. 4.4

$$U_o = \frac{p_o}{k} = \frac{10}{40} = 0.25 \text{ in.} \tag{4}$$

From Eq. 4.7

$$r = \frac{\Omega}{\omega_n} = \frac{10}{20} = 0.5 \tag{5}$$

Therefore,

$$\frac{U_o}{1 - r^2} = \frac{0.25}{1 - (0.5)^2} = \frac{0.25}{1 - 0.25} = \frac{0.25}{0.75} = 0.33 \text{ in.} \tag{6}$$

Use the initial conditions to evaluate A_1 and A_2.

$$u(0) = 0 = \frac{U_o}{1 - r^2} + A_1 \tag{7}$$

Therefore,

$$A_1 = -\frac{U_o}{1 - r^2} = -0.33 \text{ in.} \tag{8}$$

$$\dot{u}(0) = 0 = A_2\omega_n \tag{9}$$

Thus,

$$A_2 = 0 \tag{10}$$

Finally,

$$\boxed{u = 0.33[\cos(10t) - \cos(20t)] \text{ in.}} \tag{11}$$

This equation is plotted below.

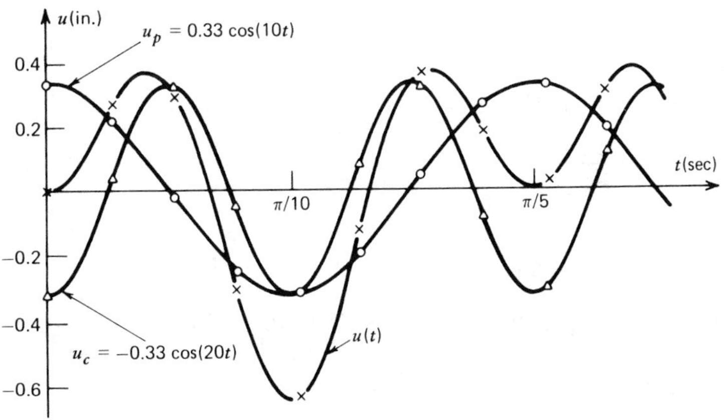

From the responses plotted in Example 4.1 observe the following:

a. The steady-state response has the same frequency as the excitation and is in-phase with the excitation since $r < 1$.

b. The forced motion and natural motion alternately reinforce each other and cancel each other giving the appearance of a *beat phenomenon*. Thus, the total response is not simple harmonic motion.

c. The maximum total response ($u = -0.66$ in. at $t = \pi/10$ s) is greater in magnitude than the maximum steady-state response ($u_p = 0.33$ in. at $t = 0$).

The *total dynamic magnification factor* is defined as

$$D = \max_t \frac{|u(t)|}{U_o} \tag{4.12}$$

In the above example it has the value $0.66/0.33 = 2$. Figure 4.2 shows a plot of the total dynamic magnification factor, D, and the steady-state magnification factor, D_s, as a function of the frequency ratio, r, for an undamped system which is initially at rest and is subjected to harmonic excitation $p_o \sin \Omega t$.

Equations 4.9 and 4.11 are not valid at $r = 1$. The condition $r = 1$, or $\Omega = \omega_n$, is called *resonance,* and it is obvious from Fig. 4.2 that at excitation frequencies near resonance the response becomes very large. The importance attached to the study of the response of structures to harmonic excitation stems, in large part, from the necessity of avoiding the resonance condition, where large amplitude motion can occur. When $r = 1$ it is necessary to replace Eq. 4.2 by the assumed solution

$$u_p = Ct \sin \Omega t, \qquad \Omega = \omega_n \tag{4.13}$$

Then, by substituting Eq. 4.13 into Eq. 4.1 we obtain

$$C = \frac{p_o}{2m\omega_n} \tag{4.14}$$

or

$$u_p = \tfrac{1}{2}(U_o\omega_n t) \sin \omega_n t \tag{4.15}$$

This is plotted in Fig. 4.3 below. Note that, although Fig. 4.2 indicates an infinite amplitude at $r = 1$, Fig. 4.3 shows that the amplitude builds as a linear function of time.

4.2 Response of Viscous-Damped SDOF Systems to Harmonic Excitation

The classical analytical model of a linear SDOF system is the spring-mass-dashpot model of Fig. 3.1. When this system is subjected to harmonic excitation, $p_o \cos \Omega t$, its equation of motion is

$$m\ddot{u} + c\dot{u} + ku = p_o \cos \Omega t \tag{4.16}$$

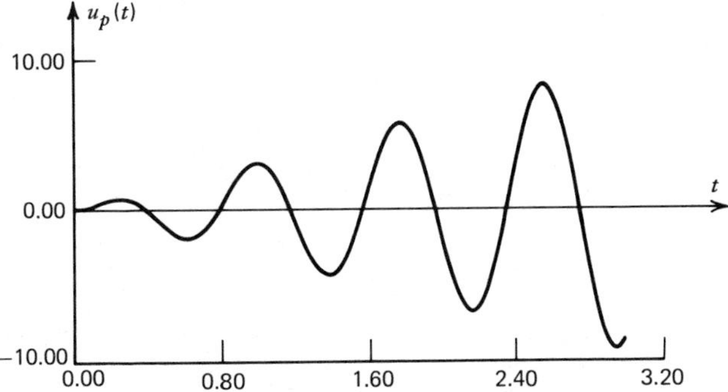

Figure 4.3. *Response* $u_p(t)$ *at resonance,* $\Omega = \omega_n$.

Due to the presence of the damping term in Eq. 4.16, the steady-state response will not be in-phase (or 180° out-of-phase) with the excitation, but will be given by

$$u_p = U \cos(\Omega t - \alpha) \tag{4.17}$$

where U is the *steady-state amplitude* and α is the *phase angle* of the steady-state response relative to the excitation.

The determination of U and α is facilitated by the use of rotating vectors. The velocity and acceleration, needed in Eq. 4.16, are given by

$$\begin{aligned} \dot{u}_p &= -\Omega U \sin(\Omega t - \alpha) \\ \ddot{u}_p &= -\Omega^2 U \cos(\Omega t - \alpha) \end{aligned} \tag{4.18}$$

Figure 4.4 shows the relationship of rotating vectors such that their projections onto the real axis are $p_o \cos \Omega t$ and the displacement, velocity, and acceleration expressions in Eqs. 4.17 and 4.18.

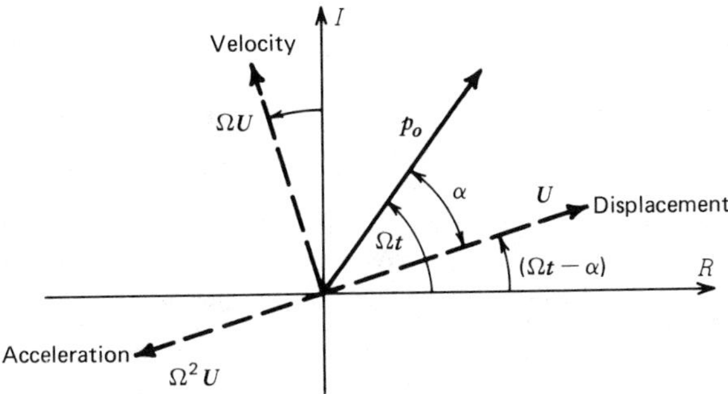

Figure 4.4. *Rotating vectors representing* p, u, u̇, and ü.

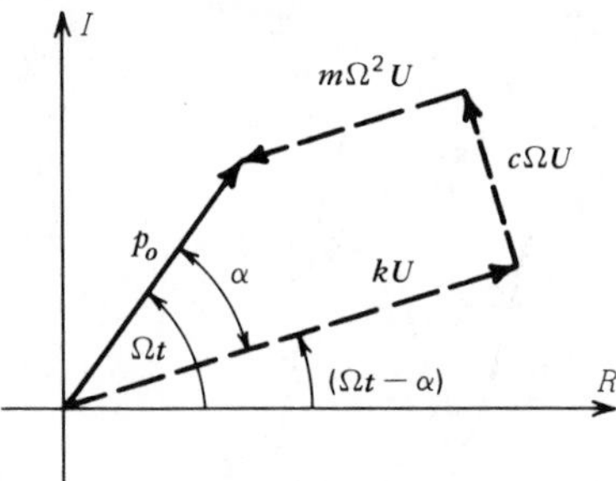

Figure 4.5. *Force vector polygon.*

When Eqs. 4.17 and 4.18 are substituted into Eq. 4.16, there results

$$-m\Omega^2 U\cos(\Omega t - \alpha) - c\Omega U\sin(\Omega t - \alpha) + kU\cos(\Omega t - \alpha) = p_o\cos\Omega t \tag{4.19}$$

This equation is conveniently represented by a *force vector polygon,* since each term in Eq. 4.19 represents a force acting on the mass of Fig. 3.1. Figure 4.5 shows a force vector polygon for the case $m\Omega^2 U < kU$, that is, $\Omega < \omega_n$. The projections of the dotted-line vectors in Fig. 4.5 onto the horizontal (real) axis are the terms on the left-hand side of Eq. 4.19; the projection of the solid vector onto the real axis gives the right-hand side of Eq. 4.19. From Fig. 4.5 it is easily seen that

$$p_o^2 = (kU - m\Omega^2 U)^2 + (c\Omega U)^2 \tag{4.20a}$$

and

$$\tan\alpha = \frac{c\Omega}{k - m\Omega^2} \tag{4.20b}$$

These can be written as

$$D_s = \frac{U}{U_o} = \frac{1}{[(1 - r^2)^2 + (2\zeta r)^2]^{1/2}} \tag{4.21a}$$

and

$$\tan\alpha = \frac{2\zeta r}{1 - r^2} \tag{4.21b}$$

where ζ, r, and U_o have been previously defined. The *steady-state magnification factor,* D_s, and phase angle, α, are plotted in Figs. 4.6. The combination of amplitude and phase information is called the *frequency response* of the system. The plots in Fig. 4.6 are called linear plots since both horizontal and vertical axes have linear scales.

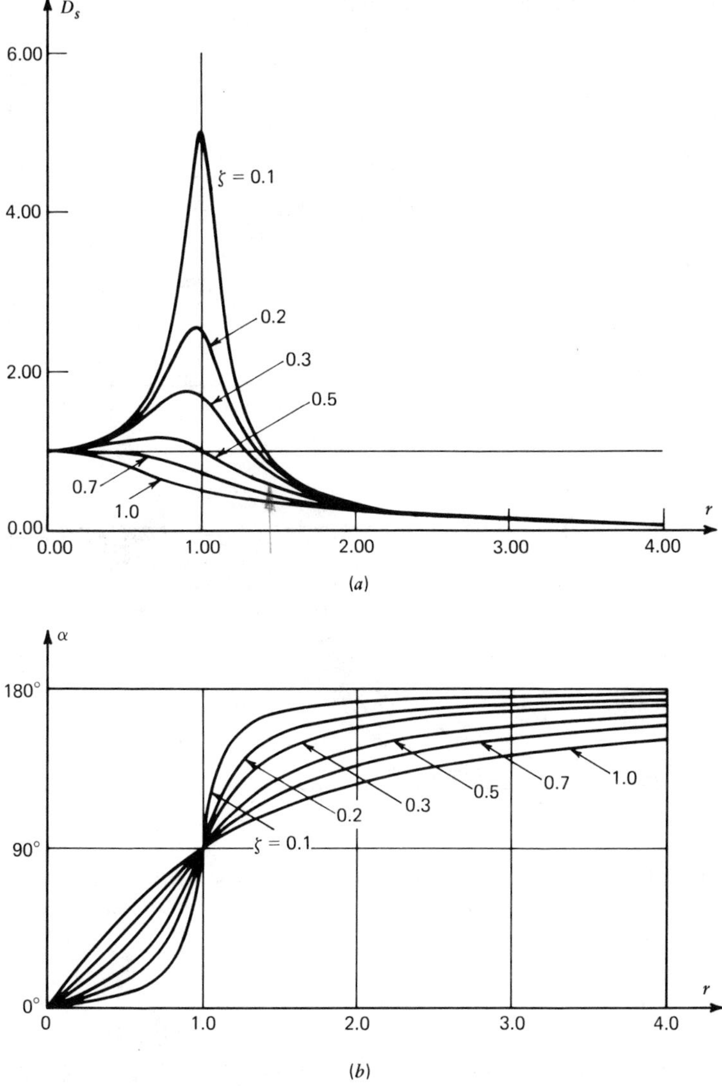

Figure 4.6. (a) *Magnification factor versus frequency ratio for various amounts of damping (linear plot).* (b) *Phase angle versus frequency ratio for various amounts of damping (linear plot).*

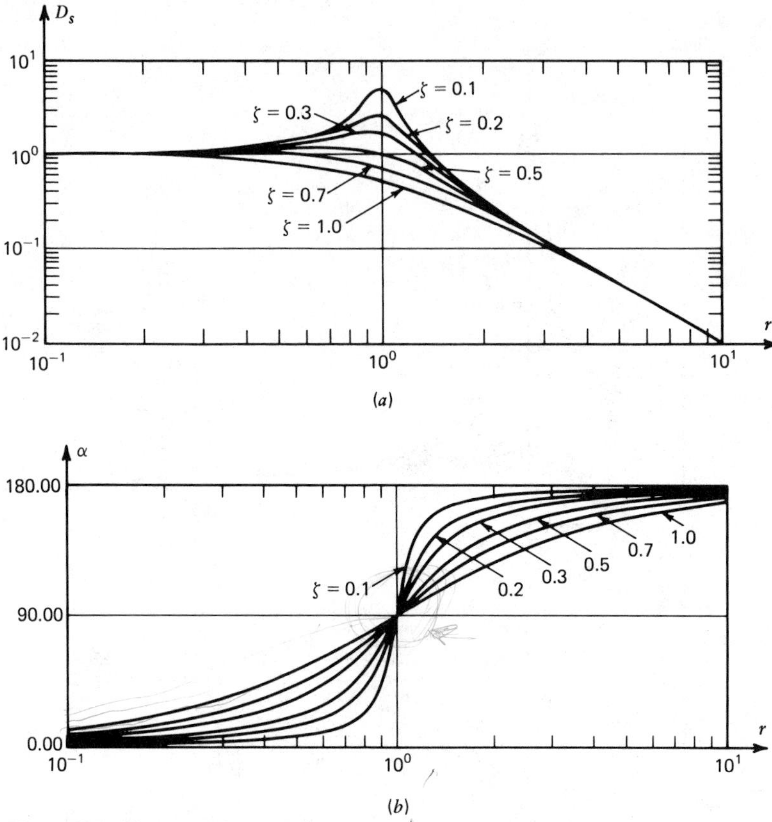

Figure 4.7. (a) *Magnification factor versus frequency ratio for various damping factors (logarithmic plot).* (b) *Phase angle versus frequency ratio for various damping factors (logarithmic frequency scale).*

From Eqs. 4.17 and 4.21 and Figs. 4.6 the following significant features of steady-state response can be observed:

a. The motion described by Eq. 4.17 is harmonic and is of the same frequency as the excitation.

b. The amplitude of the steady-state response is a function of the amplitude and frequency of the excitation as well as of the natural frequency and damping factor of the system. The steady-state magnification factor can be considerably greater than unity or less than unity.

c. The steady-state response, $u_p = U \cos(\Omega t - \alpha)$, and the excitation, $p = p_o \cos \Omega t$, are not in phase; that is, they do not attain their maximum values at the same time. The response lags the excitation by the phase angle α. This corresponds to a time lag α/Ω.

d. At resonance, $r = 1$, the amplitude is limited only by the damping force, and

$$(D_s)_{r=1} = \frac{1}{2\zeta} \tag{4.22}$$

Also, at resonance the response lags the excitation by 90°, that is, $\alpha = 90°$.

Since the dynamic magnification factor may be quite large near resonance and since excitation covering a broad range of frequencies may be of interest, the curves of Figs. 4.6 are frequently plotted to logarithmic scales as shown in Figs. 4.7. This is referred to as a *Bode plot.*

The total response is given by $u = u_p + u_c$ which, from Eqs. 4.17, 4.21, and 3.32 can be written

$$u = \frac{U_o}{[(1 - r^2)^2 + (2\zeta r)^2]^{1/2}} \cos(\Omega t - \alpha) + e^{-\zeta\omega_n t}(A_1 \cos \omega_d t + A_2 \sin \omega_d t) \tag{4.23}$$

where α is given by Eq. 4.21b, and A_1 and A_2 are constants to be determined from the initial conditions. Since the natural motion in Eq. 4.23 dies out with time, it is referred to as a *starting transient.*

Example 4.2

If damping equivalent to $\zeta = 0.2$ is added to the system in Example 4.1 and the same excitation and initial conditions prevail, determine an expression for the resulting motion. Sketch the motion.

Solution

The total response function is given by Eq. 4.23,

$$u = U \cos(\Omega t - \alpha) + e^{-\zeta\omega_n t}(A_1 \cos \omega_d t + A_2 \sin \omega_d t) \tag{1}$$

where

$$U = \frac{U_o}{[(1 - r^2)^2 + (2\zeta r)^2]^{1/2}} \tag{2}$$

ω_n, U_o, and r may be found in Example 4.1.

$$\left.\begin{aligned} \omega_n &= \left(\frac{k}{m}\right)^{1/2} = 20 \text{ rad/s} \\ U_o &= \frac{p_o}{k} = \frac{10}{40} = 0.25 \text{ in.} \\ r &= \frac{\Omega}{\omega_n} = \frac{10}{20} = 0.5 \\ \zeta\omega_n &= (0.2)(20) = 4 \text{ rad/s} \end{aligned}\right\} \tag{3}$$

Therefore, from Eqs. 2 and 3

$$\begin{aligned} U &= \frac{0.25}{\{[1-(0.5)^2]^2 + [2(0.2)(0.5)]^2\}^{1/2}} = \frac{0.25}{[(0.75)^2 + (0.2)^2]^{1/2}} \\ &= \frac{0.25}{\sqrt{0.6025}} = \frac{0.25}{0.776} = 0.32 \text{ in.} \end{aligned} \tag{4}$$

From Eq. 4.21b

$$\tan\alpha = \frac{2\zeta r}{1-r^2} = \frac{2(0.2)(5)}{1-(0.5)^2} = \frac{0.2}{0.75} = 0.267 \tag{5}$$

Therefore,

$$\alpha = 0.26 \text{ rad} \tag{6}$$

From Eq. 3.31a

$$\omega_d = \omega_n\sqrt{1-\zeta^2} = 20\sqrt{1-(0.2)^2} = 20(0.98) = 19.6 \text{ rad/sec} \tag{7}$$

Equation 1 is differentiated with respect to time to give

$$\begin{aligned} \dot{u} &= -\Omega U \sin(\Omega t - \alpha) \\ &\quad + e^{-\zeta\omega_n t}[(A_2\omega_d - A_1\zeta\omega_n)\cos\omega_d t - (A_1\omega_d + A_2\zeta\omega_n)\sin\omega_d t] \end{aligned} \tag{8}$$

Evaluate Eqs. 1 and 8 at $t = 0$ and set equal to the initial conditions.

$$u(0) = 0 = 0.32\cos(-0.26) + A_1 \tag{9}$$

Therefore,

$$A_1 = -0.32\cos(-0.26) = -0.32(0.966) = -0.31 \text{ in.} \tag{10}$$

$$\begin{aligned} \dot{u}(0) = 0 = -(0.32)(10)\sin(-0.26) \\ + [A_2(19.6) - (-0.31)(0.2)(20)] \end{aligned} \tag{11}$$

$$A_2 = -0.11 \text{ in.} \tag{12}$$

Therefore, Eqs. 1, 4, 6, 10, and 12 may be combined to give

$$\boxed{\begin{aligned} u &= 0.32\cos(10t - 0.26) \\ &-e^{-4t}[0.31\cos(19.6t) + 0.11\sin(19.6t)] \text{ in.} \end{aligned}} \tag{13}$$

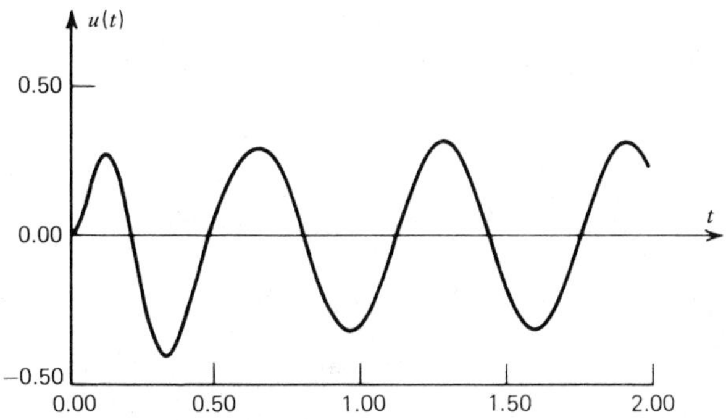

To see the effect of damping, compare the above results with those of Example 4.1.

4.3 Complex Frequency Response

In the previous section the amplitude, U, and phase angle, α, of the steady-state response to excitation, $p_o \cos \Omega t$, were determined by projecting rotating vectors onto the real axis to obtain the terms in Eq. 4.19. The use of vectors in the complex plane will be pursued further in this section because it greatly simplifies many frequency response calculations and also provides a useful way of examining experimental results, as will be seen later.

Consider again the steady-state response of the SDOF system in Fig. 3.1. Let the subscript R (for projection onto the real axis) be used to designate the steady-state, that is, forced, motion due to $\cos \Omega t$ excitation. Thus,

$$m\ddot{u}_R + c\dot{u}_R + ku_R = p_o \cos \Omega t \tag{4.24}$$

and

$$u_R = U \cos(\Omega t - \alpha) \tag{4.25}$$

Likewise, let

$$m\ddot{u}_I + c\dot{u}_I + ku_I = p_o \sin \Omega t \tag{4.26}$$

and

$$u_I = U \sin(\Omega t - \alpha) \tag{4.27}$$

be the corresponding equation of motion and steady-state solution for $\sin \Omega t$ excitation. Now, if Eq. 4.26 is multiplied by $i = \sqrt{-1}$ and added to Eq. 4.24, and Euler's formula, Eq. 3.14, is used, there results

$$m\ddot{\bar{u}} + c\dot{\bar{u}} + k\bar{u} = \bar{p} = p_o e^{i\Omega t} \tag{4.28}$$

where a bar denotes a vector in the complex plane. Equation 4.28 is called the *complex equation of motion,* and the vector

$$\bar{u} = u_R + iu_I \tag{4.29}$$

is called the *complex response.* It is understood that the actual steady-state motion will be given by either the real part of $\bar{u}$ or its imaginary part, depending on whether the excitation is of cos Ωt or sin Ωt type.

The steady-state solution of Eq. 4.28 may be assumed to have the form

$$\bar{u} = \bar{U}e^{i\Omega t} \tag{4.30}$$

where $\bar{U}$ is the complex amplitude, which may also be written

$$\bar{U} = Ue^{-i\alpha} \tag{4.31}$$

where U and α are the same amplitude and phase angle introduced in Eq. 4.17.

By substituting Eq. 4.30 into Eq. 4.28 we obtain directly

$$\bar{U} = \frac{p_o}{(k - m\Omega^2) + ic\Omega} \tag{4.32}$$

which can be written in the form

$$\bar{H}(\Omega) = \frac{\bar{U}}{U_o} = \frac{1}{(1 - r^2) + i(2\zeta r)} \tag{4.33}$$

where $\bar{H}(\Omega)$ is called the *complex frequency response.*

From Eqs. 4.31 and 4.33 it is clear that to determine the amplitude and phase of the steady-state response we need only to find the amplitude and phase of the complex expression on the right-hand side of Eq. 4.33.

Let us summarize a couple of results from the theory of complex numbers:

- *Rectangular and polar representation.* If a complex number (vector) is represented in rectangular form by

$$\bar{A} = A_R + iA_I \tag{4.34a}$$

and in polar form by

$$\bar{A} = Ae^{i\alpha} \tag{4.34b}$$

then,

$$A \equiv |\bar{A}| = \sqrt{A_R^2 + A_I^2} \tag{4.34c}$$

and

$$\tan\alpha = \frac{A_I}{A_R} \tag{4.34d}$$

• *Quotient of two complex numbers.* If $\overline{A}$ and $\overline{B}$ are two complex numbers, then

$$\frac{\overline{B}}{\overline{A}} = \left(\frac{Be^{i\beta}}{Ae^{i\alpha}}\right) = \left(\frac{B}{A}\right) e^{i(\beta-\alpha)} \tag{4.35}$$

By using Eqs. 4.34 to express the denominator of Eq. 4.33 in polar form and Eq. 4.35 to obtain the amplitude and phase of the quotient in Eq. 4.33, we obtain

$$|\overline{H}(\Omega)| = \frac{U}{U_o} = \frac{1}{[(1-r^2)^2 + (2\zeta r)^2]^{1/2}} \tag{4.36a}$$

and

$$\tan\alpha = \frac{2\zeta r}{1-r^2} \tag{4.36b}$$

which are the same results as obtained in Eqs. 4.21. Thus, the complex frequency response, $\overline{H}(\Omega)$, contains both the magnitude and phase of the steady-state response, and Eqs. 4.34 and 4.35 can be used to extract this magnitude and phase information quite easily.

In summary, the four steps employed in using complex vectors to determine the steady-state response are:

1. Write the differential equation in terms of complex excitation and complex response, Eq. 4.28.

2. Assume a solution with complex amplitude $\overline{U}$ as in Eq. 4.30.

3. Substitute the assumed response into the differential equation to get an expression for the complex frequency response $\overline{H}(\Omega)$.

4. Use Eqs. 4.34 and 4.35 to obtain the amplitude and phase of the complex frequency response.

The force vector polygon employed in Sec. 4.2 can now be related directly to the complex differential equation, Eq. 4.28. By differentiating Eq. 4.30 we obtain

$$\begin{aligned} \dot{\overline{u}} &= i\Omega\overline{U}e^{i\Omega t} = i\Omega\overline{u} \\ \ddot{\overline{u}} &= (i\Omega)^2\overline{U}e^{i\Omega t} = -\Omega^2\overline{u} \end{aligned} \tag{4.37}$$

Figures 4.4 and 4.5 can thus be relabeled in terms of complex vectors as shown in Fig. 4.8*a*, and the force vector polygon in Fig. 4.8*b* directly represents Eq. 4.28.

The results given by Eqs. 4.21a and 4.21b were plotted as amplitude and

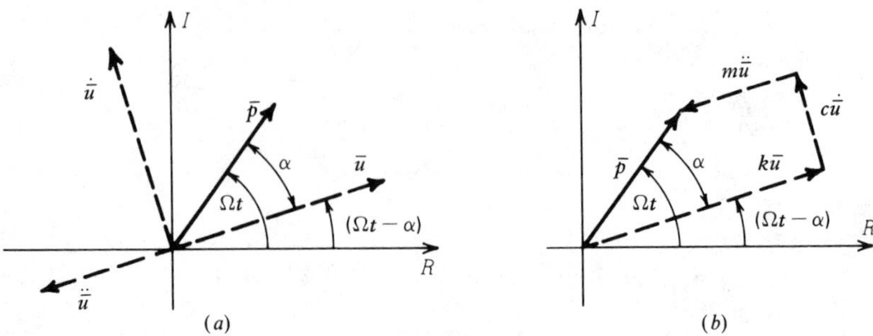

Figure 4.8. Complex vector notation for rotating vectors.

phase plots in Figs. 4.6. The amplitude and phase information can be combined in a single plot by plotting the results of Eq. 4.33 on the complex plane. The resulting plot is called a *vector response plot.* It is also called by some authors a Nyquist plot or a Argand plot. Figure 4.9 shows vector response plots for $\zeta = 0.1$ and $\zeta = 0.05$.

The vector response diagram is very useful in examining experimental results in structural dynamics as will be seen later. It can also be helpful to plot $R(\overline{H})$ versus frequency and $I(\overline{H})$ versus frequency.

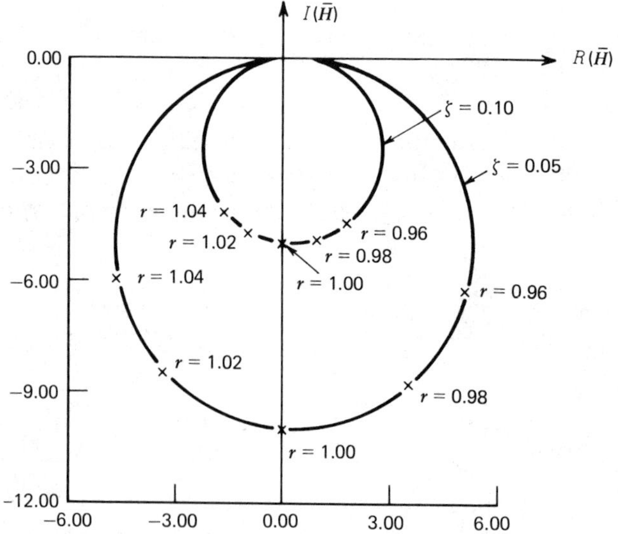

Figure 4.9. Vector response plot for steady-state vibration of a viscous-damped system.

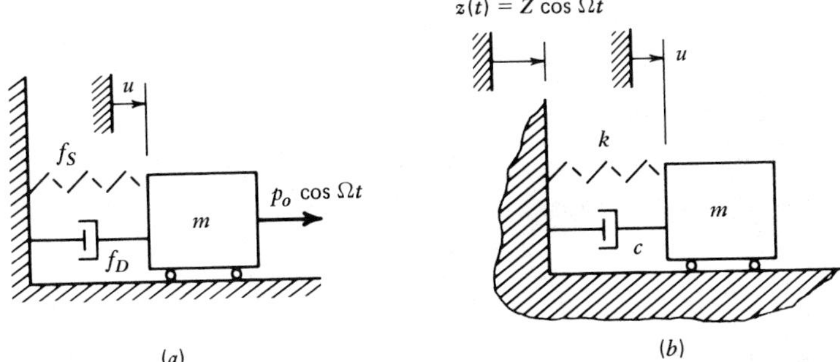

Figure 4.10. Vibration isolation situations. (a) *Force transmitted to stationary base.* (b) *Moving base.*

4.4 Vibration Isolation—Force Transmissibility and Base Motion

Having considered the basic frequency response behavior of a SDOF system, we will now consider two important topics related to frequency response: (a) the force that is transmitted through the spring and dashpot of a SDOF system to the supporting fixed base, and (b) the motion imparted to a mass when the base to which its spring and dashpot supports are attached is moving. Figure 4.10 illustrates these situations. The former situation arises when, for example, a machine is attached to a floor structure by some shock isolation mounts which may be modeled as a combination of a spring and viscous dashpot. There are numerous applications of the moving base problem: earthquake motion of a building, motion of a car over a rough road, and so forth. It is convenient, although not essential, to use the complex frequency response technique to solve these two vibration isolation problems.

Force Transmissibility

From Fig. 4.8*b* the force transmitted to the base can be written in vector form as

$$\bar{f}_{\text{tr}} = \bar{f}_S + \bar{f}_D = k\bar{u} + c\dot{\bar{u}} \tag{4.38}$$

Incorporating Eqs. 4.30 and 4.37*a*, we get

$$\bar{f}_{\text{tr}} = (k + ic\Omega)\bar{U}e^{i\Omega t} \tag{4.39}$$

The expression for $\bar{U}$ in Eq. 4.33 can be inserted into Eq. 4.39 to give

$$\bar{f}_{\text{tr}} = \left[\frac{(k + ic\Omega)U_o}{(1 - r^2) + i(2\zeta r)}\right] e^{i\Omega t} \tag{4.40}$$

or

$$\bar{f}_{tr} = \left[\frac{1 + i(2\zeta r)}{(1 - r^2) + i(2\zeta r)}\right] kU_o e^{i\Omega t} \tag{4.41}$$

Since $|e^{i\Omega t}| = 1$ and since, from Eq. 4.35, $|\bar{B}/\bar{A}| = |\bar{B}|/|\bar{A}|$, the magnitude of the transmitted force is found to be

$$|\bar{f}_{tr}| = \frac{kU_o[1 + (2\zeta r)^2]^{1/2}}{[(1 - r^2)^2 + (2\zeta r)^2]^{1/2}} \tag{4.42}$$

The *transmissibility* is defined as the ratio of the dynamic force $|\bar{f}_{tr}|$ to the force $p_o = kU_o$, which would be transmitted to the base if the force p_o were applied statically. Thus,

$$TR = \frac{|\bar{f}_{tr}|}{kU_o} = D_s[1 + (2\zeta r)^2]^{1/2} \tag{4.43}$$

Figure 4.11 shows the value of TR as a function of frequency ratio. Two important conclusions can be drawn from the curves of Fig. 4.11: (a) the force transmitted to the base dynamically is less than the static force only if $r > \sqrt{2}$, and (b) decreasing the damping decreases the transmitted force provided $r > \sqrt{2}$. Further details on the design of vibration isolation mounting may be found in References 4.1 and 4.2.

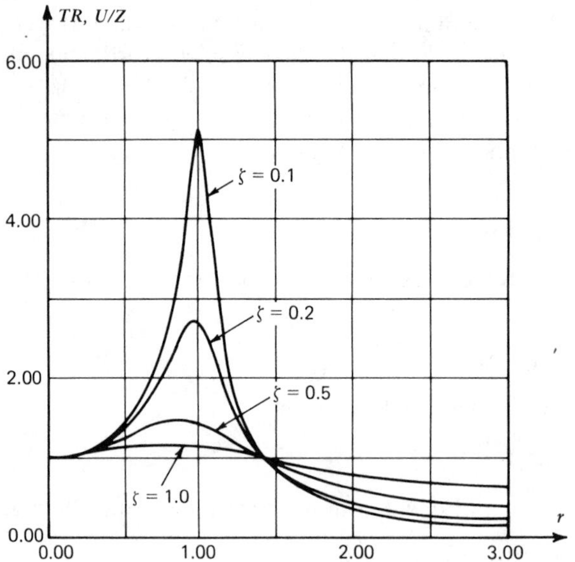

Figure 4.11. *Transmissibility, absolute response to base excitation.*

Base Motion

The equations of motion for absolute motion, u, and the relative motion, $w = u - z$, were derived in Example 2.2. We will consider the case when $p(t) = 0$ and when the base undergoes simple harmonic motion $z = Z\cos\Omega t$. The complex equations of motion can be written

$$m\ddot{\bar{u}} + c\dot{\bar{u}} + k\bar{u} = c\dot{\bar{z}} + k\bar{z} = (k + ic\Omega)Ze^{i\Omega t} \tag{4.44}$$

and

$$m\ddot{\bar{w}} + c\dot{\bar{w}} + k\bar{w} = -m\ddot{\bar{z}} = \Omega^2 mZe^{i\Omega t} \tag{4.45}$$

Assuming complex steady-state responses of the form

$$\bar{u} = \bar{U}e^{i\Omega t}, \qquad \bar{w} = \bar{W}e^{i\Omega t} \tag{4.46}$$

we get the complex frequency response functions

$$\frac{\bar{U}}{Z} = \frac{k + ic\Omega}{(k - m\Omega^2) + ic\Omega} = \frac{1 + i(2\zeta r)}{(1 - r)^2 + i(2\zeta r)} \tag{4.47}$$

and

$$\frac{\bar{W}}{Z} = \frac{m\Omega^2}{(k - m\Omega^2) + ic\Omega} = \frac{r^2}{(1 - r^2) + i(2\zeta r)} \tag{4.48}$$

From the above complex frequency response functions we can determine the magnitude of the absolute and relative responses as functions of frequency.

$$\frac{U}{Z} \equiv \frac{|\bar{U}|}{Z} = D_s[1 + (2\zeta r)^2]^{1/2} \tag{4.49}$$

and

$$\frac{W}{Z} \equiv \frac{|\bar{W}|}{Z} = r^2 D_s \tag{4.50}$$

Since U/Z given by Eq. 4.49 is equal to TR, the plot of the absolute response is given by Fig. 4.11. The relative response is shown in Fig. 4.12.

The following important conclusions can be drawn from the relative motion frequency response function shown in Fig. 4.12:

1. When $\Omega \ll \omega_n$, that is, $r \ll 1$, there is little relative motion between the mass and the base, that is, the mass moves with the base.

2. When $\Omega \doteq \omega_n$ the usual resonance phenomenon is observed. That is, for small base motion there is a large amplitude of relative motion with only the damping force limiting the amplitude.

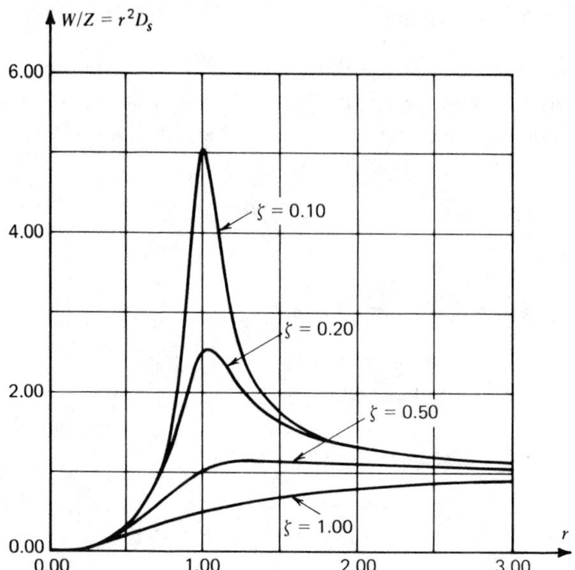

Figure 4.12. *Relative motion frequency response function.*

3. When $\Omega \gg \omega_n$ the inertia of the mass keeps it from moving much, so that the relative motion consists primarily of the base moving relative to the mass.

The above analysis of base motion will be very important to the study of motion measuring devices in Sec. 4.5.

Example 4.3

A vehicle is a complex system with many degrees of freedom. However, the following SDOF analytical model may be employed in a "first-approximation" study of ride quality of the vehicle. The steady-state magnification factor for the vehicle's absolute motion is to be determined for the vehicle when fully loaded and when empty, if the vehicle is traveling at 100 km/hr over a road whose surface has a sinusoidally varying roughness with a "period" (see sketch) of 4m. The mass of the vehicle is 1200 kg when fully loaded and 400 kg when empty, and the effective spring constant is 400 kN/m. The damping factor is $\zeta_f = 0.4$ when the vehicle is fully loaded.

Make the following assumptions:

a. As the vehicle moves forward at constant speed, only the vertical motion, $u(t)$, need be considered.

b. The tires are infinitely stiff, that is, $z(t)$ represents the motion of the axle of the vehicle.

c. The tires remain in contact with the road.

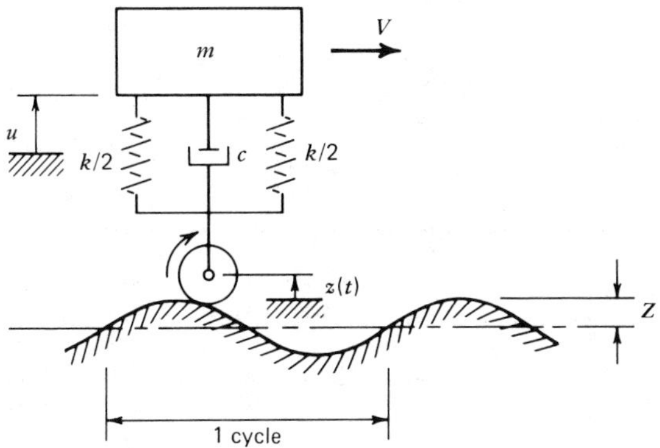

Solution

a. Compute the excitation frequency:

$$\Omega = \frac{(100{,}000 \text{ m/hr})(6.28 \text{ rad/cycle})}{(4 \text{ m/cycle})(3600 \text{ s/hr})} \tag{1}$$

$$\Omega = 43 \text{ rad/s} \tag{2}$$

b. Compute the damping factor for the empty vehicle:

From Eqs. 3.4,

$$c = 2\zeta\sqrt{km} \tag{3}$$

Since c and k do not change, but m does,

$$c = 2\zeta_f\sqrt{km_f} = 2\zeta_e\sqrt{km_e} \tag{4}$$

where the subscripts designate the full and empty vehicle configurations. Thus,

$$\zeta_e = \zeta_f\left(\frac{m_f}{m_e}\right)^{1/2} = 0.4\left(\frac{1200}{400}\right)^{1/2} \tag{5}$$

or

$$\zeta_e = 0.693 \tag{6}$$

c. Compute the magnification factors:
From Eq. 4.49 and Eq. 4.21a

$$\frac{U}{Z} = \frac{[1 + (2\zeta r)^2]^{1/2}}{[(1 - r^2)^2 + (2\zeta r)^2]^{1/2}} \tag{7}$$

The calculations are presented in the table below.

	Empty	*Full*
$\omega_n = \sqrt{k/m}$	$[400(10^3)/400]^{1/2} = 31.6$ rad/s	$[400(10^3)/1200]^{1/2} = 18.3$ rad/s
$r = \Omega/\omega_n$	$43.6/31.6 = 1.38$	$43.6/18.3 = 2.38$
$2\zeta r$	$2(0.693)(1.38) = 1.91$	$2(0.4)(2.38) = 1.90$
$(1 - r^2)^2$	$[1 - (1.38)^2]^2 = 0.818$	$[1 - (2.38)^2]^2 = 21.8$
$[1 + (2\zeta r)^2]^{1/2}$	$(1 + 3.65)^{1/2} = 2.16$	$(1 + 3.61)^{1/2} = 2.15$
$[(1 - r^2)^2 + (2\zeta r)^2]^{1/2}$	$(0.818 + 3.65)^{1/2} = 2.11$	$(21.8 + 3.61)^{1/2} = 5.04$
U/Z	$2.16/2.11 = 1.02$	$2.15/5.04 = 0.43$

Note that both damping factor and natural frequency change when the vehicle weight changes and that both of these factors enter into the magnification factor calculation.

4.5 Vibration Measuring Instruments

In Chapter 1 the importance of vibration testing was discussed. Figure 4.13 shows the major stages of a vibration measurement system[4.3]. The system consists of a motion detector-transducer, an intermediate signal modification system (e.g., amplifier), and a display system (e.g., oscilloscope). The motion quantity to be measured may be displacement, velocity, or acceleration. The purpose of the detector-transducer stage is to detect the desired motion quantity and, in most cases, to produce an output that is proportional to the input motion but of different form. The most widely used motion transducer is the accelerometer, a device that senses acceleration and produces an output voltage proportional to the input acceleration. Other motion transducers are the vibrometer for measuring displacement, the electrodynamic velocity transducer, the eddy current displacement sensor, and the optical interferometer[4.4].

Vibrometers and accelerometers are called seismic transducers. A *seismic*

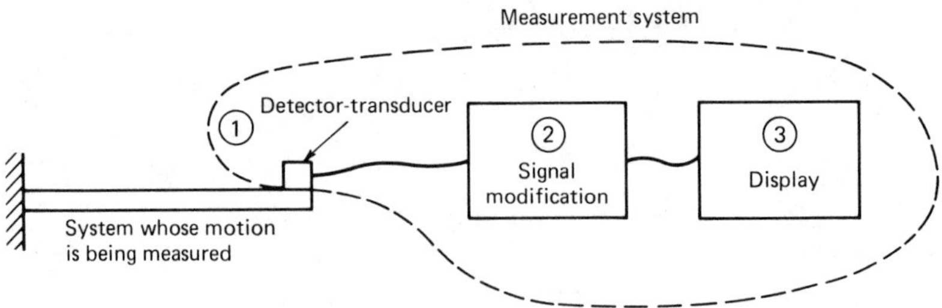

Figure 4.13. Stages in a motion measurment system.

transducer is one that employs a spring-mass system to measure motion. The behavior of seismic transducers can be analyzed using the results of Sec. 4.4. Figure 4.14 shows a schematic diagram of a seismic transducer, whose principal ingredient is a SDOF vibratory system. A mass is attached by a spring to the base of the transducer. Viscous damping is sometimes, but not always, provided. The base of the transducer is firmly attached to the specimen whose motion is to be measured. A relative motion instrument (RMI) measures, as a function of time, the relative motion between the moving mass (point a) and the base (point b). The RMI usually provides an electrical signal as its output, the signal being proportional to the relative displacement of a and b. Thus, it is necessary to consider how the relative motion

$$w = u - z \tag{4.51}$$

is related to the base motion $z(t)$. The response to harmonic base motion $z = Z \cos \Omega t$ is considered.

Vibrometer

A *vibrometer* is a seismic instrument whose output is to be proportional to the displacement of the base, that is, $w(t)$ is to be proportional to $z(t)$. From Eqs. 4.50 and 4.21 the steady-state relative displacement amplitude is related to the input displacement amplitude by

$$\frac{W}{Z} = \frac{r^2}{[(1 - r^2)^2 + (2\zeta r)^2]^{1/2}} \tag{4.52}$$

Figure 4.12 is a plot of this frequency response function. From this figure it can be seen that $W/Z \rightarrow 1$ as r becomes very large. To be useful in measuring the displacement $z(t)$, then, a vibrometer must have a low natural frequency relative to the excitation frequency. Since a system with a low natural fre-

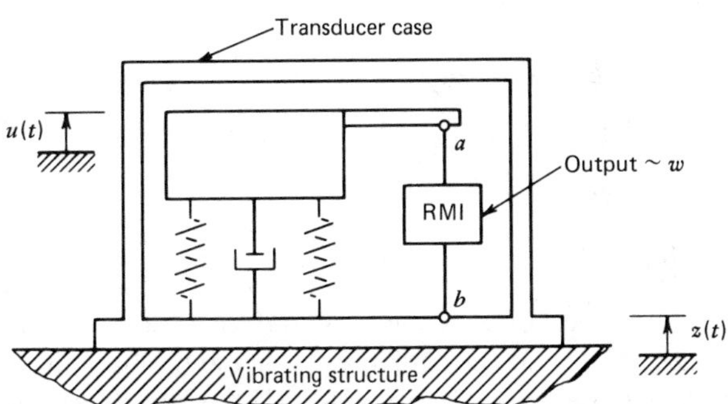

Figure 4.14. *Schematic diagram of a seismic transducer.*

quency has a (relatively) large mass and soft spring, and consequently has a large static displacement, vibrometers are seldom used other than for some seismological measurements.

Accelerometer

An *accelerometer* is a seismic instrument whose output is proportional to the base acceleration, $\ddot{z}(t)$. Consider first the steady-state output, $w_p(t)$, due to harmonic base motion. The base acceleration is

$$a_z(t) = \ddot{z} = -\Omega^2 Z \cos \Omega t \tag{4.53}$$

We need to determine how the output, $w_p(t)$, can be employed to represent the time history of the acceleration, $a_z(t)$, whose amplitude is $\Omega^2 Z$. Equation 4.52 gives the amplitude of w_p, which we can write in the form

$$W = (\Omega^2 Z)\left(\frac{1}{\omega_n^2}\right) D_s \tag{4.54}$$

where D_s is the frequency response function plotted in Fig. 4.6*a*. From Eq. 4.48 it can be seen that w_p has a phase lag given by

$$\tan \alpha = \frac{2\zeta r}{1 - r^2} \tag{4.55}$$

Thus,

$$w_p(t) = \left(\frac{1}{\omega_n^2}\right)(\Omega^2 Z) D_s \cos \Omega \left(t - \frac{\alpha}{\Omega}\right) \tag{4.56}$$

For the accelerometer to be useful over a range of frequency without having to calibrate it at each frequency of interest [in effect determine $D_s(\Omega)$] it is desirable to select an operating frequency range within which D_s is approximately constant. From Fig. 4.6*a* it can be seen that this condition holds for $r \ll 1$, where $D_s \doteq 1$.

Consider a more general acceleration input of the form

$$a_z(t) = A_1 \cos \Omega_1 t + A_2 \cos \Omega_2 t \tag{4.57}$$

The desired form of the output is

$$w_p(t) = C[A_1 \cos \Omega_1 (t - \tau) + A_2 \cos \Omega_2 (t - \tau)] \tag{4.58}$$

that is, the output may be shifted along the time axis by an amount τ and may be scaled by a constant C, but the components should not be shifted in time by different amounts or scaled by different factors. If the former occurs, the transducer is said to introduce *phase distortion;* if the latter occurs, it is said to produce *amplitude distortion.* Most modern accelerometers have very low damping. Hence, they produce little phase distortion. To minimize amplitude distortion the natural frequency of the accelerometer must be much greater than the highest input frequency, for example, $\omega_n > 10\Omega_{max}$. By introducing damping of $\zeta \doteq 0.7$ the frequency range of an accelerometer can be extended to about $0.6\omega_n$[4.5]. References 4.6 through 4.9 may be consulted for further discussion of accelerometer performance characteristics and the factors to be considered in selecting an accelerometer. Reference 4.20 is a monograph devoted to techniques for the calibration of shock and vibration measuring transducers.

4.6 Use of Frequency Response Data to Determine Natural Frequency and Damping Factor of a Lightly Damped SDOF System

In Sec. 3.3 procedures for using free vibration response to determine the natural frequency and damping factor of a lightly damped SDOF system were discussed. Forced vibration using harmonic or nonharmonic excitation may also be employed to determine these system parameters experimentally. The forced-vibration techniques may also be employed to determine system parameters for MDOF systems[4.10–4.16]. Here we will consider the viscous-damped SDOF system shown in Fig. 3.1. It is assumed that frequency response (magnitude and phase) information has been determined experimentally. This may be plotted in magnitude/ phase form as in Fig. 4.6, in vector response plot form as in Fig. 4.9, or in real/imaginary form.

Undamped Natural Frequency

As seen in Fig. 4.6*a*, when the damping is small, a sharp resonance peak occurs, with the maximum amplitude occurring at an excitation frequency just slightly below the undamped natural frequency. From Fig. 4.6*b* it is seen that the phase angle of the steady-state response is 90° when the excitation frequency is equal to the natural frequency of the system. From Fig. 4.9 it can be seen that in the vicinity of $r = 1$ (i.e., $\Omega = \omega_n$) the increment, Δs, of arc length along the frequency response curve corresponding to a given frequency increment (e.g., $\Delta r = 0.01$ or $\Delta\Omega = 0.01\omega_n$) is a maximum, that is $\Delta s/\Delta\Omega = \max$ when $\Omega = \omega_n$. Thus, the undamped natural frequency of a SDOF system may be determined quite accurately from the frequency response data by noting the excitation frequency at which any of the following occurs:

a. The response lags the input by 90° (this implies also that $R(\overline{U}) = 0.0$).

b. The response magnitude is a maximum.

c. The imaginary part of the response, $|I(\overline{U})|$, is a maximum.

d. The spacing on the vector response plot, $\Delta s/\Delta\Omega$, is a maximum.

Damping Factor

Although, from Eq. 4.22 the amplitude at resonance is given by

$$\left(\frac{U}{U_o}\right)_{r=1} = \frac{1}{2\zeta} \tag{4.59}$$

the use of this equation to determine ζ would require determination of U_o, the static deflection. Consequently, this equation is not generally employed for determining ζ. Another method, called the *half-power method,* will be described. A portion of a frequency response curve is shown in Fig. 4.15. The frequencies above and below resonance at which the response amplitude is $\sqrt{2}/2$ times the resonant response amplitude are referred to as the *half-power points.* Let their frequencies be called Ω_1 and Ω_2 with corresponding frequency ratios r_1 and r_2. These frequencies may be obtained by letting $U_i = (\sqrt{2}/2)U_{r=1}$ and using Eq. 4.21*a*. Upon squaring Eq. 4.21*a* we get

$$\left(\frac{U}{U_o}\right)^2 = \frac{1}{(1 - r^2)^2 + (2\zeta r)^2} \tag{4.60}$$

Letting $U_i = (\sqrt{2}/2)U_{r=1}$ and using Eq. 4.59 we obtain

$$\frac{1}{2}\left(\frac{1}{2\zeta}\right)^2 = \frac{1}{(1 - r_i^2)^2 + (2\zeta r_i)^2} \tag{4.61}$$

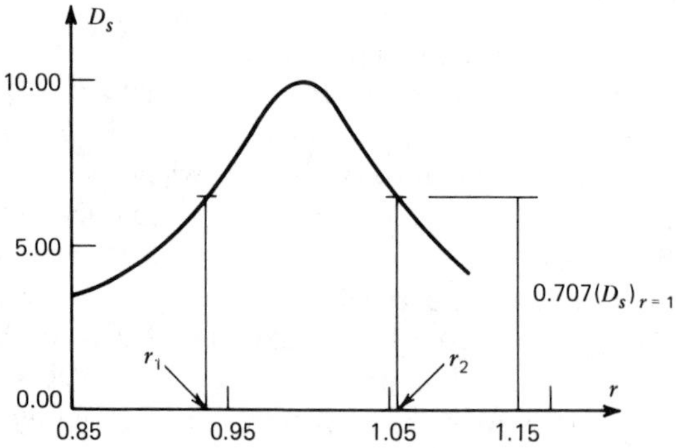

Figure 4.15. Response curve showing half-power points.

where $r_i = r_1$ or r_2. Equation 4.61 may be rewritten

$$r_i^4 - 2(1 - 2\zeta^2)r_i^2 + (1 - 8\zeta^2) = 0 \tag{4.62}$$

whose roots are given by

$$r_i^2 = (1 - 2\zeta^2) \pm 2\zeta\sqrt{1 + \zeta^2} \tag{4.63}$$

Assuming $\zeta \ll 1$ and neglecting higher order terms in ζ, we arrive at the result

$$r_i^2 = 1 \pm 2\zeta \tag{4.64}$$

Using the binomial expansion, we get

$$\begin{aligned} r_2 &= (1 + 2\zeta)^{1/2} = 1 + \tfrac{1}{2}(2\zeta) + \cdots \\ r_1 &= (1 - 2\zeta)^{1/2} = 1 - \tfrac{1}{2}(2\zeta) + \cdots \end{aligned} \tag{4.65}$$

or, since $\zeta \ll 1$,

$$r_2 - r_1 = 2\zeta \tag{4.66}$$

The half-power method for measuring the damping factor is based on Eq. 4.66. The procedure is as follows:

a. Determine the undamped natural frequency by one of the methods described above, and then determine the resonant response amplitude, $U_{r=1}$.

b. Note the points on the response amplitude curve where the amplitude is $(\sqrt{2}/2)U_{r=1}$. Call the corresponding frequencies Ω_1 and Ω_2.

c. Then, from Eq. 4.66,

$$\zeta = \frac{r_2 - r_1}{2} = \frac{1}{2}\left(\frac{\Omega_2 - \Omega_1}{\omega_n}\right) \tag{4.67}$$

The accuracy with which ζ is determined using Eq. 4.67 depends on the frequency resolution in the original frequency response data.

A related procedure using spacing of points along the vector response plot may also be used in determining damping[4.17,4.18]. The procedure is sometimes referred to as the Kennedy-Pancu method or the circle-fit method.

4.7 Equivalent Viscous Damping

Damping is present in all oscillatory systems. The primary effect of the damping is to remove energy from the system. This loss of energy from the damped system results in the decay of amplitude of free vibration, as seen in Chapter 3. In steady-state forced vibration the loss of energy is balanced by the energy that is supplied by the excitation. There are many different mechanisms which can cause damping in a system: internal friction, fluid resistance, sliding friction, and so forth. Linear viscous damping provides the simplest mathematical model of damping, namely, a force directly proportional to the velocity. Even when the true damping in a system is far more complex than linear viscous damping, it may be possible to retain the simplicity of the linear viscous damping model by introducing an "equivalent viscous damping." This section will indicate how such equivalent viscous damping may be determined.

As indicated above, one of the principal effects of damping in a system is to remove energy from the system. Hence, the concept of equivalent viscous damping is based on the equivalence of energy removed by a viscous damping mechanism and by the given nonviscous damping mechanism.

Figure 4.16 shows two forces acting on a body, the elastic spring force, f_S, which is associated with the potential energy of the system, and the damping force, f_D, associated with energy dissipation. The work done by these forces is W_S and W_D, respectively, given by

$$W_S = \int_{u_i}^{u_f} f_S \, du = \int_{t_i}^{t_f} f_S \dot{u} \, dt \tag{4.68}$$

and

$$W_D = \int_{t_i}^{t_f} f_D \dot{u} \, dt \tag{4.69}$$

Energy dissipation is usually calculated for one cycle of harmonic motion. Consider the energy dissipation in a SDOF system with a viscous dashpot. Then

$$f_D = -c\dot{u} \tag{4.70}$$

For steady-state motion the displacement and velocity are given by

$$\begin{aligned} u &= U \cos(\Omega t - \alpha) \\ \dot{u} &= -U\Omega \sin(\Omega t - \alpha) \end{aligned} \tag{4.71}$$

The work done per cycle by f_D is thus

$$\begin{aligned} W_D &= \int_0^{2\pi/\Omega} (-c\dot{u})\,\dot{u} \, dt \\ &= -c\Omega^2 U^2 \int_0^{2\pi/\Omega} \sin^2(\Omega t - \alpha) \, dt \end{aligned}$$

or

$$W_D = -\pi c \Omega U^2 \tag{4.72}$$

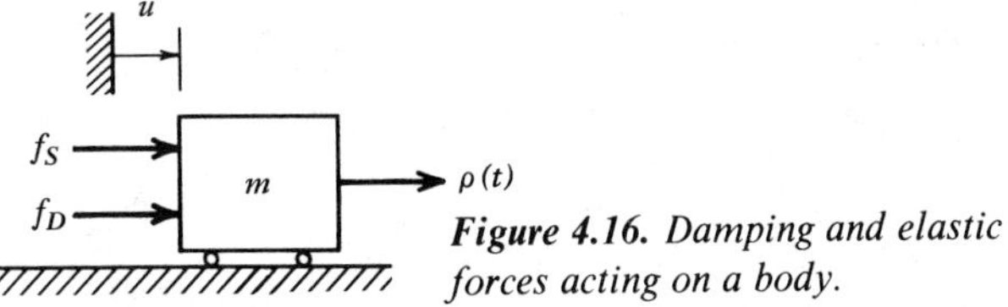

Figure 4.16. Damping and elastic forces acting on a body.

The energy dissipated per cycle by viscous damping is $(\pi c \Omega U^2)$. Thus, the energy loss per cycle due to linear viscous damping is proportional to the damping coefficient, to the excitation frequency, and to the square of the amplitude.

On Fig. 4.17 is plotted the total force $f_S + f_D$ acting on the system in Fig. 4.16, where $f_S = -ku$, $f_D = -c\dot{u}$, and where the system is executing steady-state harmonic motion. The lower portion of the curve is for motion with $\dot{u} > 0$ (i.e., the body moving to the right). The upper of the curve is for the portion of the cycle when $\dot{u} < 0$ (i.e., the body moving to the left). The triangular area marked by diagonal cross-hatching is the peak potential energy stored elastically; the area inside the elliptical curve is the energy dissipated per cycle by the damping force, that is, W_D. The latter is frequently referred to as a *hysteresis loop*. The loop is a characteristic of dissipative forces.

If the damping in a system is not of the linear viscous damping type, then an *equivalent viscous damping coefficient* may be defined by

$$c_{\text{eq}} = -\frac{W_D}{\pi \Omega U^2} \tag{4.73}$$

where W_D is the energy dissipated by the nonviscous damping mechanism. Example 4.4 illustrates the calculation of c_{eq}.

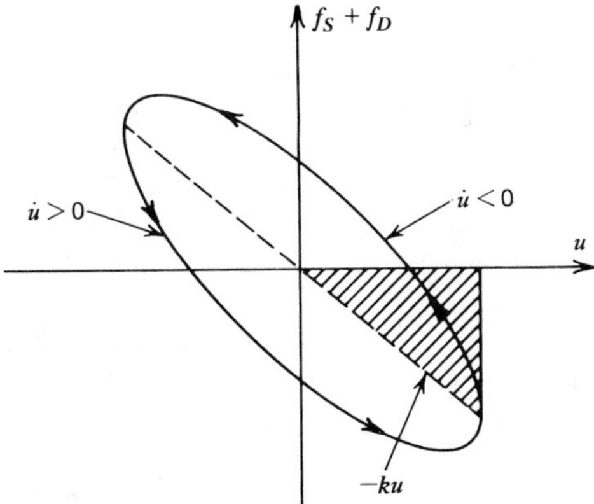

Figure 4.17. Force versus displacement for a system with linear spring and viscous dashpot.

Example 4.4

Bodies moving with moderate speed in a fluid experience a resisting force that is proportional to the square of the speed. Determine the equivalent viscous damping coefficient for this type of damping.

Solution

Let the damping force be expressed by the equation

$$f_D = \pm a\dot{u}^2 \tag{1}$$

where the negative sign is used when $\dot{u}$ is positive, and the positive sign when $\dot{u}$ is negative. Assume harmonic motion with time measured from the position of extreme negative displacement. Then

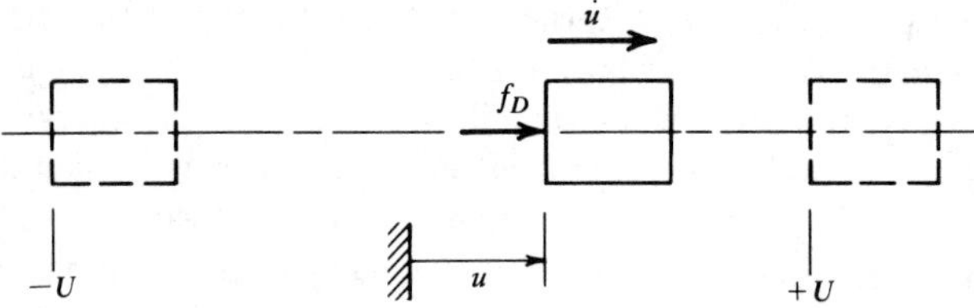

$$u = -U\cos\Omega t \tag{2}$$
$$\dot{u} = U\Omega\sin\Omega t \tag{3}$$

The energy dissipated per cycle is

$$\begin{aligned} W_D &= -2\int_{-U}^{U} a\dot{u}^2\,du = -2\int_0^{\pi/\Omega} a\dot{u}^3\,dt \\ W_D &= -2a\Omega^3U^3\int_0^{\pi/\Omega}\sin^3(\Omega t)\,dt \end{aligned} \tag{4}$$

Then,

$$W_D = -\left(\frac{8}{3}\right)a\Omega^2U^3 \tag{5}$$

From Eq. 4.73

$$c_{\text{eq}} = \frac{-W_D}{\pi\Omega U^2} = \frac{(8/3)a\Omega^2U^3}{\pi\Omega U^2}$$

or

$$\boxed{c_{\text{eq}} = \left(\frac{8}{3\pi}\right)a\Omega U} \tag{6}$$

It is convenient to define an equivalent damping factor ζ_{eq} using Eq. 4.73 and Eqs. 3.4b and 3.4c. Then,

$$\zeta_{eq} = \frac{c_{eq}}{c_c} = -\frac{W_D/\pi\Omega U^2}{2k/\omega_n} \tag{4.74}$$

But k can be determined from the elastic energy (shown cross-hatched on Fig. 4.17).

$$W_S = (\tfrac{1}{2})kU^2 \tag{4.75}$$

Thus,

$$\zeta_{eq} = \frac{-W_D}{W_S}\left(\frac{\omega_n}{4\pi\Omega}\right) \tag{4.76}$$

where Ω is the forcing frequency at which W_D is obtained.

4.8 Structural Damping

The complex notation introduced in Sec. 4.3 is particularly well suited to the introduction of a type of damping frequently employed in structural dynamics analysis, for example in aircraft vibration and flutter studies. This type of damping, called *structural damping,* is proportional to displacement but in-phase with the velocity of a harmonically oscillating system. Theodorsen and Garrick, in an early study of flutter[4.19], used this form of damping. The usual way in which it is introduced into a SDOF system's equation of motion is to write

$$m\ddot{\bar{u}} + k(1 + i\gamma)\bar{u} = p_o e^{i\Omega t} \tag{4.77}$$

where γ is the *structural damping factor.** The quantity $k(1 + i\gamma)$ is called the *complex stiffness.*

Assuming a solution of the form

$$\bar{u} = \bar{U}e^{i\Omega t} \tag{4.78}$$

and substituting this solution into Eq. 4.77, we obtain

$$\bar{U} = \frac{p_o}{(k - m\Omega^2) + ik\gamma} \tag{4.79}$$

or

$$\frac{\bar{U}}{U_o} = \frac{1}{(1 - r^2) + i\gamma} \tag{4.80}$$

*In some of the literature the symbol g is used for the structural damping factor. However, because of the confusion with the gravitational constant, the symbol γ is used here.

By comparing the denominators of Eqs. 4.80 and 4.32 we see that the factor γ in the former corresponds to the factor $(2\zeta r)$ in the latter. Since, when damping factors are small (as is generally the case in structures), damping is primarily effective at frequencies in the vicinity of resonance, it can be seen that, under harmonic excitation conditions, structural damping is essentially equivalent to viscous damping with

$$\zeta = \frac{\gamma}{2r} \doteq \frac{\gamma}{2} \tag{4.81}$$

From Eq. 4.80 the amplitude and phase angle of the response may be found to be

$$\frac{U}{U_o} = \frac{1}{[(1 - r^2)^2 + \gamma^2]^{1/2}} \tag{4.82}$$

and

$$\tan \alpha = \frac{\gamma}{1 - r^2} \tag{4.83}$$

where, again $\bar{U} = Ue^{-i\alpha}$. An excellent study of harmonic excitation of systems having structural damping was made by Bishop and Gladwell[4.14].

Figure 4.18 is a vector response plot of Eq. 4.80. The following facts may be noted:

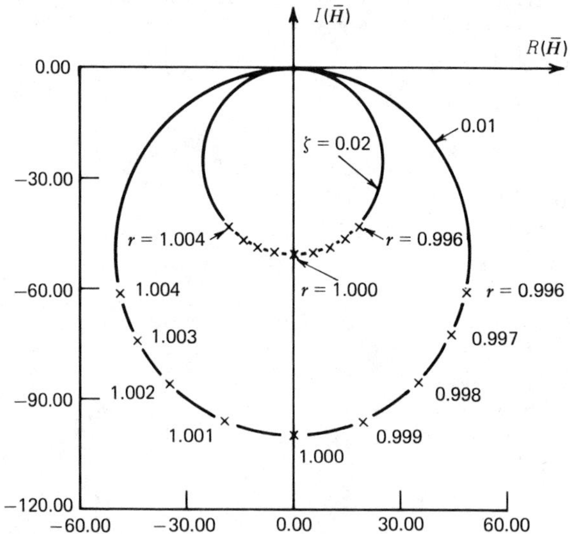

Figure 4.18. Vector response plots of steady-state response of structurally damped SDOF systems.

a. At $r = 1$ (resonance)

$$\frac{\bar{U}}{U_o} = \frac{1}{i\gamma} = -i\left(\frac{1}{\gamma}\right)$$

Therefore,

$$\frac{U}{U_o} = \frac{1}{\gamma}, \quad \text{and} \quad \alpha = 90^\circ \quad \text{at} \quad r = 1$$

b. The vector response plot for structural damping is a circle. The diameter of the circle is determined by the damping factor.

c. The spacing between points of equal frequency difference is greatest in the vicinity of $r = 1$. That is, for a given $\Delta r = r_2 - r_1$, the spacing on the circle is greatest near $r = 1$.

References

4.1 C. E. Crede, *Vibration and Shock Isolation,* Wiley, New York (1951).

4.2 C. M. Harris and C. E. Crede (eds.), *Shock and Vibration Handbook,* 2nd Ed., McGraw-Hill, New York (1976).

4.3 T. G. Beckwith and N. L. Buck, *Mechanical Measurements,* 2nd Ed., Addison-Wesley, Reading, MA (1969).

4.4 J. S. Wilson, "Instrumentation for Dynamic Measurement: Accelerometers and Other Motion Transducers," *Sound and Vibration,* v. 11, 9–13 (1977).

4.5 F. S. Tse, I. E. Morse, and R. T. Hinkle, *Mechanical Vibrations—Theory and Applications,* 2nd Ed., Allyn and Bacon, Boston (1978).

4.6 J. T. Brock, *Mechanical Vibration and Shock Measurements,* Brüel & Kjaer, Naerum, Denmark.

4.7 J. S. Wilson, "Performance Characteristics and the Selection of Accelerometers," *Sound and Vibration,* v. 12, 24–29, (1978).

4.8 N. D. Change, *General Guide to ICP Instrumentation,* Bulletin G-0001, PCB Piezotronics, Inc., Buffalo, NY.

4.9 *Instrument Notes,* Gould Inc., Statham Instruments Div., Oxnard, CA (1967).

4.10 M. Richardson, "Modal Analysis Using Digital Test Systems," *Seminar on Understanding Digital Control and Analysis in Vibration Test Systems,* The Shock and Vibration Information Center, Washington, DC, 43–64 (1975).

4.11 M. Richardson and R. Potter, "Identification of the Modal Properties of an Elastic Structure from Measured Transfer Function Data," *20th International Instrumentation Symposium,* Instrument Society of America, 239–246 (1974).

4.12 K. A. Ramsey, "Effective Measurements for Structural Dynamics Testing," *Sound and Vibration,* Pt. I, v. 9, 24–35 (1975), Pt. II, v. 10, 18–31 (1976).

4.13 S. Smith et.al., "MODALAB – A Computerized Data Acquisition and Analysis System for Structural Dynamics Testing," *21st International Instrumentation Symposium*, Instrument Society of America, 183–189 (1975).

4.14 A. L. Klosterman, *On the Experimental Determination and Use of Modal Representations of Dynamic Characteristics*, Ph.D. Dissertation, U. of Cincinnati, Cincinnati, OH (1971).

4.15 R. B. Randall, *Frequency Analysis*, Brüel & Kjaer, Naerum, Denmark (1977).

4.16 G. F. Lang, "Understanding Vibration Measurements," *Sound and Vibration*, v. 10, 26–37 (1976).

4.17 C. C. Kennedy and C. D. P. Pancu, "Use of Vectors in Vibration Measurement and Analysis," *J. Aero. Sci.*, v. 14, 603–625 (1947).

4.18 R. E. D. Bishop and G. M. L. Gladwell, "An Investigation Into the Theory of Resonance Testing," *Phil. Trans.*, v. 255, 241–280 (1963).

4.19 Th. Theodorsen and I. E. Garrick, *Mechanism of Flutter. A Theoretical and Experimental Investigation of the Flutter Problem*, NACA, Report 685 (1940).

4.20 R. R. Bouche, *Calibration of Shock and Vibration Measuring Transducers*, Monograph SUM-11, The Shock and Vibration Information Center, Naval Research Laboratory, Washington, DC (1979).

Problem Set 4.1

4.1 The system of Fig. 4.1 is at rest at $t = 0$. Subsequently, it is subjected to harmonic excitation $p(t) = p_o \sin \Omega t$.

(a) Determine an expression for the steady-state response.
(b) Determine an expression for the total response.
(c) For an excitation frequency ratio $\Omega/\omega_n = 1.5$, what is the ratio of the maximum total response to the maximum steady-state response? (See Fig. 4.2 for verification of your answer.)

4.2 What frequency range $\Omega_l/\omega_n \leq \Omega/\omega_n \leq \Omega_u/\omega_n$ must be avoided if the steady-state response of an undamped system must be less than, or equal to, five times the static deflection due to p_o?

4.3 The system shown in Fig. 4.1 has the following properties: $k = 10$ kN/m, $m = 4$ kg. The system is at rest when an excitation $120 \sin(40t)$ Newtons commences. Determine an expression for the resulting motion.

4.4 Harmonic excitation is to be used to determine the natural frequency and the mass of an "undamped" system similar to the one shown in Fig. 4.1. At a frequency $\Omega = 6$ rad/s the resonance condition is achieved, that is, the response tends to increase without bound. This frequency is therefore taken to be the undamped natural frequency, that is $\omega_n = 6$ rad/s. Next, a mass of $\Delta m = 1$ kg is attached to mass m and the resonance test is repeated. This time resonance occurs at $\Omega = 5.86$ rad/s.

(a) Determine the value of the mass m.

(b) Determine the value of the spring constant k.

4.5 (a) Starting with Eq. 4.11 determine an expression for the response of an undamped system which is at rest at $t = 0$ and is thereafter subjected to $p(t) = p_o \cos \Omega t$.

(b) Let $\Omega = \omega_n + \Delta\omega$, where $\Delta\omega/\omega_n \ll 1$. Show that the response can be written in the form

$$u(t) = A \sin(\Delta\omega t) \sin(2\omega_n + \Delta\omega)t$$

which appears to be an oscillation $A \sin(2\omega_n + \Delta\omega_n)t$ with slowly varying amplitude. (This is called a "beat" phenomenon.)

4.6 A control tab of an airplane elevator is hinged about axis A in the elevator. Although Problem 2.6 suggested a free vibration test for determining I_A, the dynamic behavior of the tab also depends on the elastic stiffness of the control linkage, which is modeled in Fig. P4.6 as a torsional spring. Since k_θ cannot be measured statically, a resonance test is to be used. The elevator is held fixed, the tab is supported by spring k_1, and harmonic excitation is applied through spring k_2 as shown.

(a) Show that ω_n for the elevator and its control linkage alone is given by $\omega_n = \sqrt{k_\theta/I_A}$.

(b) If the resonance frequency obtained in the test using the setup below is ω_r, determine an expression for k_θ in terms of ω_r, I_A and other parameters shown on the sketch.

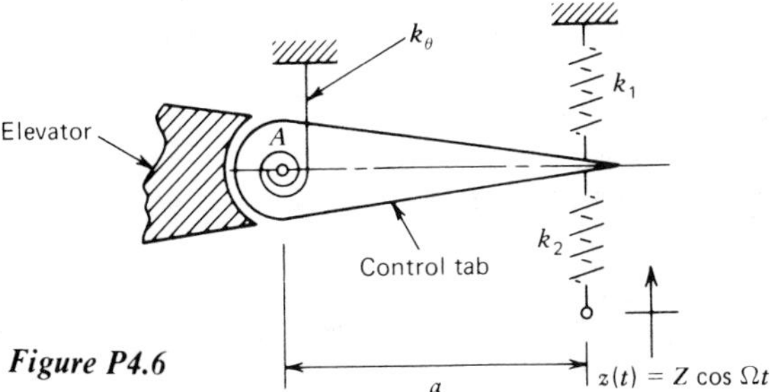

Figure P4.6

Problem Set 4.2

4.7 Figure 4.5 corresponds to harmonic excitation with $\Omega < \omega_n$. Draw the force vector polygons for the following excitation frequencies:

(a) $\Omega = \omega_n$.

(b) A typical case for which $\Omega > \omega_n$.

4.8 Unbalance in rotating machines is a common source of vibration excitation. Such a situation is illustrated schematically in Fig. P4.8. Let $(M - m)$ be the mass of the machine and m be the mass of the rotating unbalance, which rotates at Ω rad/s.

(a) Derive the equation of vertical motion of the machine.
(b) Derive an expression for the steady-state response u/u_0 of this system and show that Fig. 4.12 is the frequency response plot for this situation. Let

$$u_0 = \frac{me\Omega^2}{k}$$

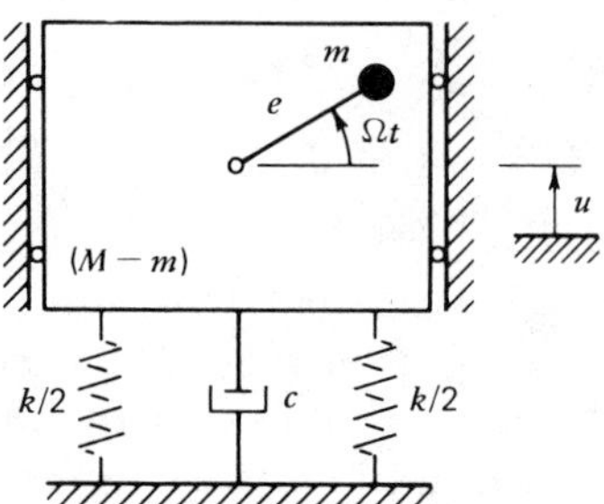

Figure P4.8

4.9 A commercial airconditioning unit weighing 1200 lb is located at the middle of two parallel, simply supported steel beams whose clear span is 16 ft. The moment of inertia of each beam is 20 in.^{4}. The motor runs at 300 rpm, and its rotor produces a rotating unbalanced force of magnitude 90 lb at this speed. Assuming a viscous damping factor of 10% and neglecting the weight of the supporting beams, determine the steady-state amplitude of the vibration which results from the rotating unbalance.

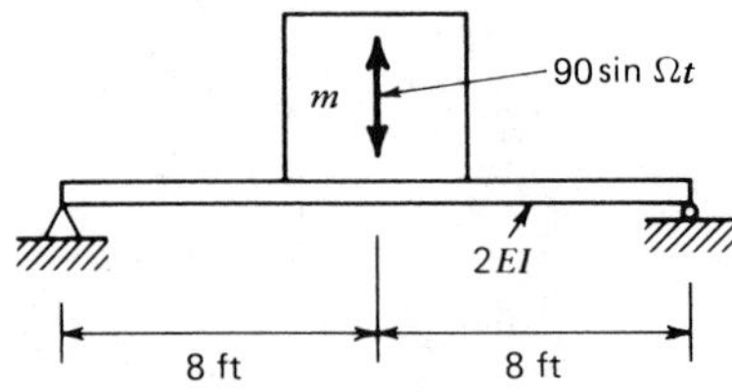

Figure P4.9

Problem Set 4.4

4.10 A SDOF system is subjected to harmonic excitation $z(t) = Z \cos \Omega t$ applied at point P. Given m, c, k, Z and Ω,

(a) Derive the equation of motion of the system with $u(t)$ as the unknown.
(b) Derive the equation of motion of the system with $w(t) = z - u$ as the unknown.
(c) Determine expressions for ω_n and ζ for this system.
(d) Determine the following complex frequency response functions: $\overline{U}/Z$, $\overline{W}/Z$.

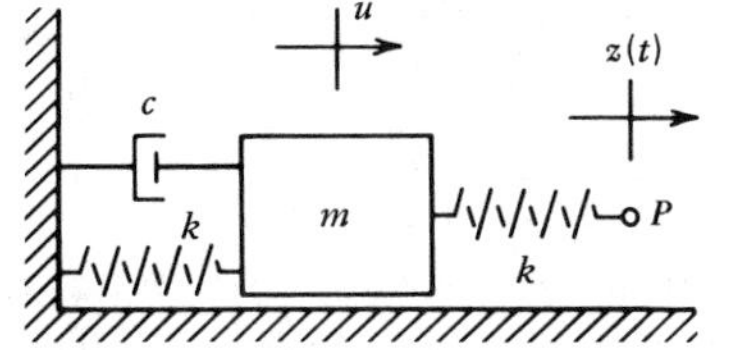

Figure P4.10

4.11 For the system shown in Fig. P2.3 use the complex frequency response method to determine expressions for the following.

(a) $\overline{U}/Z$, where $z(t) = Z \cos \Omega t$.

(b) U and α, where $u(t) = U \cos(\Omega t - \alpha)$.

4.12 For the system shown in Fig. P2.3 use the complex frequency response method to obtain an expression for the maximum steady-state force in spring BC due to harmonic excitation $z(t) = Z \cos \Omega t$.

4.13 Mechanical equipment frequently employs rotating machinery which may give rise to a force exerted on the supporting structure, for example, air conditioners on the roof of a building. Judging from Fig. 4.11, it is possible to reduce the force transmitted to the supporting structure by using vibration isolators. Assume that a machine operates at 20 Hz and that it desired to reduce the transmitted force by 90% using spring-type isolators, that is,

(a) Determine an expression giving the percent reduction in force as a function of the forcing frequency Ω and the static deflection $\delta_{st} = mg/k$.

(b) Evaluate the static deflection, $\delta_{st} = mg/k$, for the conditions listed above, that is, 90% reduction at 20 Hz. Express your answer in millimeters.

4.14 A vibration isolation block is to be installed in a laboratory so that the vibration from adjacent factory operations will not disturb certain experiments. If the isolation block weighs 2000 lb, and the surrounding floor and foundation vibrate at 24 Hz with an amplitude of 0.01 in., determine the stiffness of the isolation system such that the isolation block will have an amplitude of only 0.002 in.

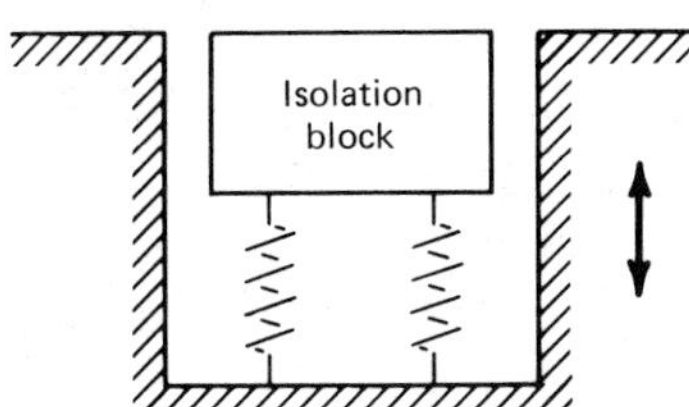

Figure P4.14

4.15 Rotating unbalance leads to harmonic forces. The nonrotating mass is $(M - m)$, while counter-rotating masses $(m/2)$ at eccentricity e rotate at an angular rate of Ω rad/s.

(a) Determine the equation of motion for the system.

(b) Use the complex frequency response method to determine an expression for $M\bar{u}/me$.

(c) Use the complex frequency response method to obtain the force transmitted to the base in the nondimensional form $\bar{f}_{tr}/me\Omega^2$.

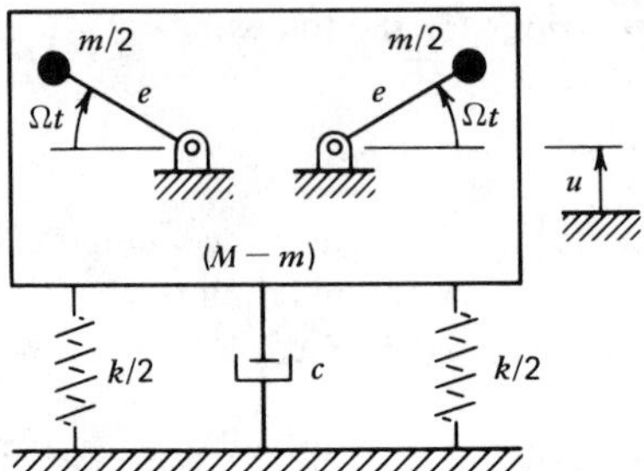

Figure P4.15

4.16 For the vehicle in Example 4.3 determine the speed of the vehicle that would produce a resonance condition:

(a) If the vehicle were empty.

(b) If the vehicle were fully loaded.

Problem Set 4.5

4.17 At a point on a vibrating structure the motion is given by

$$\begin{aligned} z &= Z_1 \cos \Omega_1 t + Z_2 \cos \Omega_2 t \\ &= 0.05 \cos(60\,\pi t) + 0.02 \cos(120\,\pi t) \end{aligned}$$

(a) Determine the vibration record $w_p(t)$ that would be obtained with an accelerometer having a damping factor $\zeta = 0.70$ and a resonant frequency of 20 kHz.

(b) Is there any significant amplitude or phase distortion produced by the accelerometer?

Problem Set 4.6

4.18 Using the data on the outer "circle" of Fig. 4.9

(a) Construct a curve of $D_s \equiv |\bar{H}|$ versus r.

(b) Using the half-power method verify that $\zeta = 0.05$, as shown on the curve. (Estimate r_2 and r_1.)

4.19 Experimental vibration data might be obtained in the form of U versus $f = \Omega/2\pi$. If a system has 1% damping, estimate the frequency resolution, Δf, with which data should be taken in order to obtain a reasonably accurate estimate of ζ for a SDOF system.

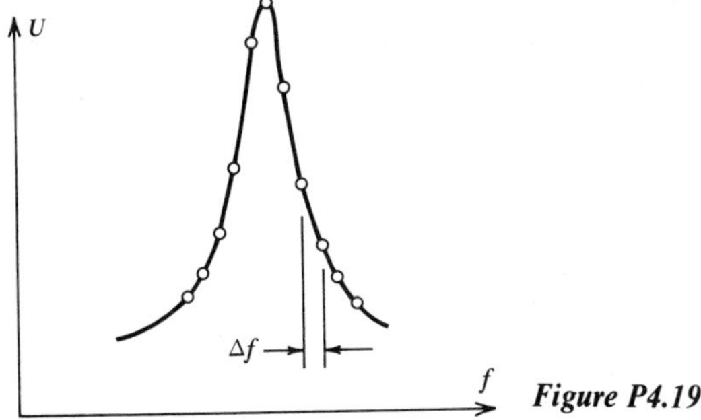

Figure P4.19

Problem Set 4.7

4.20 A force $p_o \sin \Omega t$ acts on a mass that slides on a surface having a coefficient of kinetic friction of μ_k (see Sec. 3.4).

(a) Determine the equivalent viscous damping coefficient.

(b) Determine the response amplitude, U, of the system with equivalent viscous damping.

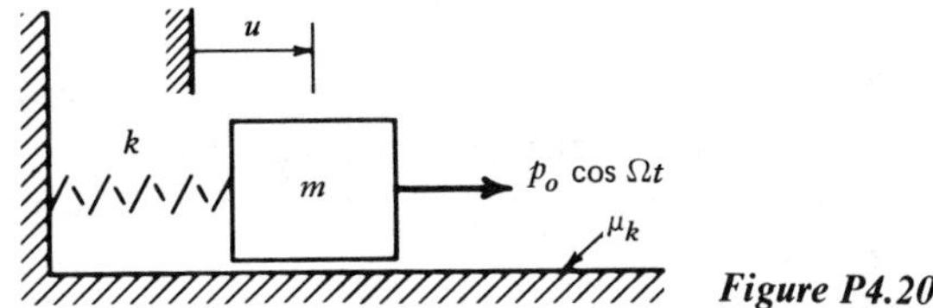

Figure P4.20

4.21 For the quadratically damped system of Example 4.4, determine the amplitude of steady-state response.

Problem Set 4.8

4.22 For harmonic motion the structural damping force can be represented by the complex form shown in Eq. 4.77, that is, $\bar{f}_d = -i\gamma k\bar{u}$, or by the real form

$$f_d = -\gamma k\dot{u}\frac{|u|}{|\dot{u}|} = -\gamma k|u|\,\mathrm{sgn}(\dot{u})$$

(a) Sketch f_d versus u for one cycle of harmonic motion, where

$$u = -U\cos(\Omega t - \alpha)$$

(b) Determine an expression for the work done per cycle.
(c) Determine an expression for c_{eq}.
(d) By substituting c_{eq} into Eq. 4.20a, determine an expression for U/U_o. Compare this with Eq. 4.82 and discuss any differences which you observe.

4.23 Using Eq. 4.80 show that the vector response plot for structural damping is a circle with center on the imaginary axis and passing through the origin as shown in Fig. 4.18 or Fig. P4.23.

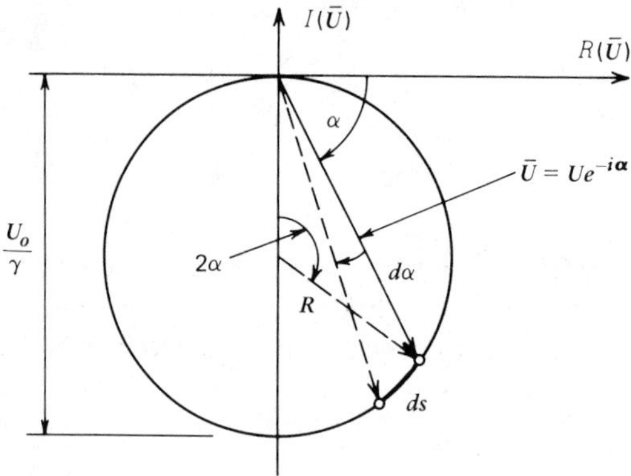

Figure P4.23

4.24 (a) Determine the rate of change of arc length along the circle with respect to frequency change by determining an expression for $ds/d(r^2)$. Use Figure P4.23.
(b) Illustrate how the damping level γ affects the above derivative by evaluating $ds/d(r^2)$ at $r = 1$.

5 RESPONSE OF SDOF SYSTEMS TO SPECIAL FORMS OF EXCITATION

In Chapter 4 you studied the response of SDOF systems to harmonic excitation and learned that the steady-state response depends primarily on the damping factor of the system and the excitation frequency ratio. Before considering the response of a SDOF system to general excitation, we will consider several simple forms of nonharmonic excitation in order to gain some insight into what system parameters and excitation parameters govern the response. Chapters 6 through 8 treat procedures for determining the response of SDOF systems to more general excitation.

Upon completion of this chapter you should be able to:

- Apply the classical differential equations method to determine the response of a SDOF system to an impulse loading at $t = 0$, a step loading applied at $t = 0$, or a ramp loading.
- Describe the difference in the response of a system loaded "rapidly" as contrasted to the response of the same system when loaded "slowly."
- Explain why it is that the maximum dynamic response to a pulse may occur after the excitation has ceased to act on the system.

5.1 Response of a Viscous-Damped SDOF System to an Ideal Step Input

Let the prototype SDOF system shown in Fig. 3.1 be subjected to an ideal step input as shown in Fig. 5.1. The equation of motion is given by Eq. 3.1. Thus,

$$m\ddot{u} + c\dot{u} + ku = p_o, \qquad t \geqslant 0 \tag{5.1}$$

Let the system be at rest at $t = 0$, that is,

$$u(0) = \dot{u}(0) = 0 \tag{5.2}$$

The solution of Eq. 5.1 consists of a particular solution which, from Eq. 5.1, can be seen to be

$$u_p = \frac{p_o}{k} \tag{5.3}$$

and a complementary solution given (for $\zeta < 1$) by Eq. 3.32. Then,

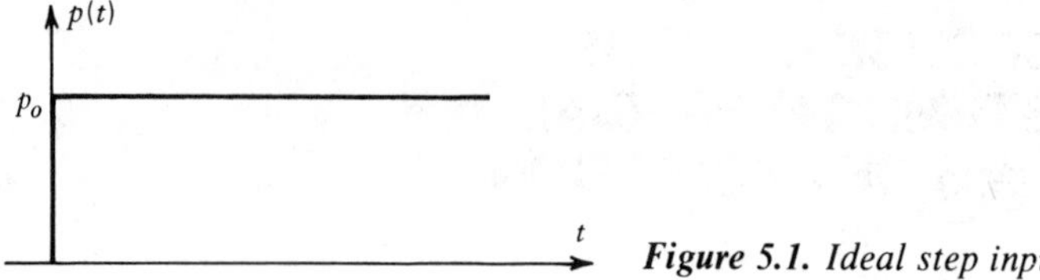

Figure 5.1. Ideal step input.

$$u = \frac{p_o}{k} + e^{-\zeta\omega_n t}(A_1 \cos \omega_d t + A_2 \sin \omega_d t) \tag{5.4}$$

Using the initial conditions to evaluate A_1 and A_2, we obtain

$$u = \frac{p_o}{k}\left\{1 - e^{-\zeta\omega_n t}\left[\cos \omega_d t + \left(\frac{\zeta\omega_n}{\omega_d}\right)\sin \omega_d t\right]\right\} \tag{5.5}$$

A useful way of examining dynamic response is to consider the *response ratio,* or dynamic load factor, defined by*

$$R(t) = \frac{ku(t)}{p_{\max}} \tag{5.6}$$

Thus, the response ratio is the ratio of dynamic response to static deformation. For the ideal step input, $R(t)$ is given by

$$R(t) = 1 - e^{-\zeta\omega_n t}\left[\cos \omega_d t + \left(\frac{\zeta\omega_n}{\omega_d}\right)\sin \omega_d t\right] \tag{5.7}$$

A typical response ratio plot is shown in Fig. 5.2. On the response ratio plot $R(t) = 1$ corresponds to the static displacement position. Because the load was instantaneously applied, there is an overshoot, and the system settles to the static value of 1 after a number of cycles of damped oscillation. The damping level determines the amount of the overshoot and the rate of decay of the oscillation about the static equilibrium position.

For an undamped system Eq. 5.5 becomes

$$u(t) = \frac{p_o}{k}(1 - \cos \omega_n t) \tag{5.8}$$

and $R_{\max} = 2$. Thus, when a load is instantaneously applied to an undamped system, a maximum displacement of twice the static displacement is attained. This is the reason that a safety factor of 2 is frequently applied to the design of structures that will be subjected to rapidly applied loads.

* $(\ )_{\max}$ will be used as an abbreviation for "maximum absolute value," that is, $\max_t |(\ )|$.

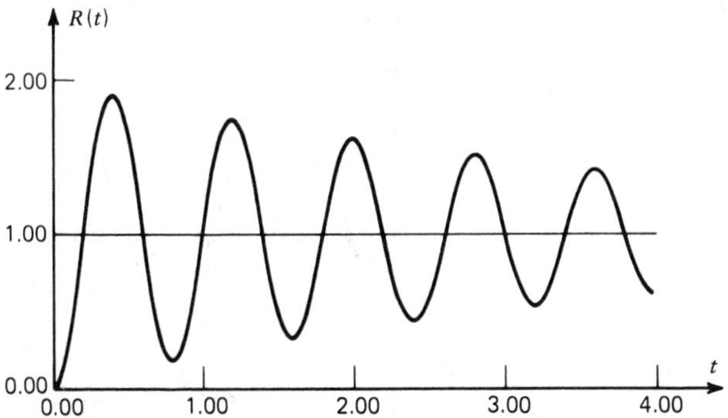

Figure 5.2. Response ratio plot for a step input.

In the next section we consider the effect of terminating the load after a duration of t_d, or of linearly increasing the load over a rise time of t_r.

5.2 Response of an Undamped SDOF System to Rectangular Pulse and Ramp Loadings

In Sec. 5.1 the response of a damped SDOF system to an ideal step input was considered. It was noted that the maximum overshoot occurs after one-half cycle of the resulting oscillation, and that the presence of damping reduces the overshoot. Now we will consider two questions—"What effect does the duration of the loading have on the response of an undamped system?" and "What effect does the rise time of the loading have on the response?"

Figure 5.3 shows a rectangular pulse input and the response ratio for an undamped system for the two cases

$$\text{(a)}\quad t_d > \frac{T_n}{2}, \qquad \text{(b)}\quad t_d < \frac{T_n}{2}$$

where t_d is the *duration* of the rectangular pulse.

From Fig. 5.3 it is clear that when $t_d \geq T_n/2$ the maximum occurs during the *forced-vibration era,* that is, prior to t_d, while the maximum occurs in the *residual vibration era,* that is, after t_d, if $t_d < T_n/2$. We can determine expressions for the maximum occurring in each case.

Case 1. Forced-Vibration Era ($0 \leq t \leq t_d$)

For this case $R(t)$ is the same as for an ideal step, namely

$$R_1(t) = 1 - \cos \omega_n t, \qquad 0 \leq t \leq t_d \tag{5.9}$$

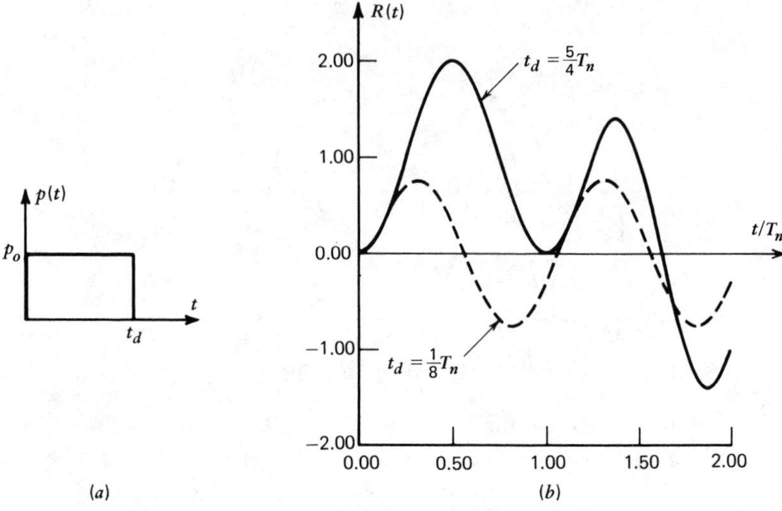

Figure 5.3. Response to a rectangular pulse input. (a) *Rectangular pulse.* (b) *Response ratios.*

and the maximum value is

$$(R_1)_{\max} = R_1\left(\frac{T_n}{2}\right) = 2 \tag{5.10}$$

Case 2. Residual-Vibration Era ($t_d < t$)

Figure 5.3*b* shows $R(t)$ for a pulse of duration $t_d = T_n/8$. Note that $R_{\max}$ occurs during the residual-vibration era. Since the response for $t > t_d$ is free vibration with "initial" conditions $R_1(t_d)$ and $\dot{R}_1(t_d)$, Eq. 3.17 can be used in the following form

$$R_2(t) = R_1(t_d)\cos\omega_n(t - t_d) + \left[\frac{\dot{R}_1(t_d)}{\omega_n}\right]\sin\omega_n(t - t_d) \tag{5.11}$$

for $t \geq t_d$, where $R_1(t_d)$ and $\dot{R}_1(t_d)$ are obtained from Eq. 5.9. Equation 3.25 can be used to determine the amplitude of this response.

$$(R_2)_{\max} = \left\{[R_1(t_d)]^2 + \left[\frac{\dot{R}_1(t_d)}{\omega_n}\right]^2\right\}^{1/2} \tag{5.12}$$

which can be simplified to the form

$$(R_2)_{\max} = 2\sin\left(\frac{\pi t_d}{T_n}\right) \tag{5.13}$$

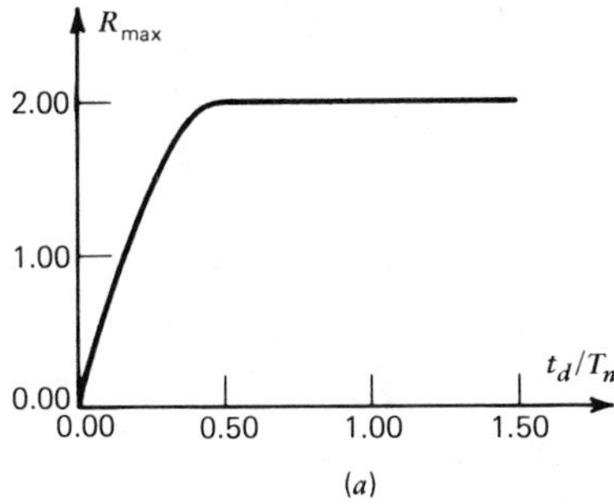

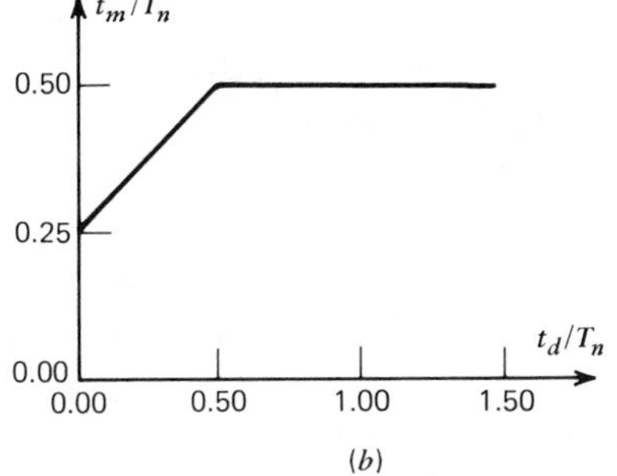

Figure 5.4. *Maximum response ratio for rectangular pulse excitation.* (a) *Maximum response amplitude.* (b) *Time at which maximum response occurs.*

Figure 5.4*a* shows the maximum response ratio as a function of pulse duration. From this figure, or from Eq. 5.13, it can be seen that any pulse of duration longer than $T_n/6$ will cause a displacement larger than the static displacement, p_o/k, and for any pulse longer than $T_n/2$, the maximum displacement will be twice the static value. Figure 5.4*b* shows the time at which the maximum response occurs. Expressions plotted in Fig. 5.4*b* are to be derived as an exercise (Problem 5.4).

Having considered the effect of the duration of a load on the maximum response, let us now examine the effect of rise time. Consider a ramp load with *rise time* t_r as shown in Fig. 5.5 applied to an undamped SDOF system that is

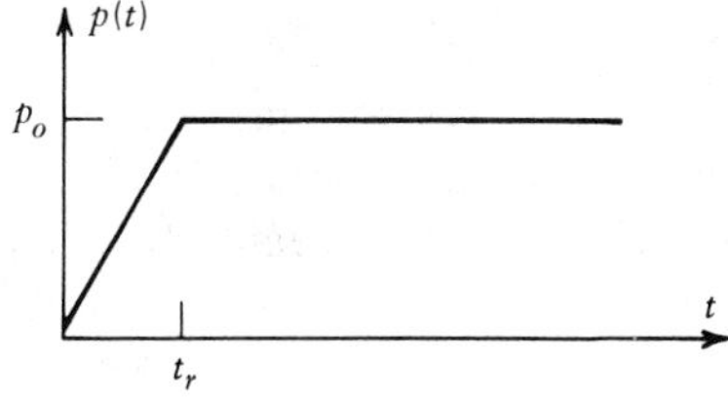

Figure 5.5. *Ramp input function.*

at rest prior to application of the load. The equation of motion and initial conditions are

$$m\ddot{u} + ku = \begin{cases} \left(\dfrac{t}{t_r}\right)p_o & 0 \le t \le t_r \\ p_o & t_r \le t \end{cases} \tag{5.14}$$

$$u(0) = \dot{u}(0) = 0 \tag{5.15}$$

For $0 \le t \le t_r$ the particular solution is seen to be

$$u_p = \left(\frac{t}{t_r}\right)\left(\frac{p_o}{k}\right) \tag{5.16}$$

Then,

$$u = \left(\frac{t}{t_r}\right)\left(\frac{p_o}{k}\right) + A_1 \cos \omega_n t + A_2 \sin \omega_n t \tag{5.17}$$

Using the initial conditions of Eq. 5.15, we get

$$u = \left(\frac{p_o}{k}\right)\left[\left(\frac{t}{t_r}\right) - \left(\frac{1}{\omega_n t_r}\right) \sin \omega_n t\right] \tag{5.18}$$

For $t \ge t_r$ the solution of Eq. 5.14b can be shown to be

$$u = \left(\frac{p_o}{k}\right)\left\{1 + \left(\frac{1}{\omega_n t_r}\right)[\sin \omega_n(t - t_r) - \sin \omega_n t]\right\} \tag{5.19}$$

Figure 5.6*a* shows the response to an input with $t_r > T_n$ and also the response to an input with $t_r < T_n$. Figure 5.6*b* summarizes the effect of rise time on maximum response.

From Fig. 5.6 it can be seen that the maximum response, $R_{max} = 2$, occurs for an ideal step input (i.e., for $t_r = 0$). For ramps with $t_r \gg T_n$ there will be little overshoot and the system will just undergo small oscillations about the pseudostatic deflection curve

$$u_{\text{pseudostatic}} = \left(\frac{t}{t_r}\right)\left(\frac{p_o}{k}\right), \quad 0 \le t \le t_r \tag{5.20}$$

Thus, a load can be considered to be "slowly applied" and dynamic effects can generally be ignored if the rise time is longer than about $3T_n$.

Figures like 5.4*a* and 5.6*b* are called response spectra. Response spectra will be discussed in greater detail in Sec. 6.2.

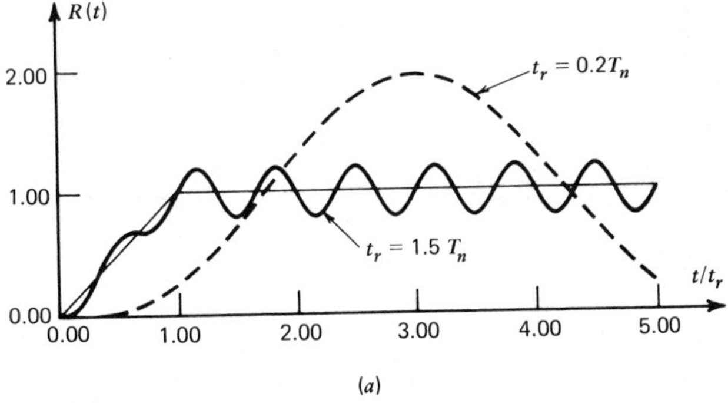

(a)

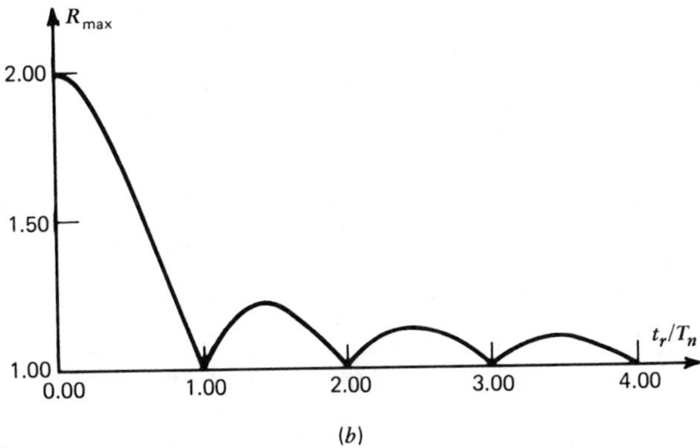

(b)

Figure 5.6. *Response of an undamped SDOF system to ramp inputs.* (a) *Response to ramp inputs.* (b) *Maximum response to ramp inputs.*

5.3 Response of an Undamped SDOF System to a Short-Duration Impulse; Unit Impulse Response

Perhaps the most important special form of excitation is the short-duration impulse, since it is also frequently used in determining the response of the system to more general forms of excitation, as will be seen in the next chapter.

Consider an undamped SDOF system subjected to a force of duration $t_d \ll T_n$ having an impulse

$$I = \int_0^{t_d} p(t)\, dt \tag{5.21}$$

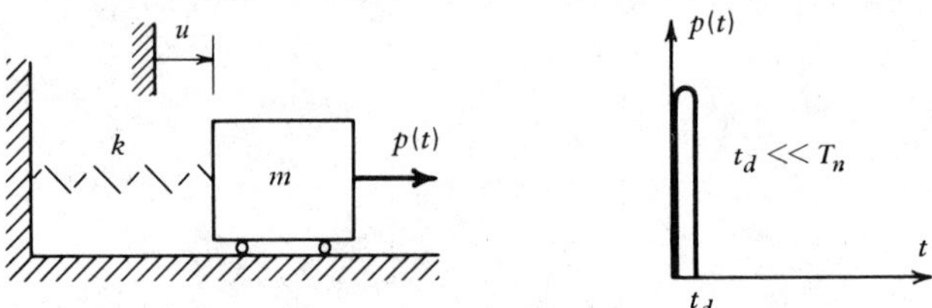

Figure 5.7. Undamped SDOF system subjected to short-duration impulse.

Let the system be at rest for $t \leq 0$, that is, prior to application of the excitation. The equation of motion and initial conditions are

$$m\ddot{u} + ku = \begin{cases} p(t) & 0 < t \leq t_d \\ 0 & t_d < t \end{cases} \tag{5.22}$$

$$u(0) = \dot{u}(0) = 0 \tag{5.23}$$

By integrating Eq. 5.22a with respect to time and incorporating the initial conditions, we get

$$m\dot{u}(t_d) + ku_{\text{avg}}t_d = I \tag{5.24}$$

where u_{avg} is the (small) average displacement in the time interval $0 < t \leq t_d$. For $t_d \to 0$, that is, $t_d \ll T_n$, the second term in Eq. 5.24 can be ignored leaving

$$m\dot{u}(0^+) = I \tag{5.25}$$

Thus, an impulse consisting of a large force acting for a very short time has the effect of giving the mass an initial velocity of

$$\dot{u}(0^+) = \frac{I}{m} \tag{5.26a}$$

but leaving it with an initial displacement of

$$u(0^+) = 0 \tag{5.26b}$$

These can be used as "initial" conditions for the free vibration problem of Eq. 5.22b. Using Eq. 3.17 we get the *impulse response*

$$u(t) = \left(\frac{I}{m\omega_n}\right) \sin \omega_n t \tag{5.27}$$

The *unit impulse response function* for an undamped SDOF system is obtained from Eq. 5.27 by letting $I = 1$. By convention, the unit impulse response function is frequently called $h(t)$. Thus,

$$h(t) = \left(\frac{1}{m\omega_n}\right) \sin \omega_n t \tag{5.28}$$

For a viscous-damped SDOF system with $\zeta < 1$ the impulse response function can be shown to be

$$u(t) = \left(\frac{I}{m\omega_d}\right) e^{-\zeta\omega_n t} \sin \omega_d t \tag{5.29}$$

and the corresponding unit impulse response function is therefore

$$h(t) = \left(\frac{1}{m\omega_d}\right) e^{-\zeta\omega_n t} \sin \omega_d t \tag{5.30}$$

Example 5.1

Assume that the impulse $I = \int p(t)\, dt$ is due to a constant force p_o applied over a time interval $0 < t \leq t_d$ to an undamped SDOF system which is at rest at $t = 0$. Show that for $t_d \ll T_n$, Eq. 5.11 reduces to Eq. 5.27.

Solution

a. Determine $u(t_d)$ and $\dot{u}(t_d)$ from Eq. 5.8:

$$u(t_d) = \left(\frac{p_o}{k}\right)(1 - \cos \omega_n t_d) \tag{1}$$

$$\dot{u}(t_d) = \left(\frac{\omega_n p_o}{k}\right) \sin \omega_n t_d \tag{2}$$

Since $\omega_n T_n = 2\pi$ and $t_d \ll T_n$, $\omega_n t_d \ll 2\pi$. Therefore,

$$(1 - \cos \omega_n t_d) \doteq \tfrac{1}{2}(\omega_n t_d)^2$$

$$\sin \omega_n t_d \doteq \omega_n t_d$$

Thus, since $I = p_o t_d$,

$$\left.\begin{aligned} u(t_d) &\doteq \frac{1}{2}\left(\frac{p_o}{k}\right)(\omega_n t_d)^2 = \left(\frac{I}{2}\right)\left(\frac{t_d}{m}\right) \\ \dot{u}(t_d) &\doteq \left(\frac{\omega_n p_o}{k}\right)(\omega_n t_d) = \frac{I}{m} \end{aligned}\right. \tag{3}$$

b. Evaluate $u(t)$ from Eq. 5.11 letting $t_d \to 0$:

$$u(t) = \lim_{t_d \to 0} \left[\left(\frac{I t_d}{2m}\right) \cos \omega_n(t - t_d) + \left(\frac{I}{m\omega_n}\right) \sin \omega_n(t - t_d)\right] \tag{4}$$

Thus, for $t_d \to 0$

$$u(t) = \left(\frac{I}{m\omega_n}\right) \sin \omega_n t \tag{5}$$

as found in Eq. 5.27.

In Chapter 6 you will learn how to determine the response of a SDOF system to general excitation. You will then be able to show that the form of $p(t)$ is not very important so long as the duration is very short, that is, $t_d \ll T_n$. Consequently, Eq. 5.27, or 5.29 for viscous-damped systems, gives a very close approximation to the actual response determined using $p(t)$.

In summary, you have learned that an excitation can be considered to be an impulse if $t_d \ll T_n$. On the other hand, a load can be considered to be applied "statically" if its rise time, t_r, is much greater than T_n. Dynamic overshoot must be considered for rapidly applied loads.

Problem Set 5.1

5.1 A very simplified model is to be used to study landing impact of a light aircraft. The airplane is modeled, as shown in Fig. P5.1, that is, as a lumped mass with a linear spring representing the landing gear. The mass m has a vertical descent speed of V when the spring touches the ground. Call the time of contact $t = 0$, and let $u(0) = 0$.

(a) Determine an expression for the vertical position of the mass as a function of time during the time that the spring remains in contact with the ground.

(b) Determine the time at which the spring loses contact with the ground upon rebound.

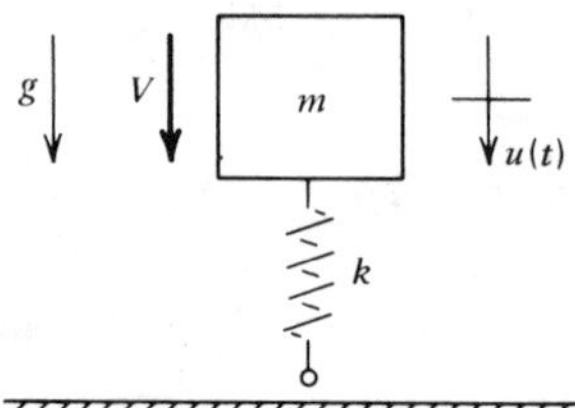

Figure P5.1

5.2 An instrument weighing 40 lb is being shipped in a container with foam packaging which has an effective vertical stiffness $k = 100$ lb/in. and which provides an effective damping factor $\zeta = 0.05$. If the container and its contents hit the ground with a vertical speed $V = 150$ in./sec, determine the maximum total force exerted on the instrument by the packaging.

(*Note:* Model this as a mass on a spring, as in Fig. P5.1, with an added viscous dashpot.)

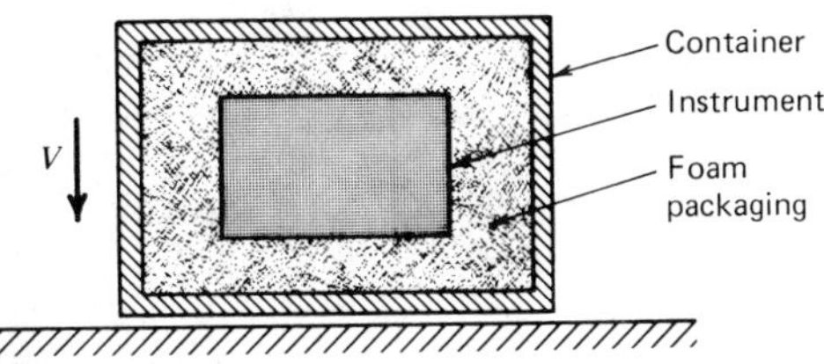

Figure P5.2

5.3 Determine an expression for the transient response of a damped SDOF system with an ideal step input as in Fig. 5.1 and with initial conditions $u(0) = 0$, $\dot{u}(0) = \dot{u}_o$.

Problem Set 5.2

5.4 Determine expressions for the time, t_m, at which maximum response occurs for an undamped SDOF system subjected to the rectangular pulse of Fig. 5.3*a*. There should be one expression for $t_d/T_n \leq 0.5$ and a second expression for $t_d/T_n \geq 0.5$ (see Fig. 5.4*b*).

5.5 An undamped SDOF system is subjected to a ramp forcing function of the form shown in Fig. 5.5. For $0 \leq t_r \leq T_n$ determine expressions for:

(a) t_m = time at which maximum response occurs.

(b) R_{max} = maximum response ratio.

Assume $u(0) = \dot{u}(0) = 0$. (See Fig. 5.6.)

5.6 Determine expressions for the response of an undamped SDOF system to the forcing function shown in Fig. P5.6 for each of the following time intervals:

(a) $0 \leq t \leq t_1$.

(b) $t_1 \leq t \leq t_2$.

(c) $t_2 \leq t$.

Assume $u(0) = \dot{u}(0) = 0$.

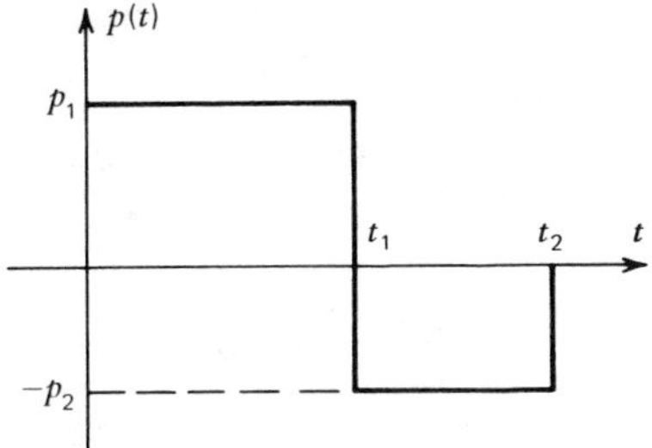

Figure P5.6

6 RESPONSE OF SDOF SYSTEMS TO GENERAL DYNAMIC EXCITATION

In Chapter 5 the classical differential equations approach was used to solve for the response of SDOF systems to three special forms of input. For these cases the particular solution was easily obtained "by inspection," and the emphasis was on noting the importance of excitation parameters such as rise time and duration. In addition to the classical approach employed above, three methods are widely employed for obtaining analytical expressions for the response of systems to general dynamic loading—the Duhamel integral method (time-domain solution), the Laplace transform method (Laplace-domain solution), and the Fourier transformation method (frequency-domain solution). The Laplace transform method provides a useful means of converting differential equations of motion to algebraic equations and of specifying system characteristics. However, we will consider only the Duhamel integral method in this chapter. In Chapter 8 the Fourier transform method will be discussed.

In designing a system to withstand a specified dynamic input it is convenient to be able to compare the response of different systems to the specified input. Response spectra, which are useful for this purpose, are described in the latter part of this chapter.

Upon completion of this chapter you should be able to:

- Employ the Duhamel integral method to obtain the response of SDOF systems to simple transient inputs.
- Employ response spectra to determine the design parameters of SDOF systems subjected to specified inputs.

6.1 Response of a SDOF System to General Dynamic Excitation—Duhamel Integral Method

The Duhamel integral method for determining the response of a SDOF system to general dynamic excitation can be developed from the impulse response function derived in Sec. 5.3. In many cases an analytical expression for the response can be obtained. For more complex forms of excitation, numerical evaluation of the response is necessary, as will be described in Chapters 7 and 8.

The Duhamel integral is based on the principle of superposition, which is valid only for linear systems. Figure 6.1 shows an undamped SDOF system that is initially at rest and is then subjected to an input $p(t)$ as shown. The response of the system to an impulse $dI = p(\tau)\, d\tau$ is called $du(t)$ and is given by

$$du(t) = \left(\frac{dI}{m\omega_n}\right) \sin \omega_n(t - \tau) \tag{6.1}$$

The total response at time t will be the sum of the response due to all incremental impulses prior to time t. Therefore,

$$u(t) = \left(\frac{1}{m\omega_n}\right) \int_0^t p(\tau) \sin \omega_n(t - \tau)\, d\tau \tag{6.2}$$

or

$$u(t) = \int_0^t p(\tau) h(t - \tau)\, d\tau \tag{6.3}$$

where $h(t - \tau)$ is obtained from Eq. 5.28 for an undamped system. Equation 6.3 is equally valid for a damped system if Eq. 5.30 is used to obtain $h(t - \tau)$. Thus, for a damped system that is initially at rest

$$u(t) = \left(\frac{1}{m\omega_d}\right) \int_0^t p(\tau) e^{-\zeta\omega_n(t-\tau)} \sin \omega_d(t - \tau)\, d\tau \tag{6.4a}$$

or

$$u(t) = \left(\frac{1}{m\omega_d}\right) I\left[\int_0^t p(\tau) e^{(i\omega_d - \zeta\omega_n)(t-\tau)}\, d\tau\right] \tag{6.4b}$$

Equations 6.2 and 6.4 are referred to as *Duhamel integral* expressions for the response of undamped and damped SDOF systems, respectively. Equation 6.3 is frequently referred to as a *convolution integral,* a more general form of which is

$$x(t) = \int_{-\infty}^{\infty} f_1(\tau) f_2(t - \tau)\, d\tau \tag{6.5}$$

Equation 6.2 or 6.4 may thus be used to determine the response of a SDOF system to general dynamic excitation if the system is initially at rest. If the system has nonzero initial conditions, then the response to the initial conditions

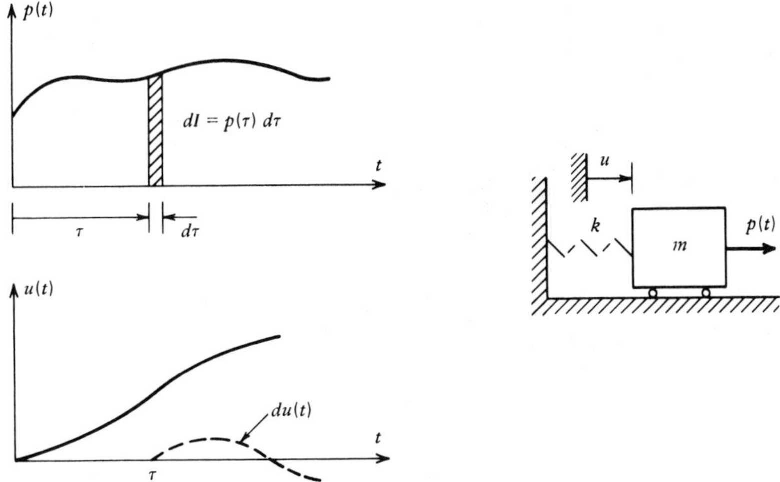

Figure 6.1. *Incremental response of an undamped SDOF system.*

is determined from Eq. 3.17 or, for $\zeta < 1$, from Eq. 3.33. Thus, for an undamped system

$$u(t) = \left(\frac{1}{m\omega_n}\right) \int_0^t p(\tau) \sin \omega_n(t - \tau)\, d\tau + u_o \cos \omega_n t + \left(\frac{\dot{u}_o}{\omega_n}\right) \sin \omega_n t \tag{6.6}$$

and for an underdamped system

$$u(t) = \left(\frac{1}{m\omega_d}\right) \int_0^t p(\tau) e^{-\zeta\omega_n(t-\tau)} \sin \omega_d(t - \tau)\, d\tau + u_o e^{-\zeta\omega_n t} \cos \omega_d t + \left(\frac{1}{\omega_d}\right) (\dot{u}_o + \zeta\omega_n u_o) e^{-\zeta\omega_n t} \sin \omega_d t \tag{6.7}$$

It is convenient to use the following trigonometric identity when evaluating Duhamel integrals.

$$\sin \omega(t - \tau) = \sin \omega t \cos \omega\tau - \cos \omega t \sin \omega\tau \tag{6.8}$$

Example 6.1

Use the Duhamel integral to determine the response of an undamped SDOF system to a "blast" loading specified by the triangular pulse shown on page 126. Obtain expressions that are valid for $t < t_d$ and for $t > t_d$. The system is initially at rest.

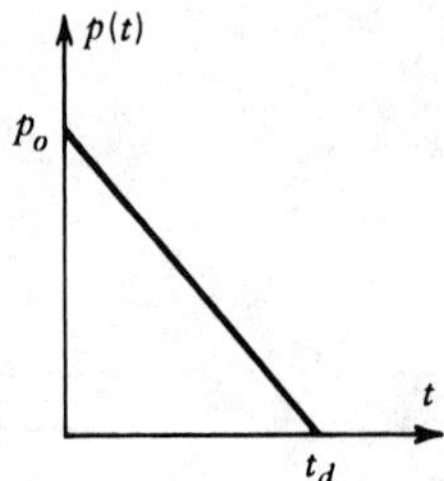

Solution

Use Eq. 6.2 with

$$p(t) = p_o\left(1 - \frac{t}{t_d}\right) \qquad 0 \leq t \leq t_d \tag{1}$$

$$p(t) = 0 \qquad t_d < t \tag{2}$$

a. For $0 \leq t \leq t_d$:

$$\begin{aligned} u(t) &= \left(\frac{1}{m\omega_n}\right)\int_0^t p_o\left(1 - \frac{\tau}{t_d}\right)\sin \omega_n(t-\tau)\, d\tau \\ &= \left(\frac{p_o}{k}\right)\sin \omega_n t \int_0^t \left(1 - \frac{\tau}{t_d}\right)\cos \omega_n\tau\, d(\omega_n\tau) \\ &\quad - \left(\frac{p_o}{k}\right)\cos \omega_n t \int_0^t \left(1 - \frac{\tau}{t_d}\right)\sin \omega_n\tau\, d(\omega_n\tau) \end{aligned} \tag{3}$$

Using integration by parts, we get

$$\begin{aligned} \int \tau \cos \omega_n\tau\, d(\omega_n\tau) &= \tau \sin \omega_n\tau - \left(\frac{1}{\omega_n}\right)\int \sin \omega_n\tau\, d(\omega_n\tau) \\ &= \tau \sin \omega_n\tau + \left(\frac{1}{\omega_n}\right)\cos \omega_n\tau \end{aligned} \tag{4}$$

Similarly,

$$\int \tau \sin \omega_n\tau\, d(\omega_n\tau) = -\tau \cos \omega_n\tau + \left(\frac{1}{\omega_n}\right)\sin \omega_n\tau \tag{5}$$

Hence,

$$\begin{aligned} u(t) = \left(\frac{p_o}{k}\right)\Bigg\{&\sin \omega_n t\left[\sin \omega_n t - \left(\frac{t}{t_d}\right)\sin \omega_n t - \left(\frac{1}{\omega_n t_d}\right)\cos \omega_n t + \left(\frac{1}{\omega_n t_d}\right)\right] \\ &- \cos \omega_n t\left[-\cos \omega_n t + 1 + \left(\frac{t}{t_d}\right)\cos \omega_n t - \left(\frac{1}{\omega_n t_d}\right)\sin \omega_n t\right]\Bigg\} \end{aligned} \tag{6}$$

Simplifying this expression, we get

$$R_1(t) = 1 - \left(\frac{t}{t_d}\right) - \cos \omega_n t + \left(\frac{1}{\omega_n t_d}\right) \sin \omega_n t \tag{7}$$

b. For $t_d < t$:

$$u(t) = \left(\frac{1}{m\omega_n}\right) \int_0^{t_d} p_o \left(1 - \frac{\tau}{t_d}\right) \sin \omega_n(t - \tau)\, d\tau \tag{8}$$

Note that this is the same as Eq. 3 except that it is evaluated at t_d, since $p(\tau) = 0$ for $t > t_d$. Equation 6 can be used by setting $t = t_d$ within the square brackets. Thus,

$$u(t) = \left(\frac{p_o}{k}\right) \left\{ \sin \omega_n t \left[-\left(\frac{1}{\omega_n t_d}\right) \cos \omega_n t_d + \left(\frac{1}{\omega_n t_d}\right) \right] - \cos \omega_n t \left[1 - \left(\frac{1}{\omega_n t_d}\right) \sin \omega_n t_d \right] \right\}$$

and, thus,

$$R_2(t) = \left(\frac{1}{\omega_n t_d}\right) [\sin \omega_n t(1 - \cos \omega_n t_d) - \cos \omega_n t(\omega_n t_d - \sin \omega_n t_d)] \tag{9}$$

(The values of R_{max} versus f_n will be plotted in Sec. 6.2.)

From Example 6.1 you can appreciate that while the response to an arbitrary input may, in principle, be obtained by use of the Duhamel integral, the work involved in evaluating the integral may be excessive. It may also be necessary to obtain the response to an input that is known only graphically, for example, a plot of ground acceleration versus time, and not in analytical form. In cases such as these, a numerical procedure is needed. Numerical procedures for computing dynamic response are discussed in Chapters 7 and 8.

6.2 Response Spectra

In the previous discussions of response calculations it has been assumed that a system has been defined, that is, k, m, and c are known, and that the response to a specified input is desired. The problem that is frequently encountered in design, particularly in preliminary design, is to select one or more of the system parameters in a manner that limits a certain response quantity, for example,

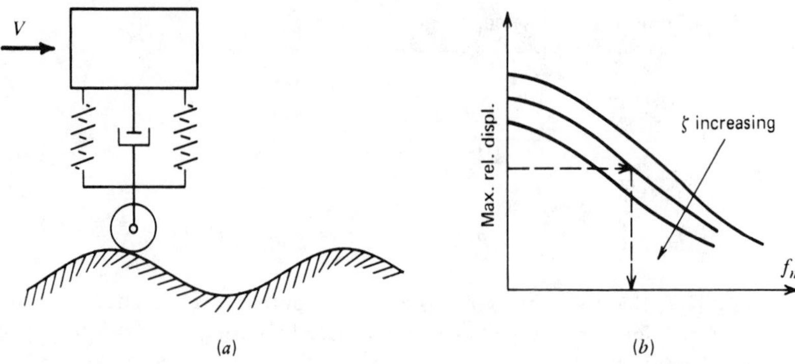

Figure 6.2. Example of a design problem amenable to solution using a response spectrum. (a) *Vehicle moving over specified bump.* (b) *Relative displacement response spectra.*

maximum absolute displacement or maximum stress, to a specified range when the system is subjected to a given input. Response spectra have been employed for this purpose in the design of buildings to withstand earthquake excitation[6.1,6.2] in the design of shock isolators[6.3,6.4], and in other design studies.

A *response spectrum* is a plot of maximum "response" (e.g., displacement, stress, acceleration, etc.) of SDOF systems to a given input versus some system parameter, generally the undamped natural frequency. A set of such curves, for example, curves plotted for various levels of system damping, may be referred to as *response spectra.*

Figure 6.2 illustrates how response spectra may be used in design. Figure 6.2*b* shows relative displacement response spectra for a system whose base excitation has a form such as that in Fig. 6.2*a*. Given the maximum permissible relative displacement and the damping level, the appropriate system natural frequency can be selected from Fig. 6.2*b* in the manner indicated by the dotted lines. This value of natural frequency, together with the given mass, can be used to determine the required spring stiffness, k.

Example 6.2

Using the response ratio expressions determined in Example 6.1,

a. Plot the response ratio versus t/t_d for $t_d/T_n = 0.25, 0.5, 1.0$ and 1.5 for $0 \leq t/t_d \leq 2.0$.

b. Determine expressions for the maximum response for $t \leq t_d$ and $t \geq t_d$. Using this information, plot a response spectrum in the form of R_{max} versus f_n.

Solution

a. Plots of the response ratio versus t/t_d are shown below.

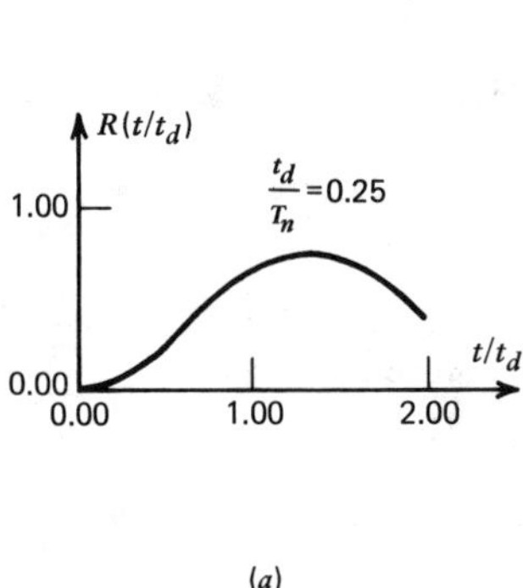

(a)

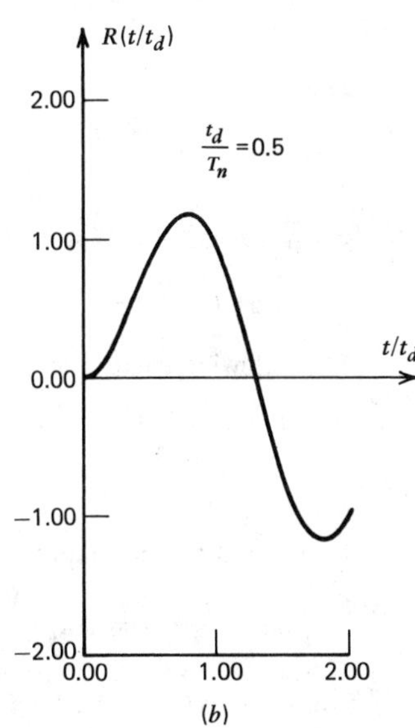

(b)

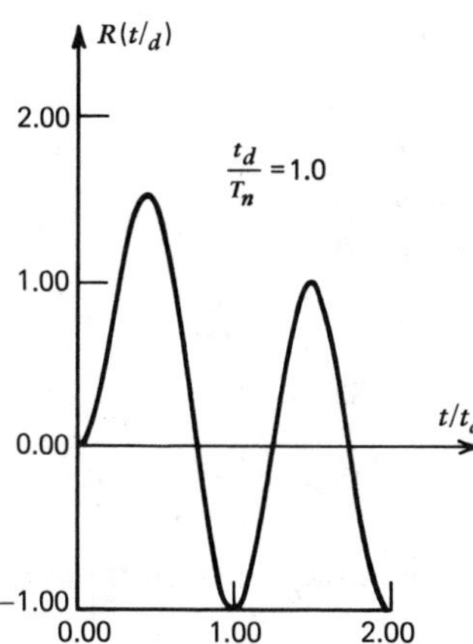

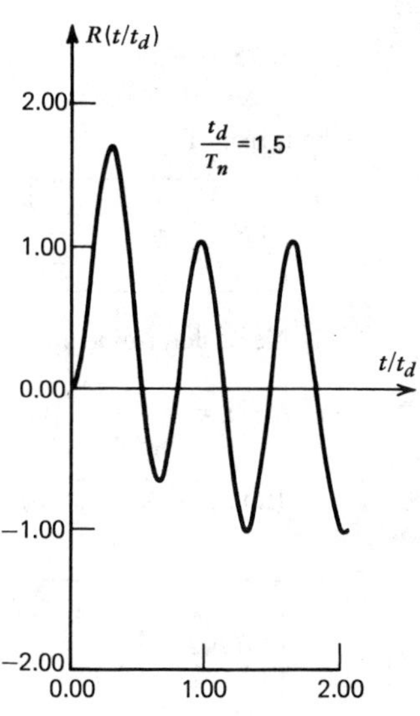

b. First determine an expression for $R_{\max}$ for $0 \le t \le t_d$. (From the curves above it can be seen that $R_{\max}$ occurs in this time interval except for short-duration loading, e.g., the $t_d/T_n = 0.25$ curve above.)

From Example 6.1,

$$R_1(t) = 1 - \left(\frac{t}{t_d}\right) - \cos \omega_n t + \left(\frac{1}{\omega_n t_d}\right) \sin \omega_n t \tag{1}$$

Then,

$$\dot{R}_1(t) = \left(\frac{1}{t_d}\right)(-1 + \omega_n t_d \sin \omega_n t + \cos \omega_n t) \tag{2}$$

Setting $\dot{R}_1 = 0$, we get

$$\omega_n t_d \sin \omega_n t_m = 1 - \cos \omega_n t_m \tag{3}$$

The following trigonometric identities are useful here:

$$\begin{aligned} (1 - \cos 2\alpha) &= 2 \sin^2 \alpha \\ \sin 2\alpha &= 2 \sin \alpha \cos \alpha \\ \cos 2\alpha &= 2 \cos^2 \alpha - 1 \end{aligned} \tag{4}$$

Then Eq. 3 can be simplified to

$$\tan\left(\frac{\omega_n t_m}{2}\right) = \omega_n t_d \tag{5}$$

From curves (b) through (d) above we can see that the maximum response occurs at the first occurrence of $\dot{R}_1 = 0$. Thus, $(\omega_n t_m/2)$ lies in the first quadrant with

$$\omega_n t_m = 2 \tan^{-1}(\omega_n t_d) \tag{6}$$

Then, from Eq. 1,

$$\boxed{(R_1)_{\max} = 1 - \frac{\omega_n t_m}{\omega_n t_d} - \cos \omega_n t_m + \left(\frac{1}{\omega_n t_d}\right) \sin \omega_n t_m} \tag{7}$$

Next, determine an expression for $R_{\max}$ for $t \ge t_d$. From Eq. 9 of Example 6.1,

$$R_2(t) = \left(\frac{1}{\omega_n t_d}\right)[(1 - \cos \omega_n t_d) \sin \omega_n t - (\omega_n t_d - \sin \omega_n t_d) \cos \omega_n t] \tag{8}$$

Equation 3.25 can be employed to determine the amplitude of this response. Thus,

$$\boxed{(R_2)_{\max} = \left(\frac{1}{\omega_n t_d}\right)[(1 - \cos \omega_n t_d)^2 + (\omega_n t_d - \sin \omega_n t_d)^2]^{1/2}} \tag{9}$$

Equation 7 is plotted as the solid curve below, and Eq. 9 is plotted as a dashed curve. For $f_n t_d < 0.371$, Eq. 9 gives the larger value of R_{max} while for $f_n t_d > 0.371$ the maximum response occurs during the forced-vibration era and is governed by Eq. 7. The "response spectrum" would be the composite maximum, or maximax, curve.

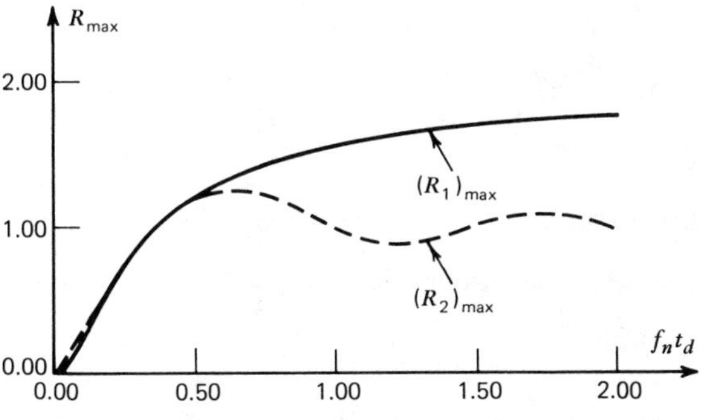

In general it is not an easy task to construct a response spectrum, since determination of the time at which the maximum occurs may involve solution of a transcendental equation and the careful selection of the maximum from among many local maxima. Consequently, determination of a response spectrum may require numerical evaluation of the response for various values of the system parameters.

In Example 6.3 we will consider an application of the response spectrum derived in Example 6.2.

Example 6.3

A building that is subjected to blast forces is modeled as a SDOF system. Determine the maximum blast force that can be sustained if the displacement is to be limited to 5 mm and if: (1) $t_d = 0.4$ s, (2) $t_d = 0.04$ s.

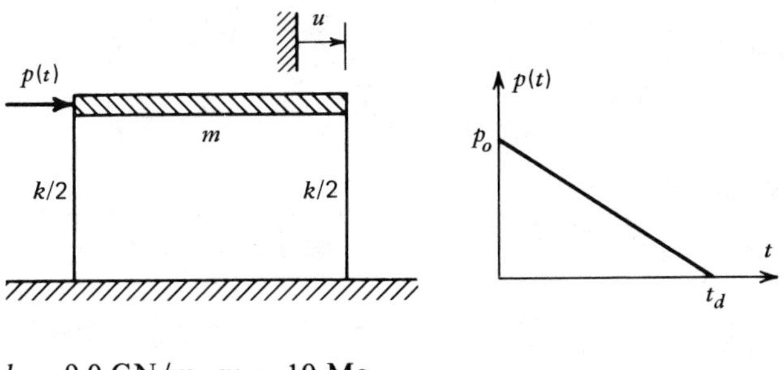

$k = 9.0$ GN/m, $m = 10$ Mg

Solution

a. Determine the natural frequency of the system

$$\omega_n = \left(\frac{k}{m}\right)^{1/2} = \left(\frac{9(10^9)}{10(10^6)}\right)^{1/2} = 30 \text{ rad/s}$$

$$f_n = \frac{\omega_n}{2\pi} = 4.77 \text{ Hz} \qquad (1)$$

b. Determine the maximum response ratio from the figure shown in Example 6.2:

For Case 1,

$$f_n t_d = 4.77(0.4) = 1.91 \qquad (2a)$$

From the response spectrum of Example 6.2

$$R_{\max} = 1.75 \qquad (2b)$$

For Case 2,

$$f_n t_d = 4.77(0.04) = 0.191 \qquad (3a)$$

and from the response spectrum of Example 6.2,

$$R_{\max} = 0.58 \qquad (3b)$$

c. Determine the static displacement for each case and then p_o:

$$u_{\max} = 5 \text{ mm} \qquad (4)$$

$$u_{\max} = R_{\max}\left(\frac{p_o}{k}\right)$$

or

$$p_o = \frac{k u_{\max}}{R_{\max}} \qquad (5)$$

Thus, for Case 1,

$$(p_o)_1 = \frac{9(10^9)(5)(10^{-3})}{1.75}$$

$$\boxed{(p_o)_1 = 25.7 \text{ MN}} \qquad (6)$$

$$(p_o)_2 = \frac{9(10^9)(5)(10^{-3})}{0.58}$$

$$\boxed{(p_o)_2 = 77.6 \text{ MN}} \qquad (7)$$

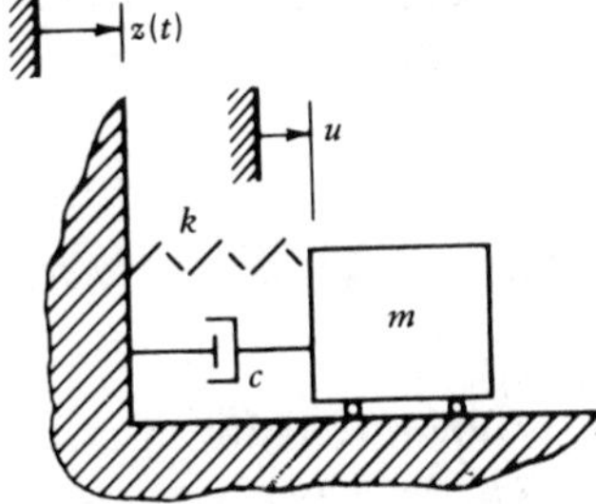

Figure 6.3. Prototype relative motion SDOF system.

Since many applications of response spectra involve relative motion, it is useful to determine appropriate response quantities for this case. Figure 6.3 shows the prototype relative motion SDOF system. As in Example 2.2, let the relative displacement be

$$w = u - z \tag{6.9}$$

Then the equation of motion can be written

$$m\ddot{w} + c\dot{w} + kw = -m\ddot{z} \tag{6.10}$$

A definition of response ratio was given in Eq. 5.6. The response ratio corresponding to relative motion should therefore be defined as

$$R(t) = \frac{-w(t)}{m\ddot{z}_{\max}/k} = \frac{-\omega_n^2\, w(t)}{\ddot{z}_{\max}} \tag{6.11}$$

By letting $\ddot{z} = \ddot{z}_{\max} f_a(t)$ and referring to Eq. 6.4 we can express $R(t)$ in Duhamel integral form. For $w(0) = \dot{w}(0) = 0$

$$R(t) = \frac{\omega_n^2}{\omega_d} \int_0^t f_a(\tau) e^{-\zeta\omega_n(t-\tau)} \sin \omega_d(t - \tau)\, d\tau \tag{6.12}$$

for damped systems and

$$R(t) = \omega_n \int_0^t f_a(\tau) \sin \omega_n(t - \tau)\, d\tau \tag{6.13}$$

for undamped systems. From Eq. 6.11 the maximum relative displacement is given by

$$w_{\max} = \left(\frac{1}{\omega_n^2}\right) R_{\max}\, \ddot{z}_{\max} \tag{6.14}$$

A second quantity which is of interest is the maximum absolute acceleration, $\ddot{u}_{\max}$. Equation 6.10 can be written

$$m\ddot{u} + c\dot{w} + kw = 0 \tag{6.15}$$

For an undamped system $\ddot{u}_{max}$ can be easily determined from Eqs. 6.14 and 6.15.

$$\ddot{u}_{max} = \omega_n^2 w_{max} \tag{6.16}$$

or, from Eqs. 6.14 and 6.16,

$$\ddot{u}_{max} = R_{max}\, \ddot{z}_{max} \qquad (c = 0) \tag{6.17}$$

Equations 6.14 and 6.17, then, relate maximum relative displacement and maximum absolute acceleration of the mass to the maximum base acceleration through R_{max}, which can be obtained from Eq. 6.12 or 6.13.

Example 6.4 illustrates a response spectrum for base motion and also indicates a typical form of response spectrum when plotted to logarithmic scales.

Example 6.4

An undamped SDOF system as in Fig. 6.2*a* experiences a base acceleration as shown below. The initial conditions are all zero. Develop expressions for w_{max} and $\ddot{u}_{max}$ and produce a log-log plot of w_{max} versus f_n.

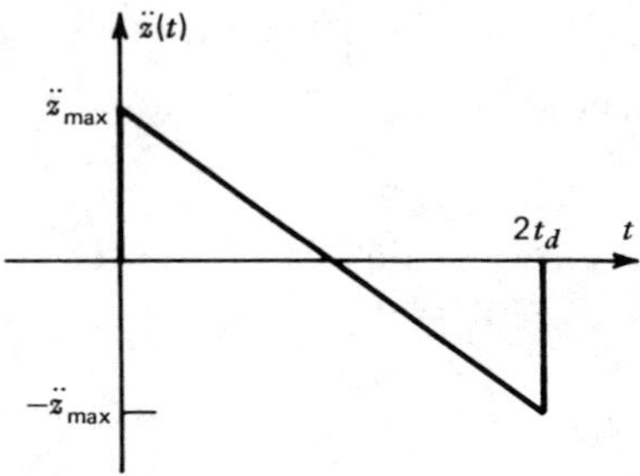

$$\ddot{z} = \ddot{z}_{max}\, f_a(t)$$

where

$$f_a(t) = \begin{cases} 1 - \dfrac{t}{t_d} & 0 \le t \le 2t_d \\ 0 & 2t_d < t \end{cases}$$

Solution

a. Determine the response for $0 \le t \le 2t_d$:

Since the excitation for $0 \le t \le 2t_d$ has the same form as that of Example 6.1, $R_1(t)$ will be the same as that given in Eq. 7 of Example 6.1, namely,

$$R_1(t) = 1 - \left(\frac{t}{t_d}\right) - \cos \omega_n t + \left(\frac{1}{\omega_n t_d}\right) \sin \omega_n t \tag{1}$$

b. Determine the response for $2t_d < t$:
The most direct way to express this response is to use Eq. 3.17 in the form

$$R_2(t) = R_1(2t_d) \cos \omega_n(t - 2t_d) + \left(\frac{\dot{R}_1(2t_d)}{\omega_n}\right) \sin \omega_n(t - 2t_d) \tag{2}$$

c. Determine the maximum response occurring in the forced-vibration era, $0 \le t \le 2t_d$:

$$\dot{R}_1(t) = \frac{1}{t_d}[-1 + (\omega_n t_d) \sin \omega_n t + \cos \omega_n t] \tag{3}$$

As in Example 6.2, setting $\dot{R}_1 = 0$ we are led to

$$\omega_n t_m = 2 \tan^{-1}(\omega_n t_d) \tag{4}$$

Unlike Example 6.2, however, the excitation now extends to negative values of $f_a(t)$, and hence the maximum absolute value of $R_1(t)$ may occur near $t = 0$ or near $t = 2t_d$. In the former case

$$\omega_n t_m = 2 \tan^{-1}(\omega_n t_d) \tag{5}$$

with $(\omega_n t_m/2)$ lying in the first quandrant. In the latter case

$$\omega_n t_m = 2 [\tan^{-1}(\omega_n t_d) + p\pi] \tag{6}$$

where p is the largest integer for which $\omega_n t_m < 2\omega_n t_d$ and $\tan^{-1}(\omega_n t_d)$ is taken in the first quadrant. Then,

$$\boxed{(R_1)_{max} = 1 - \left(\frac{\omega_n t_m}{\omega_n t_d}\right) - \cos \omega_n t_m + \left(\frac{1}{\omega_n t_d}\right) \sin \omega_n t_m} \tag{7}$$

d. Determine the maximum occurring in the residual-vibration era, $2t_d < t$:
From Eq. 2 this is simply

$$\boxed{(R_2)_{max} = \left\{[R_1(2t_d)]^2 + \left[\frac{\dot{R}_1(2t_d)}{\omega_n}\right]^2\right\}^{1/2}} \tag{8}$$

where $R_1(2t_d)$ and $\dot{R}_1(2t_d)$ are based on Eq. 1.

e. Determine expressions for w_{max} and $\ddot{u}_{max}$:
From Eq. 6.14,

$$\boxed{\frac{w_{max}}{\ddot{z}_{max} t_d^2} = \left(\frac{1}{\omega_n^2 t_d^2}\right) R_{max}} \tag{9}$$

and

$$\boxed{\frac{\ddot{u}_{max}}{\ddot{z}_{max}} = R_{max}} \qquad (10)$$

where R_{max} is the maximax response, that is, the larger of $(R_1)_{max}$ and $(R_2)_{max}$.

f. Plot w_{max} versus f_n using log-log scales:

It will be most convenient to plot the nondimensional response $w_{max}/(\ddot{z}_{max}t_d^2)$ versus the nondimensionalized natural frequency $f_n t_d$.

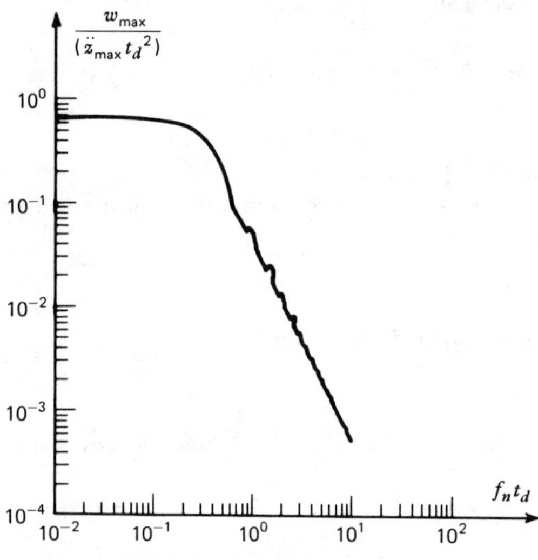

It is convenient to present response spectra in log-log form as in the above example because the spectrum can frequently be represented by two or three straight lines connected by short transition regions. For example, note from Eq. 9 of Example 6.4 that if R_{max} = const, $w_{max} \sim (1/f_n^2)$ which corresponds to a slope of -2 on the log-log plot. For large values of $f_n t_d$ the above log-log plot approximates this behavior.

Although creation of response spectra for a given input can be a tedious job, response spectra are useful tools when many systems are to be designed on the basis of their response to a given transient input, for example, a given earthquake time history. Chapter 20 presents an introduction to the use of response spectra for determining the response of linear SDOF and MDOF systems to earthquake excitation.

References

6.1 J. M. Biggs, *Introduction to Structural Dynamics,* McGraw-Hill, New York (1964).

6.2 G. W. Housner, "Design Spectrum," Chapter 5, *Earthquake Engineering* (Ed. R. L. Wiegel), Prentice-Hall, Englewood Cliffs, NJ (1970).

6.3 C. M. Harris and C. E. Crede (eds.), *Shock and Vibration Handbook,* 2nd Ed., McGraw-Hill, New York (1976).

6.4 W. T. Thomson, *Theory of Vibration with Applications,* Prentice-Hall, Englewood Cliffs, NJ (1972).

Problem Set 6.1

In Problems 6.1 through 6.4 use the Duhamel integral method to determine expressions for the response of an undamped SDOF system over the stated time intervals. Let $u(0) = \dot{u}(0) = 0$.

6.1 Determine $u(t)$ for (a) $0 \leq t \leq t_d$, and (b) $t_d \leq t$.

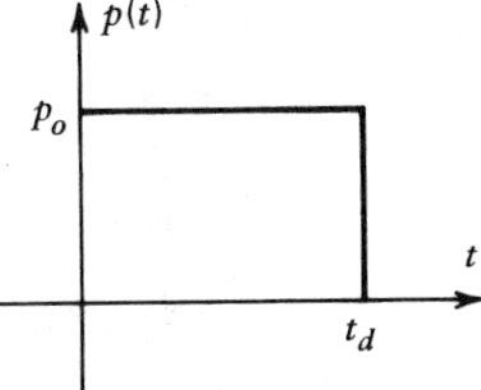

Figure P6.1

6.2 Determine $u(t)$ for (a) $0 \leq t \leq t_d$, and (b) $t_d \leq t$.

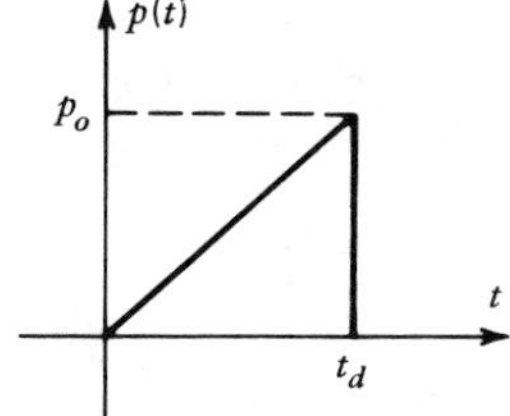

Figure P6.2

6.3 Determine $u(t)$ for (a) $0 \leq t \leq t_d$, and (b) $t_d \leq t$.

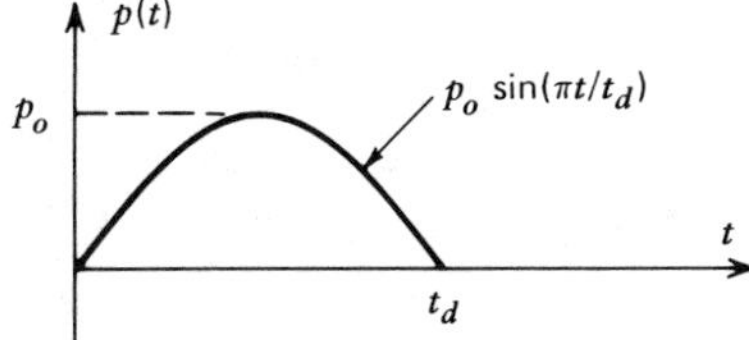

Figure P6.3

6.4 Determine $u(t)$ for (a) $0 \leq t \leq t_d$, and (b) $t_d \leq t$.

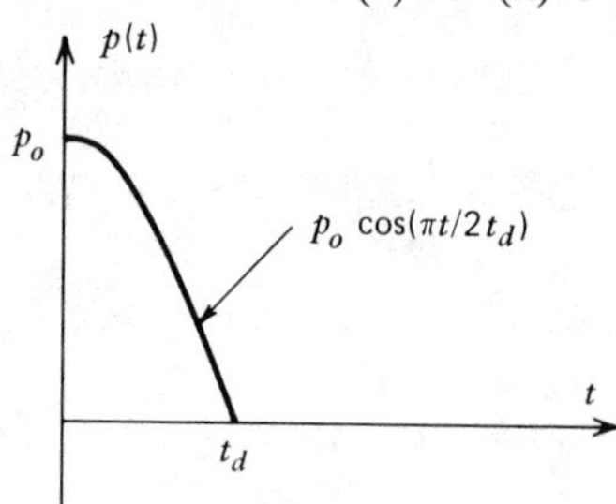

Figure P6.4

6.5 The vehicle in Example 4.3, when fully loaded, runs over a half-sine bump at 100 km/hr. The length of the bump is 2 m as shown below. Determine the maximum force exerted on the mass m by the springs.

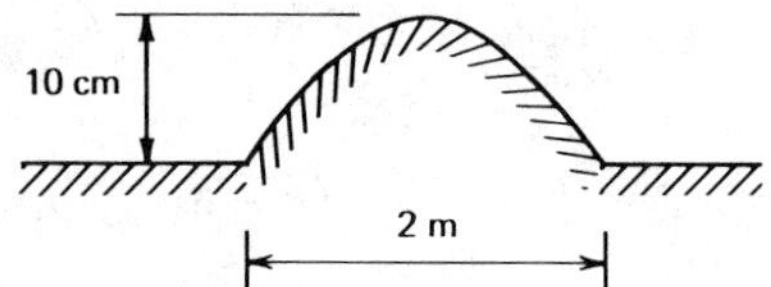

Figure P6.5

Problem Set 6.2

6.6 Plot the response spectrum $R_{\max}$ and the time for maximum response t_m/T_n versus t_d/T_n for the forcing function in Problem 6.2.

6.7 Plot the response spectrum $R_{\max}$ and the time for maximum response t_m/T_n versus t_d/T_n for the forcing function in Problem 6.3.

6.8 Plot the response spectrum $R_{\max}$ and the time for maximum response t_m/T_n versus t_d/T_n for the forcing function in Problem 6.4.

7 NUMERICAL EVALUATION OF DYNAMIC RESPONSE OF SDOF SYSTEMS

In Chapter 6, Duhamel integral expressions were obtained for the response of undamped and underdamped linear SDOF systems to arbitrary excitation. For simple forms of excitation these expressions can be evaluated in closed form. For more complex excitation a numerical solution is required. Two general approaches that may be used are described in this chapter: (1) interpolation of the excitation or interpolation of the integrand of the Duhamel integral, or (2) approximation of derivatives in the differential equation. Both approaches may be used for linear systems, while only the latter is applicable to nonlinear systems.

Upon completion of this chapter you should be able to:

- Use interpolation of the excitation to compute the response of a linear SDOF system to excitation, where the value of the excitation is given at discrete times t_i.
- Derive the computational algorithms for the average acceleration method and the linear acceleration method.
- Use the average acceleration algorithm or the linear acceleration algorithm to calculate the response of a linear SDOF system at times t_i.
- Use the average acceleration algorithm to compute the response of a SDOF system whose damping force has the form $f_D(\dot{u})$ and spring force has the form $f_S(u)$.

7.1 Numerical Solution Based on Interpolation of the Excitation Function

In many practical structural dynamics problems the excitation function, $p(t)$, is not known in the form of an analytical expression, but rather is given by a set of discrete values $p_i \equiv p(t_i)$ for $i = 0$ to N. These may be presented in tabular form or presented graphically as in Fig. 7.1. The time interval

$$\Delta t_i = t_{i+1} - t_i \tag{7.1}$$

is frequently taken to be a constant, Δt.

One approach to obtaining the response to this excitation is to use numer-

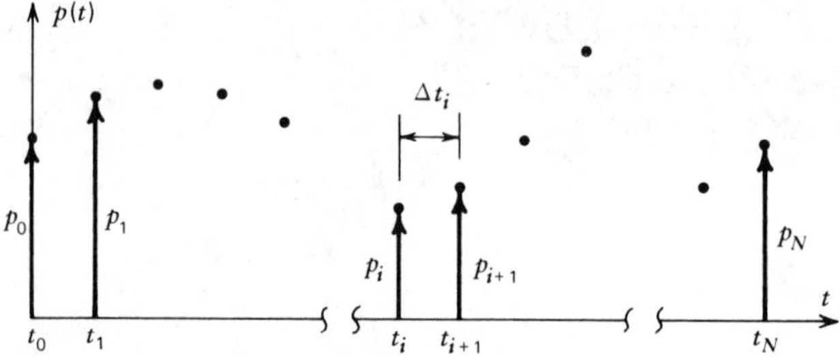

Figure 7.1. Excitation specified at discrete points.

ical quadrature formulas (e.g., the trapezoidal rule or Simpson's one-third rule) to evaluate the integrals appearing in the Duhamel integral expressions of Eqs. 6.6 or 6.7[7.1]. This involves interpolation of the integrand in these two equations.

A more direct and efficient procedure involves interpolation of the excitation function $p(t)$ and exact solution of the resulting linear response problem using results from Chapter 5 or Chapter 6. Figure 7.2 shows *piecewise-constant* and *piecewise-linear* interpolation of excitation functions. For piecewise-constant interpolation, the value of the force in the interval t_i to t_{i+1} is $\tilde{p}_i$, which could be taken to be the value p_i at the beginning of the interval, the value p_{i+1} at the end of the interval, or the average value $\tilde{p}_i = 0.5(p_i + p_{i+1})$. For piecewise-linear interpolation the interpolated force is given by

$$p(\tau) = p_i + \left(\frac{\Delta p_i}{\Delta t_i}\right)\tau \tag{7.2}$$

where

$$\Delta p_i = p_{i+1} - p_i \tag{7.3}$$

Consider the response of an undamped system. For piecewise-constant interpolation the forced response can be obtained from Eq. 5.8 and the response due to nonzero initial conditions from Eq. 3.17. Thus,

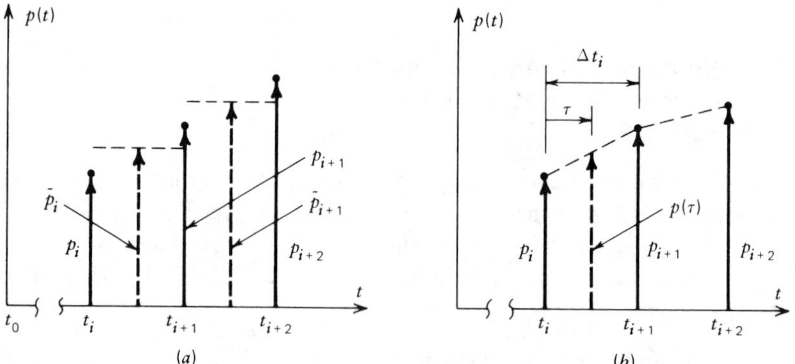

Figure 7.2. Interpolation of excitation functions. (a) *Piecewise-constant interpolation.* (b) *Piecewise-linear interpolation.*

$$u(\tau) = u_i \cos \omega_n\tau + \left(\frac{\dot{u}_i}{\omega_n}\right) \sin \omega_n\tau + \left(\frac{\tilde{p}_i}{k}\right)(1 - \cos \omega_n\tau) \tag{7.4a}$$

and

$$\frac{\dot{u}(\tau)}{\omega_n} = -u_i \sin \omega_n\tau + \left(\frac{\dot{u}_i}{\omega_n}\right) \cos \omega_n\tau + \left(\frac{\tilde{p}_i}{k}\right) \sin \omega_n\tau \tag{7.4b}$$

Evaluating these expressions at time t_{i+1} (i.e., at $\tau = \Delta t_i$), we get

$$u_{i+1} = u_i \cos(\omega_n \Delta t_i) + \left(\frac{\dot{u}_i}{\omega_n}\right) \sin(\omega_n \Delta t_i) + \left(\frac{\tilde{p}_i}{k}\right)[1 - \cos(\omega_n\Delta t_i)] \tag{7.5a}$$

$$\frac{\dot{u}_{i+1}}{\omega_n} = -u_i \sin(\omega_n \Delta t_i) + \left(\frac{\dot{u}_i}{\omega_n}\right) \cos(\omega_n \Delta t_i) + \left(\frac{\tilde{p}_i}{k}\right) \sin(\omega_n \Delta t_i) \tag{7.5b}$$

Equations 7.5 are the *recurrence formulas* for evaluating the dynamic state $(u_{i+1}, \dot{u}_{i+1})$ at time t_{i+1} given the state $(u_i, \dot{u}_i)$ at time t_i.

In contrast to piecewise-constant interpolation of the excitation, piecewise-linear interpolation permits a closer approximation. Equation 7.2 can be used to derive recurrence formulas based on piecewise-linear interpolation of the excitation. For an undamped system we get

$$u_{i+1} = u_i \cos(\omega_n \Delta t_i) + \left(\frac{\dot{u}_i}{\omega_n}\right) \sin(\omega_n \Delta t_i) + \left(\frac{p_i}{k}\right)[1 - \cos(\omega_n \Delta t_i)] + \left(\frac{\Delta p_i}{k}\right)\left(\frac{1}{\omega_n\Delta t_i}\right)[\omega_n \Delta t_i - \sin(\omega_n \Delta t_i)] \tag{7.6a}$$

$$\frac{\dot{u}_{i+1}}{\omega_n} = -u_i \sin(\omega_n \Delta t_i) + \left(\frac{\dot{u}_i}{\omega_n}\right) \cos(\omega_n \Delta t_i) + \left(\frac{p_i}{k}\right) \sin(\omega_n \Delta t_i) + \left(\frac{\Delta p_i}{k}\right)\left(\frac{1}{\omega_n \Delta t_i}\right)[1 - \cos(\omega_n \Delta t_i)] \tag{7.6b}$$

Recurrence formulas such as Eqs. 7.5 and 7.6 may be conveniently expressed in the following form.

$$u_{i+1} = Ap_i + Bp_{i+1} + Cu_i + D\dot{u}_i \tag{7.7a}$$

$$\dot{u}_{i+1} = A'p_i + B'p_{i+1} + C'u_i + D'\dot{u}_i \tag{7.7b}$$

Table 7.1. Coefficients for Recurrence Formulas for Underdamped SDOF Systems

$$A = \frac{1}{k\omega_d h}\left\{e^{-\beta h}\left[\left(\frac{\omega_d^2 - \beta^2}{\omega_n^2} - \beta h\right)\sin \omega_d h - \left(\frac{2\omega_d\beta}{\omega_n^2} + \omega_d h\right)\cos \omega_d h\right] + \frac{2\beta\omega_d}{\omega_n^2}\right\}$$

$$B = \frac{1}{k\omega_d h}\left\{e^{-\beta h}\left[-\left(\frac{\omega_d^2 - \beta^2}{\omega_n^2}\right)\sin \omega_d h + \left(\frac{2\omega_d\beta}{\omega_n^2}\right)\cos \omega_d h\right] + \omega_d h - \frac{2\beta\omega_d}{\omega_n^2}\right\}$$

$$C = e^{-\beta h}\left[\cos \omega_d h + \left(\frac{\beta}{\omega_d}\right)\sin \omega_d h\right]$$

$$D = \left(\frac{1}{\omega_d}\right)e^{-\beta h}\sin \omega_d h$$

$$A' = \frac{1}{k\omega_d h}\left\{e^{-\beta h}\left[(\beta + \omega_n^2 h)\sin \omega_d h + \omega_d \cos \omega_d h\right] - \omega_d\right\}$$

$$B' = \frac{1}{k\omega_d h}\left[-e^{-\beta h}(\beta \sin \omega_d h + \omega_d \cos \omega_d h) + \omega_d\right]$$

$$C' = -\left(\frac{\omega_n^2}{\omega_d}\right)e^{-\beta h}\sin \omega_d h$$

$$D' = e^{-\beta h}\left[\cos \omega_d h - \left(\frac{\beta}{\omega_d}\right)\sin \omega_d h\right]$$

where $\beta \equiv \zeta\omega_n$ and $h \equiv \Delta t_i$

Table 7.1 gives expressions for the coefficients A through D' for the linear force interpolation for the underdamped case ($\zeta < 1$). Coefficients can also be derived for critically damped and overdamped cases[7.2].

If the time step Δt_i is constant, the coefficients A through D' need be calculated only once, which greatly speeds the computations. Since the recurrence formulas given above are based on exact integration of the equation of motion for the various force interpolations, the only restriction on the size of the time step Δt_i is that it permit a close approximation to the excitation and that it provide output (u_i, $\dot{u}_i$) at the times where output is desired. In the latter regard, if the maximum response is desired, the time step should satisfy $\Delta t_i \leq T_n/10$ so that peaks due to natural response will not be missed.

Example 7.1

For the undamped SDOF system in Example 4.1 determine the response $u(t)$ for $0 \leq t \leq 0.2$ s:

a. By using piecewise-linear interpolation of the force with $\Delta t = 0.02$ s.

b. By evaluating the exact expression for $u(t)$ for these time steps.

Solution

From Example 4.1, $k = 40$ lb/in., $\omega_n = 20$ rad/s, and $p(t) = 10\cos(10t)$ lb.

a. For the linear interpolation method the displacement and velocity can be written in the form of Eqs. 7.7.

$$u_{i+1} = Ap_i + Bp_{i+1} + Cu_i + D\dot{u}_i \quad (1a)$$
$$\dot{u}_{i+1} = A'p_i + B'p_{i+1} + C'u_i + D'\dot{u}_i \quad (1b)$$

where the coefficients A through D' can be obtained directly from Eqs. 7.6 or by setting $\beta = 0$ and $\omega_d = \omega_n$ in Table 7.1. Thus, with $\Delta t \equiv h$,

$$\begin{aligned}
A &= \frac{1}{k\omega_n h}(\sin\omega_n h - \omega_n h\cos\omega_n h)\\
B &= \frac{1}{k\omega_n h}(\omega_n h - \sin\omega_n h)\\
C &= \cos\omega_n h\\
D &= \left(\frac{1}{\omega_n}\right)\sin\omega_n h\\
A' &= \frac{1}{kh}(\omega_n h\sin\omega_n h + \cos\omega_n h - 1)\\
B' &= \frac{1}{kh}(1 - \cos\omega_n h)\\
C' &= -\omega_n\sin\omega_n h\\
D' &= \cos\omega_n h
\end{aligned} \quad (2)$$

Thus,

$$\begin{aligned}
\omega_n h &= 20(0.02) = 0.4 \text{ rad}\\
\sin\omega_n h &= 0.38942\\
\cos\omega_n h &= 0.92106\\
A &= \frac{0.38942 - 0.4(0.92106)}{40(0.4)} = 1.312\times10^{-3}\\
B &= \frac{0.4 - 0.38942}{40(0.4)} = 6.613\times10^{-4}\\
C &= 9.211\times10^{-1}\\
D &= \frac{0.38942}{20} = 1.947\times10^{-2}\\
A' &= \frac{0.4(0.38942) + 0.92106 - 1}{40(0.02)} = 9.604\times10^{-2}\\
B' &= \frac{1.0 - 0.92106}{40(0.02)} = 9.868\times10^{-2}\\
C' &= -20.0(0.38942) = -7.788\\
D' &= 9.211\times10^{-1}
\end{aligned} \quad (3)$$

b. The exact solution is given by Eq. 11 of Example 4.1. For discrete times t_i this can be written

$$u_i = 0.3333[\cos(10t_i) - \cos(20t_i)] \quad (4)$$

Numerical Solution Based on Piecewise-Linear Interpolation of the Excitation (Exponent in Parenthesis)

i	t_i	p_i	Ap_i	Bp_{i+1}	Cu_i	$D\dot{u}_i$	u_i
0	0	10.0000	1.312 (−2)	6.481 (−3)	0.0000	0.0000	0.0000
1	0.02	9.8007	1.286 (−2)	6.091 (−3)	1.805 (−2)	3.753 (−2)	1.960 (−2)
2	0.04	9.2106	1.208 (−2)	5.458 (−3)	6.865 (−2)	6.762 (−2)	7.453 (−2)
3	0.06	8.2534	1.083 (−2)	4.607 (−3)	1.417 (−1)	8.406 (−2)	1.538 (−1)
4	0.08	6.9671	9.141 (−3)	3.573 (−3)	2.221 (−1)	8.293 (−2)	2.412 (−1)
5	0.10	5.4030	7.089 (−3)	2.396 (−3)	2.927 (−1)	6.322 (−2)	3.178 (−1)
6	0.12	3.6236	4.754 (−3)	1.124 (−3)	3.366 (−1)	2.711 (−2)	3.654 (−1)
7	0.14	1.6997	2.230 (−3)	−1.931 (−4)	3.404 (−1)	−2.040 (−2)	3.696 (−1)
8	0.16	−0.2920	−3.831 (−4)	−1.502 (−3)	2.966 (−1)	−7.221 (−2)	3.221 (−1)
9	0.18	−2.2720	−2.981 (−3)	−2.752 (−3)	2.050 (−1)	−1.203 (−1)	2.226 (−1)
10	0.20	−4.1615	—	—	—	—	7.900 (−2)

i	t_i	p_i	$A'p_i$	$B'p_{i+1}$	$C'u_i$	$D'\dot{u}_i$	$\dot{u}_i$
0	0	10.0000	9.604 (−1)	9.671 (−1)	0.0000	0.0000	0.0000
1	0.02	9.8007	9.413 (−1)	9.089 (−1)	−1.527 (−1)	1.775	1.928
2	0.04	9.2106	8.846 (−1)	8.144 (−1)	−5.805 (−1)	3.199	3.473
3	0.06	8.2534	7.927 (−1)	6.875 (−1)	−1.198	3.977	4.318
4	0.08	6.9671	6.691 (−1)	5.332 (−1)	−1.878	3.923	4.259
5	0.10	5.4030	5.189 (−1)	3.576 (−1)	−2.475	2.991	3.247
6	0.12	3.6236	3.480 (−1)	1.677 (−1)	−2.846	1.283	1.392
7	0.14	1.6997	1.632 (−1)	−2.881 (−2)	−2.878	−9.649 (−1)	−1.048
8	0.16	−0.2920	−2.804 (−2)	−2.242 (−1)	−2.508	−3.416	−3.709
9	0.18	−2.2720	−2.182 (−1)	−4.107 (−1)	−1.733	−5.689	−6.177
10	0.20	−4.1615	—	—	—	—	−8.051

Numerical Solution Based on Exact Response Function (Exponent in Parenthesis)

i	t_i	$\cos(10t_i)$	$\cos(20t_i)$	$u_i = 0.3333[\cos(10t_i) - \cos(20t_i)]$
0	0			0
1	0.02	9.801 (−1)	9.211 (−1)	1.967 (−2)
2	0.04	9.211 (−1)	6.967 (−1)	7.478 (−2)
3	0.06	8.253 (−1)	3.624 (−1)	1.543 (−1)
4	0.08	6.967 (−1)	−2.920 (−2)	2.419 (−1)
5	0.10	5.403 (−1)	−4.161 (−1)	3.188 (−1)
6	0.12	3.624 (−1)	−7.374 (−1)	3.665 (−1)
7	0.14	1.700 (−1)	−9.422 (−1)	3.707 (−1)
8	0.16	−2.920 (−2)	−9.983 (−1)	3.230 (−1)
9	0.18	−2.272 (−1)	−8.968 (−1)	2.232 (−1)
10	0.20	−4.161 (−1)	−6.536 (−1)	7.916 (−2)

By comparing the solution based on piecewise-linear interpolation of the excitation with the exact solution, we see that there is good agreement. Since the period of the excitation,

$$T = \frac{2\pi}{\Omega} = 0.0628 \text{ s} \tag{5}$$

is approximately 30 times Δt, the piecewise-linear approximation of $\cos(\Omega t)$ should be quite good.

For hand calculation of the response, the tabular form of Example 7.1 is convenient. Recurrence formulas of the form given in Eq. 7.7 are easily programmed for a digital computer or programmable calculator.

7.2 Numerical Solution Based on Approximating Derivatives; Step-by-Step Numerical Integration

In Sec. 7.1 recurrence relations based on interpolation of the excitation were obtained. These permit the response of a linear system to be obtained at dis-

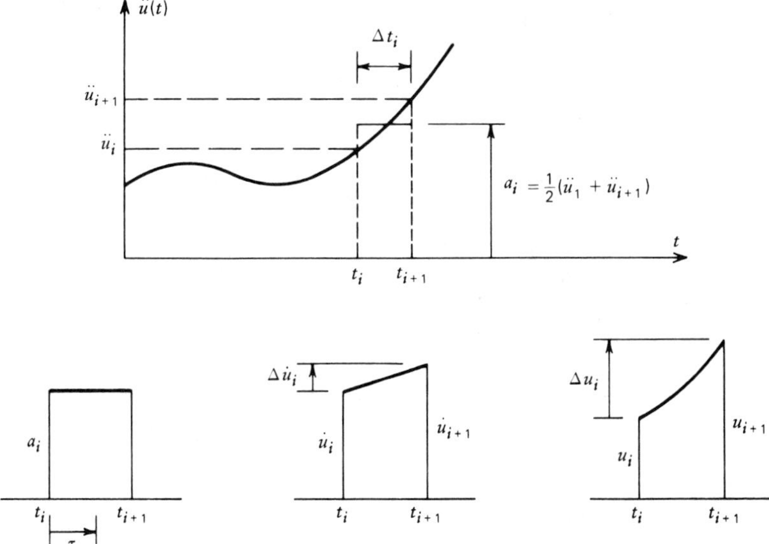

***Figure** 7.3. Numerical integration using the average acceleration method.*

crete times t_i. An alternate approach, which may be used for determining the response of both linear and nonlinear systems, is to approximate the derivatives appearing in the system equation of motion and to generate a step-by-step solution using time steps Δt_i. The procedure will be applied to linear systems in this section and will be extended to treat nonlinear systems in Sec. 7.4. Although many such procedures are available for carrying out this numerical integration (e.g., see References 7.3, 7.4, and 7.5), one of the most useful is the *average acceleration method,** which we will now consider.

The equation of motion to be integrated is

$$m\ddot{u} + c\dot{u} + ku = p(t) \tag{7.8}$$

with initial conditions u_o and $\dot{u}_o$ given. The acceleration is approximated, as shown in Fig. 7.3.

The acceleration in the time interval t_i to t_{i+1} is taken to be the average of the initial and final values of acceleration, that is,

$$\ddot{u}(\tau) = \tfrac{1}{2}(\ddot{u}_i + \ddot{u}_{i+1}) \tag{7.9}$$

*This method is also referred to as the *Newmark* $\beta = \frac{1}{4}$ *method,* the *trapezoidal rule,* or the *constant-average-acceleration method.*

Integration of Eq. 7.9 twice gives

$$\dot{u}_{i+1} = \dot{u}_i + \left(\frac{\Delta t_i}{2}\right)(\ddot{u}_i + \ddot{u}_{i+1}) \tag{7.10}$$

and

$$u_{i+1} = u_i + \dot{u}_i \, \Delta t_i + \left(\frac{\Delta t_i^2}{4}\right)(\ddot{u}_i + \ddot{u}_{i+1}) \tag{7.11}$$

In setting up the computational algorithm for this numerical integration problem, it is convenient to employ the incremental quantities Δp_i, Δu_i, $\Delta \dot{u}_i$, and $\Delta \ddot{u}_i$, where $\Delta p_i = p_{i+1} - p_i$, and so forth.* Then Eq. 7.11 can be solved for $\Delta \ddot{u}_i$, and Eqs. 7.10 and 7.11 combined to give $\Delta \dot{u}_i$ as follows:

$$\Delta \ddot{u}_i = \left(\frac{4}{\Delta t_i^2}\right)(\Delta u_i - \dot{u}_i \, \Delta t_i) - 2\ddot{u}_i \tag{7.12}$$

and

$$\Delta \dot{u}_i = \left(\frac{2}{\Delta t_i}\right)\Delta u_i - 2\dot{u}_i \tag{7.13}$$

Since Eq. 7.8 is satisfied at both t_i and t_{i+1},

$$m \, \Delta \ddot{u}_i + c \, \Delta \dot{u}_i + k \, \Delta u_i = \Delta p_i \tag{7.14}$$

Equations 7.12 through 7.14 can be combined to give the equation

$$k_i^* \, \Delta u_i = \Delta p_i^* \tag{7.15}$$

where

$$k_i^* = k + \left(\frac{2c}{\Delta t_i}\right) + \left(\frac{4m}{\Delta t_i^2}\right) \tag{7.16a}$$

and

$$\Delta p_i^* = \Delta p_i + \left[\left(\frac{4m}{\Delta t_i}\right) + 2c\right]\dot{u}_i + 2m\ddot{u}_i \tag{7.16b}$$

Once Δu_i has been determined from Eq. 7.15, $\Delta \dot{u}_i$ can be obtained from Eq. 7.13 and $\Delta \ddot{u}_i$ from Eq. 7.12, and the updated values of u, $\dot{u}$, and $\ddot{u}$ determined from

$$\begin{aligned} u_{i+1} &= u_i + \Delta u_i \\ \dot{u}_{i+1} &= \dot{u}_i + \Delta \dot{u}_i \\ \ddot{u}_{i+1} &= \ddot{u}_i + \Delta \ddot{u}_i \end{aligned} \tag{7.17}$$

*The incremental form presented here will be used in the nonlinear analysis of Sec. 7.4. An alternative "operator formulation" is presented in Chapter 18.

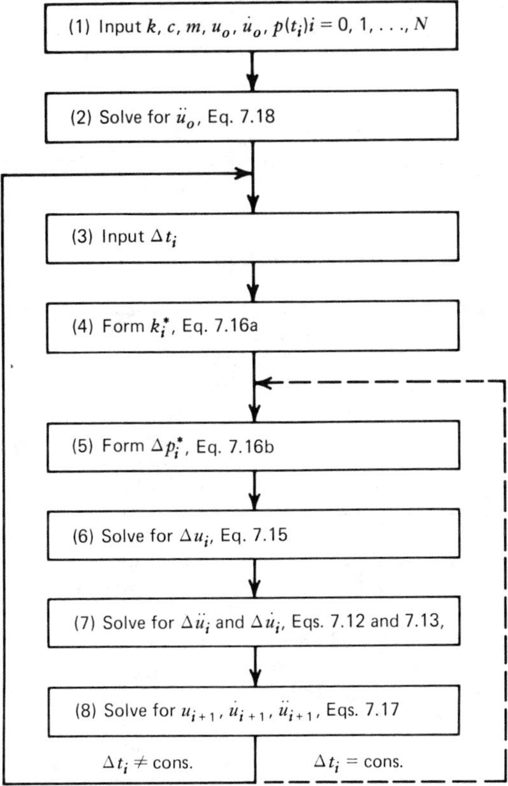

Figure 7.4. *Flowchart for step-by-step numerical integration based on the average acceleration method.*

The acceleration can also be obtained from the equation of motion.

$$\ddot{u}_i = \left(\frac{1}{m}\right)(p_i - c\dot{u}_i - ku_i) \tag{7.18}$$

rather than by Eqs. 7.12 and 7.17c. This equation is also used to obtain $\ddot{u}_o$.

A computational algorithm based on the above equations is summarized in the flowchart (Fig. 7.4). If Δt_i is constant, steps (3) and (4) can be removed from the *i*-loop. Chapter 18 contains further discussion of numerical integration methods including the average acceleration method.

Example 7.2

For the undamped SDOF system in Examples 4.1 and 7.1 use the average acceleration method to determine the response $u(t)$ for $0 \leq t \leq 0.2$ s. Use a constant time step $\Delta t = 0.02$ s and also $\Delta t = 0.005$ s. Compare the results with the exact solution of Example 7.1.

Solution

From Example 4.1, $k = 40$ lb/in., $c = 0$, $m = 0.1$ lb-sec^2/in., $\omega_n = 20$ rad/s, and $p(t) = 10\cos(10t)$ lb.

u(t$_i$) ***Determined by Average Acceleration Method (Exponent in Parenthesis)***

t_i	Exact	$\Delta t = 0.02$ s	$\Delta t = 0.005$ s
0	0	0	0
0.02	1.967 (−2)	1.904 (−2)	1.963 (−2)
0.04	7.478 (−2)	7.247 (−2)	7.464 (−2)
0.06	1.543 (−1)	1.498 (−1)	1.540 (−1)
0.08	2.419 (−1)	2.356 (−1)	2.416 (−1)
0.10	3.188 (−1)	3.116 (−1)	3.184 (−1)
0.12	3.665 (−1)	3.602 (−1)	3.662 (−1)
0.14	3.707 (−1)	3.673 (−1)	3.705 (−1)
0.16	3.230 (−1)	3.243 (−1)	3.231 (−1)
0.18	2.232 (−1)	2.303 (−1)	2.237 (−1)
0.20	7.916 (−2)	9.221 (−2)	8.002 (−2)

A "rule-of-thumb" that Δt should satisfy $\Delta t \leq T/10$, where T is the smallest period in the excitation (see Chapter 8) or the natural period T_n, whichever is smaller, is frequently stated. In Example 2.2 the natural period, $T_n = 2\pi/\omega_n = 0.314$ s, is smaller than the excitation period, $T = 0.628$ s. The time step $\Delta t = 0.02$ s satisfies the "rule-of-thumb" and, as seen in Example 7.2, gives satisfactory accuracy. The shorter time step, $\Delta t = 0.005$ s, reproduces the "exact" solution almost identically. Chapter 18 gives further information on the relationship of time step to accuracy of the solution.

7.3 Nonlinear SDOF Systems

Step-by-step numerical integration procedures such as the average acceleration method described in the preceding section are especially useful in solving for the response of nonlinear systems. Before considering the numerical solution of the nonlinear equation of motion, let us consider some of the physical phenomena that lead to nonlinearity in the equation of motion of a SDOF system.* The equation of motion of a linear SDOF system is, of course,

$$m\ddot{u} + c\dot{u} + ku = p(t) \tag{7.19a}$$

The equation of motion of a *nonlinear SDOF system* can be written in the form:

$$m\ddot{u} + f(u, \dot{u}, t) = 0 \tag{7.19b}$$

Since this is not a linear differential equation, the principle of superposition does not hold. Thus, for example, the Duhamel integral cannot be used to obtain the solution for the nonlinear response $u(t)$.

In Sec. 4.7 it was indicated that damping usually leads to nonlinear terms in the equation of motion and that equivalent viscous damping may sometimes be employed to approximate the actual nonlinear damping in a system. The damping force of a fluid acting on an object moving through it, and Coulomb damping were cited as common nonlinear damping mechanisms.

Two important classes of nonlinearity related to the displacement $u(t)$ are *geometric nonlinearity* and *material nonlinearity*. These will be illustrated briefly.

Consider the taut string (neglect bending) with attached lumped mass m as shown in Fig. 7.5. The tension in the string is given by

$$T = T_o + \left(\frac{AE}{L}\right)\delta \tag{7.20}$$

where T_o is the tension in the undeflected string and where δ is the elongation of the string given by

$$\delta = (L^2 + u^2)^{1/2} - L \tag{7.21}$$

*References 7.6, 7.7, 7.8 and many other references are available which go into far greater detail on nonlinear structural dynamics than is possible within the scope of this text. Reference 7.9, in particular, treats nonlinearities in structures subjected to earthquakes.

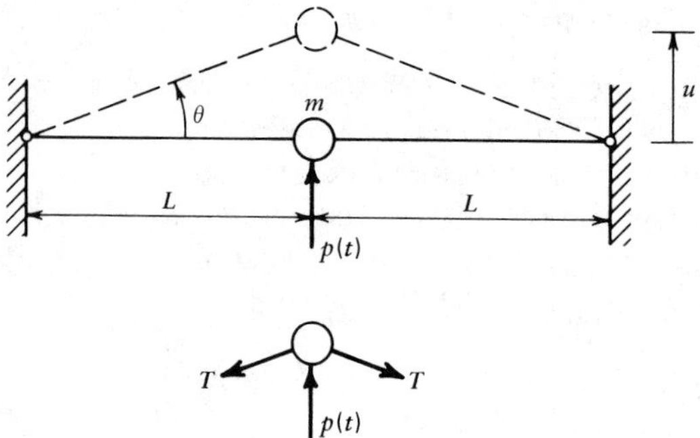

Figure 7.5. Large deflection of a mass on a taut string.

From the freebody diagram of Fig. 7.5 the equation of motion is found to be

$$\sum F_u = m\ddot{u} \tag{7.22}$$

or

$$p - 2T\sin\theta = m\ddot{u} \tag{7.23}$$

But,

$$\sin\theta = \frac{u}{(L^2 + u^2)^{1/2}} \tag{7.24}$$

Thus, the nonlinear equation of motion is

$$m\ddot{u} + 2\left\{T_o + \left(\frac{AE}{L}\right)[(L^2 + u^2)^{1/2} - L]\right\}\left[\frac{u}{(L^2 + u^2)^{1/2}}\right] = p(t) \tag{7.25}$$

This exact nonlinear differential equation has the form of Eq. 7.19 with the nonlinear term being a *geometric nonlinearity,* that is, the nonlinearity depends

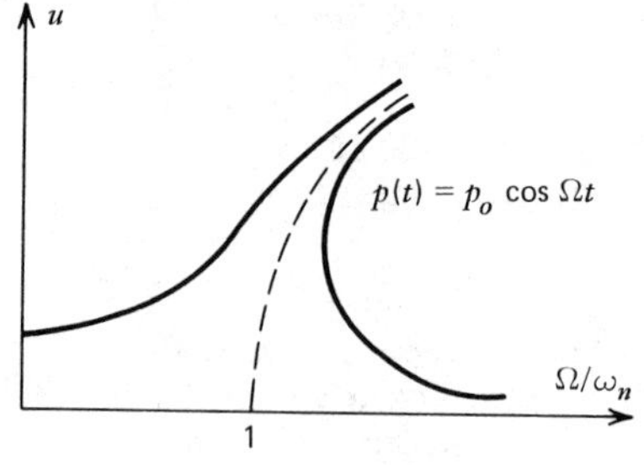

Figure 7.6. Frequency response curve for a SDOF system with hardening spring.

only on geometric quantities, length, and displacement, and not on any material properties. For "small" displacements, that is, $u \ll L$, Eq. 7.25 can be approximated by the linear equation

$$m\ddot{u} + 2T_o\left(\frac{u}{L}\right) = p(t) \tag{7.26}$$

For somewhat larger displacements

$$\delta \doteq L\left[1 + \frac{1}{2}\left(\frac{u}{L}\right)^2\right] - L$$

$$= \left(\frac{1}{2L}\right)u^2$$

$$\sin\theta \doteq \frac{u}{L}$$

so Eq. 7.25 can be approximated by

$$m\ddot{u} + \left(\frac{2T_o}{L}\right)u + \left(\frac{AE}{L^3}\right)u^3 = p(t) \tag{7.27}$$

The cubic term in Eq. 7.27, preceded by a plus sign, leads to what is referred to as a *hardening spring*. Under harmonic excitation such a system exhibits a frequency response behavior characterized by Fig. 7.6. The dashed line in Fig. 7.6 represents the locus of resonant frequencies, indicating that the resonant frequency increases with increasing amplitude of response. Note also that at some values of Ω/ω_n, three different amplitudes are possible. References 7.6, 7.7, and others give detailed studies of the response of systems with hardening or softening springs to harmonic excitation.

Next let us consider an example of a *material nonlinearity*. Figure 7.7*a* represents an idealized steel frame whose columns are assumed to be much more flexible than the horizontal "roof" member. If the load p is increased slowly, the load-deflection curve of Fig. 7.7*b* results. The load-deflection behavior is linear up to point *B*, where yielding begins at the outer fibers at the points labeled "plastic hinges," where the moment is the greatest. The load is increased further to point *C*. Thereafter the load is reduced, and the load-deflection curve from *C* to *E* follows a straight line parallel to the original slope of portion *AB*. At point *D* the direction of loading is reversed. The closed loop *ACFA* is called a *hysteresis loop*. The area inside the loop is the energy dissipated as a result of cyclic plastic deformation. The force-displacement relationship of Fig. 7.7*b* and other models of structural nonlinearities are discussed

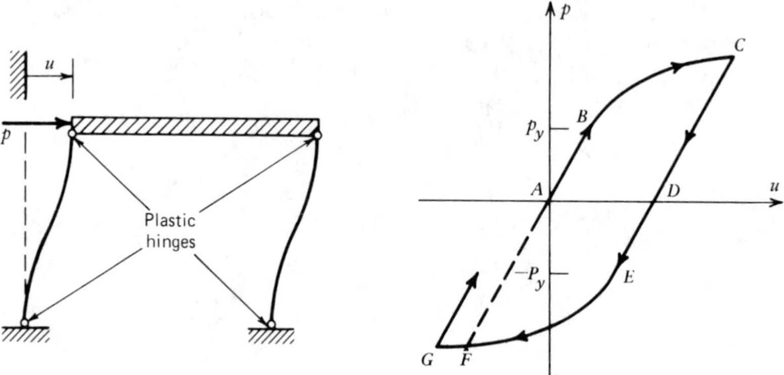

Figure 7.7. Steel frame under loading which produces yielding of columns.

in References 7.8 and 7.9. A simpler model, the elastic-perfectly plastic model, is illustrated in Example 7.3 in Sec. 7.4.

7.4 Step-by-Step Numerical Solution for Response of Nonlinear SDOF Systems

The step-by-step procedure employed in Sec. 7.2 for determining the response of linear systems may be adapted for determining the response of nonlinear systems. Figure 7.8*a* shows a SDOF system with possible nonlinear "spring"

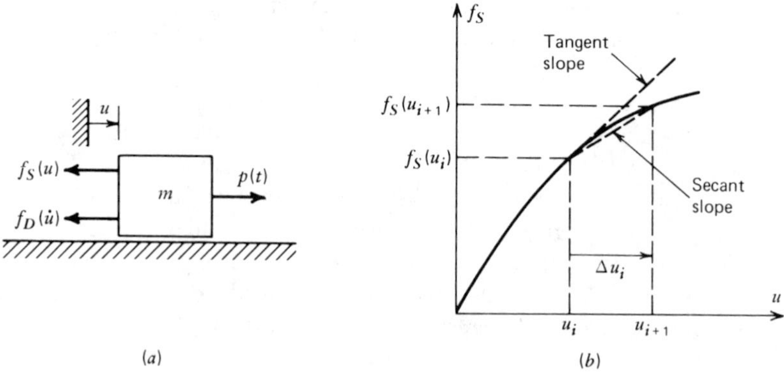

Figure 7.8. A nonlinear SDOF system. (a) *SDOF system.* (b) *Force-deformation curve.*

and "damping" elements. Figure 7.8*b* illustrates a typical nonlinear force-displacement curve for f_S expressed as a function of displacement u. It is assumed that the mass remains constant, although this could easily be generalized to be a known function of time. (The symbols f_S and f_D will be used whether referring to functions of time or to functions of displacement and velocity.) Then, the equation of motion may be written at times t_i and $t_{i+1} = t_i + \Delta t_i$ as follows:

$$m\ddot{u}_i + f_{Di} + f_{Si} = p_i \tag{7.28a}$$

$$m\ddot{u}_{i+1} + f_{D(i+1)} + f_{S(i+1)} = p_{i+1} \tag{7.28b}$$

Subtracting Eq. 7.28a from 7.28b, we get

$$m\,\Delta\ddot{u}_i + \Delta f_{Di} + \Delta f_{Si} = \Delta p_i \tag{7.29}$$

where

$$\begin{aligned} \Delta\ddot{u}_i &= \ddot{u}_{i+1} - \ddot{u}_i \\ \Delta f_{Di} &= f_{D(i+1)} - f_{Di} \\ \Delta f_{Si} &= f_{S(i+1)} - f_{Si} \\ \Delta p_i &= p_{i+1} - p_i \end{aligned} \tag{7.30}$$

Equation 7.29 differs from Eq. 7.14, which was used in the step-by-step integration algorithm, only in the nonlinear f_D and f_S terms. An approximation to the Δf_{Si} term is

$$\Delta f_{Si} \doteq k_i\,\Delta u_i \tag{7.31a}$$

where k_i is the tangent slope at u_i as shown on Fig. 7.8*b*. Similarly, an approximation to the f_D term can be written in the form

$$\Delta f_{Di} \doteq c_i\,\Delta\dot{u}_i \tag{7.31b}$$

where c_i is the tangent slope of the f_D versus $\dot{u}$ curve at $\dot{u}_i$. Then Eq. 7.29 becomes

$$m\,\Delta\ddot{u}_i + c_i\,\Delta\dot{u}_i + k_i\,\Delta u_i = \Delta p_i \tag{7.32}$$

This replaces Eq. 7.14 in the step-by-step integration algorithm of Sec. 7.2. Thus, the response of a nonlinear system can be computed using the steps outlined in the flowchart of Fig. 7.4 with c and k being replaced by c_i and k_i evaluated at the beginning of each time step. By using Eq. 7.28b to compute the acceleration at the end of the present time step (and beginning of the next time step), dynamic equilibrium is enforced at each time step. Due to the approximations in Eqs. 7.31, if $\ddot{u}_i$ were computed using Eqs. 7.12 and 7.17c, dynamic equilibrium would not be satisfied at time t_{i+1} unless c and k remained constant over the time step. In general, the time step Δt_i must be small enough that the difference in a linear solution and nonlinear solution over one time step is not great.

The following example illustrates the use of the average acceleration step-by-step integration algorithm for calculating the response of a nonlinear SDOF system.

Example 7.3

The frame shown in Fig. 7.7*a* has an elastic, perfectly plastic force-deformation behavior as shown below. (This is an idealization of the behavior shown in Fig. 7.7*b*.) The frame is subjected to a rectangular pulse loading as shown below. (Compare this problem with the corresponding linear problem in Sec. 5.2.) The frame is at rest at $t = 0$. Use the average acceleration method to compute the response $u(t)$ for $t = 0$ to $t = 0.55$ s. Use $\Delta t = 0.05$ s (constant).

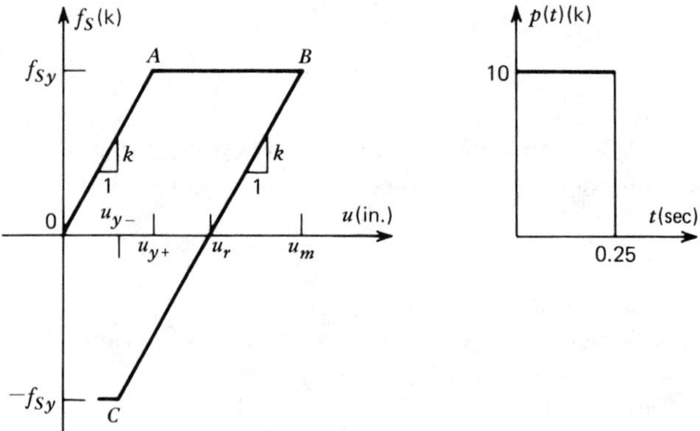

$m = 0.2$ k-sec^2/in., $\quad k = 30$ k/in. (total), $\quad f_{S_y} = 15$ k

Solution

The flowchart of Fig. 7.4 can be adapted by using the following equations.

$$\ddot{u}_i = \left(\frac{1}{m}\right)(p_i - f_{Si}) = 5.0(p_i - f_{Si}) \tag{1}$$

$$k_i^* = k_i + \left(\frac{4m}{\Delta t^2}\right) = k_i + 320.0 \tag{2}$$

$$\begin{aligned} \Delta p_i^* &= \Delta p_i + \left(\frac{4m}{\Delta t}\right)\dot{u}_i + 2m\ddot{u}_i \\ &= \Delta p_i + 16.0\dot{u}_i + 0.4\ddot{u}_i \end{aligned} \tag{3}$$

$$\Delta u_i = \frac{\Delta p_i^*}{k_i^*} \tag{4}$$

$$\Delta \dot{u}_i = \left(\frac{2}{\Delta t}\right)\Delta u_i - 2\dot{u}_i = 40.0\,\Delta u_i - 2\dot{u}_i \tag{5}$$

$$u_{i+1} = u_i + \Delta u_i \tag{6}$$

$$\dot{u}_{i+1} = \dot{u}_i + \Delta \dot{u}_i \tag{7}$$

For the present problem k_i has the value $k = 30$ k/in. or zero depending on the deformation history. The table on page 158 summarizes the calculations.

$$\begin{aligned} u_y &= \frac{f_{Sy}}{k} = 15/30 = 0.5 \text{ in.} \\ u_m &= \text{displacement at which } \dot{u} \text{ switches from } (+) \text{ to } (-). \\ u_r &= \text{residual (inelastic) displacement} = u_m - u_y = u_m - 0.5. \end{aligned} \tag{8}$$

$$f_{Si} = \begin{cases} ku_i & \text{for segment } OA \quad (9a) \\ f_{Sy} & \text{for segment } AB \quad (9b) \\ f_{Sy} - k(u_m - u_i) & \text{for segment } BC \quad (9c) \\ \text{etc.} & \end{cases}$$

The tables that follow outline the calculation of the response of the elastic-plastic system and of an elastic system (assuming $f_y > \max \mid f_S \mid$, i.e., ignoring yielding) using the average acceleration algorithm. The figure below shows the elastic-plastic response.

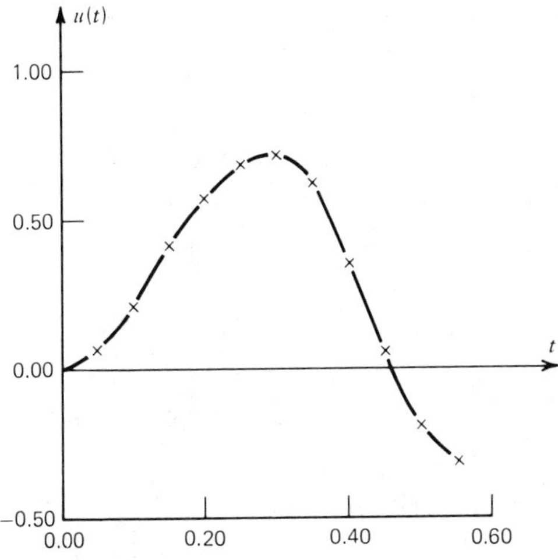

Numerical Solution for Elastic-Plastic Response Calculated by Average Acceleration Method

(1) i	(2) t_i	(3) p_i	(4) f_{Si}	(5) $\ddot{u}_i$	(6) Δp_i	(7) Δp_i^*	(8) k_i^*	(9) Δu_i	(10) $\Delta \dot{u}_i$	(11) u_i	(12) $\dot{u}_i$
			Eqs. 9	5[(3) − (4)]		(6) + 16(12) + 0.4(5)	Eq. 2	(7)/(8)	40(9) − 2(12)	(11) + (9)	(12) + (10)
0	0.00	10.0	0	50.0000	0.0	20.0000	350.0	0.05714	2.2857	0.00000	0.0000
1	0.05	10.0	1.7143	41.4286	0.0	53.1429	350.0	0.15184	1.5020	0.05714	2.2857
2	0.10	10.0	6.2694	18.6531	0.0	68.0653	350.0	0.19447	0.2034	0.20898	3.7878
3	0.15	10.0	12.1036	−10.5178	0.0	59.6511	350.0	0.17043	−1.1650	0.40345	3.9911
4	0.20	10.0	15.0000	−25.0000	0.0	35.2181	320.0	0.11006	−1.2500	0.57388†	2.8261
5	0.25	10.0	15.0000	−25.0000	0.0	15.2181	320.0	0.04756	−1.2500	0.68394	1.5761
6	0.30	0.0	15.0000	−75.0000	−10.0	−34.7819†	350.0	−0.09938	−4.6273	0.73150	0.3261
7	0.35	0.0	12.0186	−60.0929	0.0	−92.8565	350.0	−0.26530	−2.0098	0.63212	−4.3012
8	0.40	0.0	4.0595	−20.2973	0.0	−109.0943	350.0	−0.31170	0.1540	0.36682	−6.3110
9	0.45	0.0	−5.2915	26.4574	0.0	−87.9284	350.0	−0.25122	2.2650	0.05512	−6.1570
10	0.50	0.0	−12.8282	64.1410	0.0	−36.6156	350.0	−0.10462	3.5994	−0.19611	−3.8920
11	0.55	0.0	—	—	—	—	—	—	—	−0.30072	−0.2926

Numerical Solution for Elastic Response Calculated by Average Acceleration Method

(1) i	(2) t_i	(3) p_i	(4) $f_{Si} = ku_i$	(5) $\ddot{u}_i$	(6) Δp_i	(7) Δp_i^*	(8) k_i^*	(9) Δu_i	(10) $\Delta \dot{u}_i$	(11) u_i	(12) $\dot{u}_i$
			30(11)	5[(3) − (4)]		(6) + 16(12) + 0.4(5)	Eq. 2	(7)/(8)	40(9) − 2(12)	(11) + (9)	(12) + (10)
0	0.00	10.0	0.0000	50.0000	0.0	20.0000	350.0	0.05714	2.2857	0.00000	0.0000
1	0.05	10.0	1.7143	41.4286	0.0	53.1429	350.0	0.15184	1.5020	0.05714	2.2857
2	0.10	10.0	6.2694	18.6531	0.0	68.0653	350.0	0.19447	0.2034	0.20898	3.7878
3	0.15	10.0	12.1036	−10.5178	0.0	59.6511	350.0	0.17043	−1.1650	0.40345	3.9911
4	0.20	10.0	17.2165	−36.0825	0.0	30.7851	350.0	0.08796	−2.1340	0.57388	2.8261
5	0.25	10.0	19.8552	−49.2761	0.0	−8.6359	350.0	−0.02467	−2.3713	0.66184	0.6922
6	0.30	0.0	19.1150	−95.5751	−10.0	−75.0959	350.0	−0.21456	−5.2242	0.63717	−1.6791
7	0.35	0.0	12.6782	−63.3911	0.0	−135.8088	350.0	−0.38803	−1.7145	0.42261	−6.9033
8	0.40	0.0	1.0375	−5.1873	0.0	−139.9586	350.0	−0.39988	1.2402	0.03458	−8.6177
9	0.45	0.0	−10.9590	54.7949	0.0	−96.1227	350.0	−0.27464	3.7696	−0.36530	−7.3775
10	0.50	0.0	−19.1981	95.9904	0.0	−19.3304	350.0	−0.05523	5.0066	−0.63994	−3.6079
11	0.55	0.0	—	—	—	—	—	—	—	−0.69517	1.3987

Note that the expressions for k_i^* and f_{Si} change at the †-symbols; first where u_i first exceeds u_y, and next where Δp_i^* changes sign. If the computations were to continue, a further change would be required.

The above example briefly illustrates the complex nature of numerical solution of nonlinear structural dynamics problems. In Sec. 18.2 numerical solution of nonlinear MDOF problems will be discussed.

References

7.1 R. W. Clough and J. Penzien, *Dynamics of Structures,* McGraw-Hill, New York (1975).

7.2 *NASTRAN Theoretical Manual (Level 16.0),* National Aeronautics and Space Administration, Washington, DC (1976).

7.3 N. M. Newmark, "A Method of Computation for Structural Dynamics," *Trans. ASCE,* v. 127, pt. 1, 1406–1435 (1962).

7.4 K. J. Bathe and E. L. Wilson, "Stability and Accuracy Analysis of Direct Integration Methods," *Earthquake Engr. and Structural Dyn.,* v. 1, 283–291 (1973).

7.5 H. M. Hilber, T. J. R. Hughes, and R. L. Taylor, "Improved Numerical Dissipation for Time Integration Algorithms in Structural Dynamics," *Earthquake Engr. and Structural Design,* v. 5, 283–292 (1977).

7.6 J. J. Stoker, *Nonlinear Vibrations,* Wiley, New York (1950).

7.7 W. J. Cunningham, *Introduction to Nonlinear Analysis,* McGraw-Hill, New York (1958).

7.8 M. A. Sozen, "Hysteresis in Structural Elements," *Applied Mechanics in Earthquake Engineering,* AMD-v. 8, ASME, New York, 63–98 (1974).

7.9 A. K. Chopra, "Earthquake Analysis of Complex Structures," *Applied Mechanics in Earthquake Engineering,* AMD-v. 8, ASME, New York, 163–203 (1974).

Problem Set 7.1

7.1 For the undamped SDOF system in Example 4.1 determine the response $u(t)$ for $0 \leq t \leq 0.4$ s. Use piecewise-constant interpolation of the force with the magnitude of each impulse determined by the force at the beginning of each time step $\Delta t = 0.02$ s.

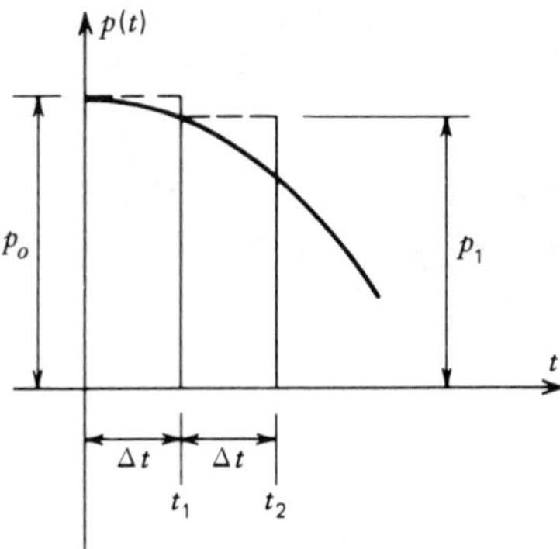

Figure P7.1

7.2 Repeat Problem 7.1, but with the magnitude of each impulse being determined by the force at the middle of each time step $\Delta t = 0.02$ s.

7.3 Repeat Problem 7.1, but with the magnitude of each impulse being determined by the force at the end of each time step $\Delta t = 0.02$ s.

7.4 (a) Repeat part (a) of Example 7.1, but double the time step to $\Delta t = 0.04$ s.

(b) Tabulate your results from (a) and compare them with the exact solution and with the numerical solution of Problem 7.1.

Problem Set 7.2

7.5 The Newmark β method, referred to in the footnote on p. 147, is embodied in the equations

$$\dot{u}_{i+1} = \dot{u}_i + \left(\frac{\Delta t_i}{2}\right)(\ddot{u}_i + \ddot{u}_{i+1})$$

$$u_{i+1} = u_i + \dot{u}_i\,\Delta t_i + \left(\frac{1}{2} - \beta\right)\ddot{u}_i(\Delta t_i)^2 + \beta\ddot{u}_{i+1}(\Delta t_i)^2$$

(a) Show that $\beta = \frac{1}{6}$ corresponds to the "linear acceleration method," wherein the acceleration approximation in Fig. 7.3 is replaced by the linear interpolation below.

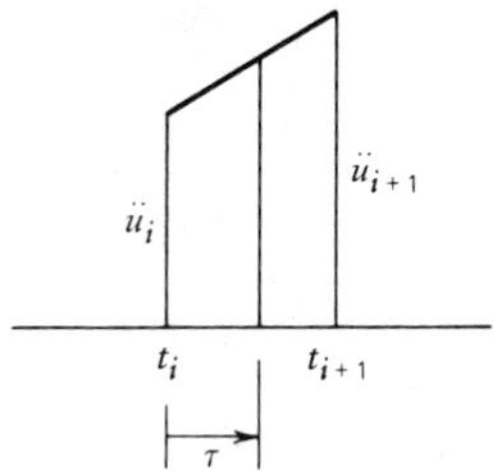

Figure P7.5

(b) Determine expressions for k_i^* and Δp_i^* for the linear acceleration method similar to those for the average acceleration method as given by Eqs. 7.16a and 7.16b.

7.6 Repeat Example 7.2 using the linear acceleration method as derived in Problem 7.5 instead of the average acceleration method.

Problem Set 7.4

7.7 Repeat Example 7.3 using the linear acceleration method as derived in Problem 7.5 instead of the average acceleration method.

8 RESPONSE OF SDOF SYSTEMS TO PERIODIC EXCITATION; FREQUENCY DOMAIN ANALYSIS

In Chapter 4 you studied the response of SDOF systems to harmonic excitation and became familiar with important concepts such as resonance. You also learned how to simplify the analysis of viscous-damped systems by using complex frequency response. These concepts from Chapter 4 will now be extended to determine the response of SDOF systems to periodic excitation. Upon completion of Chapter 8 you should be able to:

- Determine the Fourier series representation of a periodic function using the real form of the Fourier series.
- Determine the Fourier series representation of a periodic function using the complex form of the Fourier series.
- Determine the steady-state response of a SDOF system to periodic excitation using either the real or complex form.
- Determine the Fourier transform of a transient function $p(t)$.
- Discuss the Fourier transform method for computing the response of a SDOF system to transient excitation. Set up, but do not carry out, the complex integration for the inverse Fourier transform for $u(t)$.
- Define DFT and FFT and discuss the advantages of the FFT algorithm in computing numerical Fourier transforms.

8.1 Response to Periodic Excitation—Real Fourier Series

Forces acting on structures are frequently periodic, or can be approximated closely by periodic forces. For example, the forces exerted on an automobile traveling at constant speed over certain roadway surfaces can be considered to be periodic. A function of time, $p(t)$, having a period T_1, can be separated into its harmonic components by means of a Fourier series expansion. In this section we consider real Fourier series. In Sec. 8.2 complex Fourier series will be introduced. The complex form is very useful when combined with the complex frequency response function of Chapter 4 to study damped systems.

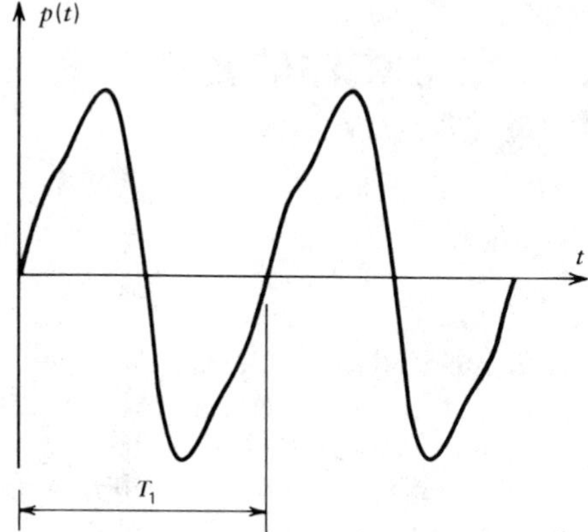

Figure 8.1. *A periodic function with period T_1.*

Figure 8.1 shows a periodic function with period T_1, that is,

$$p(t + T_1) = p(t) \tag{8.1}$$

$p(t)$ may be separated into its harmonic components by means of a *Fourier series expansion,* which may be defined as

$$p(t) = a_o + \sum_{n=1}^{\infty} a_n \cos(n\Omega_1 t) + \sum_{n=1}^{\infty} b_n \sin(n\Omega_1 t) \tag{8.2}$$

where

$$\Omega_1 = \frac{2\pi}{T_1} \tag{8.3}$$

is the fundamental frequency, and a_n and b_n are the coefficients of the nth harmonic. The coefficients a_o, a_n, and b_n are related to $p(t)$ by the equations

$$\begin{aligned} a_o &= \frac{1}{T_1}\int_{\tau}^{\tau+T_1} p(t)\, dt = \text{average value of } p(t) \\ a_n &= \frac{2}{T_1}\int_{\tau}^{\tau+T_1} p(t)\cos(n\Omega_1 t)\, dt,\ n \neq 0 \\ b_n &= \frac{2}{T_1}\int_{\tau}^{\tau+T_1} p(t)\sin(n\Omega_1 t)\, dt \end{aligned} \tag{8.4}$$

where τ is an arbitrary time.

Although theoretically a Fourier series representation of $p(t)$ may require an infinite number of terms, in actual practice $p(t)$ may generally be approximated with sufficient accuracy by a relatively small number of terms. Example 8.1 illustrates the Fourier series representation of a square wave.

Example 8.1

a. Determine expressions for the coefficients of a Fourier series representation of the square wave shown below. Write the Fourier series representation of $p(t)$.

b. Plot truncated series employing respectively one, two, and three terms of the Fourier series.

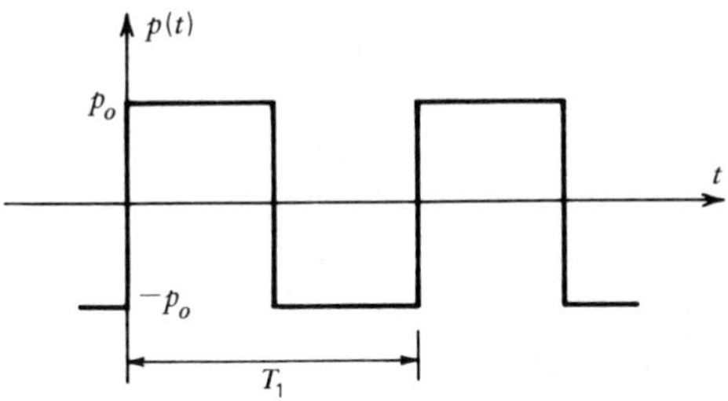

Solution

a. The integrals in Eq. 8.4 can be evaluated over the period $-T_1/2 \leq t \leq T_1/2$ and thus written in the form:

$$a_o = \frac{1}{T_1}\int_{-T_1/2}^{T_1/2} p(t)\, dt \tag{1}$$

$$a_n = \frac{2}{T_1}\int_{-T_1/2}^{T_1/2} p(t)\cos(n\Omega_1 t)\, dt \tag{2}$$

$$b_n = \frac{2}{T_1}\int_{-T_1/2}^{T_1/2} p(t)\sin(n\Omega_1 t)\, dt \tag{3}$$

where

$$p(t) = \begin{cases} -p_o & -\dfrac{T_1}{2} < t < 0 \\ p_o & 0 < t < \dfrac{T_1}{2} \end{cases} \tag{4}$$

Substituting Eqs. 4 into Eqs. 1 and 2 we get

$$\boxed{a_o = a_n = 0} \tag{5}$$

This results from the fact that $p(t)$ is an odd function of t, that is, $p(t) = -p(-t)$, while a_o and a_n are coefficients of even terms in the Fourier series.

$$b_n = \frac{4p_o}{T_1}\int_0^{T_1/2} \sin(n\Omega_1 t)\, dt \tag{6}$$

so

$$b_n = \frac{4p_o}{T_1}\left(\frac{-1}{n\Omega_1}\right)\cos(n\Omega_1 t)\Big|_0^{T_1/2} \tag{7}$$

But,

$$\Omega_1 T_1 = 2\pi \tag{8}$$

so

$$b_n = \frac{-2p_o}{n\pi}\left[\cos(n\pi) - 1\right] \tag{9}$$

or

$$\boxed{b_n = \frac{4p_o}{n\pi}} \qquad n = 1,3,5, \ldots \tag{10}$$

The Fourier series representation of the square wave is thus

$$\boxed{p(t) = \frac{4p_o}{\pi} \sum_{n=1,3,\ldots} \left(\frac{1}{n}\right) \sin(n\Omega_1 t)} \tag{11}$$

b. The plots below show the contributions of the first three nonzero terms of the Fourier series representation of the square wave.

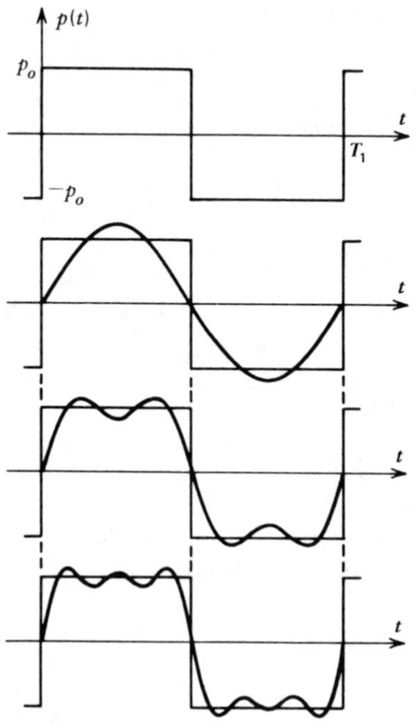

Having determined the response of SDOF systems to harmonic excitation in Chapter 4 and having determined how to represent a periodic function in terms of its harmonic components, we are now able to determine the response of SDOF systems to periodic excitation. In Example 8.2, an undamped SDOF system will be subjected to the square wave excitation of Example 8.1.

Example 8.2

The undamped SDOF system below is subjected to a square wave excitation $p(t)$ as shown in Example 8.1. Determine the steady-state response of the system if $\omega_n = 6\Omega_1$.

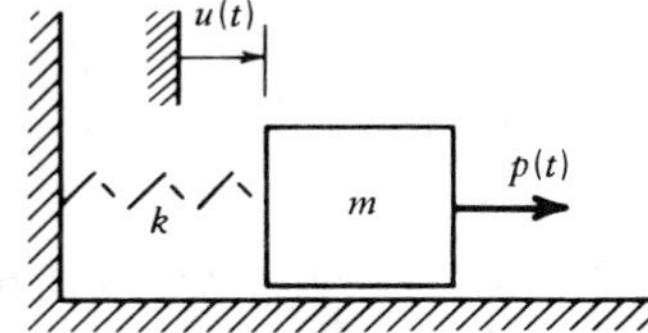

Solution

From Eq. 4.9 the steady-state response of an undamped SDOF system to harmonic excitation $p_o \cos \Omega t$ is

$$u = \left(\frac{p_o/k}{1 - r^2} \right) \cos \Omega t \tag{1}$$

where $r = \Omega/\omega_n = \Omega \sqrt{m/k}$. (2)

From Example 8.1 we can write $p(t)$ in the form

$$p(t) = \sum_{n=1}^{\infty} P_n \sin(n\Omega_1 t) \tag{3}$$

where

$$P_n = \begin{cases} \dfrac{4p_o}{n\pi} & n = \text{odd} \\ 0 & n = \text{even} \end{cases} \tag{4}$$

Hence, a typical term of the steady-state response has the form

$$U_n = \frac{P_n}{k(1 - r_n^2)} \tag{5}$$

where

$$r_n = \frac{n\Omega_1}{\omega_n} \tag{6}$$

Then, the steady-state response is

$$u = \sum_{n=1}^{\infty} U_n \sin(n\Omega_1 t) \tag{7}$$

where

$$U_n = \frac{4p_o}{nk\pi[1 - (n\Omega_1/\omega_n)^2]}, \qquad n = \text{odd} \tag{8}$$

or

$$\boxed{u = \frac{4p_o}{k\pi} \sum_{n=1,3,\ldots} \frac{\sin(n\Omega_1 t)}{n[1 - (n\Omega_1/\omega_n)^2]}} \tag{9}$$

where, for the present problem, $\Omega_1/\omega_n = \frac{1}{6}$.

It is very convenient to visualize periodic functions in terms of their spectra, that is, plots of the amplitude of each harmonic component versus frequency. The spectra of $p(t)$ and $u(t)$ from Examples 8.1 and 8.2 are plotted in Fig. 8.2. From Eqs. 4 and 8 of Example 8.2,

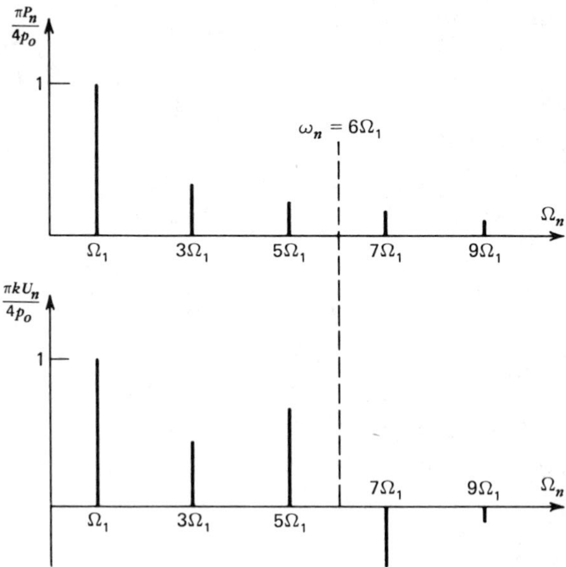

Figure 8.2. Excitation and response spectra based on Example 8.2.

$$\frac{P_n}{p_o} = \begin{cases} \dfrac{4}{n\pi} & n = \text{odd} \\ 0 & n = \text{even} \end{cases} \tag{8.5a}$$

$$\frac{U_n}{p_o/k} = \begin{cases} \dfrac{4}{n\pi[1 - (n/6)^2]} & n = \text{odd} \\ 0 & n = \text{even} \end{cases} \tag{8.5b}$$

and

$$\Omega_n = n\Omega_1 \tag{8.5c}$$

Note that for the particular ratio Ω_1/ω_n some of the Fourier components of the excitation are at frequencies below resonance while some are above. The response would have a very large Fourier component if $n\Omega_1$ were close to ω_n for some value of n.

8.2 Response to Periodic Excitation—Complex Fourier Series

In Sec. 4.3 the complex frequency response function $\overline{H}(\Omega)$ was introduced as a convenient means of representing the response of a viscous-damped system to harmonic excitation. A complex Fourier series representation of periodic functions will likewise prove to be very useful. Let

$$p(t) = \sum_{n=-\infty}^{\infty} \overline{P}_n(\Omega) e^{i(n\Omega_1 t)} \tag{8.6}$$

The bar over P_n symbolizes the fact that the coefficients of the series may be complex. To evaluate these components note that

$$\int_{\tau}^{\tau+T_1} e^{i(n\Omega_1 t)}\, e^{-i(m\Omega_1 t)}\, dt = \begin{cases} 0 & n \neq m \\ T_1 & n = m \end{cases} \tag{8.7}$$

So, multiplying Eq. 8.6 by $e^{-i(m\Omega_1 t)}$ and integrating over one period we get

$$\overline{P}_n = \frac{1}{T_1} \int_{\tau}^{\tau+T_1} p(t) e^{-i(n\Omega_1 t)}\, dt, \qquad n = 0, \pm 1, \ldots \tag{8.8}$$

Note that

$$\overline{P}_{-n} = \overline{P}_n^* = \text{complex conjugate of } \overline{P}_n \tag{8.9}$$

and

$$\overline{P}_o = \frac{1}{T_1} \int_{\tau}^{\tau+T_1} p(t)\, dt = \text{average value of } p(t) \tag{8.10}$$

[Actually, $\overline{P}_o$ is real-valued since $p(t)$ is real-valued.]

Example 8.3

a. Show that if $p(t)$ is a real-valued function, the right-hand side of Eq. 8.6 will turn out to be real-valued, as it should be.

b. If $p(t)$ is an odd function, show that $\overline{P}_n(\Omega)$ is imaginary and that $\overline{P}_{-n} = -\overline{P}_n$.

Solution

a. Expanding Eq. 8.6, we get

$$\begin{aligned} p(t) = \overline{P}_o &+ \sum_{n=1}^{\infty} \overline{P}_n[\cos(n\Omega_1 t) + i \sin(n\Omega_1 t)] \\ &+ \sum_{n=1}^{\infty} \overline{P}_{-n}[\cos(n\Omega_1 t) - i \sin(n\Omega_1 t)] \end{aligned} \tag{1}$$

From Eq. 8.10, $\overline{P}_o$ is seen to be real-valued, and from Eq. 8.9, Eq. 1 can be expressed in the form

$$p(t) = \overline{P}_o + 2 \sum_{n=1}^{\infty} [\mathcal{R}(\overline{P}_n) \cos(n\Omega_1 t) + \mathcal{I}(\overline{P}_n) \sin(n\Omega_1 t)] \tag{2}$$

which is real-valued. Q.E.D.

b. Equation 8.8 can be written out as

$$\overline{P}_n = \frac{1}{T_1} \int_{\tau}^{\tau+T_1} p(t)[\cos(n\Omega_1 t) - i \sin(n\Omega_1 t)]\, dt \tag{3}$$

Since $p(t)$ is said to be an odd function of t,

$$\overline{P}_{+n} = \frac{-i}{T_1} \int_{\tau}^{\tau+T_1} p(t) \sin(n\Omega_1 t)\, dt \tag{4}$$

Then,

$$\overline{P}_{-n} = \frac{-i}{T_1}\int_{\tau}^{\tau+T_1} p(t)\sin(-n\Omega_1 t)\,dt = \frac{i}{T_1}\int_{\tau}^{\tau+T_1} p(t)\sin(n\Omega_1 t)\,dt \tag{5}$$

Hence, $\overline{P}_n$ is purely imaginary, and $\overline{P}_{-n} = -\overline{P}_n$. Q.E.D.

By comparing Eq. 2 of the above example with Eq. 8.2 you will notice that

$$a_n = 2\mathcal{R}(\overline{P}_n), \qquad b_n = 2\mathcal{I}(\overline{P}_n) \tag{8.11}$$

Example 8.4

a. Determine $\overline{P}_n$ for the square wave of Example 8.1.

b. Sketch spectra of $\mathcal{R}(\overline{P}_n)$, $\mathcal{I}(\overline{P}_n)$, and $|\overline{P}_n|$.

Solution

a. Equation 8.8 can be evaluated over the period $0 < t < T_1$, giving

$$\overline{P}_n = \frac{1}{T_1}\int_0^{T_1/2} (p_o)e^{-i(n\Omega_1 t)}\,dt + \frac{1}{T_1}\int_{T_1/2}^{T_1} (-p_o)e^{-i(n\Omega_1 t)}\,dt \tag{1}$$

$$\overline{P}_n = \frac{-p_o}{in\Omega_1 T_1}\left[e^{-i(n\Omega_1 t)}\Big|_0^{T_1/2} - e^{-i(n\Omega_1 t)}\Big|_{T_1/2}^{T_1}\right] \tag{2}$$

But $\Omega_1 T_1 = 2\pi$, so

$$e^{-i(n\Omega_1 T_1/2)} = e^{-in\pi} = \begin{cases} +1 & n = \text{even} \\ -1 & n = \text{odd} \end{cases} \qquad e^{-i(n\Omega_1 T_1)} = e^{-i(2n\pi)} = 1 \tag{3}$$

Thus,

$$\overline{P}_n = \frac{ip_o}{2\pi n}\left[2e^{-in\pi} - 1 - e^{-i(2n\pi)}\right] \tag{4}$$

or

$$\boxed{\overline{P}_n = \frac{ip_o}{2\pi n}(2e^{-in\pi} - 2) = \begin{cases} 0 & n = \text{even} \\ \dfrac{-2p_o i}{n\pi} & n = \text{odd} \end{cases}}$$

(*Note:* $\overline{P}_n$ could also be evaluated by expressing $e^{-i(n\Omega_1 t)}$ as $[\cos(n\Omega_1 t) - i\sin(n\Omega_1 t)]$.)

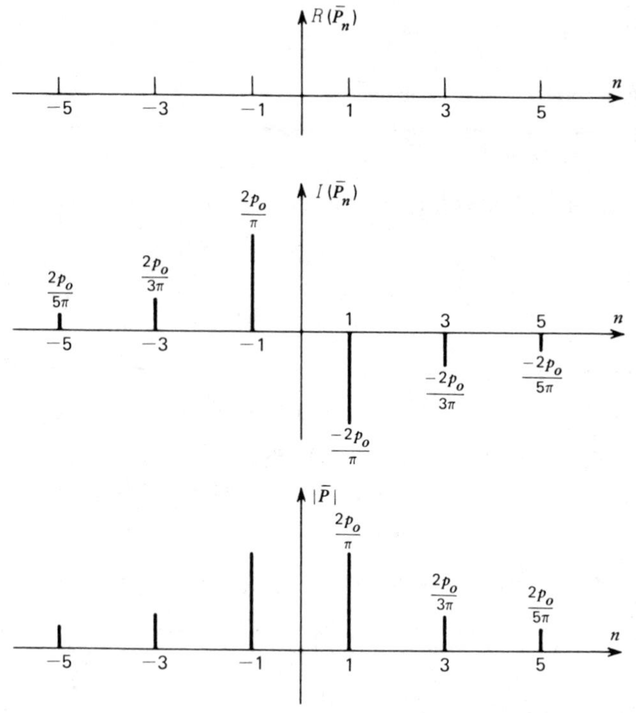

From Eqs 4.30 and 4.33 the steady-state response of a SDOF system can be written in complex form as

$$\overline{u}(t) = \overline{U}e^{i\Omega t} = \overline{H}(\Omega)p_o e^{i\Omega t} \tag{8.12}$$

where $\overline{H}(\Omega)$ is the complex frequency response function given by Eq. 4.33. That is,

$$\overline{H}(\Omega) = \frac{1/k}{[1 - (\Omega/\omega_n)^2] + i(2\zeta\Omega/\omega_n)} \tag{8.13}$$

For a periodic excitation we can use the complex Fourier series

$$p(t) = \sum_{n=-\infty}^{\infty} \overline{P}_n e^{i(n\Omega_1 t)} \tag{8.14}$$

and the steady-state response can be written as

$$u(t) = \sum_{n=-\infty}^{\infty} \overline{U}_n e^{i(n\Omega_1 t)} \tag{8.15}$$

Noting from Eq. 8.12 that for harmonic excitation $\overline{U} = \overline{H}p_o$, we see that

$$\overline{U}_n = \overline{H}_n\overline{P}_n = |\overline{H}_n|\ |\overline{P}_n|\ e^{i(\alpha_{Hn}+\alpha_{Pn})} \tag{8.16}$$

where

$$\overline{H}_n(\Omega) = \frac{1/k}{[1-(n\Omega_1/\omega_n)^2] + i(2\zeta n\Omega_1/\omega_n)} \tag{8.17}$$

Example 8.5 illustrates the use of complex Fourier series in determining the steady-state response of a SDOF system subjected to periodic excitation. The method would be even more beneficial if the system were a damped system.

Example 8.5

a. Repeat Example 8.2 by determining an expression for $\overline{U}_n$ for the undamped SDOF system subjected to a square wave excitation with $\omega_n = 6\Omega_1$.

b. Sketch $|\overline{U}_n|$ and α_{Un}.

Solution

a. From Eq. 8.16

$$\overline{U}_n = \overline{H}_n\overline{P}_n = |H_n|\,|P_n|\ e^{i(\alpha_{Hn}+\alpha_{Pn})} \tag{1}$$

and from Eq. 8.17, with $\zeta = 0$,

$$\overline{H}_n = \frac{1/k}{1-(n\Omega_1/\omega_n)^2} = \frac{1/k}{1-(n/6)^2} \tag{2}$$

From Example 8.4

$$\overline{P}_n = \begin{cases} 0 & n = \text{even} \\ \dfrac{-2p_o i}{n\pi} & n = \text{odd} \end{cases} \tag{3}$$

Hence,

$$\boxed{\overline{U}_n = \overline{H}_n\overline{P}_n = \begin{cases} \dfrac{-i(2p_o)}{n\pi k\,[1-(n/6)^2]} & n = \text{odd} \\ 0 & n = \text{even} \end{cases}} \tag{4}$$

b. From Eq. 4

$$|\overline{U}_n| = \frac{2p_o/\pi k}{|n[1 - (n/6)]^2|} \tag{5}$$

$$\alpha_{Un} = \begin{cases} -\dfrac{\pi}{2} & n = +1, +3, +5, -7, -9, \ldots \\ \dfrac{\pi}{2} & n = -1, -3, -5, +7, +9, \ldots \end{cases} \tag{6}$$

For sketching purposes, evaluate

$$\mu_n \equiv \frac{(\pi/2)|\overline{U}_n|}{p_o/k} = \frac{|1/n|}{|1 - (n/6)^2|}, \qquad n = \pm 1, \pm 3, \ldots \tag{7}$$

This gives

$$\mu_1 = \mu_{-1} = \frac{36}{35} = 1.029$$

$$\mu_3 = \mu_{-3} = \frac{4}{9} = 0.444$$

$$\mu_5 = \mu_{-5} = \frac{36}{55} = 0.655$$

$$\mu_7 = \mu_{-7} = \frac{36}{91} = 0.396$$

$$\mu_9 = \mu_{-9} = \frac{4}{45} = 0.089$$

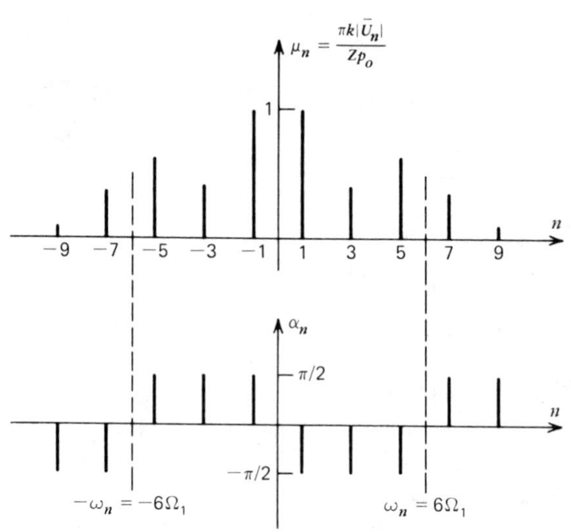

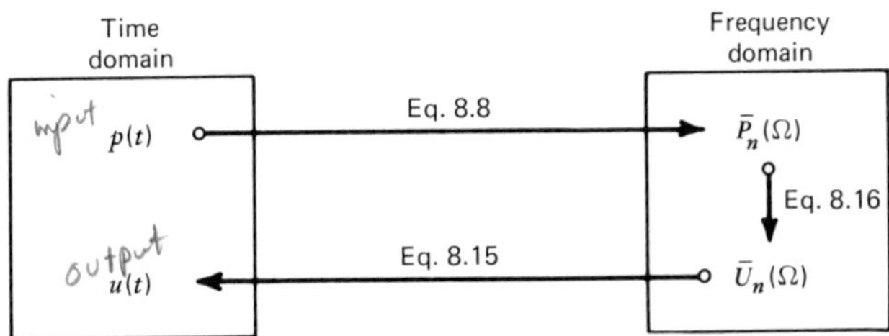

Figure 8.3. *Solution of periodic response problem by transformation to frequency domain.*

Compare the sketch of μ_n in Example 8.5 with Fig. 8.2*b* and note that the complex coefficients have an amplitude that is half that of the real coefficients. The contribution of the $(-n)$ terms in the complex Fourier series accounts for this difference.

It is convenient to think of problems such as Example 8.5 in terms of transformations from the time domain to the frequency domain (spectrum) and from the frequency domain to the time domain. Figure 8.3 illustrates this.

8.3 Response to Nonperiodic Excitation—Fourier Integral

In the previous sections you have seen that a periodic function can be represented by a Fourier series as in Eq. 8.2 or Eq. 8.6. When the function to be represented is not periodic, it can be represented by a Fourier integral, which is obtained from the Fourier series by letting the period T_1 approach infinity. It will be convenient to employ the complex Fourier series in developing expressions for Fourier integrals. Equation 8.6, repeated here, gives

$$p(t) = \sum_{n=-\infty}^{\infty} \overline{P}_n(\Omega)\, e^{i(n\Omega_1 t)} \tag{8.18}$$

where

$$\overline{P}_n = \frac{1}{T_1} \int_{\tau}^{\tau+T_1} p(t)\, e^{-i(n\Omega_1 t)}\, dt \tag{8.19}$$

provided that the integral exists.

In letting $T_1 \to \infty$, it will be convenient to introduce the following notation:

$$\begin{aligned} \Omega_1 &= \Delta\Omega \\ n\Omega_1 &= \Omega_n \\ \overline{P}_n(\Omega_n) &= T_1\overline{P}_n = \left(\frac{2\pi}{\Delta\Omega}\right)\overline{P}_n \end{aligned} \tag{8.20}$$

Then Eq. 8.18 can be written

$$p(t) = \frac{1}{2\pi}\sum_{n=-\infty}^{\infty} \overline{P}(\Omega_n)e^{i\Omega_n t}\,\Delta\Omega \tag{8.21}$$

where

$$\overline{P}(\Omega_n) = \int_{-T_1/2}^{T_1/2} p(t)e^{-i\Omega_n t}\,dt \tag{8.22}$$

The limits of integration on Eq. 8.22 have been taken as shown so that when $T_1 \to \infty$ the entire time history of $p(t)$ will be included regardless of the specific form of $p(t)$.

As $T_1 \to \infty$, Ω_n becomes a continuous variable Ω, and $\Delta\Omega$ becomes the differential $d\Omega$. Then Eqs. 8.21 and 8.22 can be written as

$$p(t) = \frac{1}{2\pi}\int_{-\infty}^{\infty} \overline{P}(\Omega)e^{i\Omega t}\,d\Omega \tag{8.23}$$

and

$$\overline{P}(\Omega) = \int_{-\infty}^{\infty} p(t)e^{-i\Omega t}\,dt \tag{8.24}$$

Equations 8.23 and 8.24 are called a *Fourier transform pair.* $\overline{P}(\Omega)$ is known as the *Fourier transform* of $p(t)$; and $p(t)$ is called the *inverse Fourier transform* of $\overline{P}(\Omega)$. The representation of $p(t)$ by its Fourier transform requires that the integral in Eq. 8.24 exist. This is guaranteed if $p(t)$ satisfies the Dirichlet conditions[8.1] in the domain $-\infty < t < \infty$ and if the integral

$$\int_{-\infty}^{\infty} |p(t)|\,dt$$

is convergent. These conditions are met by most physically realizable functions representing forces, displacements, and so forth. Finally, Eqs. 8.23 and 8.24 can be written in a more symmetric form if written in terms of the frequency $f = \Omega/2\pi$. Then,

$$p(t) = \int_{-\infty}^{\infty} \overline{P}(f)e^{i(2\pi f t)}\,df \tag{8.25}$$

and

$$\overline{P}(f) = \int_{-\infty}^{\infty} p(t) e^{-i(2\pi f t)}\, dt \tag{8.26}$$

In Eqs. 8.15 and 8.16 we found that the response of a SDOF system to periodic excitation can be expressed in the form

$$u(t) = \sum_{n=-\infty}^{\infty} \overline{H}_n \overline{P}_n e^{i(n\Omega_1 t)} \tag{8.27}$$

Following the procedure of Eqs. 8.18 through 8.24 we obtain the following Fourier transform pair for the response $u(t)$ of a SDOF system.

$$\overline{U}(f) = \int_{-\infty}^{\infty} u(t) e^{-i(2\pi f t)}\, dt \tag{8.28}$$

and

$$u(t) = \int_{-\infty}^{\infty} \overline{U}(f) e^{i(2\pi f t)}\, df \tag{8.29}$$

where

$$\overline{U}(f) = \overline{H}(f)\overline{P}(f) \tag{8.30}$$

which is the product of the complex frequency response function and the Fourier transform of the excitation.

From Eq. 8.26 it is clear that the determination of a direct Fourier transform involves a straightforward integration, as will be seen in Example 8.6.

Example 8.6

Let $p(t)$ be the rectangular pulse defined by

$$p(t) = \begin{cases} 0 & t < -T \\ p_o & -T < t < T \\ 0 & t > T \end{cases}$$

Determine the Fourier transform of this rectangular pulse which is symmetric about $t = 0$.

Solution

From Eq. 8.24

$$\overline{P}(\Omega) = \int_{-\infty}^{\infty} p(t) e^{-i\Omega t}\, dt \tag{1}$$

$$= \int_{-T}^{T} p_o e^{-i\Omega t}\, dt$$

$$= \left(\frac{p_o}{-i\Omega}\right)(e^{-i\Omega T} - e^{i\Omega T}) \tag{2}$$

This can also be written as

$$\overline{P}(\Omega) = 2p_oT\left(\frac{\sin \Omega T}{\Omega T}\right) \tag{3}$$

$\overline{P}(\Omega)$ is therefore a real function. It can be plotted versus Ω and compared with the corresponding discrete Fourier series of Example 8.4.

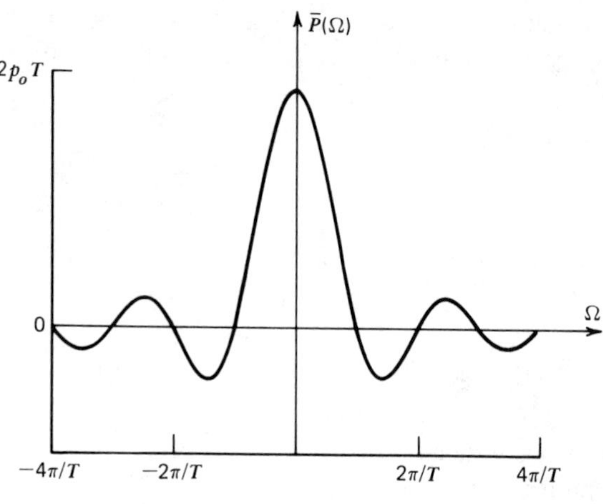

From Eqs. 8.29 and 8.30 the response can be expressed by the following inverse Fourier transform

$$u(t) = \int_{-\infty}^{\infty} \overline{H}(f)\overline{P}(f)e^{i(2\pi ft)}\,df \tag{8.31}$$

For example, for a viscous-damped system $\overline{H}(f)$ is given by

$$\overline{H}(f) = \frac{1/k}{[1 - (f/f_n)^2] + i(2\zeta f/f_n)} \tag{8.32}$$

and for a rectangular pulse as in Example 8.6

$$\overline{P}(f) = 2p_oT\left[\frac{\sin(2\pi fT)}{2\pi fT}\right] \tag{8.33}$$

Evaluation of the inverse Fourier transform of Eq. 8.31 requires contour integration in the complex plane, which is beyond the scope of this text. From the numerical standpoint the importance of the Fourier integral lies in approximating it by a discrete Fourier transform (DFT) which is evaluated by fast

Fourier transform (FFT) methods. A brief introduction to the DFT and FFT follows in Sec. 8.5.

8.4 Relationship Between Complex Frequency Response and Unit Impulse Response

The complex frequency response $\overline{H}(\Omega)$, or $\overline{H}(f)$, of a system describes its response characteristics in the frequency domain. The unit inpulse response $h(t)$ describes the system's response in the time domain. For example, for a viscous-damped system $\overline{H}(f)$ is given by Eq. 8.32 and $h(t)$ by Eq. 5.30, repeated here.

$$h(t) = \frac{1}{m\omega_d} e^{-\zeta\omega_n t} \sin \omega_d t \tag{8.34}$$

$h(t)$ is the response to a unit impulse applied at $t = 0$. From Eq. 8.26 the Fourier transform of this unit impulse is given by

$$\overline{P}(f) = \int_{-\infty}^{\infty} p(t) e^{-i(2\pi f t)}\, dt = 1 \tag{8.35}$$

Hence, Eq. 8.31 can be combined with Eq 8.35 to give

$$h(t) = \int_{-\infty}^{\infty} \overline{H}(f) e^{i(2\pi f t)}\, df \tag{8.36}$$

Thus, it can be concluded that $h(t)$ is the inverse Fourier transform of $\overline{H}(f)$ and that

$$\overline{H}(f) = \int_{-\infty}^{\infty} h(t) e^{-i(2\pi f t)}\, dt \tag{8.37}$$

The block diagram of Fig. 8.3 can thus be extended to nonperiodic functions as shown in Fig. 8.4.

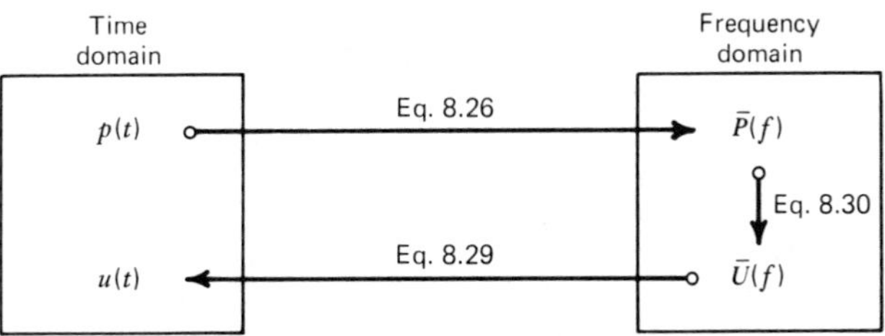

Figure 8.4. Solution of transient response problem by transformation to the frequency domain.

As pointed out above, however, the inverse transformation of Eq. 8.29 is best accomplished by numerical techniques that are discussed briefly in Sec. 8.5.

8.5 Discrete Fourier Transforms (DFT) and Fast Fourier Transforms (FFT)

Although the Fourier integral techniques discussed in Sec. 8.3 provide a means for determining the transient response of a system, numerical implementation of the Fourier integral became a practical reality only with the publication of the Cooley-Tukey algorithm for the fast Fourier transform in 1965[8.1]. Since that date the FFT has virtually led to a revolution in many areas of technology such as the area of measurements and instrumentation (e.g., References 8.2, 8.3, and 8.4).

Two steps are involved in the numerical evaluation of Fourier transforms (e.g., References 8.5 and 8.6). First, discrete Fourier transforms (DFT's), which correspond to Eqs. 8.25 and 8.26, are derived. Next, an efficient numerical algorithm (FFT) for evaluating the DFT's is described.

For numerical treatment of the Fourier transform it is necessary first to define a *discrete Fourier transform* (DFT) pair corresponding to the Fourier transform pair of Eqs 8.25 and 8.26. First, a continuous function to be transformed must be sampled at discrete time intervals, Δt. Secondly, only a finite number, N, of these sampled values may be considered due to computer memory and execution time limitations. The effect of this sampling and truncation is to approximate the continuous signal by a periodic signal of period $T_1 = N \Delta t$ sampled at times $t_m = m\,\Delta t$.

Let

$$\Omega_1 = \frac{2\pi}{T_1} \tag{8.38}$$

Since a period T_1 consists of N samples, the integral of Eq. 8.22 may be replaced by the following finite sum

$$\overline{P}(\Omega_n) = \sum_{m=0}^{N-1} p(t_m)e^{-i(n\Omega_1)(m\Delta T)}\,\Delta t \tag{8.39}$$

or

$$\overline{P}(\Omega_n) = \Delta t \sum_{m=0}^{N-1} p(t_m)e^{-i(2\pi mn/N)}, \qquad n = 0, 1, \ldots, N-1 \tag{8.40}$$

The inverse DFT can be obtained from Eq. 8.21 in a similar manner. Thus,

$$p(t_m) = \frac{1}{2\pi} \sum_{n=0}^{N-1} \overline{P}(\Omega_n) e^{i(n\Omega_1)(m\Delta T)} \Omega_1 \tag{8.41}$$

or

$$p(t_m) = \frac{1}{T_1} \sum_{n=0}^{N-1} \overline{P}(\Omega_n) e^{i(2\pi mn/N)}, \qquad m = 0, 1, \ldots, N-1 \tag{8.42}$$

Equations 8.40 and 8.42 therefore define a *discrete Fourier transform pair.* A more rigorous derivation of the Fourier transform pair may be found in Reference 8.5.

The accuracy of a DFT representation depends on the sampling interval Δt and the number of samples N, but a discussion of this topic is beyond the scope of this text.

The *fast Fourier transform* (FFT) is not a new type of transform, but is rather an efficient numerical algorithm for evaluating the DFT. Its importance lies in the fact that by eliminating most of the repetition in the calculation of a DFT it permits much more rapid computation of the DFT.

Either Eq. 8.40 or Eq. 8.42 can be cast in the form

$$A_m = \sum_{n=0}^{N-1} B_n W_N^{mn}, \qquad m = 0, 1, \ldots, N-1 \tag{8.43}$$

where

$$W_N = e^{-i(2\pi/N)} \tag{8.44}$$

A measure of the amount of computation involved in Eq. 8.43 is the number of complex products implied by the form of the equation and the range of m. It is clear that there are N sums, each of which requires N complex products, or there are N^2 products required for computing all of the A_m's. By taking advantage of the cyclical nature of powers of W_N, the total computational effort can be drastically reduced. Figure 8.5 shows the repetition cycle for

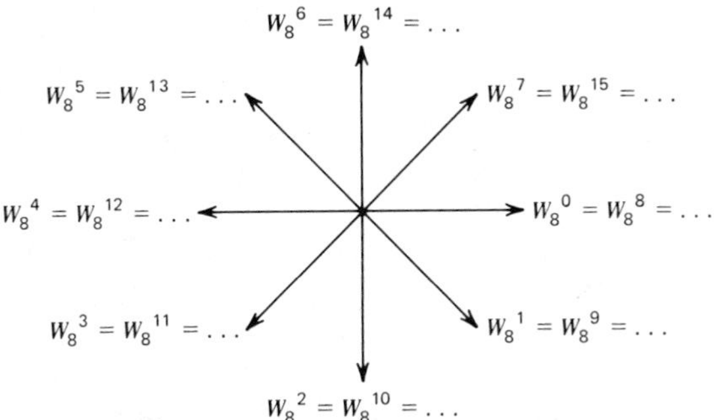

Figure 8.5. *Cyclical nature of* W_N^{mn} *for* $N = 8$.

W_8^{mn} The number of complex products for the FFT algorithm is given by $(N/2) \log_2 N$. For example, if $N = 512$, the number of FFT operations is less than 1% of the original DFT operations.

You are urged to consult further references (e.g., References 8.5, 8.6) for a thorough treatment of digital signal analysis via the fast Fourier transform (FFT).

References

8.1 J. W. Cooley and J. W. Tukey, "An Algorithm for Machine Calculation of Complex Fourier Series," *Math Computation,* v. 19, 297–301 (1965).

8.2 M. Richardson, "Modal Analysis Using Digital Test Systems," *Seminar on Understanding Digital Control and Analysis in Vibration Test Systems,* The Shock and Vibration Information Center, Washington, DC, 43–64 (1975).

8.3 K. A. Ramsey, "Effective Measurements for Structural Dynamics Testing," *Sound and Vibration,* Pt. I, v. 9, 24–35 (1975), Pt. II, v. 10, 18–31 (1976).

8.4 S. Smith et al., "MODALAB—A Computerized Data Acquisition and Analysis System for Structural Dynamics Testing," *21st International Instrumentation Symposium,* Instrument Society of America, 183–189 (1975).

8.5 E. O. Brigham, *The Fast Fourier Transform,* Prentice-Hall, Englewood Cliffs, NJ (1974).

8.6 S. D. Stearns, *Digital Signal Analysis,* Hayden Book Co., Rochelle Park, NJ (1975).

Problem Set 8.1

8.1 Determine the real Fourier series for the periodic excitation in Fig. P8.1.

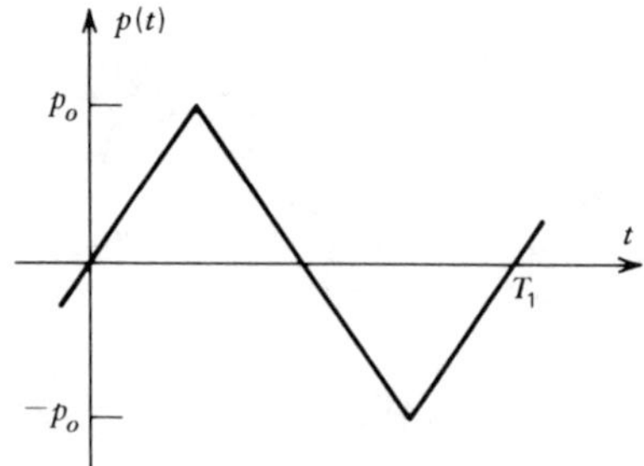

Figure P8.1

8.2 Determine the real Fourier series for the periodic excitation in Fig. P8.2.

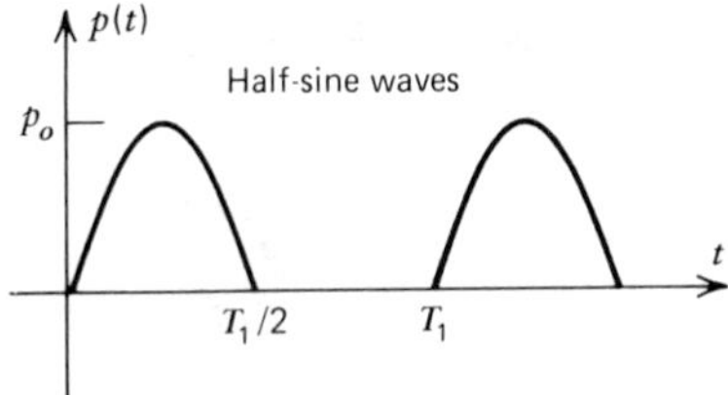

Figure P8.2

8.3 (a) Determine the real Fourier series for the square wave shown below.

(b) By comparing your results from part (a) with Example 8.1, what do you observe to be the effect(s) of the phase shift of the excitation?

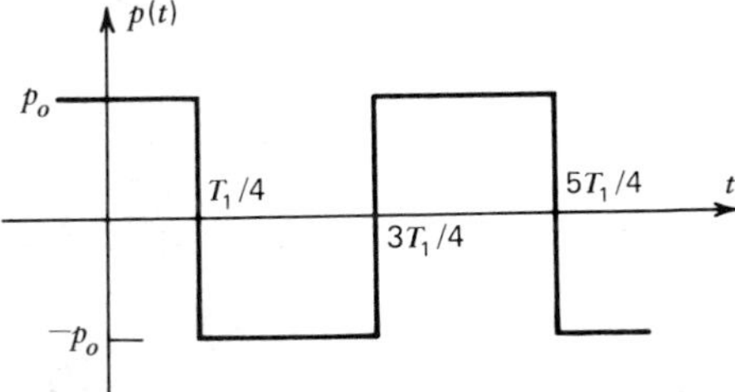

Figure P8.3

8.4 The undamped SDOF system below is subjected to the excitation $p(t)$ given in Problem 8.1.

(a) Determine the steady-state response of the system if $\omega_n = 2.5\Omega_1$.

(b) Sketch response spectra for P_n and U_n similar to those shown in Fig. 8.2.

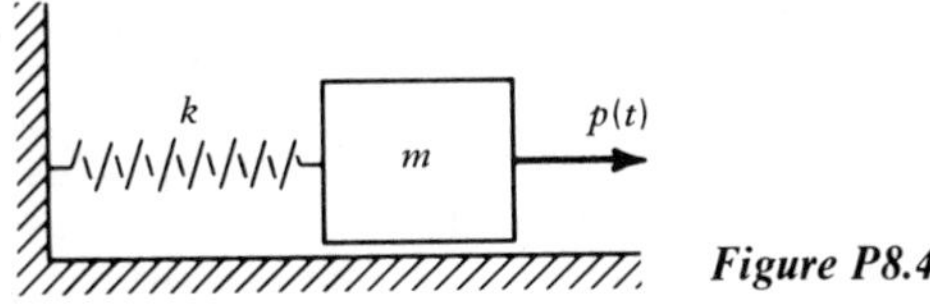

Figure P8.4

Problem Set 8.2

8.5 (a) Determine coefficients of the complex Fourier series that corresponds to the excitation $p(t)$ in Problem 8.1.

(b) Sketch the real and imaginary parts and the magnitude of the complex coefficients as in Example 8.4.

8.6 (a) Determine coefficients of the complex Fourier series that corresponds to the excitation $p(t)$ in Problem 8.3.

(b) Sketch the real and imaginary parts and the magnitude of the complex coefficients as in Example 8.4.

8.7 For the undamped base-motion system of Fig. P8.7, let

$$\overline{W}_n = \overline{H}_n \overline{Z}_n$$

where

$$z(t) = \sum_{n=-\infty}^{\infty} \overline{Z}_n e^{i(n\Omega_1 t)}$$

$$w(t) = \sum_{n=-\infty}^{\infty} \overline{W}_n e^{i(n\Omega_1 t)}$$

(a) Determine $\overline{H}_n$.

(b) Determine $\overline{Z}_n$. (This will have the same form as $\overline{P}_n$ of Problem 8.1.)

(c) Determine $\overline{W}_n$ and sketch $|\overline{W}_n|/Z$ versus frequency order as in Example 8.4.

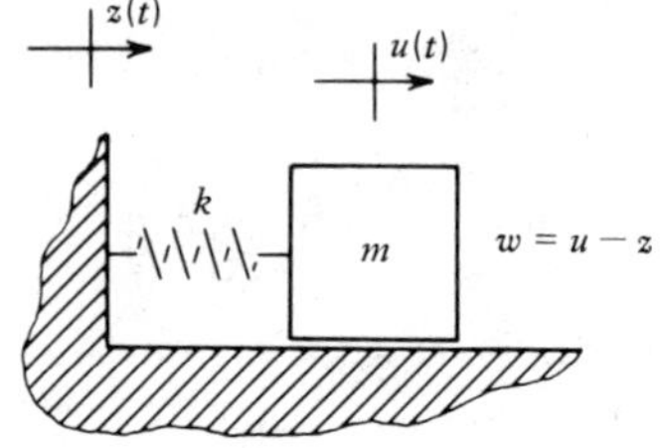

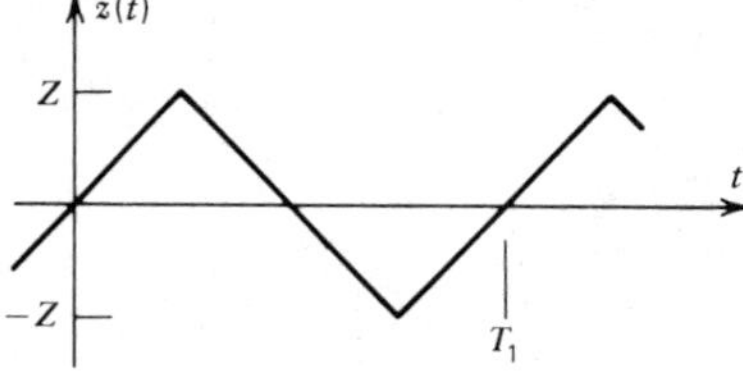

Figure P8.7

Problem Set 8.3

8.8 Determine the Fourier transform for the rectangular pulse below. (Note that this involves a time shifting of the pulse in Example 8.6.)

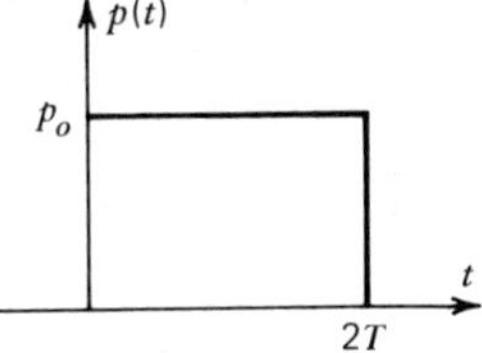

Figure P8.8

8.9 Determine the real and imaginary parts of the Fourier transform of each of the following functions.

(a) $p(t) = p_o e^{-\alpha|t|}$

(b) $$p(t) = \begin{cases} 0 & t < 0 \\ p_o \sin(2\pi f_o t) & 0 \le t \le \dfrac{1}{2f_o} \\ 0 & t > \dfrac{1}{2f_o} \end{cases}$$

PART II CONTINUOUS SYSTEMS

9 MATHEMATICAL MODELS OF CONTINUOUS SYSTEMS

All structures are, in actual fact, three dimensional bodies, and every point in such a body, unless restrained, can displace along three mutually perpendicular directions, say x, y, and z. In Chapter 1 the distinction was made between discrete-parameter models and continuous models of structures. In Chapters 2 through 8 we have considered single degree-of-freedom models and their response to various types of excitation. In the present chapter and in Chapter 10 we briefly consider structures represented by continuous models, that is, by partial differential equations. This will lay the groundwork for an extensive treatment of multiple degree-of-freedom models in the remainder of the text.

Upon completion of this chapter you should be able to:

- Use Newton's laws to derive the equation(s) of motion of "one-dimensional" bodies undergoing axial deformation, torsional deformation, bending deformation, or a combination of the above.
- State appropriate boundary conditions for axial, transverse, or torsional free vibration of members having various support conditions.
- Use Hamilton's principle to derive the equation(s) of motion of "one-dimensional" bodies undergoing axial deformation, torsional deformation, bending deformation, or a combination of the above.

9.1 Application of Newton's Laws—Axial Deformation

We will consider first the axial deformation of a long, thin member, a portion of which is shown in Fig. 9.1a. To derive the equation of motion for axial vibration we isolate a freebody diagram of an element of length Δx. Let $u(x, t)$ be the displacement of the cross section along the axial direction, and let $p(x, t)$ be the externally applied axial force per unit length with resultant lying along the centroidal axis of the member. [$A(x)$ is the cross-sectional area and $\rho(x)$ is the mass density, i.e., mass per unit volume.] The axial deformation assumptions, as treated in elementary strength of materials, are:

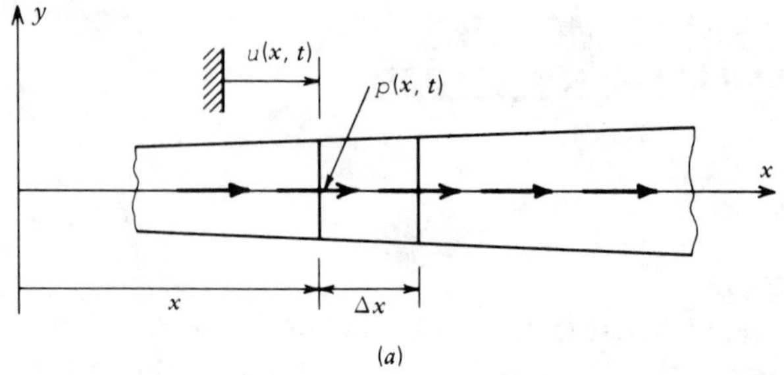

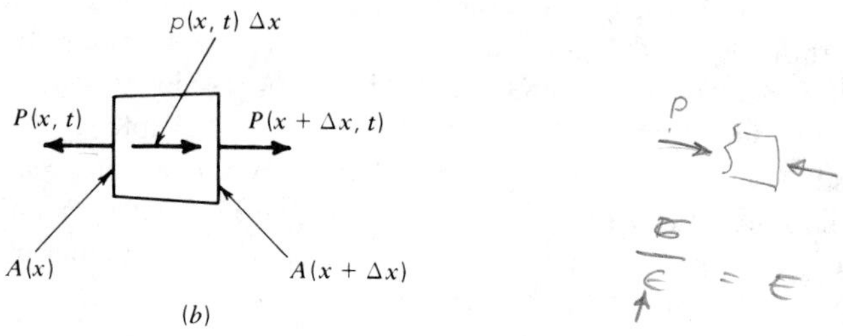

Figure 9.1. *Member undergoing axial deformation.* (a) *Portion of a member undergoing axial deformation.* (b) *Freebody diagram.*

a. Cross sections remain plane and remain perpendicular to the axis of the member.

b. The material is linearly elastic.

c. The material properties (E, ρ) are constant at a given cross section but may vary with x.

Based on these assumptions the following three equations are obtained:

$$\epsilon = \frac{\partial u}{\partial x} \tag{9.1}$$

$$\sigma = E\epsilon \tag{9.2}$$

and

$$P = A\sigma \tag{9.3}$$

If we apply Newton's second law

$$\overset{+}{\rightarrow} \sum F_x = (\Delta m)a_x \tag{9.4}$$

to the freebody diagram in Fig. 9.1b, we get

$$p\,\Delta x + P(x + \Delta x, t) - P(x, t) = \rho A\,\Delta x\,\frac{\partial^2 u}{\partial t^2} \tag{9.5}$$

where, because of the limit process to follow, the first and last terms above are taken at x rather than as averages over Δx. Taking the limit as $\Delta x \to 0$ gives

$$\lim_{\Delta x \to 0}\left(\frac{P(x + \Delta x, t) - P(x, t)}{\Delta x}\right) + p = \rho A\,\frac{\partial^2 u}{\partial t^2}$$

or

$$\frac{\partial P}{\partial x} + p = \rho A\,\frac{\partial^2 u}{\partial t^2} \tag{9.6}$$

Equations 9.1 through 9.3 may be substituted into Eq. 9.6 to give

$$\frac{\partial}{\partial x}\left(AE\,\frac{\partial u}{\partial x}\right) + p = \rho A\,\frac{\partial^2 u}{\partial t^2} \tag{9.7}$$

This is the equation of motion for axial vibration of a linearly elastic bar.

The two most common boundary conditions, or end conditions, are the force-free end and the fixed end. The appropriate boundary conditions at an end $x = x_e$ are:

$$P(x_e, t) = 0 \qquad \text{force-free end} \tag{9.8a}$$

$$u(x_e, t) = 0 \qquad \text{fixed end} \tag{9.8b}$$

By using Eqs. 9.1 through 9.3 we can write Eq. 9.8a as

$$\left.\frac{\partial u}{\partial x}\right|_{x_e} = 0 \qquad \text{free end} \tag{9.8c}$$

Only one of the above equations, that is, 9.8b or 9.8c, may be enforced at a given end.

Example 9.1

Determine the appropriate axial deformation boundary conditions at $x = 0$ for the two members shown below.

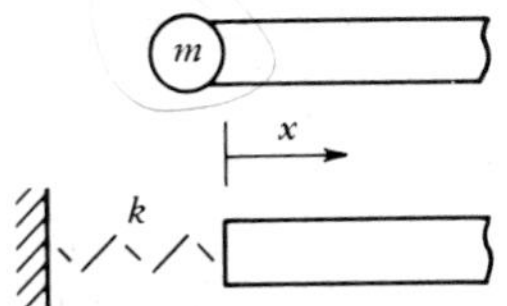

Solution

a. For the tip-mass problem draw a freebody diagram of the tip mass:

The equation of motion for the tip mass is

$$P(0, t) = m \left. \frac{\partial^2 u}{\partial t^2} \right|_{x=0} \tag{1}$$

From Eqs. 9.1 through 9.3

$$P(0, t) = AE \left. \frac{\partial u}{\partial x} \right|_{x=0} \tag{2}$$

Thus, the end condition at $x = 0$ is

$$\boxed{AE \left. \frac{\partial u}{\partial x} \right|_{x=0} = m \left. \frac{\partial^2 u}{\partial t^2} \right|_{x=0}} \tag{3}$$

which reduces to Eq. 9.8c when $m = 0$.

b. For the bar with a linear spring at $x = 0$ sketch the deflected spring:

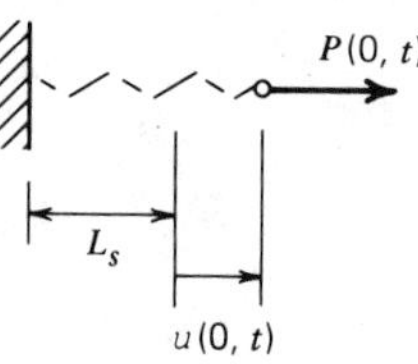

From the force-deflection equation of the linear spring

$$P(0, t) = ku(0, t) \tag{4}$$

Incorporating Eq. 2 we get

$$\boxed{AE \left. \frac{\partial u}{\partial x} \right|_{x=0} = ku(0, t)} \tag{5}$$

9.2 Application of Newton's Laws—Tranverse Vibration of Linearly Elastic Beams (Bernoulli-Euler Theory)

The equation of motion of long, thin members undergoing transverse vibration may also be derived using Newton's second law. Figure 9.2a shows a portion

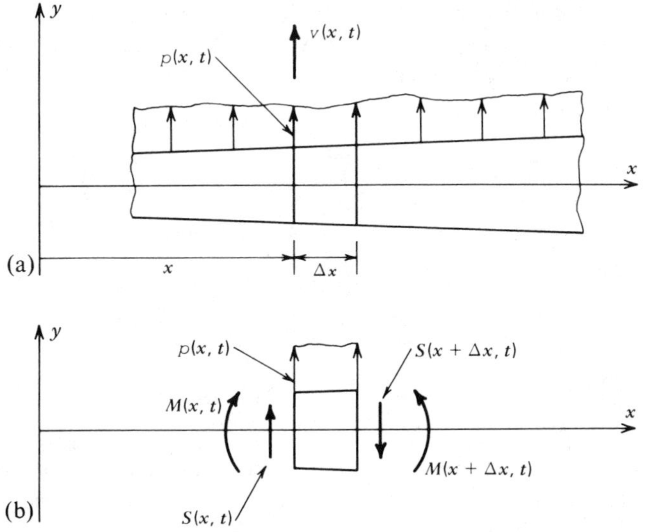

***Figure 9.2.** Member undergoing transverse vibration.*

of a member undergoing transverse motion and Fig. 9.2*b* shows an appropriate freebody diagram. $v(x, t)$ is the transverse motion of the point $(x, 0)$ on the neutral axis of the beam. The bending moment is $M(x, t)$, the transverse shear force is $S(x, t)$, and the external force per unit length is $p(x, t)$. The Bernoulli-Euler assumptions of elementary beam theory will be employed, namely:

a. There is an axis of the beam which undergoes no extension or contraction. The x axis is located along this neutral axis.

b. Cross sections perpendicular to the neutral axis in the undeformed beam remain plane and remain perpendicular to the deformed neutral axis, that is, transverse shear deformation is neglected.

c. The material is linearly elastic and the beam is homogeneous at any cross section.

d. σ_y and σ_z are negligible compared to σ_x.

e. The xy-plane is a principal plane.

In addition to the above assumptions which are made in the static analysis of beams, it will be assumed that the rotatory inertia may be neglected in the moment equation.

From kinematics the strain can be related to the curvature, $1/\mu$, of the beam by

$$\epsilon = \frac{-y}{\mu} \tag{9.9}$$

where y is distance in the cross section measured from the neutral axis. Then, for a linearly elastic beam whose properties are independent of position in the cross section, the bending moment can be related to the curvature by

$$M(x, t) = \frac{EI}{\mu} \tag{9.10}$$

where I is the moment of inertia of the cross section.

The equations of motion for the mass Δm in the above freebody diagram may be derived using Newton's laws. Thus,

$$+\uparrow \sum F_y = (\Delta m) a_y \tag{9.11}$$

and

$$\circlearrowleft + \sum M_G = (\Delta I_G)\alpha \tag{9.12}$$

where G is the mass center of Δm and α is the angular acceleration. However, it was previously stated that rotatory inertia would be neglected in the moment equation, so Eq. 9.12 reduces to

$$\sum M_G = 0 \tag{9.13}$$

Applying Eq. 9.11 to the freebody diagram, we obtain

$$S(x, t) - S(x + \Delta x, t) + p(x, t)\,\Delta x = \rho A\,\Delta x \frac{\partial^2 v}{\partial t^2} \tag{9.14}$$

By dividing Eq. 9.14 by Δx and taking the limit we obtain

$$-\frac{\partial S}{\partial x} + p(x, t) = \rho A \frac{\partial^2 v}{\partial t^2} \tag{9.15}$$

From the moment equation, Eq. 9.13, we obtain in a similar manner the equation

$$S = \frac{\partial M}{\partial x} \tag{9.16}$$

If the slope, $\partial v/\partial x$, of the beam remains small, the curvature may be approximated by $\partial^2 v/\partial x^2$, so Eq. 9.10 becomes

$$M(x, t) = EI \frac{\partial^2 v}{\partial x^2} \tag{9.17}$$

Combining Eqs. 9.15, 9.16, and 9.17 we get

$$\frac{\partial^2}{\partial x^2}\left(EI \frac{\partial^2 v}{\partial x^2} \right) + \rho A \frac{\partial^2 v}{\partial t^2} = p(x, t) \tag{9.18}$$

This is the equation of motion for transverse forced vibration of a beam which satisfies the assumptions stated above. It is valid only for beams that are relatively long and thin. In Example 9.3 the above derivation will be modified to account for an axial preload on the beam. In Sec. 9.4 shear deformation and rotatory inertia will be added.

The boundary conditions most frequently encountered in analyzing vibration of Bernoulli-Euler beams are the following.

a. Fixed end:

$$v(x_e, t) = 0 \tag{9.19a}$$

and

$$\left.\frac{\partial v}{\partial x}\right|_{x=x_e} = 0 \tag{9.19b}$$

that is, the displacement and slope vanish at a fixed end.

b. Simply supported* end:

$$v(x_e, t) = 0 \tag{9.20a}$$

and

$$M(x_e, t) = 0 \tag{9.20b}$$

that is, the displacement and bending moment vanish at a simply supported end. From Eqs. 9.20b and 9.17 we get the second boundary condition expressed in terms of v, namely

$$\left.\frac{\partial^2 v}{\partial x^2}\right|_{x=x_e} = 0 \tag{9.20c}$$

c. Force-free end:

$$S(x_e, t) = 0 \tag{9.21a}$$

and

$$M(x_e, t) = 0 \tag{9.21b}$$

*The moment is not necessarily zero at a simply supported end, but this is frequently the case. Otherwise $M(x_e, t) = M_e(t) =$ given moment.

that is, the transverse shear force and the bending moment vanish. Equations 9.16 and 9.17 may be combined with Eqs. 9.21a and 9.21b to give

$$\frac{\partial}{\partial x}\left(EI\frac{\partial^2 v}{\partial x^2}\right)\bigg|_{x=x_e} = 0 \tag{9.22a}$$

and

$$\frac{\partial^2 v}{\partial x^2}\bigg|_{x=x_e} = 0 \tag{9.22b}$$

Thus, at each end of a beam two end conditions are required. Different end conditions arise when, for example, a lumped mass or a spring is attached to the end of the beam. The best procedure to employ in such cases is to draw careful freebody diagrams employing the sign convention of Fig. 9.2*b*.

Example 9.2

Determine the appropriate boundary conditions if a point mass m is attached at the end of the beam at $x = L$.

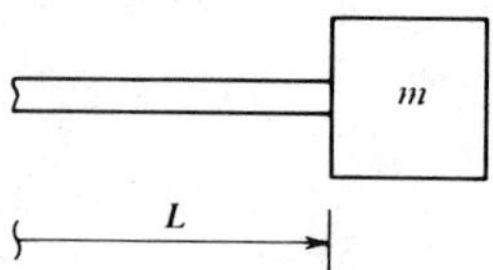

Solution

Isolate a freebody diagram of the mass m. Use the sign convention of Fig. 9.2*b*.

Apply Newton's laws to this "particle."

$$+\uparrow \sum F_y = ma_y \tag{1}$$

From the freebody diagram

$$S(L, t) = m\frac{\partial^2 v}{\partial t^2}\bigg|_{x=L} \tag{2}$$

Combine Eqs. 2, 9.16, and 9.17 to get

$$\boxed{\frac{\partial}{\partial x}\left(EI\frac{\partial^2 v}{\partial x^2}\right)\bigg|_{x=L} = m\frac{\partial^2 v}{\partial t^2}\bigg|_{x=L}} \tag{3}$$

Since the particle has no rotatory inertia

$$\sum M_G = 0 \tag{4}$$

Then,

$$M(L, t) = 0 \tag{5}$$

or

$$\boxed{\frac{\partial^2 v}{\partial x^2}\bigg|_{x=L} = 0} \tag{6}$$

Equations 3 and 6 are the appropriate end conditions for a tip mass whose rotational inertia, I_G, can be neglected.

Example 9.3

Determine the equation of motion for a beam that is subjected to a compressive end load, N, which remains parallel to the x-axis. Neglect axial strain. (*Note:* Coupled axial-bending motion will be considered in Problem 9.1.)

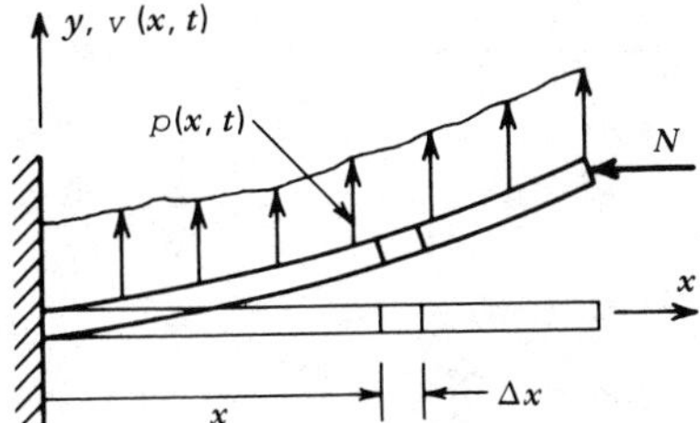

Solution

a. Draw a freebody diagram of an element of length Δx.

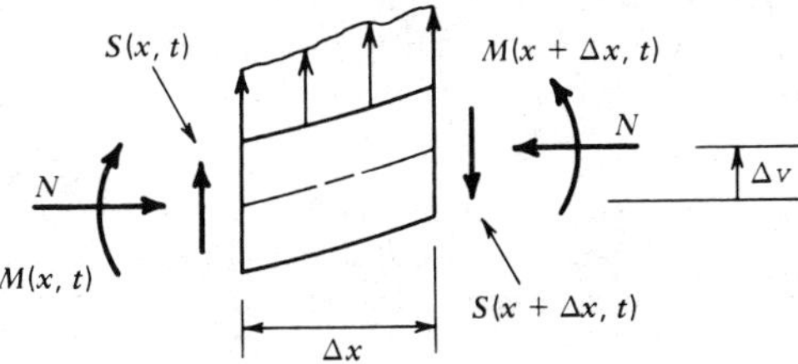

b. Write the equations of motion:
As in the previous derivation for beams without N,

$$+\uparrow \sum F_y = \Delta m\, a_y \tag{1}$$

$$\circlearrowleft\!+ \sum M_G = 0 \tag{2}$$

Equation 1 leads to

$$-\frac{\partial S}{\partial x} + p(x, t) = \rho A \frac{\partial^2 v}{\partial t^2} \tag{3}$$

as before. Equation 2 gives

$$\begin{aligned} &M(x + \Delta x, t) - M(x, t) + N[v(x + \Delta x, t) - v(x, t)] \\ &- S(x + \Delta x, t)\, \Delta x = 0 \end{aligned} \tag{4}$$

Dividing by Δx and taking the limit as $\Delta x \to 0$ we get

$$\frac{\partial M}{\partial x} + N\frac{\partial v}{\partial x} = S \tag{5}$$

Equations 3 and 5 may be combined to give

$$\frac{\partial^2 M}{\partial x^2} + N\frac{\partial^2 v}{\partial x^2} + \rho A\frac{\partial^2 v}{\partial t^2} = p(x, t) \tag{6}$$

Finally, Eq. 9.17 can be incorporated into Eq. 6 giving

$$\boxed{\frac{\partial^2}{\partial x^2}\left(EI\frac{\partial^2 v}{\partial x^2}\right) + N\frac{\partial^2 v}{\partial x^2} + \rho A\frac{\partial^2 v}{\partial t^2} = p(x, t)} \tag{7}$$

9.3 Application of Hamilton's Principle

In Chapter 2 Newton's laws and the principle of virtual displacements were employed in deriving the equations of motion of various SDOF systems. Hamilton's principle, which will be stated below, is directly related to the principle of virtual displacements* and could have been used to derive the equations of motion of the SDOF systems treated in Chapter 2. Since it employs kinetic energy and potential energy, which are scalar quantities, rather than the virtual

*The study of dynamics is frequently subdivided into Newtonian mechanics and analytical mechanics. The latter is also referred to as the study of energy methods. The principle of virtual displacements, Hamilton's principle, and Lagrange's equations are topics treated in analytical mechanics.

work of inertia forces and elastic forces, which must be obtained from vectorial quantities, Hamilton's principle may be somewhat easier to apply than is the principle of virtual displacements. For continuous systems the use of Hamilton's principle provides an important bonus, namely, the boundary conditions are handled systematically in the process of obtaining the equation(s) of motion.†

In the formulation of Hamilton's principle, virtual displacements, or virtual changes of configuration, are employed. Figure 2.5 shows a cantilever beam with a virtual change of configuration $\delta v(x, t)$. As noted in Sec. 2.4, the virtual change of configuration must satisfy all geometric boundary conditions. Hamilton also assumed that the configuration is specified at times t_1 and t_2. For the cantilever beam this would imply that $\delta v(x, t_1) = \delta v(x, t_2) = 0$.

Hamilton's principle may be stated as*

$$\int_{t_1}^{t_2} \delta(T - V)\, dt + \int_{t_1}^{t_2} \delta W_{nc}\, dt = 0 \tag{9.23}$$

where

T = total kinetic energy of the system.
V = potential energy of the system, including the strain energy and the potential energy of conservative external forces.
δW_{nc} = virtual work done by nonconservative forces, including damping forces and external forces not accounted for in V.
$\delta(\ )$ = the symbol denoting the first variation, or virtual change, in the quantity in parentheses.
t_1, t_2 = times at which the configuration of the system is known.

In this section Hamilton's principle will be employed in deriving the equation of motion and boundary conditions for a cantilevered circular rod undergoing torsional deformation. (Newton's laws could also be used to obtain the same results.)

Consider the rod in Fig. 9.3 undergoing torsional deformation due to a distributed externally applied torque $t(x, t)$ and an end torque $T_L(t)$. The rotation at section x is denoted by $\theta(x, t)$. The only geometric boundary condition is

$$\theta(0, t) = 0 \tag{9.24}$$

†See Reference 9.1, pp. 168–170, for a discussion of the important role played by energy methods in early attempts to formulate a theory for bending of flat plates.

*Hamilton's principle may be derived from the principle of virtual work. See References 9.1 and 9.2

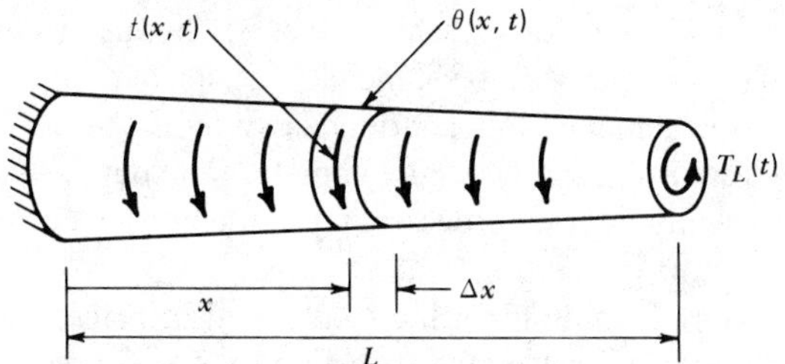

Figure 9.3. *Rod undergoing torsional deformation.*

The polar moment of area is*

$$I_p = J = \iint r^2 \, dA = \frac{\pi R^4}{2} \tag{9.25}$$

where $R(x)$ = radius of the rod at section x. The potential energy and kinetic energy are given by

$$V = \frac{1}{2} \int_0^L GJ(\theta')^2 \, dx \tag{9.26}$$

and

$$T = \frac{1}{2} \int_0^L \rho I_p (\dot{\theta})^2 \, dx \tag{9.27}$$

where

$$(\ )' = \frac{\partial (\ )}{\partial x}$$

$$(\dot{\ }) = \frac{\partial (\ )}{\partial t}$$

The virtual work of the external forces is

$$\delta W_{\text{nc}} = \int_0^L t(x, t) \, \delta\theta(x, t) \, dx + T_L(t) \, \delta\theta(L, t) \tag{9.28}$$

Substituting Eqs. 9.26 through 9.28 into Eq. 9.23 we get

$$\int_{t_1}^{t_2} \delta \left[\frac{1}{2} \int_0^L \rho I_p \, (\dot{\theta})^2 \, dx - \frac{1}{2} \int_0^L GJ(\theta')^2 \, dx \right] dt + \int_{t_1}^{t_2} \left[\int_0^L t(x, t) \, \delta\theta(x, t) \, dx + T_L(t) \, \delta\theta(L, t) \right] dt = 0 \tag{9.29}$$

*$J = I_p$ for a circular cross section. For other cross-sectional shapes $I_p = \iint r^2 \, dA \neq J$. The reader should consult texts on advanced strength of materials or elasticity to obtain expressions for J.

As in Sec. 2.4

$$\delta V = \int_0^L GJ\theta' \, \delta\theta' \, dx \tag{9.30}$$

and, similarly,

$$\delta T = \int_0^L \rho I_p \dot{\theta} \, \delta\dot{\theta} \, dx \tag{9.31}$$

It is necessary to remove the $(\ )'$ and $(\dot{\ })$ from $\delta\theta'$ and $\delta\dot{\theta}$. This is done by integration by parts. Integrating by parts with respect to x we get

$$\begin{aligned} \int_{t_1}^{t_2} \delta V \, dt &= \int_{t_1}^{t_2} \int_0^L GJ\theta' \, \delta\theta' \, dx \, dt \\ &= \int_{t_1}^{t_2} \left[(GJ\theta' \, \delta\theta \Big|_0^L - \int_0^L (GJ\theta')' \, \delta\theta \, dx \right] dt \end{aligned} \tag{9.32}$$

Integrating by parts with respect to t gives

$$\begin{aligned} \int_{t_1}^{t_2} \delta T \, dt &= \int_0^L \int_{t_1}^{t_2} \rho I_p \dot{\theta} \, \delta\dot{\theta} \, dt \, dx \\ &= \int_0^L \left[(\rho I_p \dot{\theta} \, \delta\theta) \Big|_{t_1}^{t_2} - \int_{t_1}^{t_2} \rho I_p \, \ddot{\theta} \, \delta\theta \, dt \right] dx \end{aligned} \tag{9.33}$$

In Eq. 9.33 the term evaluated at t_1 and t_2 involves $\delta\theta(x, t_1)$ and $\delta\theta(x, t_2)$. In the formulation of Hamilton's principle it is assumed that the initial and final configurations are known, that is, $\theta(x, t_1)$ and $\theta(x, t_2)$ are known. Thus, $\delta\theta(x, t_1) = \delta\theta(x, t_2) = 0$, so the integrated term on the right-hand side of Eq. 9.33 vanishes. Combining Eqs. 9.29, 9.32, and 9.33 we obtain

$$\begin{aligned} &\int_{t_1}^{t_2} \left\{ \Big[(GJ\theta') \, \delta\theta \Big]_{x=0} - \Big[(GJ\theta' - T_L) \, \delta\theta \Big]_{x=L} \right\} dt \\ &\qquad + \int_{t_1}^{t_2} \int_0^L \Big[(GJ\theta')' - \rho I_p \ddot{\theta} + t \Big] \delta\theta \, dx \, dt = 0 \end{aligned} \tag{9.34}$$

Because of the geometric boundary condition in Eq. 9.24, we must set $\delta\theta(0, t) = 0$, so the first term in Eq. 9.34 vanishes. Since $\theta(L, t)$ is not prescribed, $\delta\theta(L, t)$ is completely arbitrary. Then, the natural boundary condition

$$(GJ\theta')_{x=L} = T_L(t) \tag{9.35}$$

is obtained. A *natural boundary condition,* when the principle of virtual displacements is employed, is a boundary condition that arises when no geometric boundary condition is prescribed on a portion of the boundary. It provides a boundary condition for force-type quantities.

On the interval $0 < x < L$, $\delta\theta(x, t)$ is arbitrary. Hence, the last term in Eq. 9.34 leads to the partial differential equation of motion

$$-(GJ\theta')' + \rho I_p\ddot{\theta} = t(x, t) \tag{9.36}$$

Thus, the motion of the torsion member is governed by the partial differential equation of motion, Eq. 9.36; the geometric boundary condition at $x = 0$, Eq. 9.24; and the natural boundary condition at $x = L$, Eq. 9.35.

9.4 Application of Hamilton's Principle—Beam Flexure Including Shear Deformation and Rotatory Inertia (Timoshenko Beam Theory)

In Sec. 9.2 the equation of motion was derived for transverse vibration of a long, thin beam for which it is valid to ignore shear deformation and rotatory inertia. In this section Hamilton's principle is used in deriving the equations of motion and boundary conditions which include these effects, so that shorter, stubbier beams can be analyzed.

Figure 9.4 shows the kinematics of deformation of a beam which undergoes shear deformation in addition to pure bending. $v(x, t)$ is the total transverse displacement of the neutral axis of the beam. From Fig. 9.4

$$\beta = \alpha - \frac{\partial v}{\partial x} \tag{9.37}$$

From elementary beam theory the moment-curvature equation is

$$M = EI\alpha' \tag{9.38}$$

and the corresponding bending strain energy is

$$V_b = \frac{1}{2}\int_0^L EI(\alpha')^2\,dx \tag{9.39}$$

The shear strain energy can be expressed as

$$V_s = \frac{1}{2}\int_0^L \kappa GA\beta^2\,dx \tag{9.40}$$

where the shear coefficient κ can be obtained by computing the strain energy due to shear using

$$V_s = \frac{1}{2}\int_0^L \iint_A \tau\gamma\,dA\,dx \tag{9.41}$$

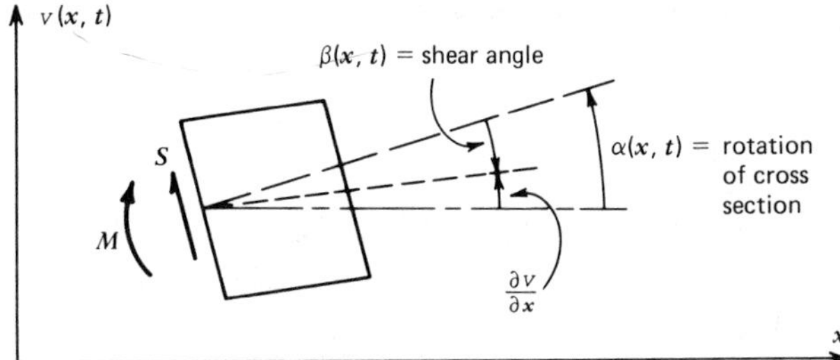

Figure 9.4. *Kinematics of deformation of a beam including shear deformation.*

For rectangular beams, $\kappa = \frac{5}{6}$.

The kinetic energy of the beam is

$$T = \frac{1}{2}\int_0^L \rho A(\dot{v})^2\,dx + \frac{1}{2}\int_0^L \rho I(\dot{\alpha})^2\,dx \tag{9.42}$$

The first term is the kinetic energy due to translation and the second term is the rotatory inertia term. For a beam with distributed transverse load $p(x, t)$

$$\delta W_{nc} = \int_0^L p(x, t)\,\delta v(x, t)\,dx \tag{9.43}$$

We might suspect that geometric boundary conditions may be imposed on the translational displacement, $v(x, t)$, and on the rotation of the cross section, $\alpha(x, t)$. Hence, v and α will be retained as the unknown displacement functions, and Eq. 9.37 can be used to eliminate the shear angle, $\beta(x, t)$. Then Eqs. 9.39, 9.40, 9.42, and 9.43 can be substituted into Hamilton's principle, Eq. 9.23, to give

$$\begin{aligned}\frac{1}{2}\int_{t_1}^{t_2}\int_0^L &\delta[\rho A\dot{v}^2 + \rho I\dot{\alpha}^2 - EI(\alpha')^2 - \kappa GA(\alpha - v')^2]\,dx \\ &+ \int_{t_1}^{t_2}\int_0^L p\,\delta v\,dx = 0\end{aligned} \tag{9.44}$$

Integrating by parts, as was done in Sec. 9.3, and noting that $\delta v(x, t_1) = \delta v(x, t_2) = \delta\alpha(x, t_1) = \delta\alpha(x, t_2) = 0$, we obtain

$$\begin{aligned}&\int_{t_1}^{t_2}\int_0^L \{-\rho A\ddot{v} - [\kappa GA(\alpha - v')]' + p\}\,\delta v\,dx\,dt \\ &+ \int_{t_1}^{t_2}\int_0^L [-\rho I\ddot{\alpha} + (EI\alpha')' - \kappa GA(\alpha - v')]\,\delta\alpha\,dx\,dt \\ &+ \int_{t_1}^{t_2}[\kappa GA\,(\alpha - v')\delta v]\Big|_0^L dt - \int_{t_1}^{t_2}[(EI\alpha')\,\delta\alpha]\Big|_0^L dt = 0\end{aligned} \tag{9.45}$$

Since δv and $\delta\alpha$ are arbitrary except where geometric boundary conditions are prescribed, Eq. 9.45 leads to the following partial differential equations of motion:

$$[\kappa GA(\alpha - v')]' + \rho A\ddot{v} = p(x, t) \tag{9.46a}$$
$$\kappa GA(\alpha - v') - (EI\alpha')' + \rho I\ddot{\alpha} = 0 \tag{9.46b}$$

The following *generalized boundary conditions* are also obtained from Eq. 9.45:

$$(\kappa GA\beta)\,\delta v = 0, \qquad \text{at } x = 0 \tag{9.47a}$$
$$(\kappa GA\beta)\,\delta v = 0, \qquad \text{at } x = L \tag{9.47b}$$
$$(EI\alpha')\,\delta\alpha = 0, \qquad \text{at } x = 0 \tag{9.47c}$$
$$(EI\alpha')\,\delta\alpha = 0, \qquad \text{at } x = L \tag{9.47d}$$

Each of the four boundary conditions above must be satisfied either as a geometric boundary condition, that is, v (or α) specified, or as a natural boundary condition, that is, $\beta = 0$ (or $\alpha' = 0$).

If the beam has uniform cross-sectional properties, Eqs. 9.46 may be combined to give a single equation in v. From Eq. 9.46a

$$\alpha' = v'' + \left(\frac{1}{\kappa GA}\right)(p - \rho A\ddot{v}) \tag{9.48}$$

Differentiating Eq. 9.46b and substituting Eq. 9.48 into the resulting equation, we get

$$\underbrace{EI\frac{\partial^4 v}{\partial x^4} - \left(p - \rho A\frac{\partial^2 v}{\partial t^2}\right)}_{\text{Bernoulli-Euler theory}} \underbrace{- \rho I\frac{\partial^4 v}{\partial x^2\,\partial t^2}}_{\text{Principal rotatory inertia term}} \tag{9.49}$$
$$\underbrace{+ \frac{EI}{\kappa GA}\frac{\partial^2}{\partial x^2}\left(p - \rho A\frac{\partial^2 v}{\partial t^2}\right)}_{\text{Principal shear deformation term}} \underbrace{- \frac{\rho I}{\kappa GA}\frac{\partial^2}{\partial t^2}\left(p - \rho A\frac{\partial^2 v}{\partial t^2}\right)}_{\text{Combined rotatory inertia and shear deformation}} = 0$$

From Eq. 9.49, it is possible to identify the terms of Bernoulli-Euler theory, Eq. 9.18, and the correction terms due to shear deformation and rotatory inertia.

References

9.1 H. L. Langhaar, *Energy Methods in Applied Mechanics,* Wiley, New York (1962).

9.2 L. Meirovitch, *Analytical Methods in Vibrations,* Macmillan, New York (1967).

Problem Set 9.2

9.1 The effect of a tip mass on the boundary conditions for axial vibration was discussed in Example 9.1 and for bending in Example 9.2. If the mass is located eccentrically, as in Fig. P9.1, it couples the axial and bending vibration. Using appropriate freebody diagrams, determine the boundary conditions for axial and transverse motion.

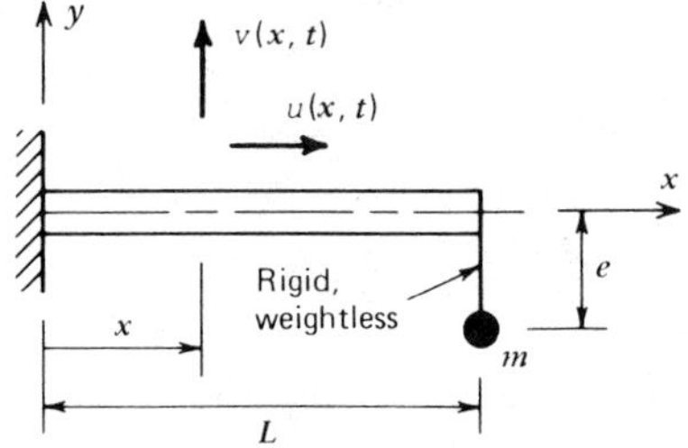

Figure P9.1

$A = \text{cons}, \quad I = \text{cons}, \quad E = \text{cons}, \quad \rho = \text{cons}$

9.2 A uniform thin rigid bar BC having mass m and length L is attached to a uniform flexible beam AB. Assume small transverse displacements. Using appropriate freebody diagrams, determine the boundary conditions at A and B.

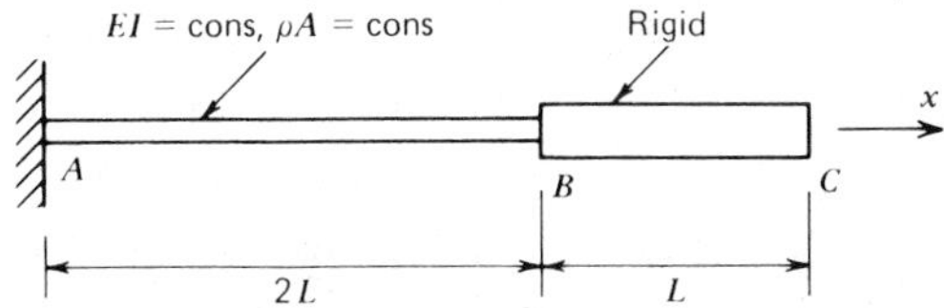

Figure P9.2

Problem Set 9.3

9.3 A small mass, $m = (\rho AL/4)$, is attached to the tip of a uniform beam by a rigid, weightless arm of length e as shown in Fig. P9.1. Consider vibration in the x-y plane only. Use Hamilton's principle to derive the differential equations of motion and the boundary conditions for the system.

9.4 A uniform cantilever bar has torsional stiffness GJ, vertical bending stiffness EI, mass per unit length ρA, and torsional mass moment of inertia ρI_p. At end B a rigid, weightless bar BC is attached parallel to the y-axis. At C a concentrated mass m is attached to bar BC, and also at C a load $P(t)$ acts parallel to the z-axis. Assume that the bar bends in the vertical (x-z) plane with deflection $v(x, t)$ and twists about its axis with angle-of-twist $\theta(x, t)$. Assume all deflection to be small (e.g., m moves approximately vertically if θ is small), and neglect gravity.

Use Hamilton's principle to derive the equations of motion and the boundary conditions for this system.

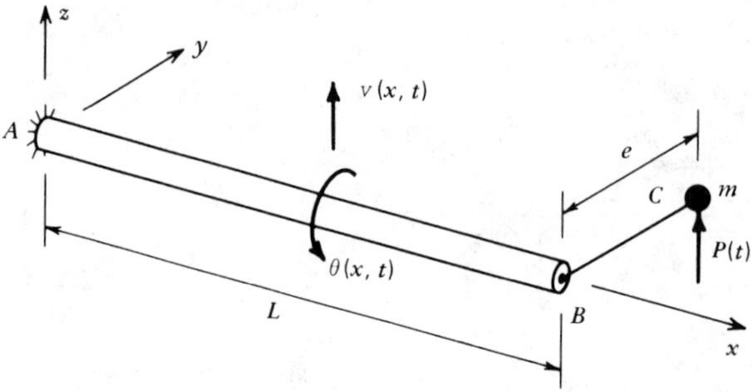

Figure P9.4

9.5 Use Hamilton's principle to derive the equation of motion and the boundary conditions for the flexible beam AB in Problem 9.2.

10 FREE VIBRATION OF CONTINUOUS SYSTEMS

In Chapter 9 the equations of motion and boundary conditions were derived for several continuous models of "one-dimensional" structures, that is, structures whose deformation depends on one spatial variable, x, and time, t. In this chapter on free vibration of continuous systems the important concepts of natural frequency and mode shape will be discussed. The results that are obtained in this chapter for uniform members will serve as "exact solutions" against which results obtained in subsequent chapters for multi-degree-of-freedom models can be compared. Forced vibration of continuous systems will not be treated in this text since very few practical structures can be conveniently modeled using uniform continuous members.

Upon completion of this chapter you should be able to:

- Solve for the natural frequencies and mode shapes for axial, transverse, or torsional free vibration of uniform members with specified boundary conditions.
- Discuss conditions under which shear deformation and/or rotatory inertia should be included in considering transverse vibration of beams.
- Discuss the relationship of frequencies of axial vibration and transverse vibration of a uniform beam.
- Use the Rayleigh method to approximate the fundamental frequency of a member, and state the relationship of the approximate frequency to the exact frequency.
- Derive the orthogonality equation for axial, transverse, or torsional vibration of bars.
- Describe three procedures for normalizing mode shapes. Define generalized stiffness and generalized mass.

10.1 Free Axial Vibration

The equation of motion for axial vibration of a bar with distributed axial load is given by Eq. 9.7. For free vibration this reduces to

$$(AEu')' = \rho A\ddot{u} \tag{10.1}$$

Assume harmonic motion given by the equation

$$u(x, t) = U(x)\cos(\omega t - \alpha) \tag{10.2}$$

Substitute this into Eq. 10.1 to obtain the *eigenvalue equation*

$$(AEU')' + \rho A\omega^2 U = 0 \tag{10.3}$$

For a given set of boundary conditions this ordinary differential equation possesses solutions only for certain values ω_r and corresponding functions $U_r(x)$.

Consider a uniform bar, that is, A = cons, E = cons, and ρ = cons. Then Eq. 10.3 reduces to

$$\frac{d^2U}{dx^2} + \left(\frac{\rho\omega^2}{E}\right)U = 0 \tag{10.4}$$

which may be written

$$\frac{d^2U}{dx^2} + \lambda^2 U = 0 \tag{10.5}$$

where

$$\lambda^2 = \frac{\rho\omega^2}{E} \tag{10.6}$$

The general solution of Eq. 10.5 is

$$U(x) = A_1 \cos \lambda x + A_2 \sin \lambda x \tag{10.7}$$

The end conditions for free vibration are obtained by substituting Eq. 10.2 into Eqs. 9.8b and 9.8c to obtain

(a) Fixed end:

$$U = 0 \tag{10.8a}$$

(b) Free end:

$$\frac{dU}{dx} = 0 \tag{10.8b}$$

Example 10.1

Derive the natural frequencies and mode shapes for a uniform cantilever bar in axial motion.

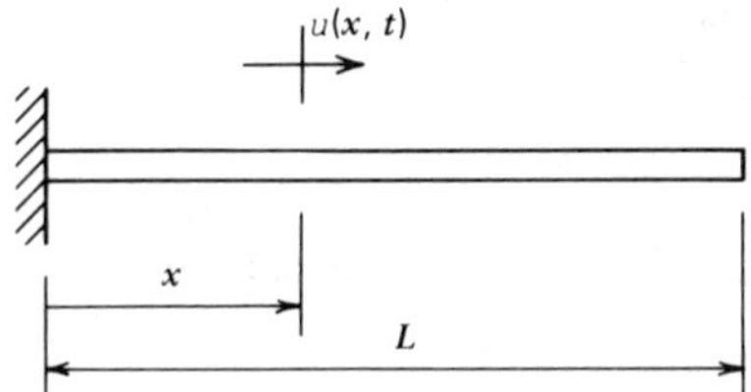

Solution

The boundary conditions are

$$U(0) = \left.\frac{dU}{dx}\right|_{x=L} = 0 \tag{1}$$

Differentiating Eq. 10.7 with respect to x gives

$$\frac{dU}{dx} = -A_1\lambda \sin \lambda x + A_2\lambda \cos \lambda x \tag{2}$$

Evaluating the two boundary conditions of Eq. 1, we obtain

$$U(0) = A_1 = 0 \tag{3}$$

$$\left.\frac{dU}{dx}\right|_{x=L} = A_2\lambda \cos \lambda L = 0 \tag{4}$$

Since $A_1 = A_2 = 0$ would represent a trivial solution, and since Eq. 3 requires that $A_1 = 0$, we must choose λ in Eq. 4 such that

$$\cos \lambda L = 0 \tag{5}$$

This is the *characteristic equation* for this problem. The roots of the characteristic equation are

$$\lambda L = \frac{\pi}{2}, \frac{3\pi}{2}, \ldots, (r - \tfrac{1}{2})\pi, \ldots \tag{6}$$

Using Eq. 10.6 we get

$$\omega_r = \lambda_r \left(\frac{E}{\rho}\right)^{1/2} = \frac{(\lambda L)_r}{L}\left(\frac{E}{\rho}\right)^{1/2} \tag{7}$$

so that the natural frequencies are

$$\boxed{\omega_r = \frac{(2r-1)\pi}{2L}\left(\frac{E}{\rho}\right)^{1/2}} \tag{8}$$

The corresponding mode shapes are obtained by combining Eqs. 10.7 and 3 to get

$$U_r(x) = C \sin \lambda_r x \tag{9}$$

or

$$U_r(x) = C \sin\left[\left(r - \frac{1}{2}\right)\frac{\pi x}{L}\right] \tag{10}$$

where C is an arbitrary scaling factor. (Section 10.5 discusses scaling.) Hence, the mode shapes are sine curves like the following.

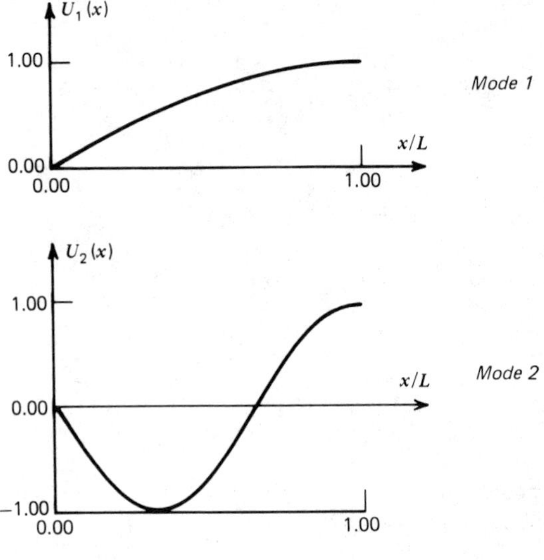

In mathematical terminology Eq. 10.5 is an *eigenvalue equation*, λ_r is an *eigenvalue*, and $U_r(x)$ is the corresponding *eigenfunction*.

Note that the curve of $U_1(x)$ in Example 10.1 exhibits no crossing of the x-axis while $U_2(x)$ has a point for which $U(x) = 0$. Such a point is called a *node point*. As the mode number r increases the number of node points increases accordingly, for example, $U_5(x)$ would have four node points.

10.2 Free Transverse Vibration of Bernoulli-Euler Beams

Transverse vibration of Bernoulli-Euler beams is governed by Eq. 9.18. For free vibration this reduces to

$$(EIv'')'' + \rho A\ddot{v} = 0 \tag{10.9}$$

Again, assume harmonic motion given by the equation

$$v(x, t) = V(x) \cos(\omega t - \alpha) \tag{10.10}$$

and substitute this into Eq. 10.9 to get the eigenvalue equation

$$(EIV'')'' - \rho A \omega^2 V = 0 \tag{10.11}$$

Since closed-form solutions are not available for this equation with variable coefficients, we will restrict our attention to free vibration of uniform beams. Free transverse vibration of uniform beams has been extensively studied [10.1,10.2], and the mode shapes of uniform beams have frequently been used in conjunction with approximation methods which will be discussed in Sec. 11.4.

For free vibration of a uniform beam, Eq. 10.11 reduces to

$$\frac{d^4 V}{dx^4} - \lambda^4 V = 0 \tag{10.12}$$

where

$$\lambda^4 = \frac{(\rho A \omega^2)}{EI} \tag{10.13}$$

The general solution of Eq. 10.12 may be written in the form

$$V(x) = A_1 e^{\lambda x} + A_2 e^{-\lambda x} + A_3 e^{i\lambda x} + A_4 e^{-i\lambda x} \tag{10.14a}$$

Two useful alternative forms are

$$V(x) = B_1 e^{\lambda x} + B_2 e^{-\lambda x} + B_3 \sin \lambda x + B_4 \cos \lambda x \tag{10.14b}$$

and

$$V(x) = C_1 \sinh \lambda x + C_2 \cosh \lambda x + C_3 \sin \lambda x + C_4 \cos \lambda x \tag{10.14c}$$

There are five constants in the general solution, that is, four amplitude constants and the eigenvalue λ. The end (boundary) conditions are used in evaluating these constants as will be illustrated in the examples below. For free vibration of uniform beams Eq. 10.10 can be substituted into the end condition equations of Sec. 9.2 to give the following:

(**a**) Fixed end:

$$V = 0 \tag{10.15a}$$

$$\frac{dV}{dx} = 0 \tag{10.15b}$$

(**b**) Simply-supported end:

$$V = 0 \tag{10.16a}$$

$$\frac{d^2 V}{dx^2} = 0 \tag{10.16b}$$

(c) Free end:

$$\frac{d^2V}{dx^2} = 0 \tag{10.17a}$$

$$\frac{d^3V}{dx^3} = 0 \tag{10.17b}$$

Example 10.2

Determine the natural frequencies and natural modes of a uniform beam simply-supported at both ends.

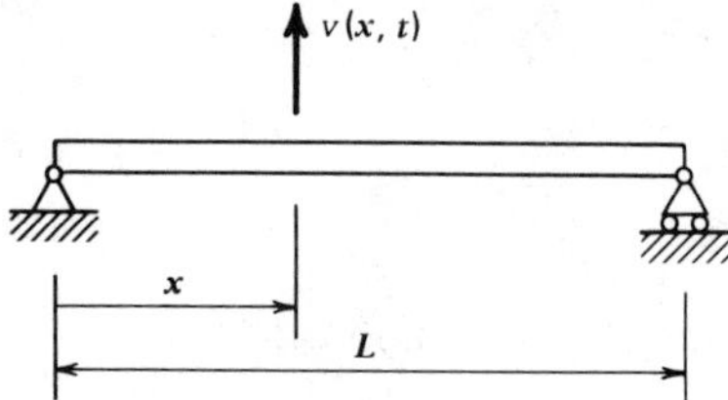

Solution

Use the general solution given in Eq. 10.14c.

$$V(x) = C_1 \sinh \lambda x + C_2 \cosh \lambda x + C_3 \sin \lambda x + C_4 \cos \lambda x \tag{1}$$

The boundary conditions are:

$$V(0) = 0 \tag{2}$$

$$\left.\frac{d^2V}{dx^2}\right|_{x=0} = 0 \tag{3}$$

$$V(L) = 0 \tag{4}$$

$$\left.\frac{d^2V}{dx^2}\right|_{x=L} = 0 \tag{5}$$

From Eq. 1,

$$\frac{d^2V}{dx^2} = \lambda^2(C_1 \sinh \lambda x + C_2 \cosh \lambda x - C_3 \sin \lambda x - C_4 \cos \lambda x) \tag{6}$$

Evaluating the boundary conditions at $x = 0$, we obtain

$$C_2 + C_4 = 0 \tag{7}$$

$$\lambda^2(C_2 - C_4) = 0 \tag{8}$$

Thus, $C_2 = C_4 = 0$. Evaluating the remaining boundary conditions, we get

$$C_1 \sinh \lambda L + C_3 \sin \lambda L = 0 \tag{9}$$

$$\lambda^2(C_1 \sinh \lambda L - C_3 \sin \lambda L) = 0 \tag{10}$$

Since this is a pair of homogeneous, linear algebraic equations in C_1 and C_3, a nontrivial solution exists only if the determinant of the coefficients vanishes, namely,

$$\begin{vmatrix} \sinh \lambda L & \sin \lambda L \\ \lambda^2 \sinh \lambda L & -\lambda^2 \sin \lambda L \end{vmatrix} = 0$$

or

$$-2\lambda^2 \sinh \lambda L \sin \lambda L = 0$$

or

$$\sinh \lambda L \sin \lambda L = 0 \tag{11}$$

Since $\sinh \lambda L = 0$ only if $(\lambda L) = 0$, the only nontrivial solutions of Eq. 11 are obtained if

$$\sin \lambda L = 0 \tag{12}$$

Equation 12 is the *characteristic equation* for this problem. It determines the eigenvalues λ_r. If Eq. 12 is substituted back into Eq. 9 (or Eq. 10), we get

$$C_1 = 0 \tag{13}$$

Thus, $C_1 = C_2 = C_4 = 0$, so

$$V(x) = C \sin \lambda x \tag{14}$$

where λ is determined from Eq. 12, and where C is an arbitrary amplitude factor. From Eq. 12,

$$\begin{aligned} \lambda_1 L &= \pi \\ \lambda_2 L &= 2\pi \\ &\;\;\vdots \\ \lambda_r L &= r\pi \end{aligned} \tag{15}$$

The natural frequency, ω_r, can be obtained by combining Eqs. 10.13 and Eq. 15 to get

$$\boxed{\omega_r = \left(\frac{r\pi}{L}\right)^2 \left(\frac{EI}{\rho A}\right)^{1/2}} \tag{16}$$

When Eq. 15 is substituted into Eq. 14, the mode shape equation becomes

$$\boxed{V_r(x) = C \sin\left(\frac{r\pi x}{L}\right)} \tag{17}$$

Several mode shapes are plotted below.

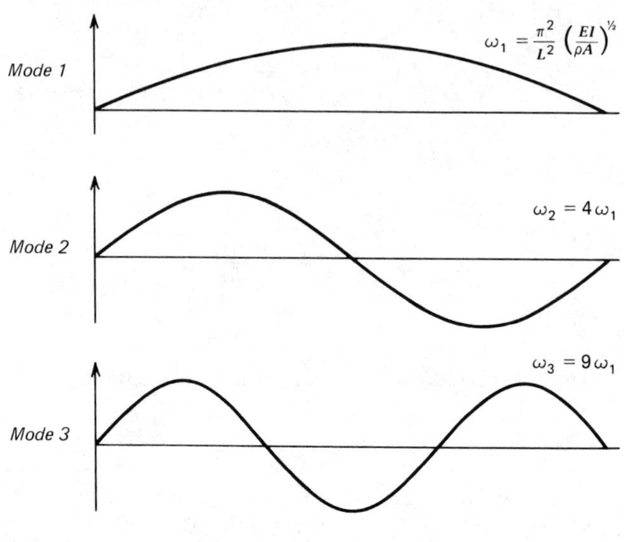

Example 10.3

Determine the natural frequencies and natural modes of a uniform cantilever beam.

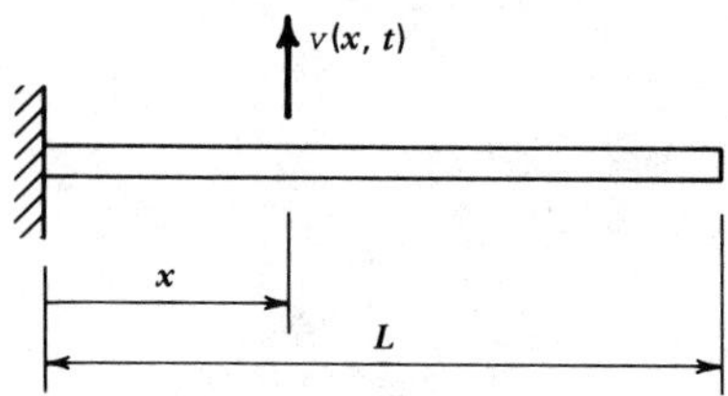

Solution

Use the general solution given in Eq. 10.14c.

$$V(x) = C_1 \sinh \lambda x + C_2 \cosh \lambda x + C_3 \sin \lambda x + C_4 \cos \lambda x \tag{1}$$

The boundary conditions are

$$V(0) = 0 \tag{2}$$

$$\left.\frac{dV}{dx}\right|_{x=0} = 0 \tag{3}$$

$$\left.\frac{d^2 V}{dx^2}\right|_{x=L} = 0 \tag{4}$$

$$\left.\frac{d^3 V}{dx^3}\right|_{x=L} = 0 \tag{5}$$

From Eq. 1,

$$\frac{dV}{dx} = \lambda(C_1 \cosh \lambda x + C_2 \sinh \lambda x + C_3 \cos \lambda x - C_4 \sin \lambda x) \tag{6}$$

$$\frac{d^2 V}{dx^2} = \lambda^2(C_1 \sinh \lambda x + C_2 \cosh \lambda x - C_3 \sin \lambda x - C_4 \cos \lambda x) \tag{7}$$

$$\frac{d^3 V}{dx^3} = \lambda^3(C_1 \cosh \lambda x + C_2 \sinh \lambda x - C_3 \cos \lambda x + C_4 \sin \lambda x) \tag{8}$$

Substituting Eqs. 1 and 6 through 8 into the boundary condition equations, Eqs. 2 through 5, we obtain the following equations

$$\begin{bmatrix} 0 & 1 & 0 & 1 \\ \lambda & 0 & \lambda & 0 \\ \lambda^2 \sinh \lambda L & \lambda^2 \cosh \lambda L & -\lambda^2 \sin \lambda L & -\lambda^2 \cos \lambda L \\ \lambda^3 \cosh \lambda L & \lambda^3 \sinh \lambda L & -\lambda^3 \cos \lambda L & \lambda^3 \sin \lambda L \end{bmatrix} \begin{Bmatrix} C_1 \\ C_2 \\ C_3 \\ C_4 \end{Bmatrix} = \begin{Bmatrix} 0 \\ 0 \\ 0 \\ 0 \end{Bmatrix} \tag{9}$$

For this set of homogeneous equations to have a nontrivial solution, the determinant of the coefficients must vanish. This leads to the *characteristic equation*

$$\cos \lambda L \cosh \lambda L + 1 = 0 \tag{10}$$

whose roots are the eigenvalues λ_r times the length L. No simple expression for the roots of the characteristic equation is available, so a numerical solution of Eq. 10 is required. Values of $(\lambda_r L)$ may be found in tables in Reference 10.2, which corrects an error in the original work of Young and Felgar[10.1]. A few values are listed below.

$$\begin{aligned} \lambda_1 L &= 1.8751 \\ \lambda_2 L &= 4.6941 \\ \lambda_3 L &= 7.8548 \\ \lambda_4 L &= 10.996 \end{aligned} \tag{11}$$

From Eq. 10.13

$$\omega_r = \frac{(\lambda_r L)^2}{L^2}\left(\frac{EI}{\rho A}\right)^{1/2} \tag{12}$$

so

$$\begin{aligned} \omega_1 &= \frac{3.516}{L^2}\left(\frac{EI}{\rho A}\right)^{1/2} \\ \omega_2 &= \frac{22.03}{L^2}\left(\frac{EI}{\rho A}\right)^{1/2} \\ \omega_3 &= \frac{61.70}{L^2}\left(\frac{EI}{\rho A}\right)^{1/2} \end{aligned} \tag{13}$$

We determine the mode shape by employing three of the four equations in Eq. 9 to express three of the constants in terms of the fourth, which remains arbitrary. The first two equations of Eq. 9 say that

$$\begin{aligned} C_4 &= -C_2 \\ C_3 &= -C_1 \end{aligned} \tag{14}$$

The third equation says that

$$C_1 \sinh(\lambda_r L) + C_2 \cosh(\lambda_r L) - C_3 \sin(\lambda_r L) - C_4 \cos(\lambda_r L) = 0 \tag{15}$$

which can be combined with Eqs. 14 to give

$$C_1 \sinh(\lambda_r L) + C_2 \cosh(\lambda_r L) + C_1 \sin(\lambda_r L) + C_2 \cos(\lambda_r L) = 0$$

or

$$C_1 = -C_2 \left[\frac{\cosh(\lambda_r L) + \cos(\lambda_r L)}{\sinh(\lambda_r L) + \sin(\lambda_r L)} \right] = -k_r C_2 \tag{16}$$

Equations 14 and 16 can be combined with Eq. 1 to give the mode shapes

$$V_r(x) = C\{\cosh(\lambda_r x) - \cos(\lambda_r x) - k_r[\sinh(\lambda_r x) - \sin(\lambda_r x]\} \tag{17}$$

where k_r is given in Eq. 16 and C is an arbitrary amplitude constant. Three mode shapes of a uniform cantilever beam are sketched below.

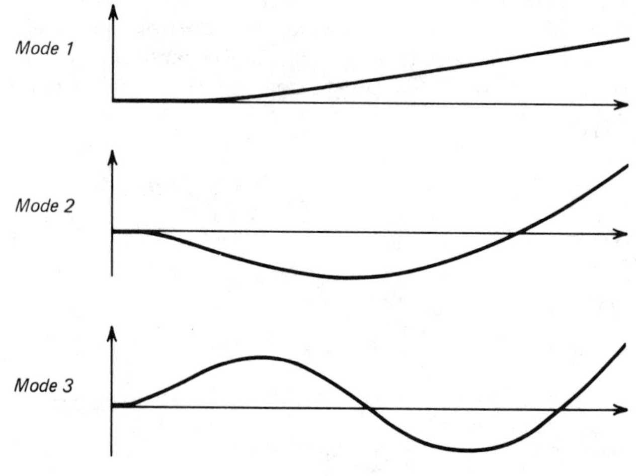

In the above examples it may be noted that the natural frequencies (eigenvalues) are determined by solving for the roots of a transcendental equation. It should also be noted that there is an arbitrary amplitude associated with each mode shape. These and other properties of modes and frequencies will be discussed further in Sec. 10.5.

10.3 Rayleigh's Method for Approximating the Fundamental Frequency of a Continuous System

In Sec. 2.4 you were introduced to the assumed-modes method for obtaining a SDOF approximation of a continuous system. This is actually a generalization of a method used by Lord Rayleigh[10.3] to estimate the fundamental frequency of an undamped continuous system. Consider a beam similar to the one in Example 2.9.

Rayleigh observed that for undamped free vibration the motion is simple harmonic motion. Thus,

$$v(x, t) = V(x) \cos(\omega_R t) = \hat{V}\psi(x) \cos(\omega_R t) \tag{10.18}$$

where ω_R will designate the Rayleigh approximation to the fundamental frequency and where $\psi(x)$ is an assumed shape function as in Sec. 2.4. Rayleigh also observed that energy is conserved. Hence, since energy is conserved, the maximum potential energy is equal to the maximum kinetic energy, that is,

$$T_{\max} = V_{\max} \tag{10.19}$$

For the beam in Fig. 10.1 the strain energy is given by

$$V = \frac{1}{2} \int_0^L EI(v'')^2 \, dx + \frac{1}{2} k_i v_i^2 \tag{10.20}$$

and the kinetic energy is given by

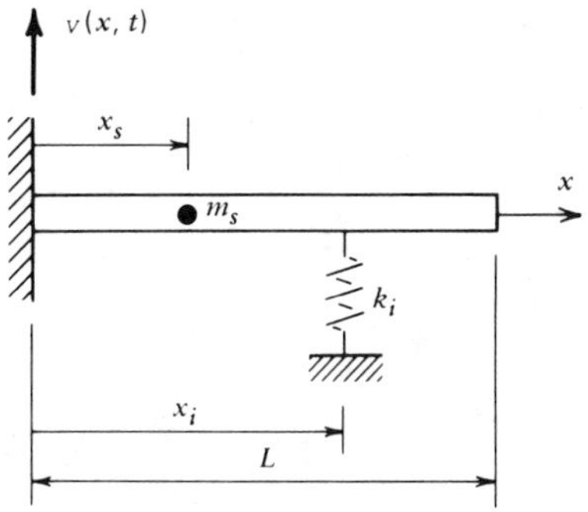

Figure 10.1. Free vibration of a beam.

$$T = \frac{1}{2}\int_0^L \rho A(\dot{v})^2\,dx + \frac{1}{2}\,m_s\dot{v}_s^2 \tag{10.21}$$

Substitution of Eq. 10.18 into these produces

$$V_{\max} = \frac{1}{2}\,k\hat{V}^2 \tag{10.22}$$

and

$$T_{\max} = \frac{1}{2}\,\omega_R^2 m\hat{V}^2 \tag{10.23}$$

where, as in Sec. 2.4,

$$k = \int_0^L EI(\psi'')^2\,dx + k_i[\psi(x_i)]^2 \tag{10.24}$$

and

$$m = \int_0^L \rho A\psi^2\,dx + m_s[\psi(x_s)]^2 \tag{10.25}$$

Finally, from Eqs. 10.19, 10.22, and 10.23 we obtain the so-called *Rayleigh quotient*

$$R(V) \equiv \omega_R^2 = \frac{k}{m} \tag{10.26}$$

Note that this same free vibration result would be obtained from using the generalized stiffness and mass of Sec. 2.4 in Eq. 3.4a. Thus, the Rayleigh method and the assumed-modes method lead to the same expression for the approximate frequency.

Example 10.4

Use the Rayleigh method to obtain an approximate fundamental frequency of a uniform cantilever beam. Use

$$\psi(x) = \left(\frac{x}{L}\right)^2$$

and compare the resulting frequency with the "exact" frequency obtained in Example 10.3.

Solution

$$\psi'' = \frac{2}{L^2} \tag{1}$$

$$k = \int_0^L EI(\psi'')^2\,dx = EIL\left(\frac{4}{L^4}\right)$$

$$k = \frac{4EI}{L^3} \tag{2}$$

$$m = \int_0^L \rho A\psi^2\,dx = \frac{\rho A}{L^4}\int_0^L x^4\,dx$$

$$m = \frac{\rho AL}{5} \tag{3}$$

From Eq. 10.26,

$$\omega_R^2 = \frac{k}{m} = \frac{20EI}{\rho AL^4} \tag{4}$$

Then,

$$\omega_R = \left(\frac{20}{L^2}\right)^{1/2}\left(\frac{EI}{\rho A}\right)^{1/2} \tag{5}$$

or

$$\boxed{\omega_R = \frac{4.472}{L^2}\left(\frac{EI}{\rho A}\right)^{1/2}} \tag{6}$$

This value compares with the "exact" value of ω_1 given in Example 10.3 as

$$\omega_1 = \frac{3.516}{L^2}\left(\frac{EI}{\rho A}\right)^{1/2}$$

It will be shown in Sec. 10.5 that the Rayleigh method always gives an upper bound to the exact fundamental frequency, as was the case in the above example.

10.4 Free Vibration of Beams Including Shear Deformation and Rotatory Inertia

The equations of motion and generalized boundary conditions for a Timoshenko beam were derived in Sec. 9.4. To gain some insight into the effect of shear deformation and rotatory inertia on the vibration of beams, consider a uniform, simply supported beam like the one in Example 10.2. The geometric boundary conditions for a simply supported beam are

$$v(0, t) = v(L, t) = 0 \tag{10.27}$$

Since the rotation α is not specified at the ends, Eqs. 9.47c and 9.47d give the natural boundary conditions

$$\alpha'(0, t) = \alpha'(L, t) = 0 \tag{10.28}$$

It will be convenient to use Eqs. 9.48 and 9.49 as the equations of motion, although Eqs. 9.46 could equally well have been chosen. For free vibration the equations of motion reduce to

$$\alpha' = v'' - \left(\frac{\rho}{\kappa G}\right)\ddot{v} \tag{10.29}$$

and

$$EI\frac{\partial^4 v}{\partial x^4} + \rho A\frac{\partial^2 v}{\partial t^2} - \rho A r_G^2 \frac{\partial^4 v}{\partial x^2\,\partial t^2}\left(1 + \frac{E}{\kappa G}\right) + \frac{\rho^2 A r_G^2}{\kappa G}\frac{\partial^4 v}{\partial t^4} = 0 \tag{10.30}$$

where r_G is the radius of gyration, $r_G^2 = I/A$.

Assume harmonic motion of the form

$$v(x, t) = V(x)\cos\omega t \tag{10.31}$$

Equation 10.29 then becomes

$$\alpha' = \left[V'' + \left(\frac{\rho\omega^2}{\kappa G}\right)V\right]\cos\omega t \tag{10.32}$$

so the boundary conditions of Eqs. 10.27 and 10.28 reduce, respectively, to

$$V(0) = V(L) = 0 \tag{10.33a}$$

and

$$V''(0) = V''(L) = 0 \tag{10.33b}$$

which are the same as those of simple beam theory. Substituting Eq. 10.31 into Eq. 10.30, we obtain

$$V^{iv} - \lambda^4 V + \lambda^4 r_G^2 V''\left(1 + \frac{E}{\kappa G}\right) + \lambda^8 r_G^4\left(\frac{E}{\kappa G}\right)V = 0 \tag{10.34}$$

where λ^4 is given by Eq. 10.13.

The simply supported beam mode shape

$$V_r(x) = C\sin\left(\frac{r\pi x}{L}\right) \tag{10.35}$$

satisfies both the boundary conditions, Eqs. 10.33, and the equation of motion,

Eq. 10.34. Therefore, it provides a solution for the present problem. Substitution of this solution into Eq. 10.34 gives

$$\underset{(a)}{\left(\frac{r\pi}{L}\right)^4} - \underset{(a)}{\lambda^4} - \lambda^4 r_G^2\left(\frac{r\pi}{L}\right)^2\left(\underset{(b)}{1} + \underset{(c)}{\frac{E}{\kappa G}}\right) + \underset{(d)}{\lambda^8 r_G^4\left(\frac{E}{\kappa G}\right)} = 0 \tag{10.36}$$

The terms denoted (a) are those of the simple beam theory characteristic equation. To estimate the relative importance of terms (c) and (d), approximate λ^4 in term (d) by the simple beam solution $(r\pi/L)^4$. Then term (d) is approximately

$$\underset{(d)}{\lambda^8 r_G^4\left(\frac{E}{\kappa G}\right)} \doteq \underbrace{\lambda^4 r_G^2\left(\frac{r\pi}{L}\right)^2\left(\frac{E}{\kappa G}\right)}_{(c)}\left(\frac{r\pi r_G}{L}\right)^2$$

Thus, when $(r\pi r_G/L) \ll 1$ the last term is small compared with term (c). To compare terms (b) and (c), observe that $E/\kappa G \doteq 3$ for a rectangular beam made of typical construction materials. Hence, the shear correction, (c), is about three times as important as the rotatory inertia correction, (b).

Neglecting term (d), we can approximate the solution of the characteristic equation, Eq. 10.36, by

$$\lambda^4 = \left(\frac{r\pi}{L}\right)^4\left[\frac{1}{1 + r_G^2(r\pi/L)^2(1 + E/\kappa G)}\right] \tag{10.37}$$

Thus, the correction due to shear and rotatory inertia increases as the mode number, r, increases and decreases as the slenderness ratio, L/r_G, increases. It is important to note that although a beam may be physically "slender," that is, have a large value of L/r_G, it is the effective slenderness ratio based on the wavelength L/r, and not on the physical length L, which governs whether correction for shear deformation and rotatory inertia are needed.

10.5 Some Properties of Natural Modes

Sections 10.1 and 10.2 present typical examples of the determination of modes and frequencies of continuous members. In the present section we consider the following properties associated with the modes: scaling (or normalization), orthogonality, the expansion theorem, and the Rayleigh quotient. These are illustrated using the Bernoulli-Euler beam equations. However, the following discussion is not limited to uniform beams, and the properties described apply to other continuous systems as well[10.4].

From Eq. 10.11, the eigenvalue equation is

$$(EIV_r'')'' - \rho A\omega_r^2 V_r = 0 \tag{10.38}$$

Homogeneous *generalized boundary conditions* can be stated as

$$(EIV'')'\,\delta V = 0 \quad \text{at} \quad x = 0 \tag{10.39a}$$
$$(EIV'')\,\delta V' = 0 \quad \text{at} \quad x = 0 \tag{10.39b}$$
$$(EIV'')'\,\delta V = 0 \quad \text{at} \quad x = L \tag{10.39c}$$
$$(EIV'')\,\delta V' = 0 \quad \text{at} \quad x = L \tag{10.39d}$$

From these, any combination of fixed, simply supported, or free ends can be obtained.

As noted previously, Eqs. 10.38 and 10.39 determine a set of *natural frequencies,* ω_r ($r = 1, 2, \ldots$), and corresponding *natural modes,* $V_r(x)$ ($r = 1, 2, \ldots$), where the latter are determined only to within a constant multiplier, that is, the *mode shape* is determined but not the amplitude. Thus, modes can be scaled in any convenient manner.

Scaling (Normalization)

Modes that have been scaled so that they have a unique amplitude will be denoted by $\phi_r(x)$. Let

$$V_r(x) = c_r\phi_r(x) \tag{10.40}$$

It is convenient to consider $\phi_r(x)$ to be dimensionless, and to let c_r carry the units of $V(x)$. Three procedures frequently employed for scaling modes are:

1. Scale the mode so that the amplitude at a particular location in the structure has a specified value, for example, make $\phi_r(x_s) = 1$ at a specified location x_s.*

2. Scale the mode so that the value of $\phi_r(x)$ at the point where $|\phi_r(x)|$ is a maximum is a specified value, for example, so $\max_x |\phi_r(x)| = 1$.

*This scaling procedure is sometimes used where the point x_s has some particular significance, for example the location of the main deck of an offshore platform, the location of the top story of a building, and so on. This method is of more limited applicability than methods (2) and (3) below, however, since the location x_s could be a node point for some mode, and hence $V_r(x_s) = 0$. Therefore, this scaling procedure could not be employed for such a mode.

3. Scale the mode so that the generalized mass, or *modal mass,* defined by

$$M_r = \int_0^L \rho A \phi_r^2 \, dx \tag{10.41}$$

has a specified value. Usually the value $M_r = 1$ is used.†

The process of rendering the amplitude of a mode to be unique is called *normalization,* and the resulting modes, $\phi_r(x)$, are called *normal modes.**

The generalized stiffness, or *modal stiffness,* for the rth mode is defined as

$$K_r = \int_0^L EI(\phi_r'')^2 \, dx \tag{10.42}$$

K_r and M_r can be related to each other in the following manner. Let Eq. 10.38 be multiplied by ϕ_r and integrated from 0 to L. Then,

$$\int_0^L (EI\phi_r'')'' \, \phi_r \, dx - \omega_r^2 \int_0^L \rho A \phi_r^2 \, dx = 0 \tag{10.43}$$

Noting that K_r contains the term $(\phi_r'')^2$ we can integrate the first term of Eq. 10.43 by parts twice to get

$$(EI\phi_r'')'\phi_r \Big|_0^L - (EI\phi_r'')\phi_r' \Big|_0^L + \int_0^L EI(\phi_r'')^2 \, dx - \omega_r^2 \int_0^L \rho A \phi_r^2 \, dx = 0 \tag{10.44}$$

From Eqs. 10.39, it will be recognized that all of the boundary terms in Eq. 10.44 vanish due to geometric or natural boundary conditions, regardless of the type of support at each end. From the definitions of M_r and K_r, then, Eq. 10.44 reduces to

$$\omega_r^2 = \frac{K_r}{M_r} \tag{10.45}$$

Orthogonality

The most important property of natural modes is the property of *orthogonality.* To derive the orthogonality property, we multiply Eq. 10.38 by $\phi_s(\neq \phi_r)$ and integrate from 0 to L.

†Since it is convenient to consider ϕ_r to be dimensionless, the units of mass should be associated with M_r, for example, $M_r = 1$ slug.

*Some authors use the term "normal mode" in a more general sense to refer to any mode shape. Others restrict the term "normal mode" to those modes normalized so that $M_r = 1$.

$$\int_0^L (EI\phi_r'')''\phi_s \, dx - \omega_r^2 \int_0^L \rho A\phi_r\phi_s \, dx = 0 \tag{10.46}$$

Integrating by parts and noting that the same boundary conditions apply to modes r and s, we can reduce Eq. 10.46 to

$$\int_0^L EI\phi_r''\phi_s'' \, dx - \omega_r^2 \int_0^L \rho A\phi_r\phi_s \, dx = 0 \tag{10.47}$$

Starting with Eq. 10.38 written for mode s, multiplying by ϕ_r, and integrating from 0 to L, we get

$$\int_0^L EI\phi_r''\phi_s'' \, dx - \omega_s^2 \int_0^L \rho A\phi_r\phi_s \, dx = 0 \tag{10.48}$$

Subtracting Eq. 10.47 from Eq. 10.48 we obtain

$$(\omega_r^2 - \omega_s^2) \int_0^L \rho A\phi_r\phi_s \, dx = 0 \tag{10.49}$$

For two modes having distinct frequencies, then, the *orthogonality property*

$$\int_0^L \rho A\phi_r\phi_s \, dx = 0 \qquad \omega_r \neq \omega_s \tag{10.50}$$

holds true. The modes are said to be *orthogonal with respect to the mass distribution.* By substituting Eq. 10.50 into either Eq. 10.47 or 10.48, we see that modes ϕ_r and ϕ_s are also *orthogonal with respect to the stiffness distribution* as defined by

$$\int_0^L EI\phi_r'' \, \phi_s'' \, dx = 0 \tag{10.51}$$

The modes $\phi_r(x)$ for which $M_r = 1$ satisfy the normalization condition

$$\int_0^L \rho A\phi_r^2 \, dx = 1 \tag{10.52}$$

and the orthogonality condition

$$\int_0^L \rho A\phi_r\phi_s \, dx = 0 \tag{10.53}$$

This set of modes $\phi_r(x)$, $r = 1, 2, \ldots$, is said to form a *set of orthonormal modes.*

Expansion Theorem

An *expansion theorem* for continuous systems can be stated as follows[10.4]: Any function $V(x)$ which satisfies the same boundary conditions as are satisfied by a set of orthonormal modes, ϕ_r $(r = 1, 2, \ldots)$, and which is such that $(d^2/dx^2)(EI\, d^2V/dx^2)$ is a continuous function, can be represented by an absolutely and uniformly convergent series of the form

$$V(x) = \sum_{r=1}^{\infty} c_r \phi_r(x) \tag{10.54}$$

where the coefficients c_r are given by

$$c_r = \frac{\int_0^L \rho A V \phi_r \, dx}{\int_0^L \rho A \phi_r^2 \, dx} \tag{10.55a}$$

or

$$c_r = \int_0^L \rho A V \phi_r \, dx, \qquad r = 1, 2, \ldots \tag{10.55b}$$

In deriving Eqs. 10.55 from Eq. 10.54, the orthogonality equation, Eq. 10.53, and the normalization condition, Eq. 10.52, were employed.

Rayleigh Quotient

Let $V(x)$ be an arbitrary function satisfying the boundary and continuity conditions mentioned in the expansion theorem above. Then the *Rayleigh quotient* is defined as

$$R(V) = \frac{k}{m} = \frac{\int_0^L EI(V'')^2 \, dx}{\int_0^L \rho A V^2 \, dx} \tag{10.56}$$

for a beam with no lumped springs or masses. The Rayleigh quotient was used in Sec. 10.3 to obtain an approximate value of the fundamental frequency. Equation 10.45 shows that $R(V) = \omega_r^2$ when $V = c_r \phi_r(x)$. Let Eq. 10.54 be substituted into Eq. 10.56. Then, using the orthonormality conditions, we get

$$R(V) = \frac{c_1^2\omega_1^2 + c_2^2\omega_2^2 + c_3^2\omega_3^2 + \cdots}{c_1^2 + c_2^2 + c_3^2 + \cdots}$$

which can be written in the form

$$R(V) = \omega_1^2 \left[\frac{1 + (c_2/c_1)^2(\omega_2/\omega_1)^2 + (c_3/c_1)^2(\omega_3/\omega_1)^2 + \cdots}{1 + (c_2/c_1)^2 \qquad\qquad + (c_3/c_1)^2 \qquad\qquad + \cdots} \right] \tag{10.57}$$

By comparing, term by term, the numerator and denominator we see that, since ω_1 is the lowest frequency, the numerator is larger than the denominator. Hence,

$$R(V) \geq \omega_1^2 \tag{10.58}$$

the equality holding when $c_2 = c_3 = \cdots = 0$. The Rayleigh quotient thus provides an upperbound to the fundamental frequency, as was seen in Example 10.4.

10.6 Vibration of Thin Flat Plates

In Chapter 9 the equations of motion were derived for several one-dimensional structural members, and so far in Chapter 10 solutions have been obtained for the modes and frequencies of these members. Although a thorough discussion of the vibration of plates and shells is beyond the scope of this text, a brief introduction is presented here so that you may note some of the features that arise in vibration of multidimensional structures. Excellent surveys of the topics of vibration of plates[10.5] and vibration of shells[10.6] have been compiled by Leissa.

The equation of motion of a flat plate, based on the Kirchhoff theory[10.5], is

$$D\nabla^4 w + \rho h w_{tt} = 0 \tag{10.59}$$

where subscripts indicate partial differentiation, and where D is the flexural rigidity of the plate as defined by

$$D = \frac{Eh^3}{12(1 - \nu^2)} \tag{10.60}$$

where h is the (constant) thickness of the plate. $w(x, y; t)$ is the transverse displacement of the point $(x, y, z = 0)$ and ∇^4 is the biharmonic operator

$$\nabla^4(\ ) = (\ )_{xxxx} + 2(\ )_{xxyy} + (\ )_{yyyy} \tag{10.61}$$

Figure 10.2 shows a rectangular plate. On an edge $y =$ cons the following generalized boundary conditions pertain:

$$[w_{yyy} + (2 - \nu)w_{xxy}]\delta w = 0 \tag{10.62a}$$

$$[w_{yy} + \nu w_{xx}]\ \delta w_y = 0 \tag{10.62b}$$

Conditions for an edge $x =$ cons are obtained by permutation of x and y.

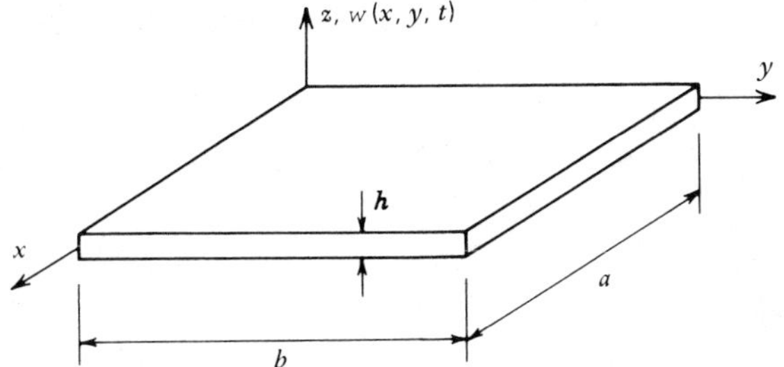

Figure 10.2. A rectangular plate.

Exact solutions for free vibration are only available for rectangular plates with a pair of opposite edges simply supported. The case of a rectangular plate with all edges simply supported leads to a particularly easy solution and is presented here to illustrate some features of plate vibration. The boundary conditions for a simply supported plate are:

$$w(0, y) = w(a, y) = w(x, 0) = w(x, b) = 0 \tag{10.63a}$$

$$w_{xx}(0, y) = w_{xx}(a, y) = w_{yy}(x, 0) = w_{yy}(x, b) = 0 \tag{10.63b}$$

Equations 10.63a are geometric boundary conditions and Eqs. 10.63b are natural boundary conditions obtained from Eq. 10.62b.

The equation of motion admits a solution which is harmonic in time, that is,

$$w(x, y; t) = W(x, y) \cos(\omega t - \alpha) \tag{10.64}$$

Then, Eq. 10.59 becomes the eigenvalue equation

$$\nabla^4 W - \lambda^4 W = 0 \tag{10.65}$$

where

$$\lambda^4 = \frac{\omega^2 \rho h}{D} \tag{10.66}$$

The spatial function

$$W_{(m,n)}(x, y) = C \sin\left(\frac{m\pi x}{a}\right) \sin\left(\frac{n\pi y}{b}\right) \tag{10.67}$$

satisfies all of the boundary conditions, and when substituted into Eq. 10.65 it gives

$$\lambda^2_{(m,n)} = \left(\frac{m\pi}{a}\right)^2 + \left(\frac{n\pi}{b}\right)^2 \tag{10.68}$$

Using Eqs. 10.66 and 10.68, we can express the natural frequency $\omega_{(m,n)}$ in the form

$$\omega_{(m,n)} = \left(\frac{\pi^4 D}{\rho h a^4}\right)^{1/2} \left[m^2 + n^2\left(\frac{a}{b}\right)^2\right] \tag{10.69a}$$

or

$$\omega_{(m,n)} = \left(\frac{\pi^4 D}{\rho h b^4}\right)^{1/2} \left[m^2\left(\frac{b}{a}\right)^2 + n^2\right] \tag{10.69b}$$

It is convenient to use the parameters (m, n) to describe the mode, rather than a single parameter as was used for one-dimensional structural members. For the simply supported plate these designate the number of half-sine waves in the x and y directions respectively, Eq. 10.67.

Modes of two-dimensional structures are frequently visualized by sketching *node lines,* as in Fig. 10.3. These are the lines along which $w(x, y; t) = 0$ for all time. The $\oplus$ and $\ominus$ indicate the relative sign of $W(x, y)$ in various portions of the plate. Note from Eq. 10.69a that if $a > b$ the mode (2, 1) has a lower frequency than the mode (1, 2). Using Fig. 10.3, this can be interpreted as saying that the plate will seek to divide itself into "squares."

One feature which is easily demonstrated on simply supported square plates, but which can also occur in other structures, is the phenomenon of repeated frequencies, that is, $\omega_{(m,n)} = \omega_{(n,m)}$ since $a = b$. Let $W_{(m,n)}$ and $W_{(n,m)}$ be distinct spatial modes corresponding to $\lambda_{(m,n)} = \lambda_{(n,m)}$. Then any linear combination $W_{(m,n)} + cW_{(n,m)}$ satisfies the equation

$$\nabla^4 W - \lambda^4 W = 0 \tag{10.70}$$

This means that there are not two unique mode shapes, but there is a two-dimensional *subspace of modes* of the form

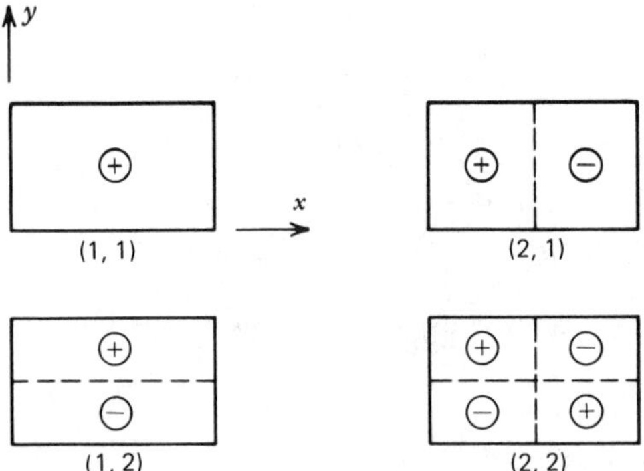

Figure 10.3. *Node lines of a rectangular plate.*

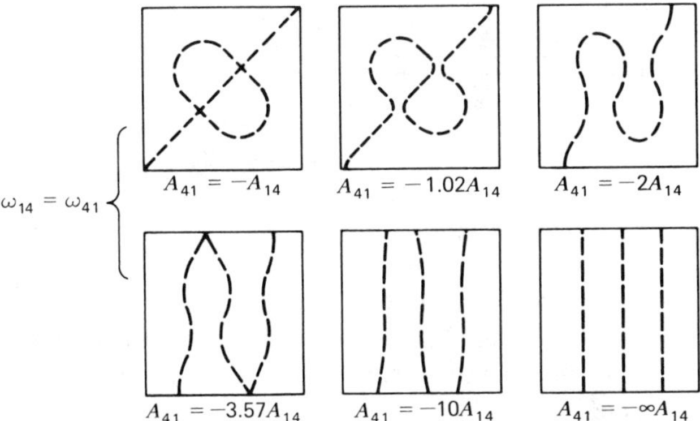

Figure 10.4. Nodal patterns for plates having a natural frequency $\omega_{(1,4)} = \omega_{(4,1)}$ *(Reference 5).*

$$W = A_{(m,n)} W_{(m,n)} + A_{(n,m)} W_{(n,m)} \tag{10.71}$$

Figure 10.4 shows some interesting nodal patterns obtained for a square plate where $W_{(1,4)}$ and $W_{(4,1)}$ are defined by Eq. 10.67.

References

10.1 D. Young and R. P. Felgar, Jr., *Tables of Characteristic Functions Representing Normal Modes of Vibration of a Beam,* Engr. Res. Series No. 44, Bureau of Engr. Research, The University of Texas, Austin, TX (1949).

10.2 T-C. Chang and R. R. Craig, Jr., "Normal Modes of Uniform Beams," *Proc. ASCE,* v. 95, n. EM 4, 1025–1031 (1969).

10.3 G. Temple and W. G. Bickley, *Rayleigh's Principle,* Oxford University Press, London (1933). Reprinted by Dover Publications, Inc., New York (1956).

10.4 L. Meirovitch, *Analytical Methods in Vibrations,* Macmillan, New York (1967).

10.5 A. W. Leissa, *Vibration of Plates,* NASA SP-160, National Aeronautics and Space Administration, Washington, DC (1969).

10.6 A. W. Leissa, *Vibration of Shells,* NASA SP-288, National Aeronautics and Space Administration, Washington, DC (1973).

Problem Set 10.1

10.1 Consider axial vibration of the uniform bar in Fig. P10.1.

(a) Obtain the characteristic equation from which eigenvalues can be determined.

(b) Solve for the fundamental axial frequency of the bar.

(c) Sketch the fundamental mode shape by plotting values at increments of $L/5$ along the bar.

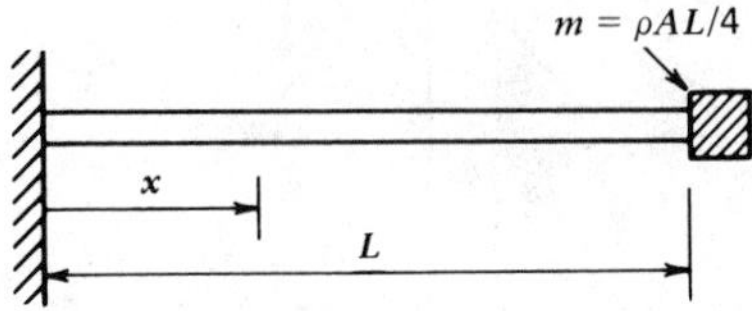

Figure P10.1

10.2 The axial frequencies of a free-free structure are sometimes of great importance, as in the "Pogo" problem experienced by some rockets. Consider axial motion of the uniform free-free bar in Fig. P10.2.

(a) Show that the system has a zero-frequency, or rigid-body, mode.

(b) Determine the characteristic equation which governs the nonzero frequencies.

(c) Solve for the fundamental frequency.

(d) Sketch the fundamental (flexible) mode by plotting values at increments of $L/5$ along the bar.

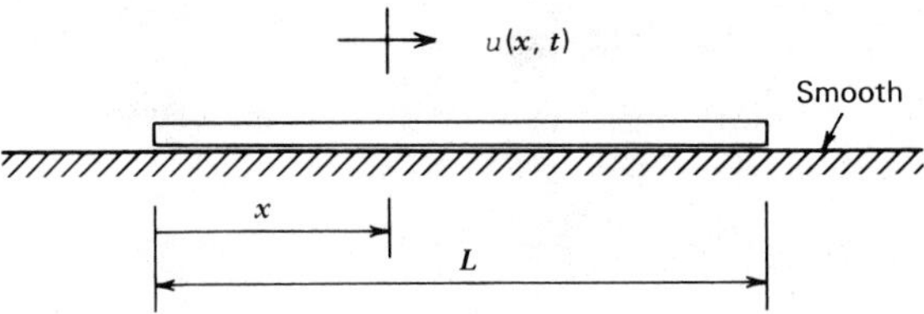

Figure P10.2

Problem Set 10.2

10.3 Consider transverse vibration of the uniform clamped-clamped beam shown in Fig. P10.3.

(a) Determine the characteristic equation.

(b) Solve for the fundamental frequency.

(c) Sketch the fundamental mode by evaluating and plotting values at increments of $L/8$ along the beam.

(d) Using the results of Example 10.2, compare the fundamental frequency of the clamped-clamped beam with that of a simply supported beam.

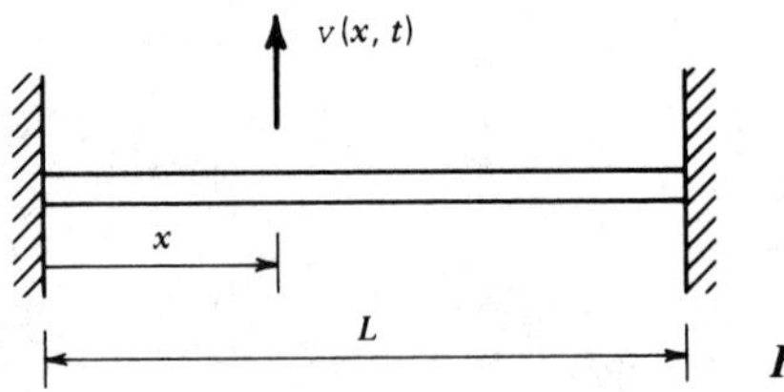

Figure P10.3

10.4 Flight vehicles such as airplanes and rockets are unconstrained, and therefore they can undergo rigid-body motion. The free-free uniform beam gives some insight into the behavior of such free-free structures. Consider transverse vibration of the uniform beam shown below.

(a) Show that the beam has two zero-frequency, or rigid-body, modes, one mode in translation and one in rotation.

(b) Determine the characteristic equation governing the nonzero frequencies. (Note that this is the same equation as obtained in Problem 10.3 for a clamped-clamped beam.)

(c) Solve for the fundamental frequency.

(d) Sketch the fundamental (flexible) mode by evaluating and plotting values at increments of $L/8$ along the beam.

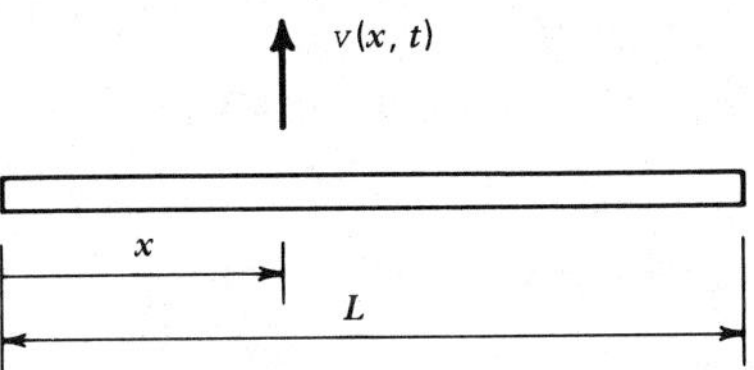

Figure P10.4

10.5 Most structural beams are neither clamped-clamped nor simply supported, but can be considered to have partial fixity. Determine the characteristic equation for the beam below, where $0 \leq \beta \leq \infty$ is a parameter controlling the amount of rotational restraint.

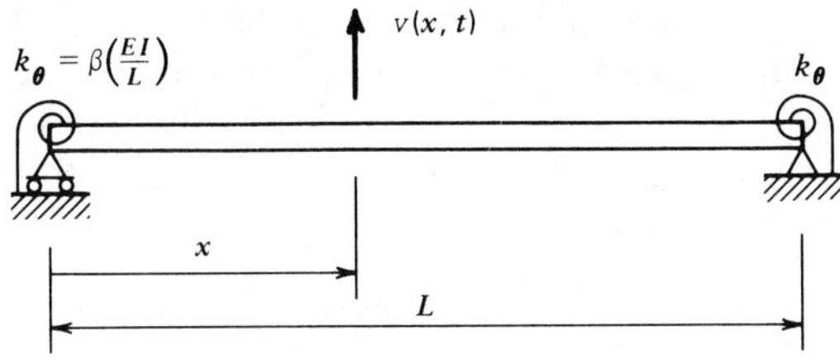

Figure P10.5

10.6 The statement "neglect axial motion" is frequently encountered in modeling frame structures such as the one shown in Fig. P10.6a. The justification for this lies in the natural frequencies associated with axial and transverse vibration. Using the results of Examples 10.1 and 10.3, tabulate the first four axial and first four bending frequencies of a cantilever beam. Use the 12 ft, W10 $\times$ 45 steel beam (I = 248.6 in.^{4}, A = 13.24 in.^{2}, γ = 0.284 lb/in.^{3}) shown in Fig. P10.6*b*.

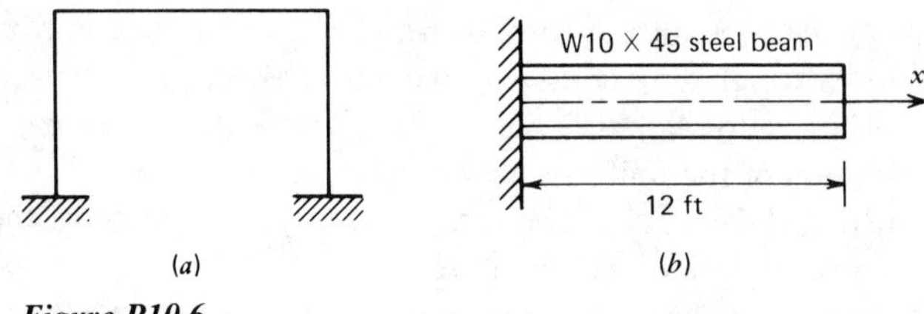

Figure P10.6

10.7 Obtain the characteristic equation for determining the frequencies of the coupled bending-axial motion of the system in Problem 9.3.

10.8 Consider *transverse* vibration, $v(x, t)$ of the beam in Fig. P10.1.

(a) Obtain the characteristic equation.

(b) Solve for the fundamental transverse bending frequency.

Problem Set 10.3

10.9 In Example 10.4 the assumed mode $\psi(x) = (x/L)^2$ gave only a crude estimate of the fundamental frequency of a uniform cantilever beam.

(a) Determine an expression for the static deflection curve of a uniform beam carrying a uniformly distributed load.

(b) Use this static deflection curve as $\psi(x)$ in estimating the fundamental frequency of a uniform cantilever beam by the Rayleigh method.

10.10 Consider *transverse* vibration of the beam with tip mass in Fig. P10.1. Use the Rayleigh method to estimate the fundamental bending frequency. Let $\psi(x)$ be the static deflection curve of a uniform beam with concentrated tip force as shown below. (The "exact" fundamental frequency was determined in Problem 10.8.)

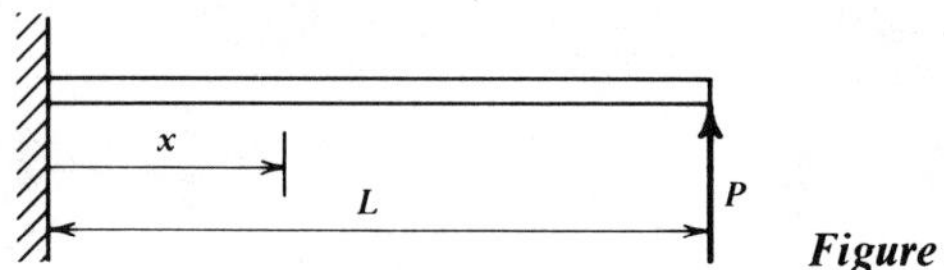

Figure P10.10

10.11 A heavy air conditioner is located at the center of a simply supported roof beam as shown below. Use the Rayleigh method to obtain an expression for estimating the effect of the air conditioner mass on the fundamental beam frequency. For $\psi(x)$ use the fundamental mode shape of the uniform beam alone, as found in Example 10.2.

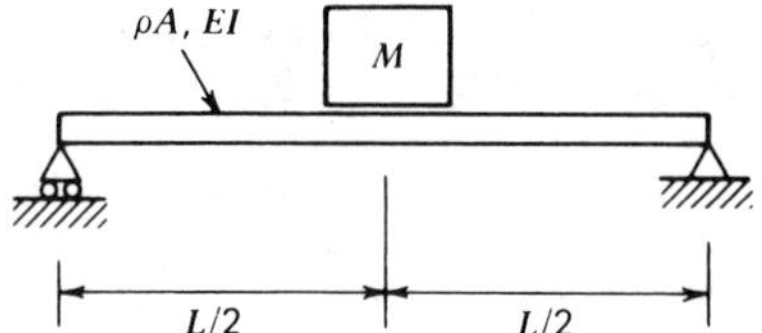

Figure P10.11

10.12 A transmission tower consists of a uniform column topped by an equipment platform. Consider the platform to be a lumped mass with mass $M_p = \mu M_c$, where $M_c = \rho AL$ = the column mass. Using the Rayleigh method with $\psi(x) = (x/L)^2$, estimate the fundamental frequency of this system. Include the geometric stiffness effect of M_p (see Example 2.9).

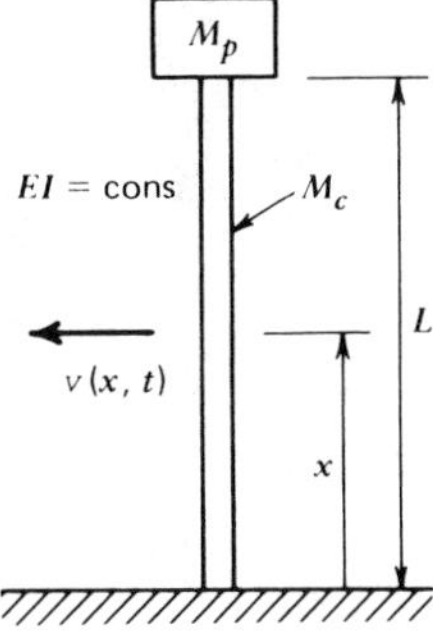

Figure P10.12

Problem Set 10.5

10.13 Consider *axial* vibration of the uniform bar in Figure P10.1. Let

$$M_r = \int_0^L \rho A \phi_r^2 \, dx + m[\phi_r(L)]^2$$

and let

$$K_r = \int_0^L EA(\phi'_r)^2 \, dx$$

Show that Eq. 10.45 holds when M_r and K_r are defined in this way.

10.14 (a) Derive an equation which axial vibration modes ϕ_r and ϕ_s of the system in Fig. P10.1 must satisfy if they are "orthogonal with respect to the mass distribution."

(b) What equation must they satisfy in order to be "orthogonal with respect to the stiffness distribution?"

10.15 For the beam in the Fig. P10.5 derive equations that define:

(a) "Orthogonality of modes ϕ_r and ϕ_s with respect to the mass distribution."

(b) "Orthogonality of modes ϕ_r and ϕ_s with respect to the stiffness distribution."

(c) K_r and M_r.

PART III
MULTIPLE-DEGREE-OF-FREEDOM SYSTEMS

11 MATHEMATICAL MODELS OF MDOF SYSTEMS

In Chapter 2 Newton's laws and the principle of virtual displacements were employed to derive the equations of motion of SDOF models. Although SDOF models may adequately describe the dynamical behavior of some systems, in most cases it is necessary to employ more "sophisticated" models, for example, MDOF models. To illustrate this, consider the axial vibration of a uniform cantilever rod as shown in Fig. 11.1. In Sec. 10.1 it was shown that the continuous model of the rod would have an infinite number of natural frequencies and corresponding mode shapes. Only if $P(t)$ is a "slowly-varying" function of time will the dynamics of the rod be adequately described by a SDOF model, such as the spring-mass model shown in Fig. 11.1. While continuous models, such as those considered in Chapter 10, give valuable insight into the dynamics of a few systems with simple geometry (e.g., uniform bars and beams), the dynamical analysis of most real structures is based on multiple-degree-of-freedom (MDOF) models. In this and subsequent chapters you will gain further insight into this important topic of modeling.

Upon completion of Chapter 11 you should be able to:

- Use Newton's laws to derive the equations of motion of systems of particles and of rigid bodies in plane motion.
- Use Lagrange's equations to derive the equations of motion of systems of particles and of rigid bodies in plane motion.
- Use Lagrange's equations to derive the equations of motion of assumed-modes models of continuous systems.

11.1 Application of Newton's Laws to Lumped-Parameter Models

In Chapter 2 applications to SDOF models were considered. Systems of connected particles and rigid bodies, and rigid bodies in general plane motion are systems that lead to MDOF models. Several examples of the application of Newton's laws to such systems are now considered.

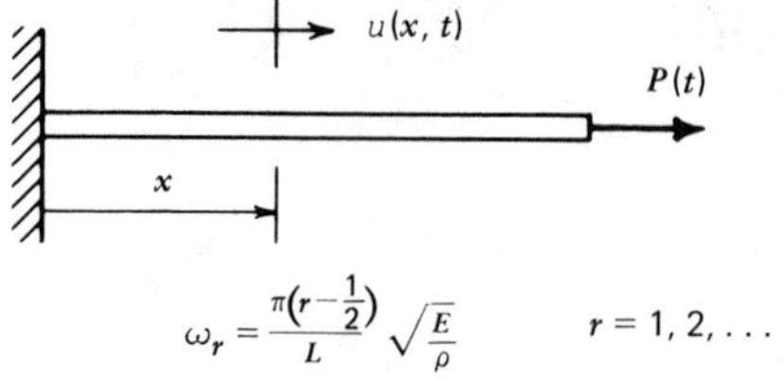

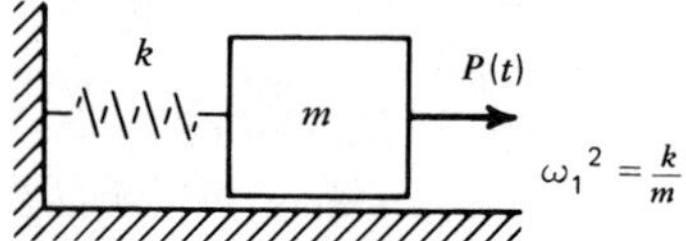

Figure 11.1. Models of a cantilevered uniform rod.

Example 11.1

All of the three systems shown below have essentially the same set of equations of motion. Derive the three equations of motion for the spring-mass system at the lower right.

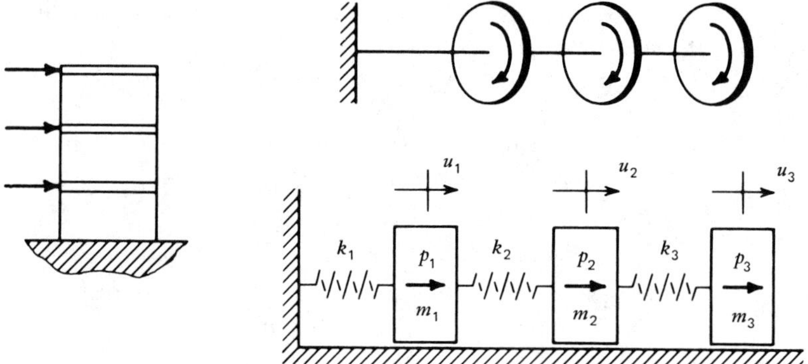

Solution

a. Draw a *freebody diagram* of each particle and label all unknown forces.

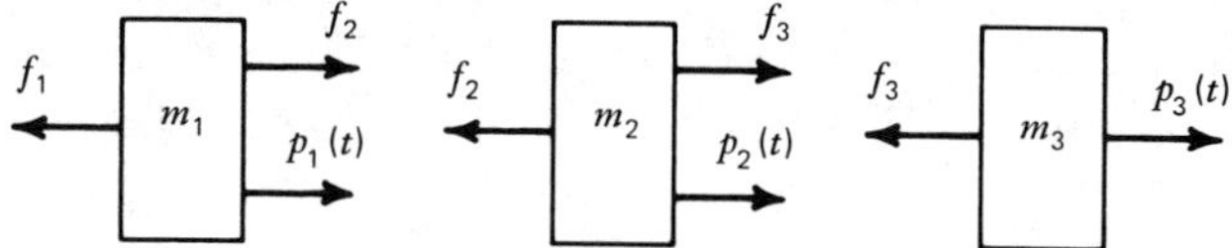

b. Write Newton's second law for each particle.

$$\overset{+}{\rightarrow} \sum F_1 = m_1\ddot{u}_1 = p_1 + f_2 - f_1 \tag{1a}$$
$$\overset{+}{\rightarrow} \sum F_2 = m_2\ddot{u}_2 = p_2 + f_3 - f_2 \tag{1b}$$
$$\overset{+}{\rightarrow} \sum F_3 = m_3\ddot{u}_3 = p_3 - f_3 \tag{1c}$$

c. Relate elastic forces to displacements.

f_1 k_1 f_1 f_2 k_2 f_2 f_3 k_3 f_3

$$f_1 = k_1e_1 = k_1u_1 \tag{2a}$$
$$f_2 = k_2e_2 = k_2(u_2 - u_1) \tag{2b}$$
$$f_3 = k_3e_3 = k_3(u_3 - u_2) \tag{2c}$$

d. Combine and simplify the equations.

$$m_1\ddot{u}_1 + (k_1 + k_2)u_1 - k_2u_2 = p_1(t) \tag{3a}$$
$$m_2\ddot{u}_2 - k_2u_1 + (k_2 + k_3)u_2 - k_3u_3 = p_2(t) \tag{3b}$$
$$m_3\ddot{u}_3 - k_3u_2 + k_3u_3 = p_3(t) \tag{3c}$$

These three equations of motion can be written in matrix form, that is,

$$\begin{bmatrix} m_1 & 0 & 0 \\ 0 & m_2 & 0 \\ 0 & 0 & m_3 \end{bmatrix} \begin{Bmatrix} \ddot{u}_1 \\ \ddot{u}_2 \\ \ddot{u}_3 \end{Bmatrix} + \begin{bmatrix} (k_1 + k_2) & -k_2 & 0 \\ -k_2 & (k_2 + k_3) & -k_3 \\ 0 & -k_3 & k_3 \end{bmatrix} \begin{Bmatrix} u_1 \\ u_2 \\ u_3 \end{Bmatrix} = \begin{bmatrix} p_1(t) \\ p_2(t) \\ p_3(t) \end{bmatrix} \tag{4}$$

or

$$\boxed{\mathbf{m}\ddot{\mathbf{u}} + \mathbf{k}\mathbf{u} = \mathbf{p}(t)} \tag{5}$$

The matrix notation employed in Eqs. 4 and 5 above, while not essential to the solution of this 3-DOF problem, is the "language" used in describing large MDOF systems, particularly when the computer is employed in the dynamical analysis of such systems. **m** is called the *mass matrix,* **k** the *stiffness matrix,* **u** the *displacement vector,* and **p**(t) the *load vector.*

In Eq. 4 of the above example it should be noted that the mass matrix, **m**, is diagonal. Since there are off-diagonal terms in the stiffness matrix, **k**, the system is said to have *stiffness coupling.*

In Example 11.1 the equations of motion were written in terms of the absolute displacements u_1, u_2, and u_3. As noted in Sec. 2.2, when the system is subjected to base motion it is convenient to employ the relative motion between the base and the structural masses. Consider the following 2-DOF system.

Example 11.2

Use Newton's laws to derive the equations of motion of the system shown below. Express the equations of motion in terms of the displacements relative to the base.

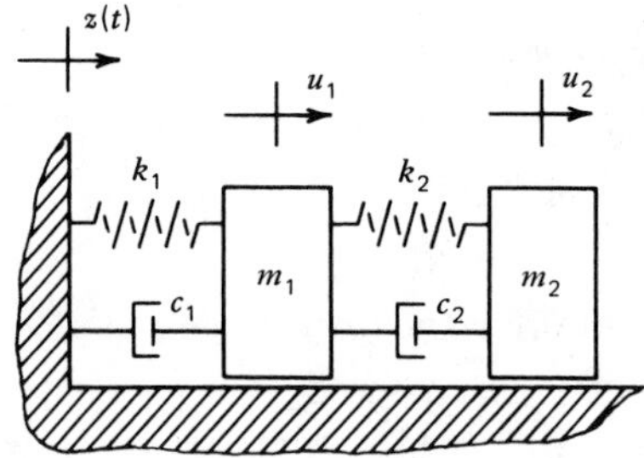

Solution

Let the relative displacements be defined by

$$w_1 = u_1 - z \tag{1a}$$

$$w_2 = u_2 - z \tag{1b}$$

a. Draw freebody diagrams of both masses and label all unknowns.

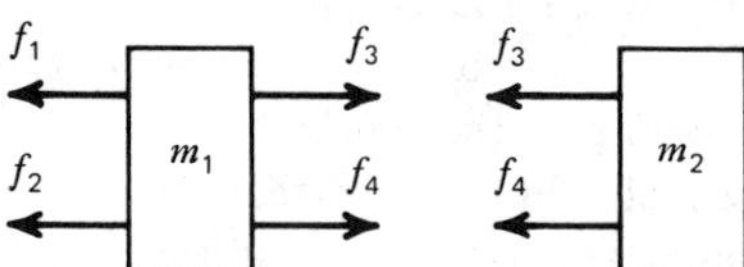

b. Write the equations of motion.

$$\overset{+}{\rightarrow}\sum F_1 = m_1\ddot{u}_1 = -f_1 - f_2 + f_3 + f_4 = m_1(\ddot{w}_1 + \ddot{z}) \tag{2a}$$

$$\overset{+}{\rightarrow}\sum F_2 = m_2\ddot{u}_2 = -f_3 - f_4 = m_2(\ddot{w}_2 + \ddot{z}) \tag{2b}$$

c. Relate elastic forces to displacements and damping forces to velocities.

$$f_1 = k_1(u_1 - z) = k_1 w_1 \tag{3a}$$

$$f_2 = c_1(\dot{u}_1 - \dot{z}) = c_1\dot{w}_1 \tag{3b}$$

$$f_3 = k_2(u_2 - u_1) = k_2(w_2 - w_1) \tag{3c}$$

$$f_4 = c_2(\dot{u}_2 - \dot{u}_1) = c_2(\dot{w}_2 - \dot{w}_1) \tag{3d}$$

d. Combine and simplify the above equations.

$$m_1\ddot{w}_1 + c_1\dot{w}_1 + k_1 w_1 - c_2(\dot{w}_2 - \dot{w}_1) - k_2(w_2 - w_1) = -m_1\ddot{z}$$

$$m_2\ddot{w}_2 + c_2(\dot{w}_2 - \dot{w}_1) + k_2(w_2 - w_1) = -m_2\ddot{z}$$

In matrix form,

$$\begin{bmatrix} m_1 & 0 \\ 0 & m_2 \end{bmatrix}\begin{Bmatrix} \ddot{w}_1 \\ \ddot{w}_2 \end{Bmatrix} + \begin{bmatrix} (c_1 + c_2) & -c_2 \\ -c_2 & c_2 \end{bmatrix}\begin{Bmatrix} \dot{w}_1 \\ \dot{w}_2 \end{Bmatrix} + \begin{bmatrix} (k_1 + k_2) & -k_2 \\ -k_2 & k_2 \end{bmatrix}\begin{Bmatrix} w_1 \\ w_2 \end{Bmatrix} = \begin{Bmatrix} -m_1\ddot{z} \\ -m_2\ddot{z} \end{Bmatrix} \tag{4}$$

or

$$\boxed{\mathbf{m}\ddot{\mathbf{w}} + \mathbf{c}\dot{\mathbf{w}} + \mathbf{k}\mathbf{w} = \mathbf{p}_{\text{eff}}} \tag{5}$$

Note that writing the equations of motion in terms of relative displacements rather than absolute displacements leads to several advantages: (1) the right-hand side has the form of an effective force that is related directly to acceleration, $\ddot{z}$, which is simpler to measure than either z or $\dot{z}$; and (2) the relative displacements can be used directly in calculating elastic forces, as seen in Eq. 3a.

The next example treats a system that is modeled as a rigid body in plane motion.

Example 11.3

The motion of a building subjected to earthquake excitation is to be studied using the lumped-parameter model shown below. Use Newton's laws (extended to rigid bodies) to derive the equations of motion of this system. Consider θ to be small and consider the foundation mass, m, to be a particle.

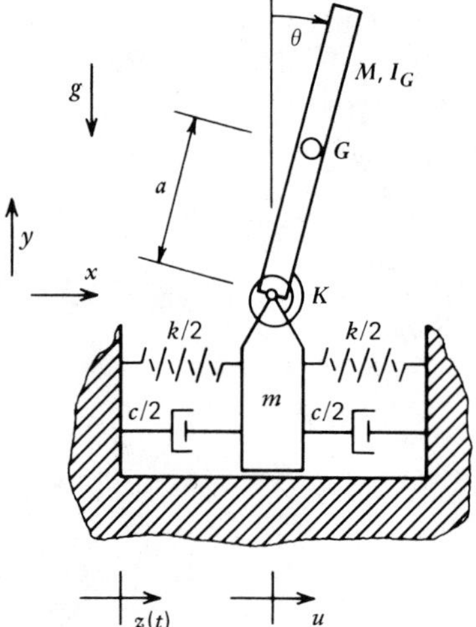

Solution

a. Draw freebody diagrams of the two components of the system and label all unknown forces.

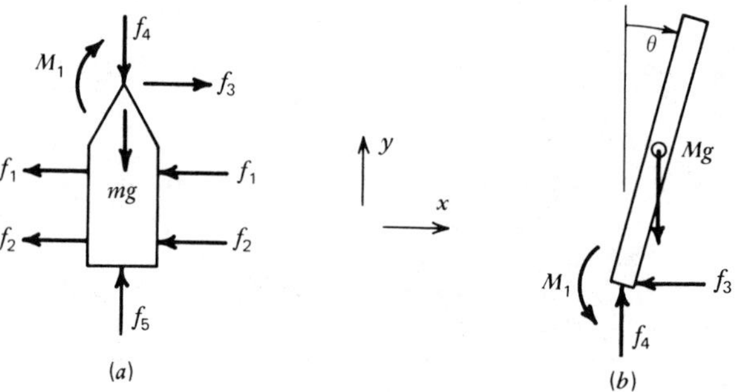

b. Write the basic equations of motion for each freebody diagram.
For (*a*):

$$\overset{+}{\rightarrow} \sum F_x = m\ddot{u} = -2f_1 - 2f_2 + f_3 \tag{1}$$

(The vertical equation of motion and moment equation are not needed.)
For (*b*):

$$\overset{+}{\rightarrow} \sum F_x = M\ddot{x}_G = -f_3 \tag{2a}$$

$$+\uparrow \sum F_y = M\ddot{y}_G = f_4 - Mg \tag{2b}$$

$$\circlearrowleft^{+} \sum M_G = I_G\ddot{\theta} = -M_1 + f_4 a \sin\theta + f_3 a \cos\theta \tag{2c}$$

c. Relate elastic forces to displacements and damping forces to velocities.

$$f_1 = \frac{k}{2}(u - z) \tag{3a}$$

$$f_2 = \frac{c}{2}(\dot{u} - \dot{z}) \tag{3b}$$

$$M_1 = K\theta \tag{3c}$$

d. Use kinematics to relate x_G and y_G to u and θ.
For small θ, $\sin\theta \doteq \theta$ and $\cos\theta \doteq 1$.

$$x_G = u + a\sin\theta \doteq u + a\theta \tag{4a}$$

$$y_G = a\cos\theta \doteq a \tag{4b}$$

The resulting two equations of motion, in matrix form, are

$$\begin{bmatrix} (M + m) & Ma \\ Ma & (I_G + Ma^2) \end{bmatrix} \begin{Bmatrix} \ddot{u} \\ \ddot{\theta} \end{Bmatrix} + \begin{bmatrix} c & 0 \\ 0 & 0 \end{bmatrix} \begin{Bmatrix} \dot{u} \\ \dot{\theta} \end{Bmatrix} + \begin{bmatrix} k & 0 \\ 0 & (K - Mga) \end{bmatrix} \begin{Bmatrix} u \\ \theta \end{Bmatrix} = \begin{Bmatrix} c\dot{z} + kz \\ 0 \end{Bmatrix} \tag{5}$$

or

$$\boxed{\mathbf{m}\ddot{\mathbf{u}} + \mathbf{c}\dot{\mathbf{u}} + \mathbf{k}\mathbf{u} = \mathbf{p}_{\text{eff}}} \tag{6}$$

Note that the base motion $z(t)$ leads to the "force" terms on the right-hand side, and note that this system has *inertia coupling* due to the off-diagonal terms in **m**.

As illustrated by the above example, the use of Newton's laws to derive the equations of motion of systems of coupled rigid bodies can become very tedious. The task of deriving the equations of motion of MDOF systems is greatly simplified in many cases by the use of Lagrange's equations, discussed below.

11.2 Lagrange's Equations

The difficulties that arise when Newton's laws are employed to derive the equations of motion of connected bodies were illustrated in Example 11.3. Separate freebodies were required for each component, and the forces of interaction had to be eliminated to arrive at the final set of equations of motion. In Example 2.7 it was shown that use of the principle of virtual displacements eliminates the necessity of using interaction forces directly. Although the principle of virtual displacements can be extended to permit the derivation of equations of motion of MDOF systems, it is far simpler to use Lagrange's equations for this purpose. This permits the use of the scalar quantities, work and kinetic energy, instead of the vector quantities, force and displacement. Lagrange's equations can be derived from the principle of virtual displacements[11.1] or from Hamilton's principle. The latter approach will be employed here.

Consider Example 11.3 again. Note that although x_G and y_G were required in writing the equations of motion of the mass M, Eqs. 4 permitted x_G and y_G to be related to u and θ, which were the only coordinates appearing in the final equations of motion, Eqs. 5. Kinematical equations, such as Eqs. 4, are called *equations of constraint.* The system considered in Example 11.3 is a 2-DOF system, and any arbitrary configuration of the system can be specified by giving the values of u and θ. Note that it is possible to vary θ while holding u constant, and vice versa. Coordinates of this type are called generalized coordinates. For an N-DOF system, *generalized coordinates* are defined as any set of N independent quantities which are sufficient to completely specify the position of every point within the system.

Hamilton's principle, stated in Eq. 9.23, is repeated here for convenience.

$$\delta \int_{t_1}^{t_2} (T - V)\, dt + \int_{t_1}^{t_2} \delta W_{\text{nc}}\, dt = 0 \tag{11.1}$$

where T is the total kinetic energy of the system, V is its potential energy, and δW_{nc} is the virtual work of nonconservative forces acting on the system.

For most mechanical and structural systems the kinetic energy can be expressed in terms of the generalized coordinates and their first time derivatives, and the potential energy can be expressed in terms of generalized coordinates alone. Also, the virtual work of nonconservative forces, as they act through virtual displacements caused by arbitrary variations in the generalized coordinates, can be expressed as a linear function of those variations. Thus,

$$T = T(q_1, q_2, \ldots, q_N, \dot{q}_1, \dot{q}_2, \ldots, \dot{q}_N, t) \tag{11.2a}$$

$$V = V(q_1, q_2, \ldots, q_N, t) \tag{11.2b}$$

$$\delta W_{\text{nc}} = Q_1\, \delta q_1 + Q_2\, \delta q_2 + \cdots + Q_N\, \delta q_N \tag{11.2c}$$

where $Q_1, Q_2, \ldots, Q_N$ are called the *generalized forces.* The generalized forces have units such that each term $Q_i\, \delta q_i$ has the units of work.

Example 11.4

A particle of mass m slides along a weightless rod. Write an expression for the kinetic energy of the particle in terms of the q_1, q_2, and their time derivatives.

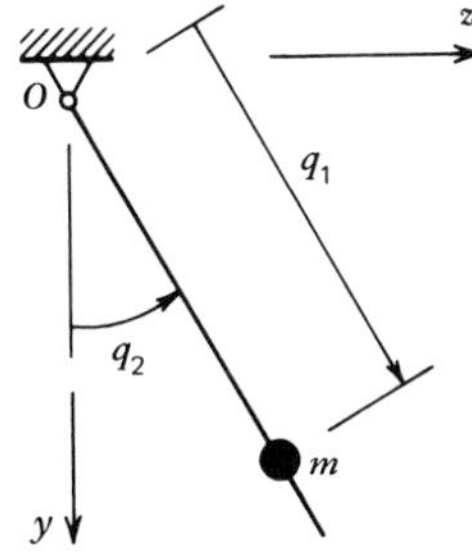

Solution

a. The kinetic energy is given by

$$T = \tfrac{1}{2}m(\dot{y}^2 + \dot{z}^2) \tag{1}$$

b. From kinematics

$$y = q_1 \cos q_2 \tag{2a}$$

$$z = q_1 \sin q_2 \tag{2b}$$

Therefore,

$$\dot{y} = \dot{q}_1 \cos q_2 - q_1\dot{q}_2 \sin q_2 \tag{3a}$$

$$\dot{z} = \dot{q}_1 \sin q_2 + q_1\dot{q}_2 \cos q_2 \tag{3b}$$

c. Combine and simplify to get

$$\boxed{T = \tfrac{1}{2}m(\dot{q}_1^2 + q_1^2\dot{q}_2^2)} \tag{4}$$

Note that Eq. 4 has the form of Eq. 11.2a, that is, both generalized coordinates and generalized velocities appear in T.

Example 11.5

A force P acts tangent to the path of a particle of weight W, which is attached to a rigid bar of length L. Obtain expressions for the potential energy of W and the virtual work done by P. Also determine an expression for the generalized force Q_θ.

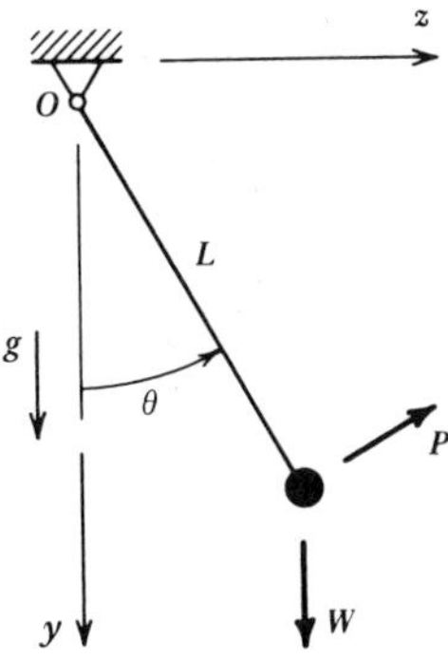

Solution

a. Let the potential energy be zero when $\theta = \pi/2$. Then,

$$\boxed{V = -Wy = -WL\cos\theta} \tag{1}$$

b. Determine δW_P due to a variation in θ as shown below.

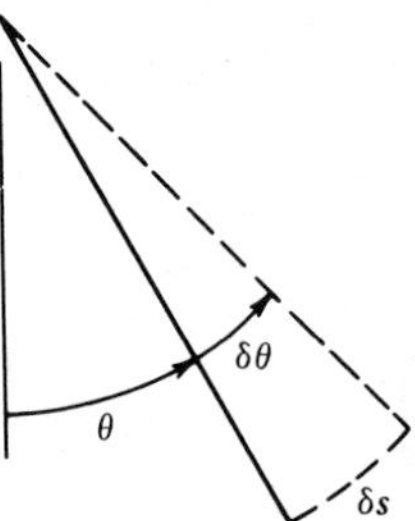

$$\delta s = L\,\delta\theta \tag{2}$$

is the tangential distance moved by P. Thus,

$$\boxed{\delta W_P = P\,\delta s = (PL)\,\delta\theta} \tag{3}$$

Then, from Eq. 11.2c,

$$\boxed{Q_\theta = PL} \tag{4}$$

Since θ is the generalized coordinate for this problem, Equation 1 has the form specified by Eq. 11.2b and Eq. 3 has the form of Eq. 11.2c.

Continuing now with the derivation of Lagrange's equations, substitute Eqs. 11.2 into 11.1 to give

$$\int_{t_1}^{t_2} \left(\frac{\partial T}{\partial q_1}\delta q_1 + \frac{\partial T}{\partial q_2}\delta q_2 + \ldots + \frac{\partial T}{\partial q_N}\delta q_n + \frac{\partial T}{\partial \dot{q}_1}\delta \dot{q}_1 + \frac{\partial T}{\partial \dot{q}_2}\delta \dot{q}_2 \right.$$
$$+ \ldots + \frac{\partial T}{\partial \dot{q}_N}\delta \dot{q}_N - \frac{\partial V}{\partial q_1}\delta q_1 - \frac{\partial V}{\partial q_2}\delta q_2 - \ldots - \frac{\partial V}{\partial q_N}\delta q_N + Q_1\,\delta q_1$$
$$\left. + Q_2\,\delta q_2 + \ldots + Q_N\,\delta q_N \right) dt = 0 \tag{11.3}$$

The terms involving $\delta\dot{q}_i$ may be integrated by parts in the following manner.

$$\int_{t_1}^{t_2} \frac{\partial T}{\partial \dot{q}_i}\delta \dot{q}_i\,dt = \left[\frac{\partial T}{\partial \dot{q}_i}\delta q_i \right|_{t_1}^{t_2} - \int_{t_1}^{t_2} \frac{d}{dt}\left(\frac{\partial T}{\partial \dot{q}_i}\right)\delta q_i\,dt \tag{11.4}$$

The first term on the right-hand side of the above equation is zero for each coordinate, since $\delta q_i(t_1) = \delta q_i(t_2) = 0$ is a basic condition imposed to make Hamilton's principle valid. When Eq. 11.4 is substituted into Eq. 11.3, the resulting equation can be written in the form:

$$\int_{t_1}^{t_2} \left\{ \sum_{i=1}^{N} \left[-\frac{d}{dt}\left(\frac{\partial T}{\partial \dot{q}_i}\right) + \frac{\partial T}{\partial q_i} - \frac{\partial V}{\partial q_i} + Q_i \right] \delta q_i \right\} dt = 0 \tag{11.5}$$

Since the variations δq_i ($i = 1, 2, \ldots, N$) must be independent, Eq. 11.5 can be satisfied in general only when the bracketed expression in Eq. 11.5 vanishes for each value of i, that is, when

$$\frac{d}{dt}\left(\frac{\partial T}{\partial \dot{q}_i}\right) - \frac{\partial T}{\partial q_i} + \frac{\partial V}{\partial q_i} = Q_i, \qquad i = 1, 2, \ldots, N \tag{11.6}$$

Equations 11.6 are known as Lagrange's equations.

The restrictions imposed in deriving Eqs. 11.6 were that the coordinates q_i be independent and that T, V, and δW_{nc} have the forms shown in Eqs. 11.2. Thus, Eqs. 11.6 are valid for nonlinear systems as well as linear systems. An extension of Lagrange's equations to systems represented by constrained (i.e., not independent) coordinates will be presented in Sec. 11.5.

11.3 Application of Lagrange's Equations to Lumped-Parameter Models

The following examples will illustrate the application of Lagrange's equations to systems represented by lumped-parameter models.

Example 11.6

Symmetrical vibration of an airplane wing-body combination is modeled by a "fuselage" M to which are attached "wing" masses m by rigid beams of length L. The elastic behavior of the wings is modeled by the torsional springs of constant k which are connected between the fuselage and wings. Use Lagrange's equations to derive the equations of motion of this system. Assume small θ. Neglect gravity.

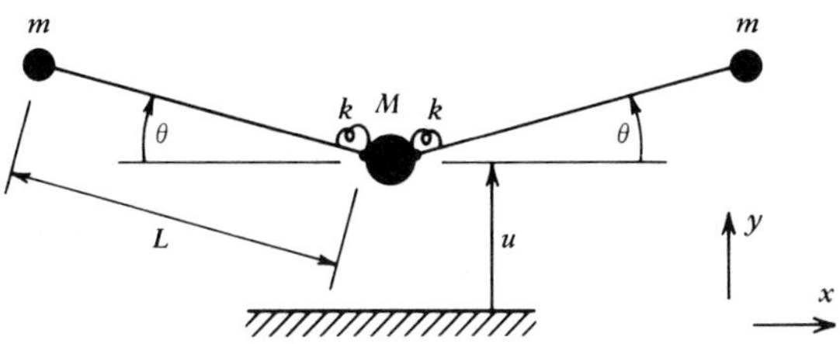

Solution

Let $q_1 \to u$, $q_2 \to \theta$.

a. Write expressions for T and V:

$$T = 2\,[\tfrac{1}{2}m\dot{y}_m^2] + \tfrac{1}{2}M\dot{u}^2 \tag{1}$$

For small θ,

$$y_m \doteq u + L\theta \tag{2}$$
$$T = m(\dot{u} + L\dot{\theta})^2 + \tfrac{1}{2}M\dot{u}^2 \tag{3}$$
$$V = 2[\tfrac{1}{2}k\theta^2] \tag{4}$$

b. Apply Lagrange's equations, Eqs. 11.6:

$$\begin{aligned}
\frac{\partial T}{\partial \dot{u}} &= 2m(\dot{u} + L\dot{\theta}) + M\dot{u}\\
\frac{\partial T}{\partial \dot{\theta}} &= 2mL(\dot{u} + L\dot{\theta})\\
\frac{\partial T}{\partial u} &= \frac{\partial T}{\partial \theta} = 0\\
\frac{\partial V}{\partial u} &= 0, \qquad \frac{\partial V}{\partial \theta} = 2k\theta\\
Q_u &= Q_\theta = 0
\end{aligned} \tag{5}$$

$$\frac{d}{dt}\left(\frac{\partial T}{\partial \dot{q}_i}\right) - \frac{\partial T}{\partial q_i} + \frac{\partial V}{\partial q_i} = Q_i \tag{6}$$

Therefore,

$$2m(\ddot{u} + L\ddot{\theta}) + M\ddot{u} = 0 \tag{7a}$$
$$2mL(\ddot{u} + L\ddot{\theta}) + 2k\theta = 0 \tag{7b}$$

Equations 7 can be written in matrix form as

$$\begin{bmatrix} (M + 2m) & 2mL \\ 2mL & 2mL^2 \end{bmatrix} \begin{Bmatrix} \ddot{u} \\ \ddot{\theta} \end{Bmatrix} + \begin{bmatrix} 0 & 0 \\ 0 & 2k \end{bmatrix} \begin{Bmatrix} u \\ \theta \end{Bmatrix} = \begin{Bmatrix} 0 \\ 0 \end{Bmatrix} \tag{8}$$

Note that the above equations of motion were obtained without the necessity of employing freebody diagrams and interaction forces between the wings and fuselage.

Example 11.7

Use Lagrange's equations to derive the equations of motion of the system shown in Example 11.3.

Solution

Let $q_1 \to u$, $q_2 \to \theta$.

a. Write expressions for T, V, and δW_{nc}:

$$T = \tfrac{1}{2}m\dot{u}^2 + \tfrac{1}{2}M(\dot{x}_G^2 + \dot{y}_G^2) + \tfrac{1}{2}I_G\dot{\theta}^2 \tag{1}$$

From kinematics

$$x_G = u + a \sin \theta \tag{2a}$$
$$y_G = a \cos \theta \tag{2b}$$

For small θ, $\sin \theta \doteq \theta$, $\cos \theta \doteq 1 - \frac{1}{2}\theta^2$. Therefore,

$$\dot{x}_G \doteq \dot{u} + a\dot{\theta} \tag{3a}$$
$$\dot{y}_G \doteq 0 \tag{3b}$$

where the nonlinear $\theta\dot{\theta}$ term in $\dot{y}_G$ has been neglected. So

$$T = \tfrac{1}{2}m\dot{u}^2 + \tfrac{1}{2}M(\dot{u} + a\dot{\theta})^2 + \tfrac{1}{2}I_G\dot{\theta}^2 \tag{4}$$

The potential energy is stored as elastic strain energy in the springs and as gravitational potential energy.

$$V = 2\left[\frac{1}{2}\left(\frac{k}{2}\right)(u - z)^2\right] + \frac{1}{2}K\theta^2 + Mga\cos\theta \tag{5}$$

For small θ,

$$V = 2\left[\frac{1}{2}\left(\frac{k}{2}\right)(u - z)^2\right] + \frac{1}{2}K\theta^2 + Mga\left(1 - \frac{1}{2}\theta^2\right) \tag{6}$$

The nonconservative forces in this problem are the damping forces exerted on the foundation mass. These forces are shown in the sketch below. (*Note:* This is not a freebody diagram! It is just a sketch to show damping forces only.)

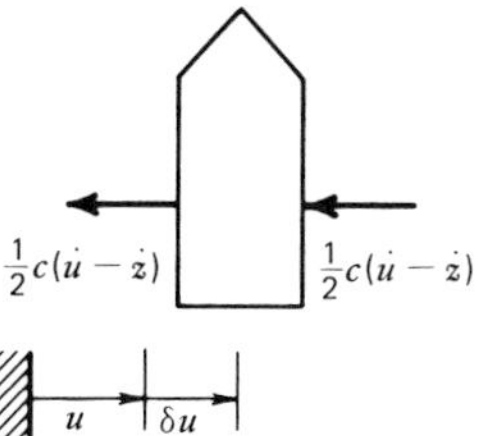

$$\delta W_{nc} = -2[\tfrac{1}{2}c(\dot{u} - \dot{z})]\,\delta u \tag{7}$$

b. Apply Lagrange's equations:

$$\frac{\partial T}{\partial \dot{u}} = m\dot{u} + M(\dot{u} + a\dot{\theta})$$
$$\frac{\partial T}{\partial \dot{\theta}} = Ma(\dot{u} + a\dot{\theta}) + I_G\dot{\theta}$$

$$\frac{\partial T}{\partial u} = \frac{\partial T}{\partial \theta} = 0$$
$$\frac{\partial V}{\partial u} = k(u - z), \qquad \frac{\partial V}{\partial \theta} = K\theta - Mga\theta \tag{8}$$
$$Q_u = -c(\dot{u} - \dot{z}), \qquad Q_\theta = 0$$
$$\frac{d}{dt}\left(\frac{\partial T}{\partial \dot{q}_i}\right) - \frac{\partial T}{\partial q_i} + \frac{\partial V}{\partial q_i} = Q_i \tag{9}$$

Therefore,

$$(M + m)\ddot{u} + Ma\ddot{\theta} + k(u - z) = -c(\dot{u} - \dot{z}) \tag{10a}$$
$$Ma\ddot{u} + (Ma^2 + I_G)\ddot{\theta} + (K - mga)\theta = 0 \tag{10b}$$

It may be observed that Eqs. 10 are the same as Eqs. 5 of Example 11.3.

As a final application of Lagrange's equations to lumped-parameter systems consider the following nonlinear problem.

Example 11.8

A particle of mass m slides along a uniform thin rod of mass M and length L. Use Lagrange's equations to derive the equations of motion of the system.

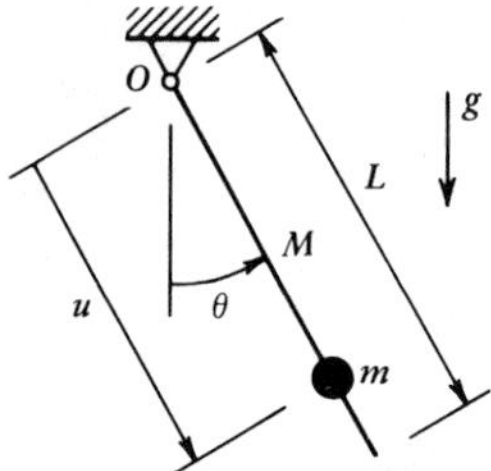

Solution

a. Write expressions for T and V (The system has no nonconservative forces): From Example 11.4, the kinetic energy of m can be obtained. Then,

$$T = \tfrac{1}{2}(\tfrac{1}{3}ML^2)\dot{\theta}^2 + \tfrac{1}{2}m(\dot{u}^2 + u^2\dot{\theta}^2) \tag{1}$$
$$V = -\left[Mg\left(\frac{L}{2}\right) + mgu\right]\cos\theta \tag{2}$$

b. Apply Lagrange's equations:

$$\begin{aligned}
\frac{\partial T}{\partial \dot{u}} &= m\dot{u} \\
\frac{\partial T}{\partial \dot{\theta}} &= \frac{1}{3} ML^2\dot{\theta} + mu^2\dot{\theta} \\
\frac{\partial T}{\partial u} &= mu\dot{\theta}^2 \\
\frac{\partial T}{\partial \theta} &= 0 \\
\frac{\partial V}{\partial u} &= -mg \cos \theta \\
\frac{\partial V}{\partial \theta} &= \left[Mg\left(\frac{L}{2}\right) + mgu \right] \sin \theta \\
Q_u &= Q_\theta = 0
\end{aligned} \tag{3}$$

$$\frac{d}{dt}\left(\frac{\partial T}{\partial \dot{q}_i}\right) - \frac{\partial T}{\partial q_i} + \frac{\partial V}{\partial q_i} = Q_i \tag{4}$$

Therefore,

$$m\ddot{u} - mu\dot{\theta}^2 - mg\cos\theta = 0 \tag{5a}$$

$$\left(\frac{1}{3} ML^2 + mu^2\right)\ddot{\theta} + 2mu\dot{u}\dot{\theta} + \left[Mg\left(\frac{L}{2}\right) + mgu\right] \sin\theta = 0 \tag{5b}$$

Observe that these equations of motion are highly nonlinear.

11.4 Application of Lagrange's Equations to Continuous Models: The Assumed-Modes Method

In Sec. 2.4 the displacement of a continuous system was assumed to have the form

$$u(x, t) = \psi(x)u(t) \tag{11.7}$$

This assumption produced a single-DOF model of the continuous system with $u(t)$ being the generalized coordinate. A similar procedure led to the Rayleigh method, which was discussed in Sec. 10.3.

To generate an N-DOF model of a continuous system, Eq. 11.7 is expanded to include N functions, $\psi_i(x)$. Thus, $u(x, t)$ is approximated by

$$u(x, t) = \sum_{i=1}^{N} \psi_i(x)u_i(t) \tag{11.8}$$

The *assumed-modes method*[11.2] consists in substituting Eq. 11.8 into expressions for T, V, and δW_{nc} and then applying Lagrange's equations to derive equations of motion of the resulting N-DOF model. In this chapter the *global assumed-modes method* will be considered, that is, the functions $\psi_i(x)$ will each represent a displacement shape for the entire structure under consideration. In Chapter 16 the relationship of the assumed-modes method to the finite element method will be noted.

It is by choosing the functions $\psi_i(x)$ that the analyst defines the N-DOF model. The functions $\psi_i(x)$ must form a linearly independent set. In addition, each $\psi_i(x)$ must possess derivatives up to the order appearing in V and must satisfy all prescribed boundary conditions, that is, displacement-type boundary conditions. Functions that satisfy these conditions are called *admissible functions.*

For the propped cantilever beam in Fig. 11.2, each $\psi_i(x)$ must satisfy the boundary conditions

$$\psi_i(0) = \psi_i'(0) = \psi_i(L) = 0$$

since

$$v(0, t) = v'(0, t) = v(L, t) = 0$$

for all t. Since the strain energy expression for a Bernoulli-Euler beam contains $v''(x, t)$, each $\psi_i(x)$ must be a continuous function of x, and its first derivative with respect to x must be continuous (i.e., the beam can have no abrupt changes of displacement or slope).

It is not necessary that the ψ_i's also satisfy the *natural boundary conditions,* that is, force-type boundary conditions such as $M(L, t) = 0$ in Fig. 11.2. However, in cases where it is possible to obtain functions $\psi_i(x)$ that satisfy both prescribed and natural boundary conditions, such functions may be used in Eq. 11.8.

To introduce the assumed-modes method, we will apply it first to the problem of determining approximate solutions for the axial vibration of a linearly elastic bar. In this case, $u(x, t)$ represents the axial motion of the plane cross section at x, as shown in Fig. 11.3.

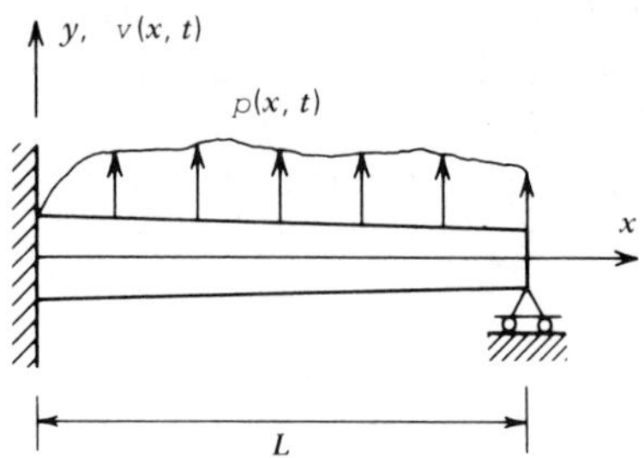

Figure 11.2. *Boundary condition example.*

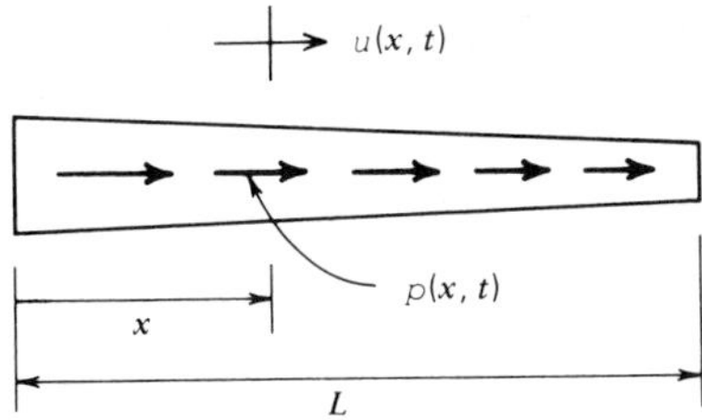

Figure 11.3. Axial motion of a thin bar.

The strain energy in the bar is given by

$$V = \frac{1}{2} \int_0^L EA(u')^2 \, dx \tag{11.9}$$

and the kinetic energy of the bar due to the axial displacement $u(x, t)$ is given by

$$T = \frac{1}{2} \int_0^L \rho A(\dot{u})^2 \, dx \tag{11.10}$$

where $\rho(x)$ is the mass per unit volume and $A(x)$ is the cross-sectional area.

When Eq. 11.8 is substituted into Eq. 11.9, the resulting expression for potential energy may be written in the form

$$V = \frac{1}{2} \sum_{i=1}^{N} \sum_{j=1}^{N} k_{ij} u_i u_j \tag{11.11}$$

where

$$k_{ij} = \int_0^L EA\psi_i'\psi_j' \, dx \tag{11.12}$$

In other words, V is a quadratic function of the generalized coordinates, with the coefficients in the quadratic expression being defined by Eq. 11.12. The quadratic expression may be written conveniently in matrix form, namely

$$V = \frac{1}{2} \mathbf{u}^T \mathbf{k} \mathbf{u} \tag{11.13}$$

where

$$\mathbf{u} = \begin{Bmatrix} u_1 \\ u_2 \\ \cdot \\ \cdot \\ \cdot \\ u_N \end{Bmatrix} \qquad \mathbf{k} = \begin{bmatrix} k_{11} & k_{12} & \cdots & k_{1N} \\ k_{21} & k_{22} & \cdots & k_{2N} \\ & & \cdot & \\ & & \cdot & \\ & & \cdot & \\ k_{N1} & k_{N2} & \cdots & k_{NN} \end{bmatrix} \tag{11.14}$$

In a similar manner, Eqs. 11.8 and 11.10 may be combined to give

$$T = \frac{1}{2} \sum_{i=1}^{N} \sum_{j=1}^{N} m_{ij} \dot{u}_i \dot{u}_j \tag{11.15}$$

where

$$m_{ij} = \int_0^L \rho A \psi_i \psi_j \, dx \tag{11.16}$$

In matrix form

$$T = \frac{1}{2} \dot{\mathbf{u}}^T \mathbf{m} \dot{\mathbf{u}} \tag{11.17}$$

For reasons explained in Sec. 16.2, when the mass matrix of a system is determined by using Eq. 11.16, the mass matrix is called a *consistent mass matrix*.

If the bar is subjected to external forces, as shown in Fig. 11.3, the corresponding generalized forces are determined by employing virtual work. Thus,

$$\delta W = \int_0^L p(x, t) \, \delta u(x, t) \, dx = \sum_{i=1}^{N} p_i \, \delta u_i \tag{11.18}$$

From Eq. 11.8, $\delta u(x, t)$ must be approximated by

$$\delta u(x, t) = \sum_{i=1}^{N} \psi_i(x) \, \delta u_i \tag{11.19}$$

Hence, upon combining Eqs. 11.18 and 11.19 we obtain

$$p_i(t) = \int_0^L p(x, t) \psi_i(x) \, dx \tag{11.20}$$

Lagrange's equations may now be used to determine the equations of motion of the N-DOF model defined by Eq. 11.8. When Eqs. 11.11 and 11.15 are substituted into Eqs. 11.6 we obtain

$$\sum_{j=1}^{N} m_{ij} \ddot{u}_j + \sum_{j=1}^{N} k_{ij} u_j = p_i, \qquad i = 1, 2, \ldots, N \tag{11.21}$$

which can be written in the familiar matrix form

$$\mathbf{m}\ddot{\mathbf{u}} + \mathbf{k}\mathbf{u} = \mathbf{p} \tag{11.22}$$

Although we have gone through Lagrange's equations to arrive at Eq. 11.22, it may be seen that the only steps that are actually required in using the assumed-modes method to arrive at the equations of motion of an N-DOF model are:

1. Select a set of N admissible functions, $\psi_i(x)$.

2. Compute the coefficients of the stiffness matrix by using Eq. 11.12.

3. Compute the coefficients of the mass matrix by using Eq. 11.16.

4. Compute the generalized forces corresponding to applied forces by using Eq. 11.20.

5. Form the equations of motion using Eq. 11.22.

This procedure will now be illustrated.

Example 11.9

Use the assumed-modes method with a polynomial approximation of $u(x, t)$ to obtain a 2-DOF model for axial vibration of a uniform cantilever bar subjected to an end force $P(t)$.

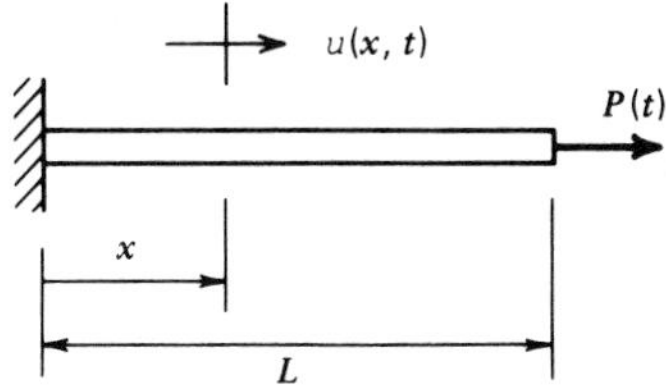

Solution

a. Select the shape functions, $\psi_i(x)$.
The only prescribed boundary condition is

$$u(0, t) = 0 \tag{1}$$

Thus, the functions $\psi_i(x)$ must satisfy

$$\psi_1(0) = \psi_2(0) = 0 \tag{2}$$

Therefore, let

$$\psi_1(x) = \frac{x}{L} \tag{3a}$$

$$\psi_2(x) = \left(\frac{x}{L}\right)^2 \tag{3b}$$

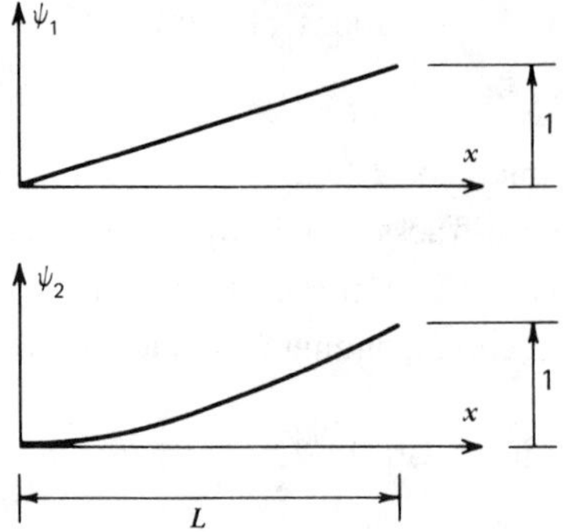

(It is convenient, although not essential, to nondimensionalize the ψ functions as has been done in Eqs. 3.)

b. Compute the stiffness coefficients, k_{ij}.
From Eqs. 3,

$$\psi_1' = \frac{1}{L} \tag{4a}$$

$$\psi_2' = \left(\frac{2}{L}\right)\left(\frac{x}{L}\right) \tag{4b}$$

Then, from Eq. 11.12,

$$k_{11} = \int_0^L EA(\psi_1')^2\, dx = \frac{EA}{L} \tag{5a}$$

$$k_{12} = k_{21} = \frac{EA}{L} \tag{5b}$$

$$k_{22} = \frac{4EA}{3L} \tag{5c}$$

Compute the mass coefficients, m_{ij}. From Eq. 11.16,

$$m_{11} = \int_0^L \rho A(\psi_1)^2\, dx = \frac{\rho AL}{3} \tag{6a}$$

$$m_{12} = m_{21} = \frac{\rho AL}{4} \tag{6b}$$

$$m_{22} = \frac{\rho AL}{5} \tag{6c}$$

c. Compute the generalized forces, p_i.
Equation 11.20 was derived for the distributed forces shown on Fig. 11.3. To determine the generalized forces corresponding to the concentrated end force, $P(t)$, we can go back to virtual work and write

$$\delta W = P\, \delta u(L, t) = p_1\, \delta u_1 + p_2\, \delta u_2 \tag{7}$$

But,

$$\delta u(L, t) = \psi_1(L)\,\delta u_1 + \psi_2(L)\,\delta u_2 \tag{8}$$

Therefore,

$$p_1 = P\psi_1(L) = P \tag{9a}$$
$$p_2 = P\psi_2(L) = P \tag{9b}$$

d. Assemble the equations of motion in matrix form.

$$\boxed{\rho AL \begin{bmatrix} \frac{1}{3} & \frac{1}{4} \\ \frac{1}{4} & \frac{1}{5} \end{bmatrix} \begin{Bmatrix} \ddot{u}_1 \\ \ddot{u}_2 \end{Bmatrix} + \frac{EA}{L} \begin{bmatrix} 1 & 1 \\ 1 & \frac{4}{3} \end{bmatrix} \begin{Bmatrix} u_1 \\ u_2 \end{Bmatrix} = \begin{Bmatrix} P(t) \\ P(t) \end{Bmatrix}} \tag{10}$$

In the same manner, the assumed-modes method may be used to derive the equations of motion of other elastic systems. For the Bernoulli-Euler beam the strain energy is given by

$$V = \frac{1}{2}\int_0^L EI(v'')^2\,dx \tag{11.23}$$

where $v(x, t)$ refers to transverse motion, as in Fig. 11.2, and where EI may be a function of x. The kinetic energy is given by

$$T = \frac{1}{2}\int_0^L \rho A(\dot{v})^2\,dx \tag{11.24}$$

Equations 11.23 and 11.24 lead to expressions for stiffness and mass coefficients of the following form

$$k_{ij} = \int_0^L EI\psi_i''(x)\psi_j''(x)\,dx \tag{11.25}$$

$$m_{ij} = \int_0^L \rho A\psi_i(x)\psi_j(x)\,dx \tag{11.26}$$

The generalized forces due to a distributed transverse force, $p(x, t)$, as shown in Fig. 11.2, are given by

$$p_i = \int_0^L p(x, t)\psi_i(x)\,dx \tag{11.27}$$

Example 11.10

An offshore drilling platform is modeled as a uniform flexible beam of length L with a lumped mass M at the top and a rotational spring k at the bottom. Use the assumed-modes method to derive the equations of motion of a 2-DOF model of this structure. Assume small rotation of the beam at $x = 0$.

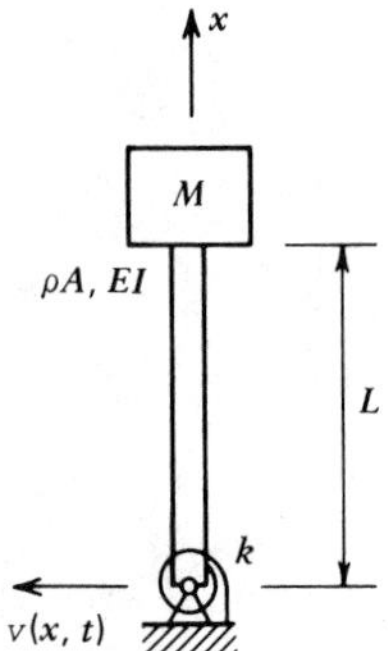

Solution

a. Choose the shape functions, $\psi_i(x)$.
The only prescribed boundary condition is

$$v(0, t) = 0 \tag{1}$$

Thus, the functions $\psi_i(x)$ must satisfy

$$\psi_1(0) = \psi_2(0) = 0 \tag{2}$$

The simplest functions to use are terms of a polynomial in x. Therefore, as in Example 11.9, choose

$$\psi_1 = \frac{x}{L} \tag{3a}$$

$$\psi_2 = \left(\frac{x}{L}\right)^2 \tag{3b}$$

b. Compute the stiffness coefficients, k_{ij}.
When the beam is deflected as shown, there is potential energy stored in the base spring as well as in the beam. Equation 11.23 must be supplemented by the potential energy in the spring. Thus,

$$V = \int_0^L EI(v'')^2\, dx + \frac{1}{2} k\theta_o^2 \tag{4}$$

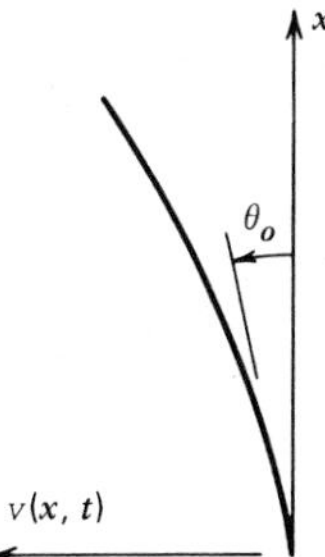

For small rotations at the base,

$$\theta_o \doteq v'(0, t) = \sum_{i=1}^{N} \psi_i'(0)\, v_i(t) \tag{5}$$

When Eq. 5 is substituted into the second term of Eq. 4, we see that Eq. 11.25 is modified to the form

$$k_{ij} = \int_0^L EI\psi_i''\psi_j''\, dx + k\psi_i'(0)\psi_j'(0) \tag{6}$$

From Eqs. 3,

$$\psi_1' = \frac{1}{L}, \qquad \psi_1'' = 0 \tag{7a}$$

$$\psi_2' = \frac{2}{L}\left(\frac{x}{L}\right), \qquad \psi_2'' = \frac{2}{L^2} \tag{7b}$$

Therefore,

$$k_{11} = 0 + k\left(\frac{1}{L}\right)\left(\frac{1}{L}\right) = \frac{k}{L^2} \tag{8a}$$

$$k_{12} = k_{21} = 0 + 0 = 0 \tag{8b}$$

$$k_{22} = \frac{4EI}{L^3} + 0 = \frac{4EI}{L^3} \tag{8c}$$

c. Compute the mass coefficients, m_{ij}.

Due to the presence of the mass M at $x = L$, the expression for T in Eq. 11.24 must be modified, which results in changing Eq. 11.26 to

$$m_{ij} = \int_0^L \rho A\psi_i\psi_j\, dx + M\psi_i(L)\psi_j(L) \tag{9}$$

Thus,

$$m_{11} = \rho A \int_0^L \left(\frac{x}{L}\right)^2 dx + M(1)(1)$$

or

$$m_{11} = \frac{\rho AL}{3} + M \tag{10a}$$

Similarly,

$$m_{12} = m_{21} = \frac{\rho AL}{4} + M \tag{10b}$$

$$m_{22} = \frac{\rho AL}{5} + M \tag{10c}$$

d. Assemble the equations of motion.
There are no external forces, so $p_1 = p_2 = 0$.

$$\begin{bmatrix} \left(M + \frac{\rho AL}{3}\right) & \left(M + \frac{\rho AL}{4}\right) \\ \left(M + \frac{\rho AL}{4}\right) & \left(M + \frac{\rho AL}{5}\right) \end{bmatrix} \begin{Bmatrix} \ddot{q}_1 \\ \ddot{q}_2 \end{Bmatrix} + \begin{bmatrix} \left(\frac{k}{L^2}\right) & 0 \\ 0 & \left(\frac{4EI}{L^3}\right) \end{bmatrix} \begin{Bmatrix} q_1 \\ q_2 \end{Bmatrix} = \begin{Bmatrix} 0 \\ 0 \end{Bmatrix} \tag{11}$$

In a situation like Example 11.10, where a member is subjected to axial loading (e.g., due to the weight of M) and also undergoes transverse deflection, the axial load has an effect on the stiffness of the member. An expression for the geometric stiffness coefficient for a 1-DOF model was given in Eq. 2.28. For an N-DOF system the corresponding expression is

$$k_{G_{ij}} = \int_0^L N(x)\psi_i'(x)\psi_j'(x)\,dx \tag{11.28}$$

Example 11.11

The damping effect of soil on the vertical motion of a foundation is modeled by distributed damping such that a damping force proportional to velocity is exerted on the foundation, that is,

$$p(x, t) = -\xi(x)\dot{v}(x, t)$$

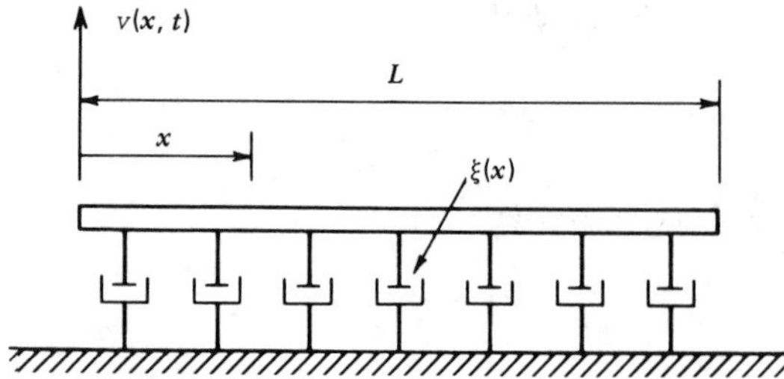

where $\xi(x)$ is the distributed damping coefficient. Determine an expression for the generalized force p_i corresponding to this damping.

Solution

Equation 11.27 can be used directly. Thus,

$$p_i = \int_0^L p(x, t)\psi_i(x)\, dx$$
$$= \int_0^L \left[-\xi(x) \sum_{j=1}^{N} \psi_j(x)\dot{v}_j(t) \right] \psi_i(x)\, dx$$

or

$$\boxed{p_i = -\sum_{j=1}^{N} \dot{v}_j \left[\int_0^L \xi(x)\psi_i(x)\psi_j(x)\, dx \right]}$$

When a system is influenced by viscous damping, as in Example 11.11, it is possible to write the generalized forces corresponding to this as

$$p_i = -\sum_{j=1}^{N} c_{ij}\dot{v}_j \tag{11.29}$$

Since these generalized forces depend on the unknown generalized velocities, the p_i due to viscous damping may be placed on the left-hand side of Eq. 11.21. Then Eq. 11.22 may be written

$$\mathbf{m}\ddot{\mathbf{q}} + \mathbf{c}\dot{\mathbf{q}} + \mathbf{k}\mathbf{q} = \mathbf{Q} \tag{11.30}$$

where $\mathbf{c}$ is called the *damping matrix*.

From a practical standpoint, the damping properties of a system are seldom known in the same way that the inertia and stiffness properties of the system are known, so it is not practical to seek the coefficients of the damping matrix $\mathbf{c}$. Other procedures for treating damping will be introduced in Chapter 15 and in Chapter 18.

11.5 Constrained Coordinates and Lagrange Multipliers

In Sec. 11.2 Lagrange's equations were derived using a set of N generalized coordinates, $q_1, q_2, \ldots, q_N$. A crucial step in the derivation, from Eq. 11.5 to 11.6, required that the coordinates be independent. Occasionally it is desirable to employ a set of coordinates that are not independent. Let these be denoted by $g_1, g_2, \ldots, g_M$, where $M > N$. Associated with this set of constrained coordinates will be a set of $C = (M - N)$ constraint equations. Although texts on advanced dynamics[11.3,11.4] handle the case of quite general constraints, it will suffice here for us to consider constraint equations that involve only the coordinates g_i (and not derivatives). Let these constraint equations be written in the form

$$f_j(g_1, g_2, \ldots, g_M) = 0, \qquad j = 1, 2, \ldots, C \tag{11.31}$$

Let each coordinate g_r be given a variation δg_r. Then

$$\delta f_j = \frac{\partial f_j}{\partial g_1}\delta g_1 + \frac{\partial f_j}{\partial g_2}\delta g_2 + \ldots + \frac{\partial f_j}{\partial g_M}\delta g_M = 0 \tag{11.32}$$

or

$$\sum_{i=1}^{M} \frac{\partial f_j}{\partial g_i}\delta g_i = 0, \qquad j = 1, 2, \ldots, C \tag{11.33}$$

Thus, the δg's are not independent but are related to each other by the C equations of Eq. 11.33.

Now let us return to Hamilton's principle by extending Eq. 11.5 to include the M coordinates, that is,

$$\int_{t_1}^{t_2} \left\{ \sum_{i=1}^{M} \left[-\frac{d}{dt}\left(\frac{\partial T}{\partial \dot{g}_i}\right) + \frac{\partial T}{\partial g_i} - \frac{\partial V}{\partial g_i} + Q_i \right] \delta g_i \right\} dt = 0 \tag{11.34}$$

Since the δg_i are not independent, we cannot just set the expression in square brackets to zero as we did in Eq. 11.6. At this point, however, we can introduce Lagrange multiplier functions, or *Lagrange multipliers*, $\lambda_j(t)$, $j = 1, 2, \ldots, C$. Multiply each of the C equations in Eqs. 11.33 by a corresponding Lagrange multiplier, λ_j, and sum these, that is

$$\sum_{j=1}^{C} \lambda_j \sum_{i=1}^{M} \frac{\partial f_j}{\partial g_i}\delta g_i = 0 \tag{11.35}$$

Since this sum is still equal to zero, it may be inserted into Eq. 11.34 to give

$$\int_{t_1}^{t_2} \left\{ \sum_{i=1}^{M} \left[-\frac{d}{dt}\left(\frac{\partial T}{\partial \dot{g}_i}\right) + \frac{\partial T}{\partial g_i} - \frac{\partial V}{\partial g_i} + Q_i + \sum_{j=1}^{C} \lambda_j \frac{\partial f_j}{\partial g_i} \right] \delta g_i \right\} dt = 0 \tag{11.36}$$

While the δg_i are still not independent, we can choose the Lagrange multipliers λ_j so as to make the bracketed expressions for δg_i ($i = 1, 2, \ldots, C$) equal to zero. The remaining $N = M - C$ coordinates can be considered to be independent, so the expression in square brackets must also vanish for δg_i ($i = C + 1, \ldots, M$). Thus, the bracketed expression must vanish for all g_i's, giving the following modified Lagrange equations.

$$\frac{d}{dt}\left(\frac{\partial T}{\partial \dot{g}_i}\right) - \frac{\partial T}{\partial g_i} + \frac{\partial V}{\partial g_i} - \sum_{j=1}^{C} \lambda_j \frac{\partial f_j}{\partial g_i} = Q_i, \; i = 1, 2, \ldots, M \tag{11.37}$$

Alternatively, Lagrange's equations may be written in the form

$$\frac{d}{dt}\left(\frac{\partial T}{\partial \dot{g}_i}\right) - \frac{\partial T}{\partial g_i} + \frac{\partial V^*}{\partial g_i} = Q_i, \qquad i = 1, 2, \ldots, M \tag{11.38}$$

where V^* is a modified potential energy function given by

$$V^* = V - \sum_{j=1}^{C} \lambda_j f_j \tag{11.39}$$

In conclusion, it can be seen that Eqs. 11.31 and 11.38 provide a set of $M + C$ equations in the $M + C$ unknowns g_i and λ_j.

A very simple example of the use of Lagrange multipliers will now be given. A more sophisticated example is given in Chapter 19.

Example 11.12

Use the assumed-modes method with Lagrange multipliers to derive the equation of motion for axial free vibration of a fixed-fixed bar. Use

$$u(x, t) = \left(\frac{x}{L}\right) g_1 + \left(\frac{x}{L}\right)^2 g_2$$

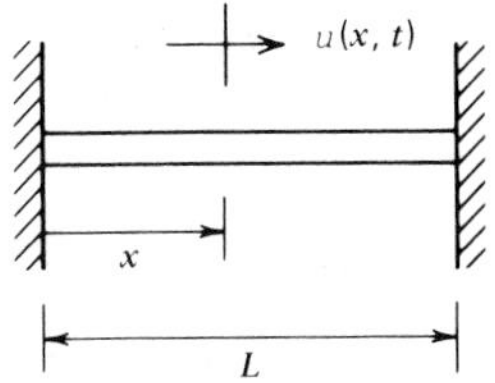

Solution

a. Establish the constraint equation.
The prescribed boundary conditions for the bar are

$$u(0, t) = u(L, t) = 0 \tag{1}$$

The first of these is satisfied by the choice of the shape functions (x/L) and $(x/L)^2$. Then, we must enforce the other boundary condition by setting

$$f(g_1, g_2) \equiv u(L, t) = g_1 + g_2 = 0 \tag{2}$$

This is the constraint equation in the form shown in Eq. 11.31.

b. Determine the mass and stiffness coefficients.

Since the structure is the same as in Example 11.9 and the two shape functions are the same, we can use the k_{ij} and m_{ij} derived there.

$$\mathbf{k} = \left(\frac{EA}{L}\right)\begin{bmatrix} 1 & 1 \\ 1 & \frac{4}{3} \end{bmatrix}, \qquad \mathbf{m} = \rho AL \begin{bmatrix} \frac{1}{3} & \frac{1}{4} \\ \frac{1}{4} & \frac{1}{5} \end{bmatrix} \tag{3}$$

c. Use Eq. 11.37 to obtain the equations of motion.

Since there is only one constraint equation, we need only to append $\lambda(\partial f/\partial g_1)$ and $\lambda(\partial f/\partial g_2)$ to their respective equations. But, from Eq. 2,

$$\frac{\partial f}{\partial g_1} = \frac{\partial f}{\partial g_2} = 1 \tag{4}$$

Therefore, the equations of motion become

$$\rho AL \begin{bmatrix} \frac{1}{3} & \frac{1}{4} \\ \frac{1}{4} & \frac{1}{5} \end{bmatrix} \begin{Bmatrix} \ddot{g}_1 \\ \ddot{g}_2 \end{Bmatrix} + \left(\frac{EA}{L}\right)\begin{bmatrix} 1 & 1 \\ 1 & \frac{4}{3} \end{bmatrix} \begin{Bmatrix} g_1 \\ g_2 \end{Bmatrix} - \begin{Bmatrix} \lambda \\ \lambda \end{Bmatrix} = \begin{Bmatrix} 0 \\ 0 \end{Bmatrix} \tag{5}$$

Equations 2 and 5 are thus to be solved for g_1, g_2, and λ.

References

11.1 H. L. Langhaar, *Energy Methods in Applied Mechanics,* Wiley, New York (1962).

11.2 L. Meirovitch, *Analytical Methods in Vibrations,* Macmillan, New York (1967).

11.3 D. T. Greenwood, *Principles of Dynamics,* Prentice-Hall, Englewood Cliffs, NJ (1965).

11.4 L. Meirovitch, *Methods of Analytical Dynamics,* McGraw-Hill, New York (1970).

Problem Set 11.1

Use Newton's laws (with extension to rigid bodies where required) to determine the equations of motion of the MDOF systems below. Show all necessary freebody diagrams for each problem.

11.1 Use as coordinates the absolute motion of mass 1 and the motion of mass 2 relative to mass 1.

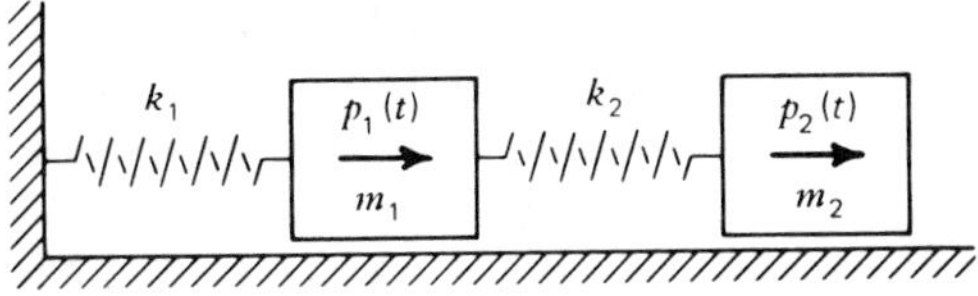

Figure P11.1

11.2 The motion of an airplane wing section is modeled by the 2-DOF system shown below. Use the vertical displacement of C and the angle of rotation as coordinates. Assume small angle of rotation. Neglect gravity.

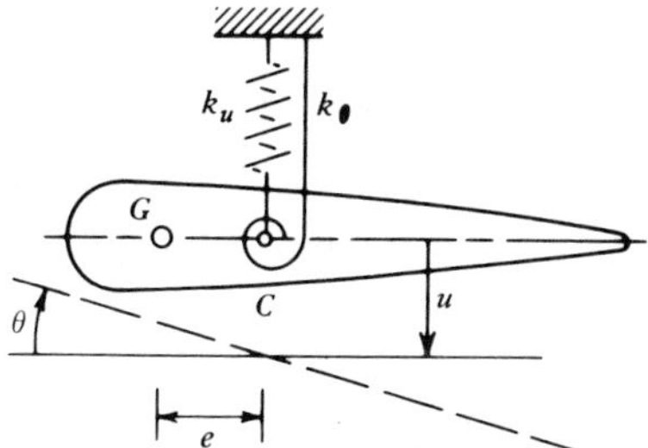

Figure P11.2

11.3 Use as coordinates for the system below the vertical displacements of the two masses. Assume that the rotation of the bar remains small. Neglect gravity.

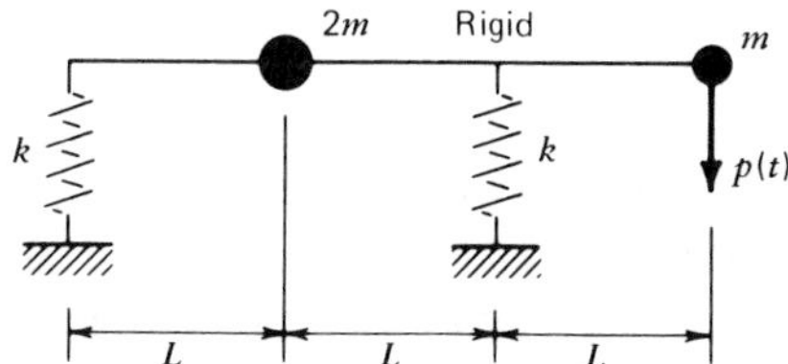

Figure P11.3

11.4 A three-story shear building is subjected to earthquake motion in addition to wind forces. Derive the equations of motion using the displacements of the concentrated masses relative to the ground as coordinates.

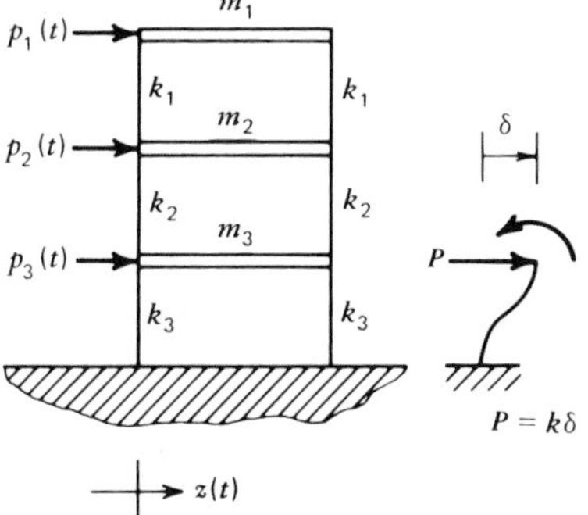

Figure P11.4

11.5 A vehicle is modeled as a rigid body *AB* having mass M and mass moment of inertia I_G about its mass center. The mass of the axles and wheels is modeled by lumped masses m, the stiffness of the springs by k_1, and the stiffness of the tires by k_2. The shock absorbers are modeled by viscous dashpots as shown. Use as coordinates the vertical displacements of the smaller masses and the displacements at *A* and *B*. Assume a small angle of rotation of the rigid body *AB*.

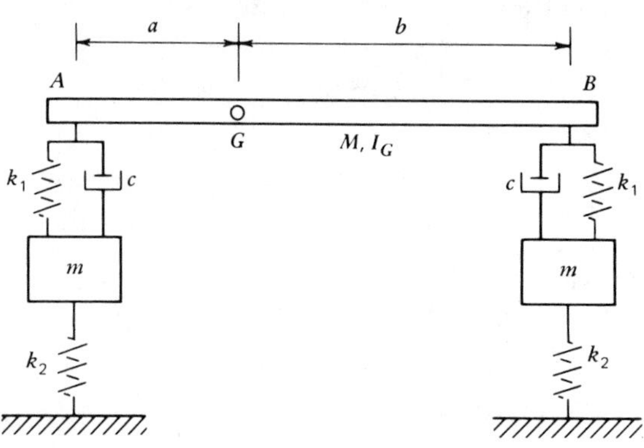

Figure P11.5

11.6 The rotating unbalance of a machine causes it to move vertically on its spring supports. The mass of the machine is m_1 and the rotating unbalance has mass m. A mass m_2 attached to the machine mass m_1 by a spring k_2 can be used as a *vibration absorber*. First determine the form of the vertical force $P(t)$ exerted at C by the arm of the rotating unbalance if the arm rotates counterclockwise at Ω rad/s as indicated. Then determine the equations of motion for m_1 and m_2. Neglect gravity.

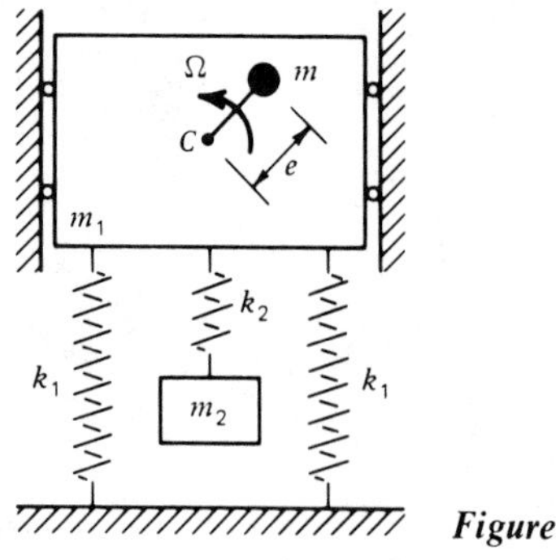

Figure P11.6

11.7 A coupled pendulum system is modeled by two lumped masses attached to rigid weightless rods, which are connected by a spring whose stiffness is k. Use the angles θ_1 and θ_2 as coordinates.

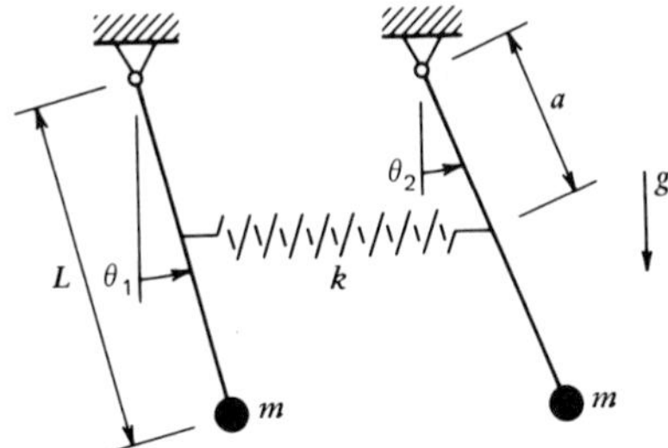

Figure P11.7

11.8 The 2-DOF system below is a model of a high-speed monorail train whose two cars are coupled by a spring and dashpot. Use as coordinates the absolute motion of m_1 and the motion of m_2 relative to m_1.

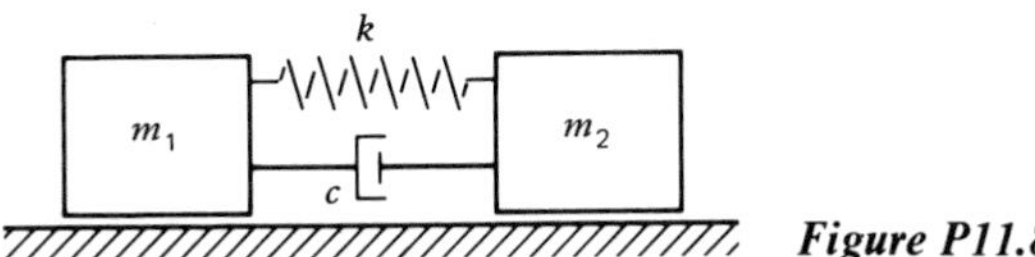

Figure P11.8

11.9 Two gears are attached to a shaft as shown. Determine the equations of motion using the absolute angles of rotation of the gears as coordinates.

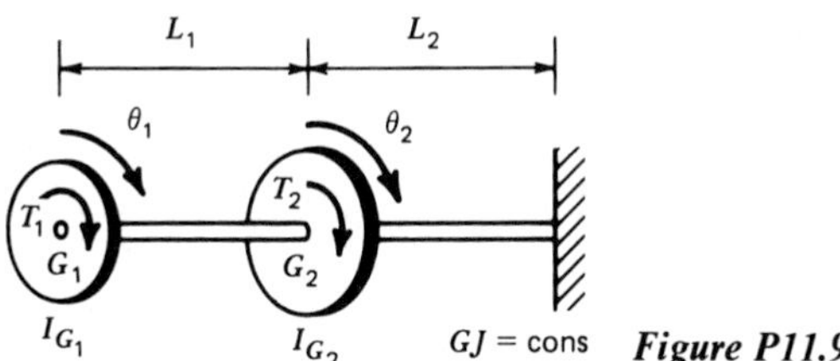

Figure P11.9

11.10 A restaurant and observation deck are situated atop a tall tower. The system is modeled as a lumped mass m supported by a rigid, massless column of length L. The elasticity of the tower and of the soil are modeled by springs as shown. The base is subjected to horizontal excitation, $z(t)$. Include the gravitational effect of the "restaurant" mass, and employ as coordinates the horizontal motion of the "foundation" and the angle the column makes with the vertical. Assume that the angle remains small.

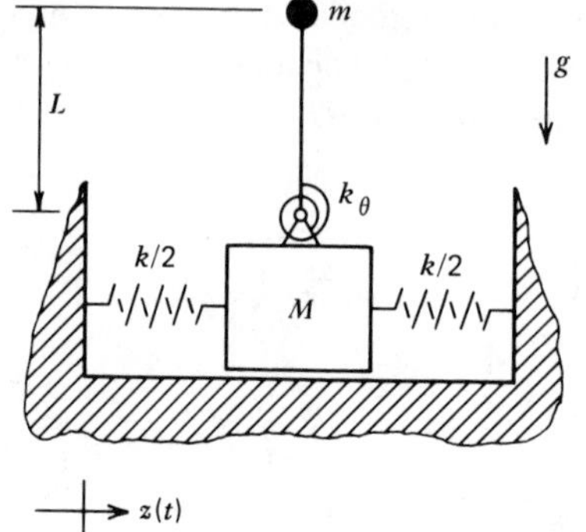

Figure P11.10

Problem Set 11.3

Use Lagrange's equations to derive the equations of motion of the following systems.

11.11 The system described in Problem 11.1.

11.12 See Problem 11.2.

11.13 See Problem 11.3.

11.14 See Problem 11.4.

11.15 See Problem 11.5.

11.16 See Problem 11.8.

11.17 See Problem 11.9.

11.18 See Problem 11.10.

11.19 Assume small rotations of the two rigid bars.

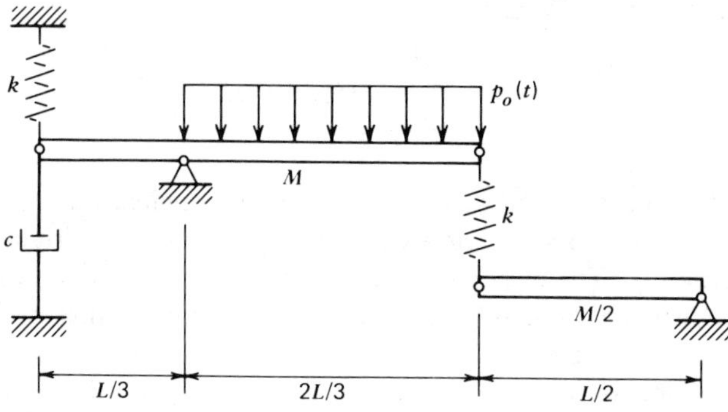

Figure P11.19

11.20 A double pendulum is modeled by lumped masses and massless, rigid rods as shown. Neglect friction. $P(t)$ remains horizontal. Use θ_1 and θ_2 as the system generalized coordinates. θ_1 and θ_2 are not necessarily small.

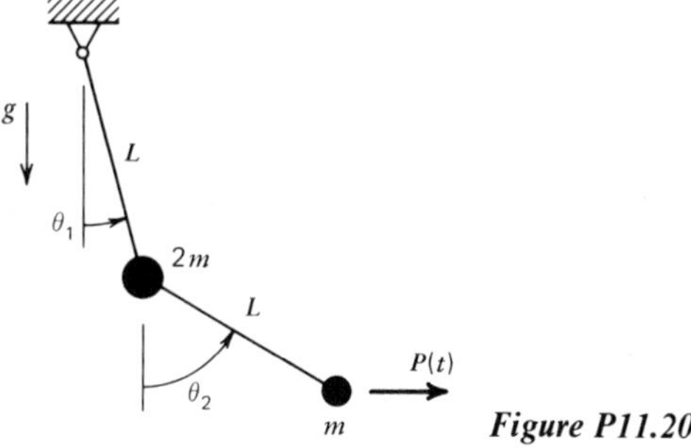

Figure P11.20

Problem Set 11.4

11.21 A uniform bar is attached at both ends to rigid supports.

(a) Show that the following shape functions satisfy the prescribed boundary conditions.

$$\psi_1(x) = \sin\left(\frac{\pi x}{L}\right)$$

$$\psi_2(x) = \sin\left(\frac{2\pi x}{L}\right)$$

(b) Derive the equations of motion for a 2-DOF assumed-modes model based on the above shape functions.

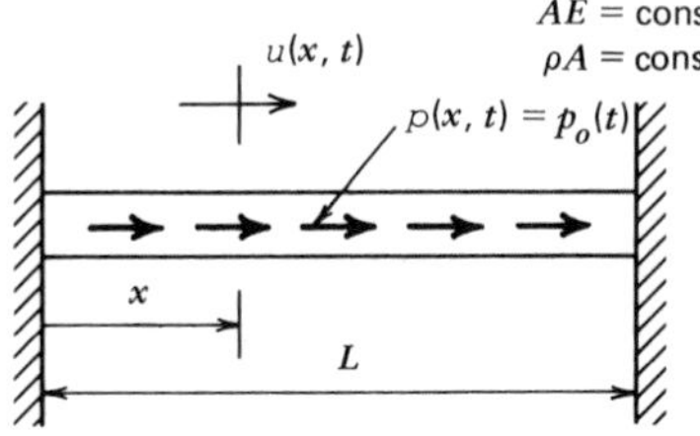

Figure P11.21

11.22 For the clamped-clamped bar in Fig. P11.21 determine two shape functions $\psi_1(x)$ and $\psi_2(x)$ based on a polynomial in x of the form

$$\psi(x) = a + b\left(\frac{x}{L}\right) + c\left(\frac{x}{L}\right)^2 + d\left(\frac{x}{L}\right)^3$$

and the two prescribed boundary conditions.

11.23 A uniform floor beam has a total mass M and supports a heavy machine of mass m at its center. Use the shape functions

$$\psi_1(x) = \sin\left(\frac{\pi x}{L}\right)$$

$$\psi_2(x) = \sin\left(\frac{2\pi x}{L}\right)$$

to derive a 2-DOF assumed-modes model of this system.

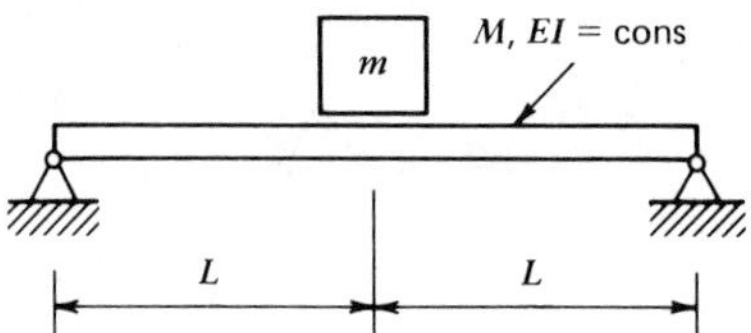

Figure P11.23

11.24 A uniform beam is subjected to a constant horizontal compressive force N as shown. Using a 2-DOF assumed-modes model based on a polynomial in x, determine the equations of motion for transverse vibration.

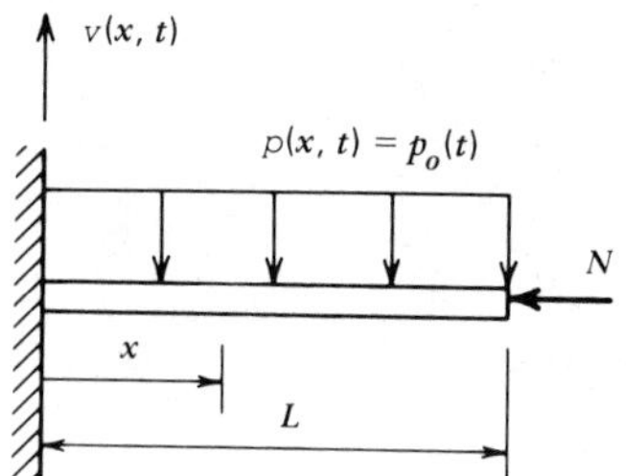

Figure P11.24

11.25 A tapered shaft with circular cross section is to be represented by a 2-DOF assumed-modes model based on the shape functions

$$\psi_1(x) = \left[1 - \left(\frac{x}{L}\right)\right]$$

$$\psi_2(x) = \left[1 - \left(\frac{x}{L}\right)^2\right]$$

Derive the two equations of motion for rotation of the shaft. The diameter is given by $d(x) = a\left(\frac{x}{L}\right)$.

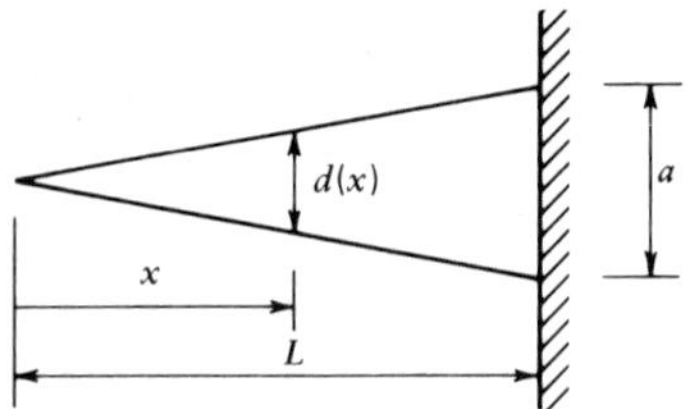

Figure P11.25

11.26 A cantilever beam is to be modeled by a 2-DOF assumed-modes model for which the generalized coordinates can be interpreted as the deflection and slope (small) at the free end, that is, $v(t)$ and $\theta(t)$. The corresponding shape functions should have the shapes shown below.

(a) Derive polynomial type shape functions $\psi_1(x)$ and $\psi_2(x)$ based on the general polynomial

$$\psi(x) = a + b\left(\frac{x}{L}\right) + c\left(\frac{x}{L}\right)^2 + d\left(\frac{x}{L}\right)^3$$

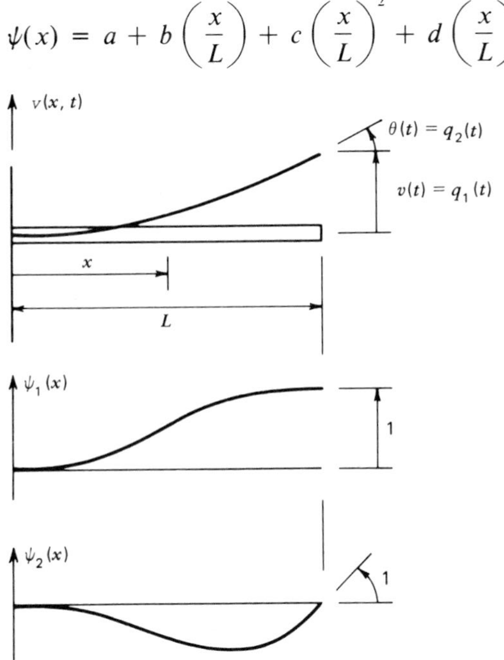

Figure P11.26

(b) Derive the equations of motion for this 2-DOF model.

11.27 A large sign is modeled as a uniform rectangular block of mass M atop a massless flexible pole. Determine the equations of motion of the sign using as coordinates the lateral translation and slope at B. (*Hint:* Use the results of Problem 11.26.)

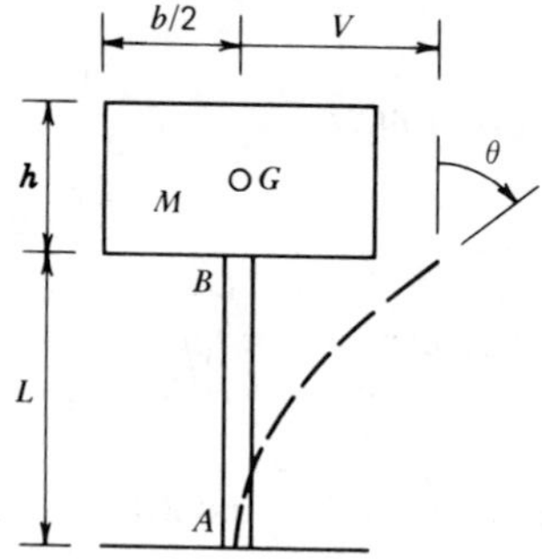

Figure P11.27

12 VIBRATION OF UNDAMPED 2-DOF SYSTEMS

Having the equations of motion of MDOF systems, as derived in Chapter 11, we could proceed immediately to a study of the dynamic response of systems with many degrees of freedom. It is very instructive, however, to begin this study with undamped systems having only 2 degrees-of-freedom. The equations of motion will thus have the form

$$\begin{bmatrix} m_{11} & m_{12} \\ m_{21} & m_{22} \end{bmatrix} \begin{Bmatrix} \ddot{u}_1 \\ \ddot{u}_2 \end{Bmatrix} + \begin{bmatrix} k_{11} & k_{12} \\ k_{21} & k_{22} \end{bmatrix} \begin{Bmatrix} u_1 \\ u_2 \end{Bmatrix} = \begin{Bmatrix} p_1(t) \\ p_2(t) \end{Bmatrix} \tag{12.1}$$

where the coordinates u_i may be physical displacements or may be generalized coordinates. The solution of Eq. 12.1 will consist of a complementary solution plus a particular solution. The complementary solution, obtained by setting the right-hand side of Eq. 12.1 to zero, involves the dynamical properties of the system called natural frequencies and natural modes. By examining the response of the 2-DOF system to harmonic excitation, you will be introduced to another very important topic in structural dynamics, namely mode-superposition.

Upon completion of this chapter you should be able to:

- Calculate the natural frequencies and natural modes of a 2-DOF system.
- Write expressions for the response of a 2-DOF system to nonzero initial conditions.
- Calculate the natural frequencies and natural modes of 2-DOF or 3-DOF systems having rigid-body modes.
- Determine the generalized mass matrix and generalized stiffness matrix of a 2-DOF system.
- Write expressions for the steady-state response of a 2-DOF system both in principal coordinates and in physical coordinates.

12.1 Free Vibration of 2-DOF Systems

We now consider the solution of the equations of motion of a 2-DOF system. The following steps may be followed in obtaining the solution for free vibration, that is, the solution of the set of equations

$$\begin{bmatrix} m_{11} & m_{12} \\ m_{21} & m_{22} \end{bmatrix} \begin{Bmatrix} \ddot{u}_1 \\ \ddot{u}_2 \end{Bmatrix} + \begin{bmatrix} k_{11} & k_{12} \\ k_{21} & k_{22} \end{bmatrix} \begin{Bmatrix} u_1 \\ u_2 \end{Bmatrix} = \begin{Bmatrix} 0 \\ 0 \end{Bmatrix} \tag{12.2}$$

1. Assume a *harmonic solution* of the form

$$\begin{aligned} u_1 &= U_1 \cos(\omega t - \alpha) \\ u_2 &= U_2 \cos(\omega t - \alpha) \end{aligned} \tag{12.3}$$

2. Substitute the assumed solution into the equations of motion to obtain the *algebraic eigenvalue problem.*

$$\left[\begin{bmatrix} k_{11} & k_{12} \\ k_{21} & k_{22} \end{bmatrix} - \omega^2 \begin{bmatrix} m_{11} & m_{12} \\ m_{21} & m_{22} \end{bmatrix} \right] \begin{Bmatrix} U_1 \\ U_2 \end{Bmatrix} = \begin{Bmatrix} 0 \\ 0 \end{Bmatrix} \tag{12.4}$$

3. Since this is a set of homogeneous linear algebraic equations, the only nontrivial solutions of Eq. 12.4 correspond to values of ω^2 which satisfy the *characteristic equation*

$$\left| \begin{bmatrix} k_{11} & k_{12} \\ k_{21} & k_{22} \end{bmatrix} - \omega^2 \begin{bmatrix} m_{11} & m_{12} \\ m_{21} & m_{22} \end{bmatrix} \right| = 0 \tag{12.5}$$

That is, for which the determinant of the coefficients of Eq. 12.4 is equal to zero. This is a polynomial equation of order two in ω_i^2.

4. Solve for the two roots of the characteristic equation. Label these roots ω_1^2 and ω_2^2 with $\omega_1 \leq \omega_2$. ω_1 and ω_2 are called the *circular natural frequencies,* or simply the natural frequencies. (In mathematical terminology ω_1^2 and ω_2^2 are called *eigenvalues.*)

5. Substitute ω_1^2 back into the first (or the second, but not both) of the equations in Eq. 12.4 and obtain the ratio $\beta_1 = (U_2/U_1)^{(1)}$. This ratio defines the *natural mode,* or *mode shape,* corresponding to the natural frequency ω_1. Do the same for ω_2 to determine $\beta_2 = (U_2/U_1)^{(2)}$. (In mathematical terminology natural modes are called *eigenvectors.* The natural modes for an undamped system are sometimes referred to as *real modes* to distinguish them from *complex modes* which can occur in some damped linear systems.) Sketch the natural modes.

6. Free vibration can occur only at frequencies ω_1 and ω_2 as determined in step 4 and with the respective ratios β_1 and β_2 determined in step 5. The general solution of Eq. 12.2 is a linear combination of these two. The constants A_1, A_2, α_1, and α_2 are determined from initial conditions.

$$\begin{aligned} u_1 &= A_1 \cos(\omega_1 t - \alpha_1) + A_2 \cos(\omega_2 t - \alpha_2) \\ u_2 &= \beta_1 A_1 \cos(\omega_1 t - \alpha_1) + \beta_2 A_2 \cos(\omega_2 t - \alpha_2) \end{aligned} \tag{12.6}$$

An alternate form of the general solution is

$$u_1 = A_1 \cos \omega_1 t + B_1 \sin \omega_1 t + A_2 \cos \omega_2 t + B_2 \sin \omega_2 t \tag{12.7}$$
$$u_2 = \beta_1 A_1 \cos \omega_1 t + \beta_1 B_1 \sin \omega_1 t + \beta_2 A_2 \cos \omega_2 t + \beta_2 B_2 \sin \omega_2 t$$

The initial conditions determine A_1, B_1, A_2, B_2 when Eq. 12.7 is used.

Example 12.1

Obtain the natural frequencies and mode shapes of the system shown. The equations of motion are

$$m\ddot{u}_1 + 2ku_1 - ku_2 = 0 \tag{1a}$$
$$m\ddot{u}_2 - ku_1 + 2ku_2 = 0 \tag{1b}$$

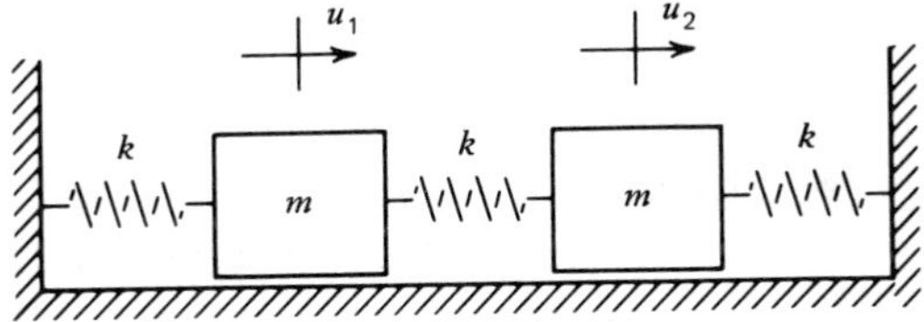

Solution

Step 1: Assume the harmonic solution:

$$u_1 = U_1 \cos(\omega_i t - \alpha) \tag{2a}$$
$$u_2 = U_2 \cos(\omega_i t - \alpha) \tag{2b}$$

Note that this means that u_1 and u_2 have the same time dependence and that, at all times, the amplitude ratio of $u_2/u_1 = U_2/U_1$ has the same value.

Step 2. Substitute Eqs. 2 into Eqs. 1 to get the algebraic eigenvalue problem:

$$\begin{bmatrix} (2k - m\omega_i^2) & -k \\ -k & (2k - m\omega_i^2) \end{bmatrix} \begin{Bmatrix} U_1 \\ U_2 \end{Bmatrix} = \begin{Bmatrix} 0 \\ 0 \end{Bmatrix} \tag{3}$$

Step 3. Set up the characteristic equation, which is the determinant of the coefficient matrix in Eq. 3:

$$\begin{vmatrix} (2k - m\omega_i^2) & -k \\ -k & (2k - m\omega_i^2) \end{vmatrix} = 0 \tag{4}$$

Expand the determinant to get

$$(2k - m\omega_i^2)^2 - k^2 = 0$$

or

$$m^2\omega_i^4 - 4km\omega_i^2 + 3k^2 = 0 \tag{5}$$

Step 4. Obtain the roots of the characteristic equation:
In the present case the equation can be factored easily to obtain

$$(m\omega_i^2 - k)(m\omega_i^2 - 3k) = 0$$

or

$$\omega_1^2 = \frac{k}{m} \tag{6a}$$

$$\omega_2^2 = \frac{3k}{m} \tag{6b}$$

The (circular) natural frequencies are thus

$$\boxed{\omega_1 = \left(\frac{k}{m}\right)^{1/2} \qquad \omega_2 = \left(\frac{3k}{m}\right)^{1/2}} \tag{7}$$

Note that the lower frequency is labeled ω_1.

Step 5. Substitute the eigenvalues given by Eq. 6 into the first of Eqs. 3 to obtain the mode shape ratios:

$$(2k - m\omega_i^2)U_1^{(i)} - kU_2^{(i)} = 0$$

or

$$\beta_i = \left(\frac{U_2}{U_1}\right)^{(i)} = \frac{2k - m\omega_i^2}{k} = 2 - \frac{m\omega_i^2}{k} \tag{8}$$

Therefore,

$$\beta_1 = 2 - 1 = 1 \tag{9a}$$

$$\beta_2 = 2 - 3 = -1 \tag{9b}$$

Sketch the mode shapes by letting $U_1^{(i)} = 1$. Then, $U_2^{(i)} = \beta_i U_1^{(i)} = \beta_i$.

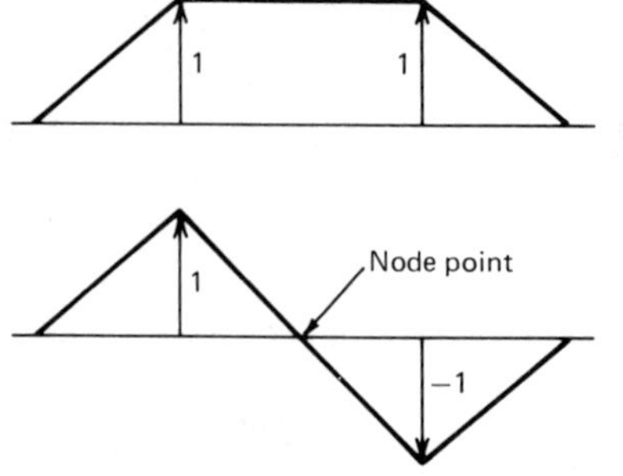

Mode 1

$$\omega_1 = \left(\frac{k}{m}\right)^{1/2}, \qquad \mathbf{U}_1 = \begin{Bmatrix} 1 \\ 1 \end{Bmatrix}$$

Mode 2

$$\omega_2 = \left(\frac{3k}{m}\right)^{1/2}, \qquad \mathbf{U}_2 = \begin{Bmatrix} 1 \\ -1 \end{Bmatrix}$$

Notice that the original system is symmetric about the center of the middle spring. For this reason the mode shapes turn out to be symmetric (mode 1) and antisymmetric (mode 2). This is an important result since many structures possess such physical symmetry. For example, an airplane is symmetric (nominally) about a vertical plane through the axis of the fuselage.

Also note that mode 2 has a *node point*, that is, a point that always remains stationary.

The general solution for free vibration of an undamped 2-DOF system is given by either Eqs. 12.6 or 12.7. Note that in either case the motion is a combination of motion of the system with mode shape one at frequency ω_1 and of motion with mode shape 2 at frequency ω_2. In the following example you can see how the four arbitrary constants in Eqs. 12.7 are determined.

Example 12.2

The following initial conditions are imposed on the system studied in Example 12.1.

$$u_1(0) = \dot{u}_1(0) = \dot{u}_2(0) = 0$$
$$u_2(0) = u_o$$

Determine the subsequent free vibration motion.

Solution

Use the general solution form given in Eqs. 12.7.

$$u_1 = A_1 \cos \omega_1 t + B_1 \sin \omega_1 t + A_2 \cos \omega_2 t + B_2 \sin \omega_2 t \quad (1a)$$
$$u_2 = A_1\beta_1 \cos \omega_1 t + B_1\beta_1 \sin \omega_1 t + A_2\beta_2 \cos \omega_2 t + B_2\beta_2 \sin \omega_2 t \quad (1b)$$

Differentiate Eqs. 1 to get

$$\dot{u}_1 = -A_1\omega_1 \sin \omega_1 t + B_1\omega_1 \cos \omega_1 t - A_2\omega_2 \sin \omega_2 t + B_2\omega_2 \cos \omega_2 t \quad (2a)$$
$$\dot{u}_2 = -A_1\beta_1\omega_1 \sin \omega_1 t + B_1\beta_1\omega_1 \cos \omega_1 t - A_2\beta_2\omega_2 \sin \omega_2 t + B_2\beta_2\omega_2 \cos \omega_2 t \quad (2b)$$

From the mode shape information obtained in Example 12.1,

$$\beta_1 = 1, \qquad \beta_2 = -1 \quad (3)$$

Therefore,

$$\begin{aligned} u_1(0) &= A_1 + A_2 = 0 \\ u_2(0) &= A_1\beta_1 + A_2\beta_2 = u_o \\ \dot{u}_1(0) &= B_1\omega_1 + B_2\omega_2 = 0 \\ \dot{u}_2(0) &= B_1\beta_1\omega_1 + B_2\beta_2\omega_2 = 0 \end{aligned} \quad (4)$$

Solve these equations to get

$$A_1 = \frac{u_o}{2}, \qquad A_2 = \frac{-u_o}{2}$$
$$B_1 = 0, \qquad B_2 = 0 \tag{5}$$

Therefore,

$$u_1 = \left(\frac{u_o}{2}\right)(\cos \omega_1 t - \cos \omega_2 t) \tag{6a}$$

$$u_2 = \left(\frac{u_o}{2}\right)(\cos \omega_1 t + \cos \omega_2 t) \tag{6b}$$

To see the motion of the two masses when the system is given initial conditions as stated above, take $u_o = 2$ in. and $\omega_1 = 1$ rad/s. From Example 12.1 $\omega_2 = \omega_1 \sqrt{3}$. Therefore, $\omega_2 = \sqrt{3} = 1.732$ rad/s. The resulting motions, $u_1(t)$ and $u_2(t)$, are shown below.

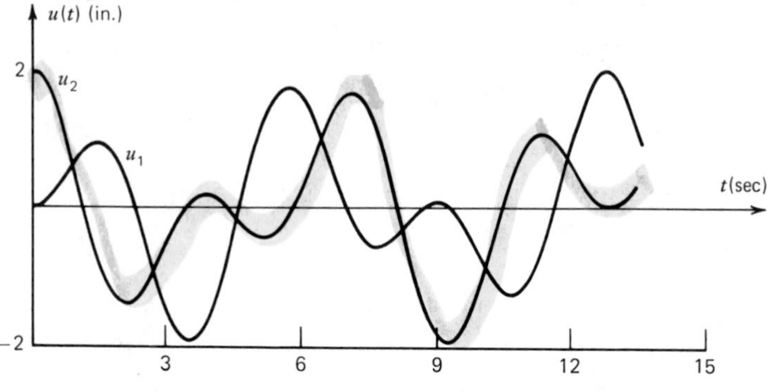

By a proper choice of initial conditions it is possible to initiate free vibration in a single mode. This is considered in a homework problem.

12.2 Further Examples of Modes and Frequencies

Examples 12.1 and 12.2 of the previous section fully illustrate the basic steps required in determining the free vibration of a 2-DOF system. This section reviews the process of determining natural frequencies and modes, indicates how this process applies to assumed-modes models, and presents a comparison of various models that represent the same physical system.

Example 12.3

A 2-DOF assumed-modes model of a cantilever bar was obtained in Example 11.9. Solve for the natural frequencies and modes of this model, and sketch the modes. Use the notation $q_1 \rightarrow u_1$, $q_2 \rightarrow u_2$.

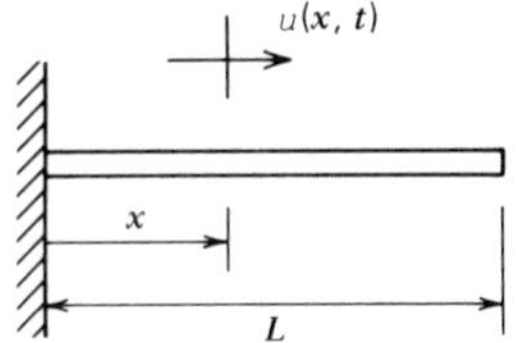

Solution

a. Write the equation of motion for free vibration.
From Eq. 10 of Example 11.9,

$$\frac{\rho AL}{60}\begin{bmatrix} 20 & 15 \\ 15 & 12 \end{bmatrix}\begin{Bmatrix} \ddot{u}_1 \\ \ddot{u}_2 \end{Bmatrix} + \frac{EA}{3L}\begin{bmatrix} 3 & 3 \\ 3 & 4 \end{bmatrix}\begin{Bmatrix} u_1 \\ u_2 \end{Bmatrix} = \begin{Bmatrix} 0 \\ 0 \end{Bmatrix} \tag{1}$$

b. Assume harmonic motion.

$$\begin{Bmatrix} u_1 \\ u_2 \end{Bmatrix} = \begin{Bmatrix} U_1 \\ U_2 \end{Bmatrix}\cos(\omega t - \alpha) \tag{2}$$

c. Obtain the algebraic eigenvalue problem

$$\left[\begin{bmatrix} 3 & 3 \\ 3 & 4 \end{bmatrix} - \mu_i^2\begin{bmatrix} 20 & 15 \\ 15 & 12 \end{bmatrix}\right]\begin{Bmatrix} U_1 \\ U_2 \end{Bmatrix} = \begin{Bmatrix} 0 \\ 0 \end{Bmatrix} \tag{3}$$

where $\mu_i^2 = \left(\dfrac{\rho L^2}{20E}\right)\omega_i^2$

d. From the determinant of the coefficients, obtain the characteristic equation.

$$(3 - 20\mu_i^2)(4 - 12\mu_i^2) - (3 - 15\mu_i^2)^2 = 0$$

or

$$15(\mu_i^2)^2 - 26(\mu_i^2) + 3 = 0 \tag{5}$$

e. Solve for the roots of the characteristic equation.

$$\mu_i^2 = \frac{26 \pm \sqrt{496}}{30} = 0.1243, \qquad 1.6090 \tag{6}$$

Then, from Eqs. 4 and 6,

$$(\omega_1 L)^2 = \left(\frac{20E}{\rho}\right)\mu_1^2 = 2.486\left(\frac{E}{\rho}\right) \tag{7a}$$

$$(\omega_2 L)^2 = \left(\frac{20E}{\rho}\right)\mu_2^2 = 32.18\left(\frac{E}{\rho}\right) \tag{7b}$$

In Sec. 10.1 it was found that the "exact" natural frequencies of this system are

$$(\omega_1 L)^2_{\text{exact}} = 2.467\left(\frac{E}{\rho}\right) \tag{8a}$$

$$(\omega_2 L)^2_{\text{exact}} = 22.21\left(\frac{E}{\rho}\right) \tag{8b}$$

Note that in both cases the approximate frequencies are higher than the exact ones, and that the fundamental (i.e., lowest) frequency is calculated with much greater accuracy than the second frequency. These phenomena are discussed further in Sec. 17.1. It is characteristic of assumed-modes models that frequencies are too high and that the higher frequencies which are calculated are inaccurate.

f. Determine the mode shapes.

Substitute μ_i^2 into the first of the equations of Eq. 3.

$$(3 - 20\mu_i^2)U_1^{(i)} + (3 - 15\mu_i^2)U_2^{(i)} = 0 \tag{9}$$

or

$$\beta_i \equiv \left(\frac{U_2}{U_1}\right)^{(i)} = -\frac{(3 - 20\mu_i^2)}{(3 - 15\mu_i^2)} \tag{10}$$

Therefore,

$$\beta_1 = -\frac{0.514}{1.136} = -0.453 \tag{11a}$$

$$\beta_2 = -1.381 \tag{11b}$$

To draw the mode shapes we must recall from Example 11.9 that

$$u(x, t) = \left(\frac{x}{L}\right)u_1(t) + \left(\frac{x}{L}\right)^2 u_2(t) \tag{12}$$

From Eqs. 2, 10, and 12 we find that motion in the ith mode is given by

$$u^{(i)}(x, t) = \left[\left(\frac{x}{L}\right) + \beta_i\left(\frac{x}{L}\right)^2\right]\cos(\omega_i t - \alpha_i) \tag{13}$$

or

$$u^{(i)}(x, t) = \phi_i(x)\cos(\omega_i t - \alpha_i) \tag{14}$$

where $\phi_i(x)$ is the mode shape, which is given by

$$\phi_i(x) = \left(\frac{x}{L}\right) + \beta_i\left(\frac{x}{L}\right)^2 \tag{15}$$

The mode shapes are plotted below.

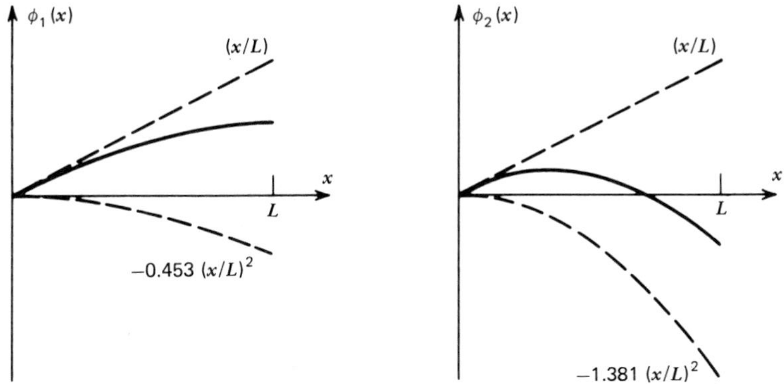

Compare these shapes with the "exact" modes shown in Example 10.1.

The above example indicates that good estimates of natural frequencies are attainable by the use of assumed-modes models. Consider now how the assumed-modes model compares with a 2-DOF lumped-mass model.

Example 12.4

The cantilever bar of Example 11.9 is now to be modeled by a massless uniform bar to which are attached two lumped masses representing the mass of the original system. This is "equivalent" to the spring-mass system with $k = 2AE/L$, $m = \rho AL$.

Determine the natural frequencies and normal modes of this model.

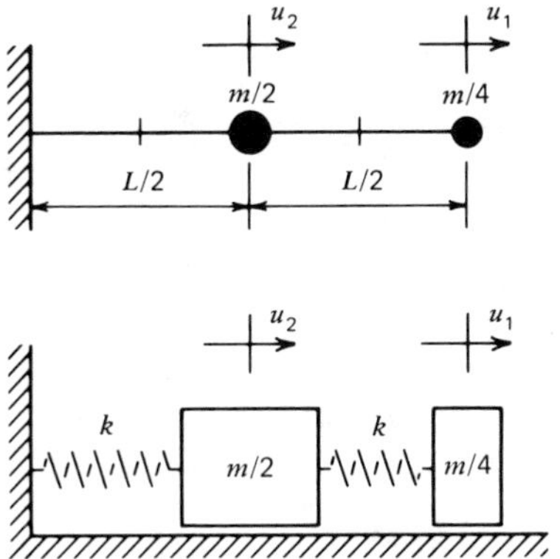

Solution

The same steps will be followed as in previous examples.

$$\frac{\rho AL}{4}\begin{bmatrix}1 & 0\\ 0 & 2\end{bmatrix}\begin{Bmatrix}\ddot{u}_1\\ \ddot{u}_2\end{Bmatrix} + \frac{2AE}{L}\begin{bmatrix}1 & -1\\ -1 & 2\end{bmatrix}\begin{Bmatrix}u_1\\ u_2\end{Bmatrix} = \begin{Bmatrix}0\\ 0\end{Bmatrix} \tag{1}$$

$$\mathbf{u} = \mathbf{U}\cos(\omega_i t - \alpha) \tag{2}$$

$$\left[\begin{bmatrix}1 & -1\\ -1 & 2\end{bmatrix} - \mu_i^2\begin{bmatrix}1 & 0\\ 0 & 2\end{bmatrix}\right]\begin{Bmatrix}U_1\\ U_2\end{Bmatrix} = \begin{Bmatrix}0\\ 0\end{Bmatrix} \tag{3}$$

where

$$\mu_i^2 = \left(\frac{\rho L^2}{8E}\right)\omega_i^2 \tag{4}$$

$$(1 - \mu_i^2)(2 - 2\mu_i^2) - (-1)(-1) = 0$$

$$2\mu_i^4 - 4\mu_i^2 + 1 = 0 \tag{5}$$

$$\mu_i^2 = \frac{4 \pm \sqrt{8}}{4} = 1 \pm \sqrt{2}/2 \tag{6}$$

$$\mu_1^2 = 0.2929, \qquad \mu_2^2 = 1.707$$

$$(\omega_1 L)^2 = 8(0.2929)\left(\frac{E}{\rho}\right) = 2.343\left(\frac{E}{\rho}\right) \tag{7a}$$

$$(\omega_2 L)^2 = 8(1.707)\left(\frac{E}{\rho}\right) = 13.66\left(\frac{E}{\rho}\right) \tag{7b}$$

Recall that the corresponding "exact" values are 2.467 and 22.21. Hence, it can be seen that this lumped-mass model gives frequencies that are below the exact value in each case. Further examples are given in Sec. 17.1.

For the mode shapes, Eq. 3 gives

$$(1 - \mu_i^2)\, U_1^{(i)} - U_2^{(i)} = 0 \tag{8}$$

or

$$\beta_i = \left(\frac{U_2}{U_1}\right)^{(i)} = 1 - \mu_i^2 \tag{9}$$

Therefore,

$$\boxed{\beta_1 = 0.707, \qquad \beta_2 = -0.707}$$

and the mode shapes can be plotted as below.

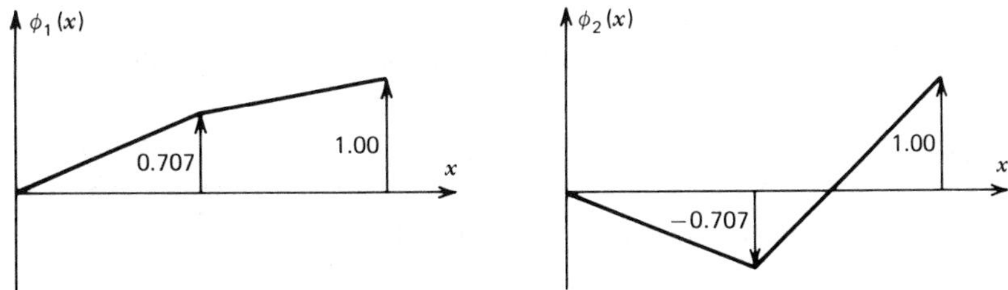

By comparing this example with Example 12.3 and with Sec. 10.1, you can observe that the choice of model is quite important. Although the lumped-mass approximation did not give as accurate results for the fundamental frequency as did the assumed-modes model, lumped-mass models may be quite acceptable if the number of lumped masses is large and the number of frequencies required is a fraction of the total DOF of the system.

12.3 Systems with Rigid-Body Modes

Frequently systems are encountered which have one or more *rigid-body modes,* that is, modes in which none of the elastic elements is deformed. This is true for aerospace vehicles in flight, since the structure is free to move as a rigid body without deformation. For example, an airplane wing and body modeled by three lumped masses along a flexible beam has the mode shapes shown in Fig. 12.1. The first two modes are rigid-body modes (a "plunge" mode and a "roll" mode in airplane terminology). Rigid-body modes have a frequency of zero, as will be seen in Example 12.5.

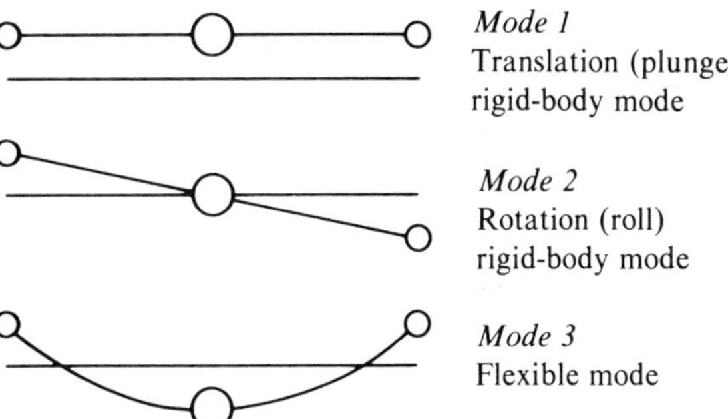

Figure 12.1. *Mode shapes of a three-mass model of an airplane wing illustrating rigid-body modes.*

To illustrate how rigid-body modes are treated, we will consider the system in Example 12.5. This system has only one rigid-body mode.

Example 12.5

Determine the natural frequencies and mode shapes of the system shown below.

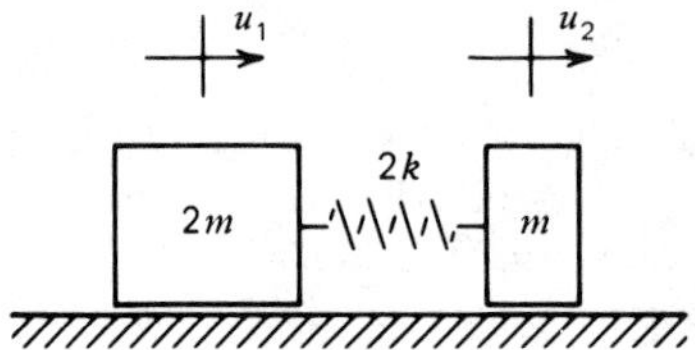

(Note that the system is free to translate as a rigid-body with the spring remaining unstretched, that is, the masses can move with $u_1 = u_2$ without any external forcing.)

Solution

Proceed in exactly the same manner as before, that is, follow the steps outlined in Sec. 12.1. The equations of motion are

$$\begin{aligned} 2m\ddot{u}_1 + 2ku_1 - 2ku_2 &= 0 \\ m\ddot{u}_2 - 2ku_1 + 2ku_2 &= 0 \end{aligned} \tag{1}$$

Step 1. Assume

$$\begin{aligned} u_1 &= U_1 \cos \omega_i t \\ u_2 &= U_2 \cos \omega_i t \end{aligned} \tag{2}$$

Step 2. Form the eigenvalue equation

$$\begin{bmatrix} (2k - 2m\omega_i^2) & -2k \\ -2k & (2k - m\omega_i^2) \end{bmatrix} \begin{Bmatrix} U_1 \\ U_2 \end{Bmatrix} = \begin{Bmatrix} 0 \\ 0 \end{Bmatrix} \tag{3}$$

Step 3. Obtain the characteristic equation by setting the determinant of coefficients in Eq. 3 equal to zero.

$$(2k - 2m\omega_i^2)(2k - m\omega_i^2) - (2k)^2 = 0$$

or

$$m\omega_i^2(2m\omega_i^2 - 6k) = 0 \tag{4}$$

Step 4. Solve for the roots (eigenvalues).

$$\boxed{\begin{aligned} \omega_1^2 &= 0 \\ \omega_2^2 &= \frac{3k}{m} \end{aligned}} \tag{5}$$

Step 5. Calculate the mode shapes. From Eq. 3,

$$(2k - 2m\omega_i^2)U_1^{(i)} - 2kU_2^{(i)} = 0$$

or

$$\beta_i = \left(\frac{U_2}{U_1}\right)^{(i)} = \frac{2k - 2m\omega_i^2}{2k} = 1 - \omega_i^2\left(\frac{m}{k}\right) \tag{6}$$

Therefore,

$$\boxed{\begin{aligned}\beta_1 &= 1 - 0 = 1\\ \beta_2 &= 1 - 3 = -2\end{aligned}} \tag{7}$$

Sketch the mode shapes. Let $U_1 = 1$ in each case.

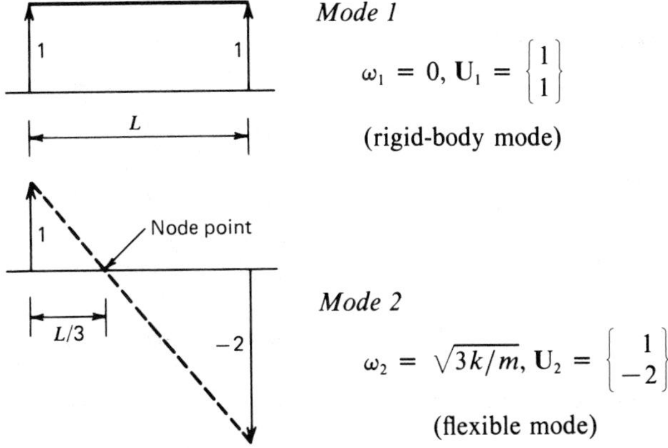

Mode 1

$$\omega_1 = 0,\ \mathbf{U}_1 = \begin{Bmatrix}1\\1\end{Bmatrix}$$

(rigid-body mode)

Mode 2

$$\omega_2 = \sqrt{3k/m},\ \mathbf{U}_2 = \begin{Bmatrix}1\\-2\end{Bmatrix}$$

(flexible mode)

Note that in the rigid-body mode the frequency is zero and the spring is undeformed. In mode 2 the masses move in opposite directions and there is a *node* point (i.e., a point which remains stationary) as shown above.

An eigenvalue problem that results in one or more zero eigenvalues, as in Example 12.5, is called a *semidefinite eigenvalue problem.* In structural dynamics, zero eigenvalues result when the structure can move as a rigid body. Mathematically, the zero eigenvalues result from the fact that the stiffness matrix is singular, that is, the determinant of the coefficients of the stiffness matrix is zero.

12.4 Response of an Undamped 2-DOF System to Harmonic Excitation: Mode-Superposition

In Chapter 4 you studied the response of single-DOF systems to harmonic excitation. You learned that when $\Omega = \omega_n$, resonance occurs, that is, the displacement of the undamped system becomes unbounded. In the preceding sections of this chapter you have seen that MDOF systems possess multiple natural frequencies, so it is reasonable to expect resonance to be an important factor to consider in MDOF systems. In this section you will consider the response of a 2-DOF system to harmonic excitation. In addition, you will be introduced to the mode-superposition method for solving for the dynamic response of MDOF systems. This important topic will be considered in greater detail in Chapter 15.

Example 12.6

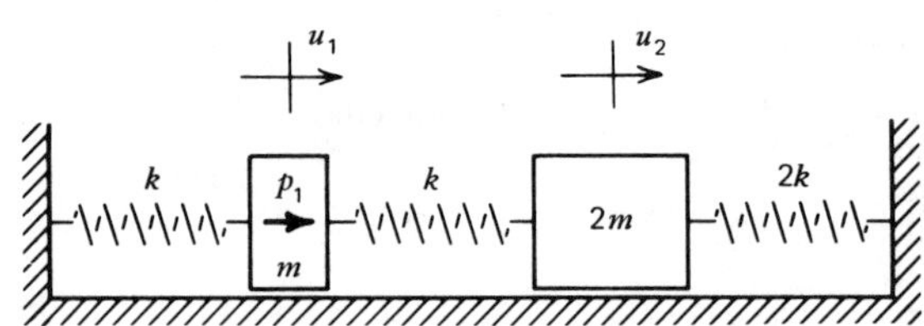

The 2-DOF system shown above is subjected to a single harmonic force, $p_1 = P_1 \cos \Omega t$. Determine the steady-state response of each of the masses as a function of frequency. Use mode-superposition in solving this problem. The equation of motion of the system is

$$m\begin{bmatrix} 1 & 0 \\ 0 & 2 \end{bmatrix}\begin{Bmatrix} \ddot{u}_1 \\ \ddot{u}_2 \end{Bmatrix} + k\begin{bmatrix} 2 & -1 \\ -1 & 3 \end{bmatrix}\begin{Bmatrix} u_1 \\ u_2 \end{Bmatrix} = \begin{Bmatrix} P_1 \\ 0 \end{Bmatrix} \cos \Omega t$$

and the natural frequencies and modes are

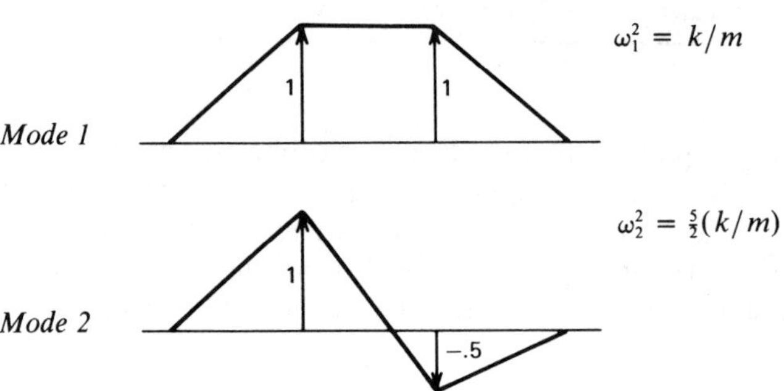

Solution

a. Form the *modal matrix.*
This is a matrix whose respective columns are the natural modes.

$$\boldsymbol{\Phi} = [\boldsymbol{\phi}_1, \boldsymbol{\phi}_2] = \begin{bmatrix} 1 & 1 \\ 1 & -\frac{1}{2} \end{bmatrix} \tag{1}$$

b. Define the set of principal coordinates, η_1 and η_2, by the matrix transformation

$$\mathbf{u} = \boldsymbol{\Phi}\boldsymbol{\eta} \tag{2}$$

$$\begin{Bmatrix} u_1 \\ u_2 \end{Bmatrix} = \begin{bmatrix} 1 & 1 \\ 1 & -\frac{1}{2} \end{bmatrix} \begin{Bmatrix} \eta_1 \\ \eta_2 \end{Bmatrix} = \begin{Bmatrix} \eta_1 + \eta_2 \\ \eta_1 - \frac{1}{2}\eta_2 \end{Bmatrix} \tag{3}$$

c. Transform the equations of motion to principal coordinates.

$$\boldsymbol{\Phi}^T [\mathbf{m}(\boldsymbol{\Phi}\ddot{\boldsymbol{\eta}}) + \mathbf{k}(\boldsymbol{\Phi}\boldsymbol{\eta}) = \mathbf{p}(t)] \tag{4}$$

or

$$(\boldsymbol{\Phi}^T\mathbf{m}\boldsymbol{\Phi})\ddot{\boldsymbol{\eta}} + (\boldsymbol{\Phi}^T\mathbf{k}\boldsymbol{\Phi})\boldsymbol{\eta} = \boldsymbol{\Phi}^T\mathbf{p}$$

or

$$\boldsymbol{M}\ddot{\boldsymbol{\eta}} + \boldsymbol{K}\boldsymbol{\eta} = \boldsymbol{P} \tag{5}$$

where

$$\boldsymbol{M} = \boldsymbol{\Phi}^T\mathbf{m}\boldsymbol{\Phi}, \qquad \boldsymbol{K} = \boldsymbol{\Phi}^T\mathbf{k}\boldsymbol{\Phi}, \qquad \boldsymbol{P} = \boldsymbol{\Phi}^T\mathbf{p} \tag{6}$$

(***M*** and ***K*** are usually referred to as the *modal mass matrix* and *modal stiffness matrix,* and ***P*** is called the *modal force vector.**) Use Eqs. 1 and 6 to calculate these.

$$\boldsymbol{M} = \begin{bmatrix} 1 & 1 \\ 1 & -\frac{1}{2} \end{bmatrix} \begin{bmatrix} m & 0 \\ 0 & 2m \end{bmatrix} \begin{bmatrix} 1 & 1 \\ 1 & -\frac{1}{2} \end{bmatrix} = m\begin{bmatrix} 3 & 0 \\ 0 & \frac{3}{2} \end{bmatrix} \tag{7}$$

$$\boldsymbol{K} = \begin{bmatrix} 1 & 1 \\ 1 & -\frac{1}{2} \end{bmatrix} \begin{bmatrix} 2k & -k \\ -k & 3k \end{bmatrix} \begin{bmatrix} 1 & 1 \\ 1 & -\frac{1}{2} \end{bmatrix} = k\begin{bmatrix} 3 & 0 \\ 0 & \frac{15}{4} \end{bmatrix} \tag{8}$$

$$\boldsymbol{P} = \begin{bmatrix} 1 & 1 \\ 1 & -\frac{1}{2} \end{bmatrix} \begin{Bmatrix} P_1 \\ 0 \end{Bmatrix} \cos \Omega t = \begin{Bmatrix} P_1 \\ P_1 \end{Bmatrix} \cos \Omega t \tag{9}$$

Therefore, the equation of motion in principal coordinates is

$$m\begin{bmatrix} 3 & 0 \\ 0 & \frac{3}{2} \end{bmatrix} \begin{Bmatrix} \ddot{\eta}_1 \\ \ddot{\eta}_2 \end{Bmatrix} + k\begin{bmatrix} 3 & 0 \\ 0 & \frac{15}{4} \end{bmatrix} \begin{Bmatrix} \eta_1 \\ \eta_2 \end{Bmatrix} = \begin{Bmatrix} P_1 \\ P_1 \end{Bmatrix} \cos \Omega t \tag{10}$$

But, the transformation to principal coordinates has *uncoupled the equations of motion* (compare Eq. 10 with the original equation of motion). This leads to two

*The term generalized mass matrix is frequently used instead of modal mass matrix, and so forth.

separate single-DOF equations which can be solved by the methods of Chapter 4. Thus, the equations to solve are

$$3m\ddot{\eta}_1 + 3k\eta_1 = P_1 \cos \Omega t \tag{11a}$$

$$(3/2)m\ddot{\eta}_2 + (15/4)k\eta_2 = P_1 \cos \Omega t \tag{11b}$$

d. Solve the uncoupled equations of motion for the steady-state response. Assume harmonic motion.

$$\eta_1 = Y_1 \cos \Omega t \tag{12a}$$

$$\eta_2 = Y_2 \cos \Omega t \tag{12b}$$

Then,

$$Y_1 = \frac{P_1}{3k - 3m\Omega^2} = \frac{(1/3k)P_1}{1 - (\Omega/\omega_1)^2} \tag{13a}$$

$$Y_2 = \frac{P_1}{(15/4)k - (3/2)m\Omega^2} = \frac{(4/15k)P_1}{1 - (\Omega/\omega_2)^2} \tag{13b}$$

e. Transform back to physical coordinates.
Equations 3, 12, and 13 can be combined to give

$$u_1 = U_1 \cos \Omega t \tag{14a}$$

$$u_2 = U_2 \cos \Omega t \tag{14b}$$

where

$$U_1 = \left[\left(\frac{P_1/3k}{1 - (\Omega/\omega_1)^2}\right) + \left(\frac{4P_1/15k}{1 - (\Omega/\omega_2)^2}\right)\right] \tag{15a}$$

$$U_2 = \left[\left(\frac{P_1/3k}{1 - (\Omega/\omega_1)^2}\right) - \frac{1}{2}\left(\frac{4P_1/15k}{1 - (\Omega/\omega_2)^2}\right)\right] \tag{15b}$$

The figures below show $U_1(\Omega)$ and $U_2(\Omega)$. Note the two resonance "peaks" in these figures at $\Omega = \omega_1$ and $\Omega = \omega_2$.

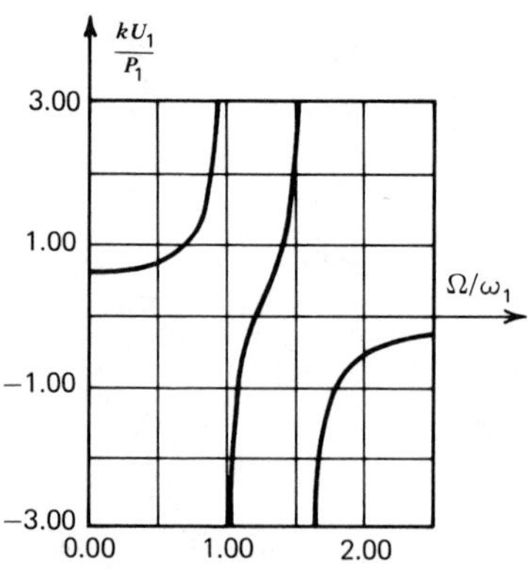

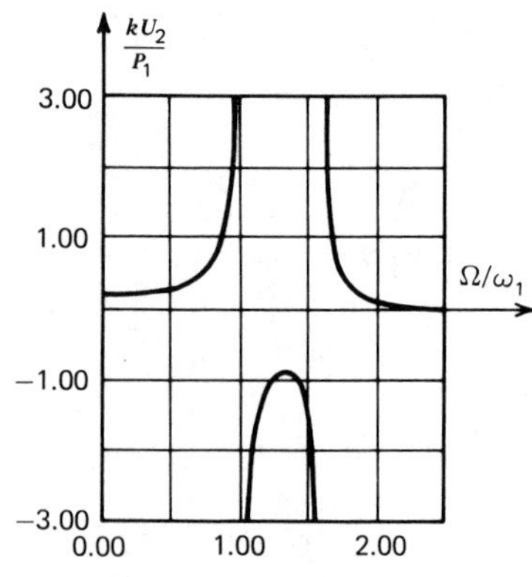

The above example could easily have been solved by substituting Eqs. 14 directly into the original equations of motion. This would not have been the case, however, with general excitation $\mathbf{P}(t)$, as you will see in Chapter 15.

From Example 12.6 you should have discovered two important reasons for solving for the modes and frequencies of MDOF systems: (1) the natural frequencies and modes determine how the system will respond to dynamic excitation, and (2) the modal matrix can be used to uncouple the equations of motion of undamped (and some damped) systems.

In the next chapter you will study some of the important properties of modes and frequencies, and in Chapter 14 you will learn about some of the numerical procedures available for solving for modes and frequencies of systems with many degrees of freedom.

Problem Set 12.1

12.1 The stiffness and mass matrices of the two-story building shown below are

$$\mathbf{k} = 600\begin{bmatrix} 1 & -1 \\ -1 & 3 \end{bmatrix} \text{ k/in.,} \qquad \mathbf{m} = 2\begin{bmatrix} 1 & 0 \\ 0 & 1 \end{bmatrix} \text{ k-sec}^2\text{/in.}$$

(a) Determine the two natural frequencies of this structure.
(b) Determine the two corresponding mode shapes. Scale them so that the maximum displacement is 1.0. Sketch the two modes.

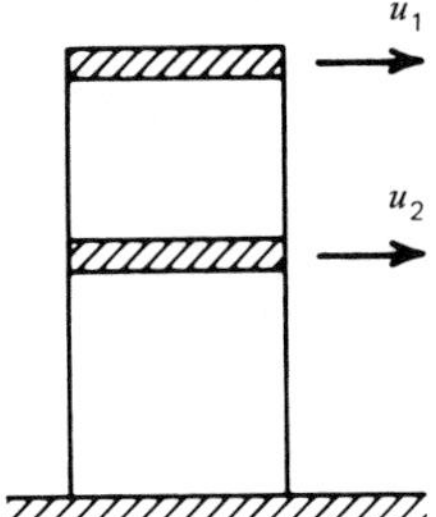

Figure P12.1

12.2 The lumped-mass beam model shown below has the following stiffness and mass matrices.

$$\mathbf{k} = k\begin{bmatrix} 2 & -5 \\ -5 & 16 \end{bmatrix} \text{ lb/in.,} \qquad \mathbf{m} = m\begin{bmatrix} 1 & 0 \\ 0 & 4 \end{bmatrix} \text{ lb-sec}^2\text{/in.}$$

(a) Determine the two natural frequencies.
(b) Determine the two mode shapes. Scale them so that the maximum displacement is 1.0. Sketch the two modes.

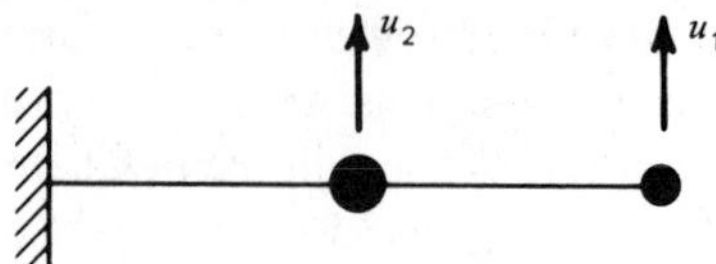

Figure P12.2

12.3 A slender rigid uniform beam of mass m is supported by two linear springs as shown below.

(a) Write the equations of motion in terms of the end displacements u_1 and u_2.
(b) Determine the two natural frequencies.
(c) Determine the two mode shapes. Scale them so that the maximum displacement is 1.0. Sketch the mode shapes.

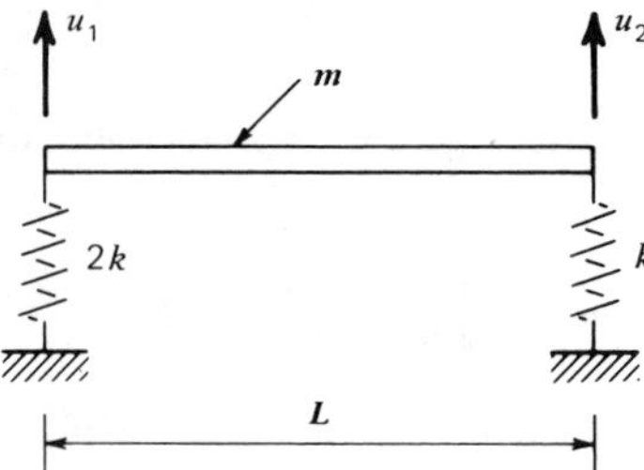

Figure P12.3

12.4 For the lumped mass system below:
(a) Determine the two natural frequencies.
(b) Determine the mode shapes.

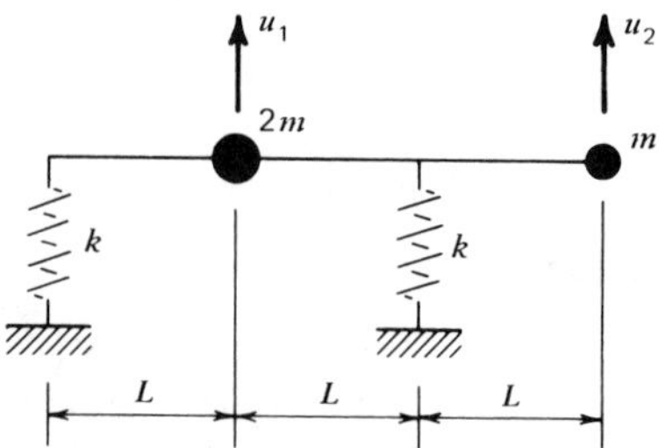

Figure P12.4

12.5 The frequencies and mode shapes of the structure in Fig. P12.1 are $\omega_1^2 = 175.7 \text{ rad}^2/\text{s}^2$, $\omega_2^2 = 1024 \text{ rad}^2/\text{s}^2$, and

$$\mathbf{U}_1 = \begin{Bmatrix} 1.000 \\ 0.414 \end{Bmatrix} \qquad \mathbf{U}_2 = \begin{Bmatrix} -0.414 \\ 1.000 \end{Bmatrix}$$

If the initial conditions are

$$u_1(0) = 2 \text{ in.}, \qquad u_2(0) = 1 \text{ in.}, \qquad \dot{u}_1(0) = \dot{u}_2(0) = 0$$

determine the expressions for $u_1(t)$ and $u_2(t)$.

12.6 Replace the spring between masses m_1 and m_2 in Example 12.1 by a weaker coupling spring having a stiffness $k/10$.

(a) Determine the two natural frequencies ω_1 and ω_2.

(b) Determine the two mode shapes and scale each so that the largest displacement is 1.0.

(c) For the following initial conditions, determine expressions for $u_1(t)$ and $u_2(t)$:

$$u_1(0) = u_o, \qquad \dot{u}_1(0) = u_2(0) = \dot{u}_2(0) = 0$$

(d) Show that $u_1(t)$ and $u_2(t)$ can be written in the form

$$u_1(t) = u_o \cos\left(\frac{\omega_2 - \omega_1}{2}\right)t \cos\left(\frac{\omega_1 + \omega_2}{2}\right)t$$

$$u_2(t) = u_o \sin\left(\frac{\omega_2 - \omega_1}{2}\right)t \sin\left(\frac{\omega_1 + \omega_2}{2}\right)t$$

(*Note:* This form indicates that the system will exhibit what is known as a "beat phenomenon." That is, the motion appears to be harmonic motion at the faster frequency $\left(\frac{\omega_1 + \omega_2}{2}\right)$ with a slowly varying amplitude at frequency $\left(\frac{\omega_2 - \omega_1}{2}\right)$

Problem Set 12.2

12.7 Axial vibration of the system in Fig. P10.1 is to be modeled by a 2-DOF system based on

$$\psi_1(x) = \frac{x}{L}, \qquad \psi_2(x) = \left(\frac{x}{L}\right)^2$$

(a) Determine the equations of motion for the 2-DOF model.

(b) Determine the two natural frequencies and compare the fundamental frequency with the "exact" frequency determined in Problem 10.1.

(c) Determine the two mode shapes and sketch them as in Example 12.3 using values computed for increments of $L/5$.

12.8 A 2-DOF model for transverse vibration of a uniform cantilever beam is to be created using

$$\psi_1(x) = \left(\frac{x}{L}\right)^2, \qquad \psi_2(x) = \left(\frac{x}{L}\right)^3$$

(a) Derive the equations of motion for the 2-DOF model.
(b) Determine the natural frequencies. Compare both frequencies with the "exact" values from Example 10.3 and compare the fundamental with the value from Example 10.4.

12.9 For the beam in Fig. P10.11 let $M = \rho AL/2$.
(a) Derive the equations of motion for a 2-DOF model of this system using

$$\psi_1 = \sin\left(\frac{\pi x}{L}\right), \qquad \psi_2 = \sin\left(\frac{3\pi x}{L}\right)$$

(b) Determine the natural frequencies of this 2-DOF model and compare the fundamental frequency with that obtained by the Rayleigh method in Example 10.11.

Problem Set 12.3

12.10 Determine the natural frequencies and mode shapes of the system shown below. Sketch the three mode shapes. (*Note:* $\omega_1 = 0$.)

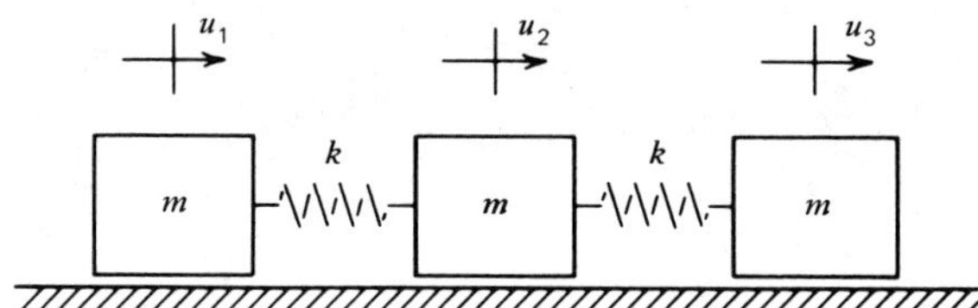

Figure P12.10

12.11 The axial vibration of a free-free uniform bar is to be studied by using a lumped-mass model as shown below. The resulting equations of motion, in matrix form, are

$$m\begin{bmatrix} 1 & 0 & 0 \\ 0 & 2 & 0 \\ 0 & 0 & 1 \end{bmatrix}\begin{Bmatrix} \ddot{u}_1 \\ \ddot{u}_2 \\ \ddot{u}_3 \end{Bmatrix} + k\begin{bmatrix} 1 & -1 & 0 \\ -1 & 2 & -1 \\ 0 & -1 & 1 \end{bmatrix}\begin{Bmatrix} u_1 \\ u_2 \\ u_3 \end{Bmatrix} = \begin{Bmatrix} 0 \\ 0 \\ 0 \end{Bmatrix}$$

where $m = \rho AL/4$, $k = 2AE/L$.
(a) Solve for the three natural frequencies of this system. (*Note*: $\omega_1 = 0$.)

(b) Solve for the three mode shapes. Scale each mode so that the maximum displacement is 1.0 and sketch the modes.

(c) Compare the frequencies with "exact" values from Problem 10.2.

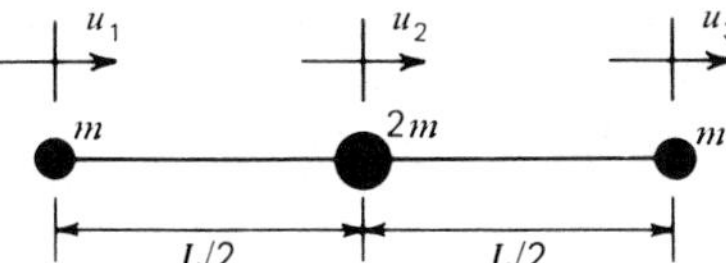

Figure P12.11

Problem Set 12.4

12.12 Example 12.6 illustrates the method of mode-superposition, which will be employed extensively in Chapter 15. A more direct solution for the steady-state response of the system in Example 12.6 is obtained by substituting Eqs. 14 directly into the equation of motion. Show that this leads to the frequency response expressions in Eqs. 15.

12.13 One method of reducing the vibration amplitude of a SDOF system subjected to harmonic excitation is to attach a "tuned vibration absorber," which is a second spring-mass system. For the resulting 2-DOF system shown below:

(a) Determine the equations of motion.

(b) Let the steady-state response be given by

$$u_1(t) = U_1 \cos \Omega t$$
$$u_2(t) = U_2 \cos \Omega t$$

and show that

$$U_1(\Omega) = \frac{(k_2 - \Omega^2 m_2) P_1}{D(\Omega)}$$
$$U_2(\Omega) = \frac{k_2 P_1}{D(\Omega)}$$

where $D(\Omega) = (k_1 + k_2 - \Omega^2 m_1)(k_2 - \Omega^2 m_2) - k_2^2$

(c) The absorber is "tuned" so that $k_2/m_2 = k_1/m_1$. Thus, when $\Omega^2 = k_1/m_1$, that is, the original system is excited at resonance, the response amplitude U_1 is reduced to zero. Let $m_2/m_1 = 0.25$ for a particular "tuned" absorber system. Plot $k_1 U_1/P_1$ and $k_1 U_2/P_1$ versus the frequency ratio $r = \Omega/\sqrt{k_1/m_1}$.

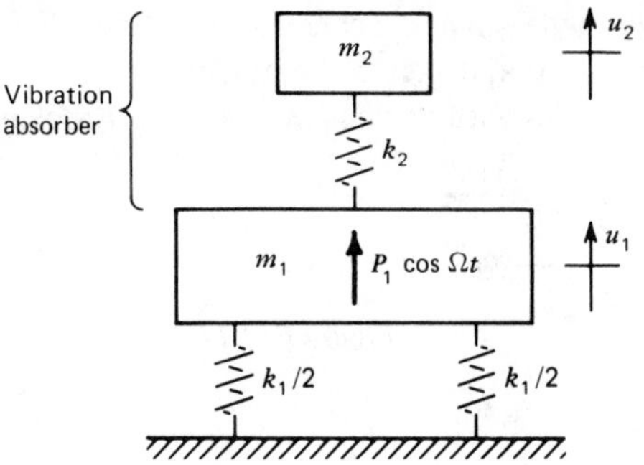

Figure P12.13

12.14 Using information from Problems 12.1 and 12.5, and using the mode-superposition method of Example 12.6, determine the steady-state response of the two-story building of Fig. P12.1 to the following harmonic excitation

$$\mathbf{P}(t) = \begin{Bmatrix} 10 \\ 0 \end{Bmatrix} \sin(25t) \text{ k}$$

12.15 Using the mode-superposition method of Example 12.6, show that symmetric harmonic excitation given by

$$\begin{Bmatrix} P_1 \\ P_2 \end{Bmatrix} = \begin{Bmatrix} P \\ P \end{Bmatrix} \cos \Omega t$$

produces steady-state response in the symmetric mode only for the symmetric system of Example 12.1.

13 FREE VIBRATION OF MDOF SYSTEMS

In Chapter 11 the equations of motion of MDOF systems were derived, and in Chapter 12 you were introduced to techniques for obtaining natural frequencies and natural modes as you considered free vibration of 2-DOF systems. In the present chapter a number of properties of natural frequencies and natural modes of MDOF systems are discussed.

Upon completion of this chapter you should be able to:

- Define the following terms: positive definite matrix, singular matrix, natural mode, normal mode, modal stiffness, modal mass, orthogonality, modal matrix, Rayleigh quotient, flexibility coefficient.
- Determine the mode shapes of a MDOF system given the distinct and/or repeated natural frequencies of the system.
- State the condition that leads to rigid-body modes of a system.
- Derive the orthogonality equations that must be satisfied by modes that correspond to distinct frequencies.
- Obtain an estimate to the fundamental frequency of an MDOF system by using the Rayleigh method, and state the fundamental property of the Rayleigh quotient.
- Obtain an estimate of $\hat{N}$ frequencies of an N-DOF system ($\hat{N} \leq N$) by using the Rayleigh-Ritz method. Describe how these frequency estimates could be improved.

13.1 Some Properties of Natural Frequencies and Natural Modes

The purpose of the present section is to discuss some of the more important properties of natural frequencies and natural modes. In the interest of conciseness most of the equations will be written in matrix form. Before considering properties of frequencies and modes it will be useful for us to consider some of the properties of the stiffness matrices and mass matrices encountered in structural dynamics problems.

Properties of k and m

In Chapter 11, **k** and **m** were related to strain energy and kinetic energy, respectively, by the quadratic forms

$$V = \tfrac{1}{2}\mathbf{u}^T\mathbf{k}\mathbf{u} \tag{13.1a}$$

and

$$T = \tfrac{1}{2}\dot{\mathbf{u}}^T\mathbf{m}\dot{\mathbf{u}} \tag{13.1b}$$

and **k** and **m** were seen to be symmetric,* that is, $\mathbf{k}^T = \mathbf{k}$ and $\mathbf{m}^T = \mathbf{m}$. For most structures **k** and **m** are *positive definite matrices*, that is, when arbitrary vectors **u** and $\dot{\mathbf{u}}$ are chosen and V and T are computed from Eqs. 13.1, the resulting values of V and T are positive except for the trivial cases $\mathbf{u} = 0$ and $\dot{\mathbf{u}} = 0$. The physical significance of this is as follows:

a. For any arbitrary displacement of an MDOF system from its undeformed configuration the strain enegy will be positive.

b. For any arbitrary velocity distribution of an MDOF system a positive kinetic energy will result.

Exceptions to the positive definiteness of **k** are systems having rigid-body freedoms. Then the displacement **u** can be taken as a rigid-body mode. In this case **k** is said to be positive *semidefinite,* that is, V can be either zero (for rigid-body modes) or greater than zero (for deformable modes). When **k** is positive semidefinite, $|\mathbf{k}| \equiv \det(\mathbf{k}) = 0$. A matrix whose determinant vanishes is called a *singular* matrix. An example of a semidefinite system is given as Example 13.2.

Exceptions to the positive definiteness of **m** are systems for which there are degrees of freedom having no associated inertia. Such a model is sometimes generated by the lumped mass approach as illustrated in Fig. 13.1 and discussed further in Sec. 16.2.

$$\mathbf{m} = \begin{bmatrix} m_1 & 0 & 0 & 0 \\ 0 & m_2 & 0 & 0 \\ 0 & 0 & 0 & 0 \\ 0 & 0 & 0 & 0 \end{bmatrix}$$

*The symmetry of stiffness coefficients, that is, $k_{ij} = k_{ji}$, for linear structures is often referred to as *Maxwell's reciprocity relationship.* The form of k_{ij} obtained by the assumed-modes method [e.g., Eq. (11.12)] shows that $k_{ij} = k_{ji}$. A homework problem (Problem 13.1) illustrates how a nonsymmetric stiffness or mass matrix might arise.

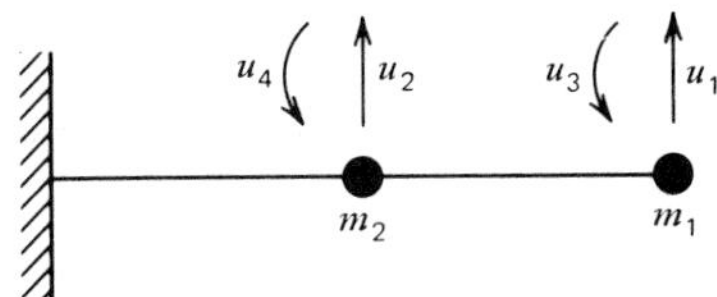

Figure 13.1. System with positive semidefinite mass matrix.

Eigensolution

The equation of motion of an undamped MDOF system, from Eq. 11.22, can be written

$$\mathbf{m}\ddot{\mathbf{u}} + \mathbf{k}\mathbf{u} = \mathbf{0} \tag{13.2}$$

where $\mathbf{m}$ and $\mathbf{k}$ are $(N \times N)$ matrices and $\mathbf{u}(t)$ is an $(N \times 1)$ vector of physical or generalized displacement coordinates.

Harmonic motion given by

$$\mathbf{u} = \mathbf{U}\cos(\omega t - \alpha) \tag{13.3}$$

may be substituted into Eq. 13.2 to give the Nth-order *algebraic eigenvalue problem*

$$(\mathbf{k} - \omega^2\mathbf{m})\mathbf{U} = \mathbf{0} \tag{13.4}$$

For there to be a nontrivial solution of Eq. 13.4 it is necessary that

$$\det(\mathbf{k} - \omega^2\mathbf{m}) = 0 \tag{13.5}$$

This is called the *characteristic equation.* When the determinant of Eq. 13.5 is expanded, there results a polynomial equation of degree N in ω^2 whose roots are the *eigenvalues,* or squared *natural frequencies,* ω_r^2. These can be ordered from lowest to highest

$$0 \leq \omega_1^2 \leq \omega_2^2 \leq \ldots \leq \omega_r^2 \leq \ldots \leq \omega_N^2 \tag{13.6}$$

Corresponding to each eigenvalue, ω_r^2, there will be an *eigenvector,* or *natural mode,* $\mathbf{U}_r$, where

$$\mathbf{U}_r = \begin{Bmatrix} U_1 \\ U_2 \\ \cdot \\ \cdot \\ \cdot \\ U_N \end{Bmatrix}_r \quad r = 1, 2, \ldots, N \tag{13.7}$$

The modes are determined only to within a constant multiplier. Thus, modes can be scaled in any convenient manner.

Scaling (Normalizing)

If the value of one of the elements of a natural mode vector $\mathbf{U}_r$ is assigned a specified value, then the remaining $(N - 1)$ elements are determined uniquely. The process of scaling a natural mode so that each of its elements has a unique value is called *normalization,* and the resulting modal vectors are called *normal modes.*

We will denote by $\boldsymbol{\phi}_r$ a mode that has been scaled to make its amplitude unique, and we will assume $\boldsymbol{\phi}_r$ to be "dimensionless," that is, an arbitrary modal vector corresponding to ω_r can be written in the form

$$\mathbf{U}_r = c_r \boldsymbol{\phi}_r \tag{13.8}$$

where c_r is a scaling constant whose units are such that $\boldsymbol{\phi}_r^T \mathbf{m} \boldsymbol{\phi}_r$ has the dimensions of mass.†

Three procedures for normalizing a mode were noted for continuous systems in Sec. 10.5. The corresponding procedures may be applied to MDOF systems.

1. Scale the rth mode so that $(\phi_i)_r = 1$ at a specified coordinate i.

2. Scale the rth mode so that $(\phi_i)_r = 1$, where $|(\phi_i)_r| = \max_j |(\phi_j)_r|$, that is, the maximum displacement is at coordinate i.

3. Scale the rth mode so that the generalized mass, or *modal mass,* defined by

$$M_r = \boldsymbol{\phi}_r^T \mathbf{m} \boldsymbol{\phi}_r \tag{13.9}$$

has a specified value. The value $M_r = 1$ is frequently used.*

The generalized stiffness, or *modal stiffness,* for the rth mode is defined as

$$K_r = \boldsymbol{\phi}_r^T \mathbf{k} \boldsymbol{\phi}_r \tag{13.10}$$

If Eq. 13.4 is written for rth mode and premultiplied by $\boldsymbol{\phi}_r^T$, we obtain

$$(\boldsymbol{\phi}_r^T \mathbf{k} \boldsymbol{\phi}_r) = \omega_r^2 (\boldsymbol{\phi}_r^T \mathbf{m} \boldsymbol{\phi}_r) \tag{13.11}$$

†The reason for this particular definition of "dimensionless" is that some vectors $\mathbf{U}$ may contain a mixture of types of coordinates, for example, translations and rotations. Thus, it would not be possible to simultaneously make all components of $\boldsymbol{\phi}$ dimensionless in the usual sense of the word.

*As previously noted, it is convenient to scale $\boldsymbol{\phi}_r$ so that the product $\boldsymbol{\phi}_r^T \mathbf{m} \boldsymbol{\phi}_r$ has the units of mass. Thus $M_r = 1$ kg (or 1 slug) is applied.

Then,

$$\omega_r^2 = \frac{K_r}{M_r} \tag{13.12}$$

as was seen for continuous systems in Eq. 10.44.

Mode Shapes—Distinct Frequency Case

If ω_r is a distinct eigenvalue, $\boldsymbol{\phi}_r$ can be determined in the following manner. Let

$$\mathbf{D}(\omega_r) \equiv \mathbf{k} - \omega_r^2\mathbf{m} \tag{13.13}$$

Assume that coordinate 1 is not a node point of the rth mode, that is, it is not a point whose displacement is zero, and partition Eq. 13.4 is as follows

$$\begin{bmatrix} D_{aa}(\omega_r) & \mathbf{D}_{ab}(\omega_r) \\ \mathbf{D}_{ba}(\omega_r) & \mathbf{D}_{bb}(\omega_r) \end{bmatrix} \begin{Bmatrix} 1 \\ \boldsymbol{\phi}_b \end{Bmatrix}_r = \begin{Bmatrix} 0 \\ \mathbf{0} \end{Bmatrix} \tag{13.14}$$

where $\boldsymbol{\phi}_r$ has been scaled by setting $\phi_1 = 1$, and where

$$(\boldsymbol{\phi}_b)_r = \begin{Bmatrix} \phi_2 \\ \phi_3 \\ \cdot \\ \cdot \\ \cdot \\ \phi_N \end{Bmatrix}_r \tag{13.15}$$

Since ω_r is an eigenvalue, Eq. 13.4 states that $\det[\mathbf{D}(\omega_r)] = 0$, that is, $\mathbf{D}(\omega_r)$ is singular. The *rank* of a matrix is defined as the largest submatrix having a nonzero determinant. If ω_r is a distinct eigenvalue, the rank of $\mathbf{D}(\omega_r)$ is $(N - 1)$ and the coordinates can be labeled so that $\mathbf{D}_{bb}(\omega_r)$ is nonsingular. Therefore $[\mathbf{D}_{bb}(\omega_r)]^{-1}$ exists, and the lower partition of Eq. 13.14 can be solved for the remainder of the mode shape $\boldsymbol{\phi}_r$, that is,

$$(\boldsymbol{\phi}_b)_r = -[\mathbf{D}_{bb}(\omega_r)]^{-1}\mathbf{D}_{ba}(\omega_r) \tag{13.16}$$

The example below illustrates the characteristic polynomial and the use of Eq. 13.16.

Example 13.1

a. For the system shown below determine the values of $\omega_r^2 (r = 1, 2, 3)$ and sketch the characteristic polynomial.

b. Solve for the mode shape corresponding to ω_2.

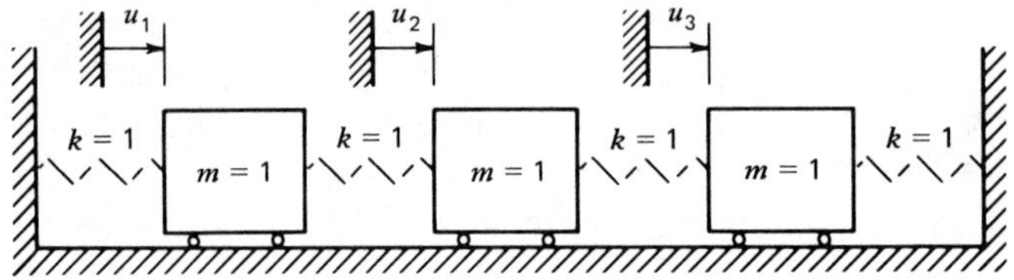

Solution

a. The stiffness matrix and mass matrix are

$$\mathbf{k} = \begin{bmatrix} 2 & -1 & 0 \\ -1 & 2 & -1 \\ 0 & -1 & 2 \end{bmatrix}, \qquad \mathbf{m} = \begin{bmatrix} 1 & 0 & 0 \\ 0 & 1 & 0 \\ 0 & 0 & 1 \end{bmatrix} \tag{1}$$

Then,

$$\mathbf{D}(\omega) = \begin{bmatrix} (2 - \omega^2) & -1 & 0 \\ -1 & (2 - \omega^2) & -1 \\ 0 & -1 & (2 - \omega^2) \end{bmatrix} \tag{2}$$

Expanding $\det[\mathbf{D}(\omega)]$ using the first row (or column), we get

$$\det[\mathbf{D}(\omega)] = (2 - \omega^2) \begin{vmatrix} (2 - \omega^2) & -1 \\ -1 & (2 - \omega^2) \end{vmatrix} + 1 \begin{vmatrix} -1 & -1 \\ 0 & (2 - \omega^2) \end{vmatrix}$$

The characteristic polynomial is

$$\det[\mathbf{D}(\omega)] = (2 - \omega^2)(\omega^4 - 4\omega^2 + 2) \tag{3}$$

Then,

$$\boxed{\omega_1^2 = 2 - \sqrt{2}, \qquad \omega_2^2 = 2, \qquad \omega_3^2 = 2 + \sqrt{2}} \tag{4}$$

The characteristic polynomial can be computed for various values of ω and plotted as shown below.

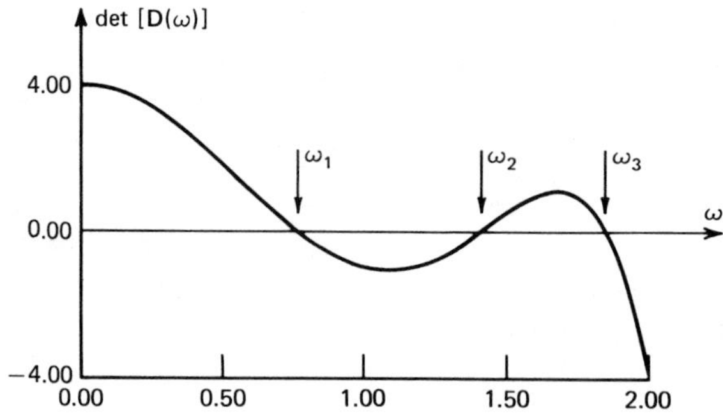

b. $\mathbf{D}(\omega_2)$ can be partitioned as in Eq. 13.14.

$$\begin{bmatrix} D_{aa}(\omega_2) & \mathbf{D}_{ab}(\omega_2) \\ \mathbf{D}_{ba}(\omega_2) & \mathbf{D}_{bb}(\omega_2) \end{bmatrix} = \left[\begin{array}{c|cc} 0 & -1 & 0 \\ \hline -1 & 0 & -1 \\ 0 & -1 & 0 \end{array}\right] \tag{5}$$

Then,

$$\mathbf{D}_{bb}(\omega_2) = \begin{bmatrix} 0 & -1 \\ -1 & 0 \end{bmatrix}, \qquad \mathbf{D}_{ba}(\omega_2) = \begin{Bmatrix} -1 \\ 0 \end{Bmatrix} \tag{6}$$

The inverse of $\mathbf{D}_{bb}$ is

$$[\mathbf{D}_{bb}(\omega_2)]^{-1} = \begin{bmatrix} 0 & -1 \\ -1 & 0 \end{bmatrix} \tag{7}$$

From Eq. 13.16 we get

$$\begin{aligned} (\boldsymbol{\phi}_b)_2 &= -[\mathrm{D}_{bb}(\omega_2)]^{-1}\mathrm{D}_{ba}(\omega_2) \\ &= -\begin{bmatrix} 0 & -1 \\ -1 & 0 \end{bmatrix}\begin{Bmatrix} -1 \\ 0 \end{Bmatrix} = \begin{Bmatrix} 0 \\ -1 \end{Bmatrix} \end{aligned} \tag{8}$$

Then, since $\phi_1 = 1$, mode 2 becomes

$$\boxed{\boldsymbol{\phi}_2 = \begin{Bmatrix} 1 \\ 0 \\ -1 \end{Bmatrix}} \tag{9}$$

as shown below.

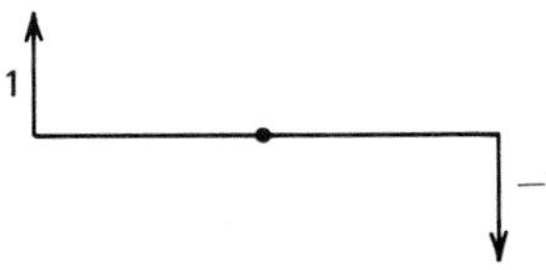

In the next example some features of semidefinite systems, that is, ones with rigid-body modes, will be illustrated.

Example 13.2

The assumed-modes method of Sec. 11.4 (or the finite element method to be discussed in Chapter 16) may be used to derive the following 3-DOF model of a uniform bar that undergoes unrestrained axial motion.

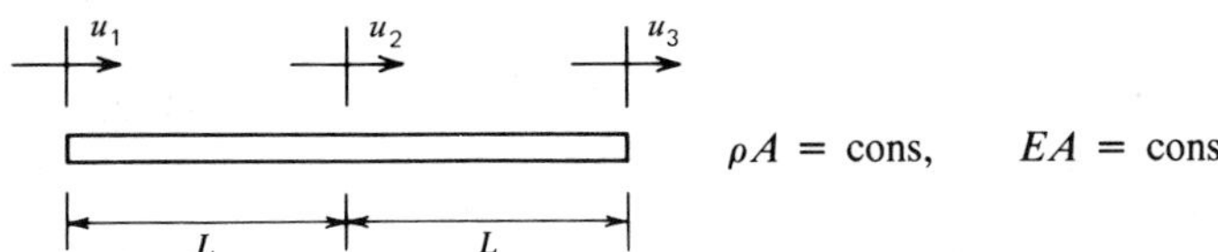

$$\frac{\rho AL}{6}\begin{bmatrix} 2 & 1 & 0 \\ 1 & 4 & 1 \\ 0 & 1 & 2 \end{bmatrix}\begin{Bmatrix} \ddot{u}_1 \\ \ddot{u}_2 \\ \ddot{u}_3 \end{Bmatrix} + \frac{EA}{L}\begin{bmatrix} 1 & -1 & 0 \\ -1 & 2 & -1 \\ 0 & -1 & 1 \end{bmatrix}\begin{Bmatrix} u_1 \\ u_2 \\ u_3 \end{Bmatrix} = \begin{Bmatrix} 0 \\ 0 \\ 0 \end{Bmatrix}$$

a. Show that **k** is singular.
b. Solve for the natural frequencies, and thus show that $\omega_1^2 = 0$.
c. Use the procedure of Eq. 13.16 to determine the rigid-body mode.

Solution

a. Evaluate det(**k**).

$$\det(\mathbf{k}) = \frac{EA}{L}\begin{vmatrix} 1 & -1 & 0 \\ -1 & 2 & -1 \\ 0 & -1 & 1 \end{vmatrix}$$

Therefore, expanding on the first row we get

$$\det(\mathbf{k}) = \begin{vmatrix} 2 & -1 \\ -1 & 1 \end{vmatrix} + \begin{vmatrix} -1 & -1 \\ 0 & 1 \end{vmatrix} = 0 \tag{1}$$

Since $\det(\mathbf{k}) = 0$, **k** is singular.

b. Set up the characteristic polynomial and determine its roots. Let

$$\mu = \omega^2\left(\frac{\rho L^2}{6E}\right) \tag{2}$$

Then, $\mathbf{D}(\omega)$ can be written in partitioned form as

$$\mathbf{D}(\mu) = \left[\begin{array}{c|cc} (1-2\mu) & (-1-\mu) & 0 \\ \hline (-1-\mu) & 2(1-2\mu) & (-1-\mu) \\ 0 & (-1-\mu) & (1-2\mu) \end{array}\right] \tag{3}$$

Expanding $\det[\mathbf{D}(\mu)]$ we obtain the characteristic polynomial

$$\det[\mathbf{D}(\mu)] = \mu(1-2\mu)(-2+\mu) = 0 \tag{4}$$

The roots of the characteristic polynomial are thus

$$\mu_1 = 0, \qquad \mu_2 = \tfrac{1}{2}, \qquad \mu_3 = 2 \tag{5}$$

so

$$\boxed{\omega_1^2 = 0, \qquad \omega_2^2 = \frac{3E}{\rho L^2}, \qquad \omega_3^2 = \frac{12E}{\rho L^2}} \tag{6}$$

d. For a rigid-body mode, $\omega = 0$. Therefore, use Eq. 13.16 to determine ϕ_1, which is a rigid-body mode. From Eq. 3,

$$\mathbf{D}_{bb}(\omega_1) = \begin{bmatrix} 2 & -1 \\ -1 & 1 \end{bmatrix}, \qquad \mathbf{D}_{ba} = \begin{Bmatrix} -1 \\ 0 \end{Bmatrix} \tag{7}$$

The inverse is

$$[\mathbf{D}_{bb}(\omega_1)]^{-1} = \begin{bmatrix} 1 & 1 \\ 1 & 2 \end{bmatrix} \tag{8}$$

Then, from Eq. 13.16,

$$\begin{aligned} (\boldsymbol{\phi}_b)_1 &= -[\mathbf{D}_{bb}(\omega_1)]^{-1}\mathbf{D}_{ba}(\omega_1) \\ &= -\begin{bmatrix} 1 & 1 \\ 1 & 2 \end{bmatrix}\begin{Bmatrix} -1 \\ 0 \end{Bmatrix} = \begin{Bmatrix} 1 \\ 1 \end{Bmatrix} \end{aligned} \tag{9}$$

Since $\phi_1 = 1$, the rigid-body mode is

$$\boxed{\boldsymbol{\phi}_1 = \begin{Bmatrix} 1 \\ 1 \\ 1 \end{Bmatrix}} \tag{10}$$

This is the expected rigid body mode shape, since Eq. 10 says that all points on the bar move the same distance in this mode. Thus, there is no deformation along the bar.

Orthogonality

The most important property of natural modes is the *orthogonality property*. We begin by writing Eq. 13.4 for mode r and premultiplying the equation by $\boldsymbol{\phi}_s^T$ to get

$$(\boldsymbol{\phi}_s^T\mathbf{k}\boldsymbol{\phi}_r) - \omega_r^2(\boldsymbol{\phi}_s^T\mathbf{m}\boldsymbol{\phi}_r) = 0 \tag{13.17}$$

Then, writing Eq. 13.4 for the sth mode and premultiplying it by $\boldsymbol{\phi}_r^T$, we get

$$(\boldsymbol{\phi}_r^T\mathbf{k}\boldsymbol{\phi}_s) - \omega_s^2(\boldsymbol{\phi}_r^T\mathbf{m}\boldsymbol{\phi}_s) = 0 \tag{13.18}$$

Since $\mathbf{k}$ and $\mathbf{m}$ are symmetric, Eq. 13.18 can be transposed and written

$$(\boldsymbol{\phi}_s^T\mathbf{k}\boldsymbol{\phi}_r) - \omega_s^2(\boldsymbol{\phi}_s^T\mathbf{m}\boldsymbol{\phi}_r) = 0 \tag{13.19}$$

Equation 13.19 may be subtracted from Eq. 13.17 to give

$$(\omega_s^2 - \omega_r^2)(\boldsymbol{\phi}_s^T\mathbf{m}\boldsymbol{\phi}_r) = 0 \tag{13.20}$$

For modes with distinct frequencies, that is, $\omega_r \neq \omega_s$, it is necessary that

$$\boldsymbol{\phi}_s^T\mathbf{m}\boldsymbol{\phi}_r = 0, \qquad (\omega_r \neq \omega_s) \tag{13.21}$$

The rth and sth modes are said to be *orthogonal with respect to the mass matrix*. Equation 13.21 can be substituted into Eq. 13.17 to show that the rth mode and sth mode are also *orthogonal with respect to the stiffness matrix,* that is,

$$\boldsymbol{\phi}_s^T \mathbf{k} \boldsymbol{\phi}_r = 0 \tag{13.22}$$

Mode Shapes—Repeated Frequency Case

It frequently happens in complex systems that there are "closely spaced" frequencies, that is, cases in which ω_r and ω_{r+1} differ by only 1% or so. It occasionally happens that a system has a *repeated frequency,* that is, $\omega_r = \omega_{r+1} = \cdots = \omega_{r+p-1}$. A theorem of linear algebra* states that if the eigenvalue is repeated p times, there will be p linearly independent eigenvectors associated with this repeated eigenvalue. Although it is not necessary that these eigenvectors be orthogonal to each other, it is possible to choose the eigenvectors such that they will, in fact, satisfy the orthogonality relationships of Eqs. 13.21 and 13.22, even though $\omega_r = \omega_s$.

The procedure for determining mode shapes corresponding to a repeated frequency differs slightly from that established in Eq. 13.16 and illustrated in Examples 13.1 and 13.2. The rank of $\mathbf{D}(\omega_r)$ is $(N - p)$ if the frequency ω_r is repeated p times. Let Eq. 13.4 be written in partitioned form

$$\begin{bmatrix} \underset{p \times p}{\mathbf{D}_{aa}(\omega_r)} & \underset{p \times (N-p)}{\mathbf{D}_{ab}(\omega_r)} \\ \underset{(N-p) \times p}{\mathbf{D}_{ba}(\omega_r)} & \underset{(N-p) \times (N-p)}{\mathbf{D}_{bb}(\omega_r)} \end{bmatrix} \begin{Bmatrix} \boldsymbol{\phi}_a \\ \boldsymbol{\phi}_b \end{Bmatrix}_r = \begin{Bmatrix} \mathbf{0} \\ \mathbf{0} \end{Bmatrix} \tag{13.23}$$

where $\mathbf{D}_{bb}$ is nonsingular and where

$$(\boldsymbol{\phi}_a)_r = \begin{Bmatrix} \phi_1 \\ \phi_2 \\ \cdot \\ \cdot \\ \cdot \\ \phi_p \end{Bmatrix}_r, \qquad (\boldsymbol{\phi}_b)_r = \begin{Bmatrix} \phi_{p+1} \\ \phi_{p+2} \\ \cdot \\ \cdot \\ \cdot \\ \phi_N \end{Bmatrix}_r \tag{13.24}$$

The lower partition can be solved for $(\boldsymbol{\phi}_b)_r$, giving

$$(\boldsymbol{\phi}_b)_r = -[\mathbf{D}_{bb}(\omega_r)]^{-1}\mathbf{D}_{ba}(\omega_r)(\boldsymbol{\phi}_a)_r \tag{13.25}$$

According to the theorem stated above there will be p linearly independent vectors corresponding to the repeated frequency ω_r. Hence, we must pick p linearly independent vectors $(\boldsymbol{\phi}_a)_r$, $(\boldsymbol{\phi}_a)_{r+1}$, . . . , $(\boldsymbol{\phi}_a)_{r+p-1}$, and Eq. 13.25 will

*The theorem is usually proved for the standard eigenvalue problem $\mathbf{Ax} = \lambda\mathbf{x}$, where $\mathbf{A}$ is symmetric. However, the generalized eigenvalue problem $\mathbf{k}\boldsymbol{\phi} = \omega^2\mathbf{m}\boldsymbol{\phi}$ can be cast into the required standard, symmetric form.

determine the remaining components of the modal vectors ϕ_r, and so forth. The vectors

$$(\phi_a)_r = \begin{Bmatrix} \phi_1 \\ \phi_2 \\ \phi_3 \\ \cdot \\ \cdot \\ \cdot \\ \phi_p \end{Bmatrix} = \begin{Bmatrix} 1 \\ 0 \\ 0 \\ \cdot \\ \cdot \\ \cdot \\ 0 \end{Bmatrix}, \quad (\phi_a)_{r+1} = \begin{Bmatrix} 0 \\ 1 \\ 0 \\ \cdot \\ \cdot \\ \cdot \\ 0 \end{Bmatrix}, \ \ldots, \ (\phi_a)_{r+p-1} = \begin{Bmatrix} 0 \\ 0 \\ 0 \\ \cdot \\ \cdot \\ \cdot \\ 1 \end{Bmatrix} \tag{13.26}$$

form a convenient linearly independent set of vectors to use as ϕ_a vectors. Example 13.3 illustrates how $\mathbf{D}(\omega_r)$ is partitioned for a system with $\omega_1 = \omega_2 = 0$, that is, $p = 2$ for the frequency $\omega_r = 0$.

Example 13.3

A model of a uniform beam is created by lumping mass at three nodes as shown.*

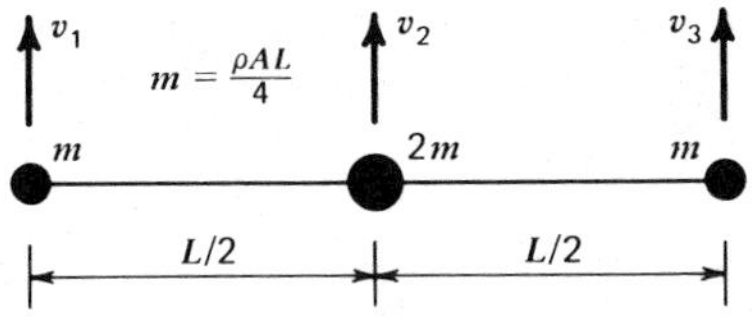

The resulting equation of motion is

$$\frac{\rho AL}{4}\begin{bmatrix} 1 & 0 & 0 \\ 0 & 2 & 0 \\ 0 & 0 & 1 \end{bmatrix}\begin{Bmatrix} \ddot{v}_1 \\ \ddot{v}_2 \\ \ddot{v}_3 \end{Bmatrix} + \frac{12EI}{L^3}\begin{bmatrix} 1 & -2 & 1 \\ -2 & 4 & -2 \\ 1 & -2 & 1 \end{bmatrix}\begin{Bmatrix} v_1 \\ v_2 \\ v_3 \end{Bmatrix} = \begin{Bmatrix} 0 \\ 0 \\ 0 \end{Bmatrix}$$

a. Solve for the three natural frequencies of this system.
b. Solve for the three normal modes of the system.
c. Evaluate $\phi_1^T \mathbf{m} \phi_2$ and $\phi_1^T \mathbf{m} \phi_3$.

Solution

a. Set up and solve the characteristic equation.
Let

$$\mu = \omega^2 \left(\frac{\rho AL^4}{48EI}\right) \tag{1}$$

*The names *grid point* and *node* are used for locations on a structure where displacement coordinates, for example, v_1, v_2, are assigned. This usage of "node" should not be confused with a "node" of a mode shape as noted in Example 12.1.

Then the eigenvalue equation

$$(\mathbf{k} - \omega^2\mathbf{m})\boldsymbol{\phi} = \mathbf{0} \tag{2}$$

can be written in the form

$$\mathbf{D}(\mu)\boldsymbol{\phi} = \begin{bmatrix} (1-\mu) & -2 & 1 \\ -2 & 2(2-\mu) & -2 \\ 1 & -2 & (1-\mu) \end{bmatrix} \begin{Bmatrix} \phi_1 \\ \phi_2 \\ \phi_3 \end{Bmatrix} = \begin{Bmatrix} 0 \\ 0 \\ 0 \end{Bmatrix} \tag{3}$$

The characteristic equation is $\det[\mathbf{D}(\mu)] = 0$, which simplifies to

$$\mu^2(\mu - 4) = 0 \tag{4}$$

so

$$\mu_1 = \mu_2 = 0, \qquad \mu_3 = 4 \tag{5}$$

Then,

$$\boxed{\omega_1 = \omega_2 = 0, \qquad \omega_3 = 4\left(\frac{48EI}{\rho AL^4}\right)} \tag{6}$$

b. Solve first for mode 3, which corresponds to the unique frequency ω_3.

Use the same procedure as in Examples 13.1 and 13.2. Partition $\mathbf{D}(\mu_3)$ by isolating the top row and the left-most column

$$\begin{bmatrix} D_{aa}(\mu_3) & \mathbf{D}_{ab}(\mu_3) \\ \mathbf{D}_{ba}(\mu_3) & \mathbf{D}_{bb}(\mu_3) \end{bmatrix} = \left[\begin{array}{c|cc} -3 & -2 & 1 \\ \hline -2 & -4 & -2 \\ 1 & -2 & -3 \end{array}\right] \tag{7}$$

Then, $\mathbf{D}_{bb}$ is nonsingular and

$$[\mathbf{D}_{bb}(\mu_3)]^{-1} = \frac{1}{8}\begin{bmatrix} -3 & 2 \\ 2 & -4 \end{bmatrix} \tag{8}$$

From Eq. 13.16,

$$\begin{aligned} (\boldsymbol{\phi}_b)_3 &= -[\mathbf{D}_{bb}(\mu_3)]^{-1}\mathbf{D}_{ba}(\mu_3) \\ &= -\frac{1}{8}\begin{bmatrix} -3 & 2 \\ 2 & -4 \end{bmatrix}\begin{Bmatrix} -2 \\ 1 \end{Bmatrix} = \begin{Bmatrix} -1 \\ 1 \end{Bmatrix} \end{aligned} \tag{9}$$

Since $(\phi_a)_3$ is assumed to be 1,

$$\boxed{\boldsymbol{\phi}_3 = \begin{Bmatrix} 1 \\ -1 \\ 1 \end{Bmatrix}} \tag{10}$$

Solve for the two modes corresponding to $\omega = 0$. For $\mu = 0$, $\mathbf{D}(\mu)$ becomes

$$\mathbf{D}(\mu_1) = \mathbf{D}(\mu_2) = \mathbf{D}(0) = \begin{bmatrix} 1 & -2 & 1 \\ -2 & 4 & -2 \\ 1 & -2 & 1 \end{bmatrix} \tag{11}$$

It is easily seen that each row (column) is a multiple of the first row (column) and that there is no 2×2 submatrix of $\mathbf{D}(0)$ which has a nonzero determinant. Thus, the rank of $\mathbf{D}(0)$ is $(N - p) = (3 - 2) = 1$. For $\mu = 0$, let Eq. 3 be written in partitioned form:

$$\begin{bmatrix} \mathbf{D}_{aa}(0) & \mathbf{D}_{ab}(0) \\ \mathbf{D}_{ba}(0) & D_{bb}(0) \end{bmatrix} = \left[\begin{array}{cc|c} 1 & -2 & 1 \\ -2 & 4 & -2 \\ \hline 1 & -2 & 1 \end{array}\right] \tag{12}$$

From Eq. 13.25,

$$(\phi_3)_r = -1[1 \quad -2] \begin{Bmatrix} \phi_1 \\ \phi_2 \end{Bmatrix}_r, \qquad r = 1, 2 \tag{13}$$

Choose the linearly independent vectors

$$\begin{Bmatrix} \phi_1 \\ \phi_2 \end{Bmatrix}_1 = \begin{Bmatrix} 1 \\ 0 \end{Bmatrix} \quad \text{and} \quad \begin{Bmatrix} \phi_1 \\ \phi_2 \end{Bmatrix}_2 = \begin{Bmatrix} 0 \\ 1 \end{Bmatrix} \tag{14}$$

From Eqs. 13 and 14,

$$(\phi_3)_1 = [-1 \quad 2] \begin{Bmatrix} 1 \\ 0 \end{Bmatrix} = -1 \tag{15a}$$

$$(\phi_3)_2 = [-1 \quad 2] \begin{Bmatrix} 0 \\ 1 \end{Bmatrix} = 2 \tag{15b}$$

Then

$$\boxed{\boldsymbol{\phi}_1 = \begin{Bmatrix} 1 \\ 0 \\ -1 \end{Bmatrix}, \qquad \boldsymbol{\phi}_2 = \begin{Bmatrix} 0 \\ 1 \\ 2 \end{Bmatrix}} \tag{16}$$

which are the rigid-body modes sketched below.

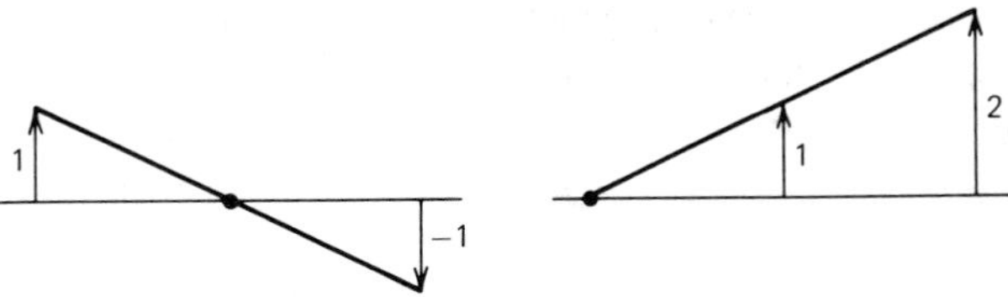

c. Evaluate $\boldsymbol{\phi}_1^T \mathbf{m} \boldsymbol{\phi}_2$ and $\boldsymbol{\phi}_1^T \mathbf{m} \boldsymbol{\phi}_3$

$$\boldsymbol{\phi}_1^T \mathbf{m} \boldsymbol{\phi}_2 = [1 \quad 0 \quad -1] \begin{bmatrix} m & 0 & 0 \\ 0 & 2m & 0 \\ 0 & 0 & m \end{bmatrix} \begin{Bmatrix} 0 \\ 1 \\ 2 \end{Bmatrix} = -2m \neq 0 \tag{17}$$

Therefore, the two rigid-body modes, although linearly independent, are not orthogonal. Let $\hat{\phi}_2 = \phi_2 + \phi_1$, which is a particular linear combination of the above rigid-body modes. Then

$$\boldsymbol{\phi}_1^T\mathbf{m}\hat{\boldsymbol{\phi}}_2 = [1 \quad 0 \quad -1]\begin{bmatrix} m & 0 & 0 \\ 0 & 2m & 0 \\ 0 & 0 & m \end{bmatrix}\begin{Bmatrix} 1 \\ 1 \\ 1 \end{Bmatrix} = 0 \tag{18}$$

Thus, the modes ϕ_1 and $\hat{\phi}_2$ form a set of rigid body modes that are orthogonal with respect to the mass matrix.

Now evaluate $\boldsymbol{\phi}_1^T\mathbf{m}\boldsymbol{\phi}_3$.

$$\boldsymbol{\phi}_1^T\mathbf{m}\boldsymbol{\phi}_3 = [1 \quad 0 \quad -1]\begin{bmatrix} m & 0 & 0 \\ 0 & 2m & 0 \\ 0 & 0 & m \end{bmatrix}\begin{Bmatrix} 1 \\ -1 \\ 1 \end{Bmatrix} = 0 \tag{19}$$

Thus, orthogonality is satisfied since $\omega_1 \neq \omega_3$.

As noted in the example above, systems that possess repeated frequencies, or even very closely spaced frequencies, require special consideration when natural modes are to be determined.

Modal Matrix

It will be convenient in the remainder of this section to assume that all the modes are orthogonal, that is, Eqs. 13.21 and 13.22 hold for $r \neq s$, including cases where $\omega_r = \omega_s$. The *modal matrix* of a system is a matrix whose columns are the respective normal modes, that is,

$$\boldsymbol{\Phi} = [\boldsymbol{\phi}_1, \boldsymbol{\phi}_2 \ldots \boldsymbol{\phi}_N] \tag{13.27}$$

Generalized Mass Matrix and Generalized Stiffness Matrix

By using the definitions of modal mass and modal stiffness given in Eqs. 13.9 and 13.10 together with the orthogonality equations, Eqs. 13.21 and 13.22, we obtain a diagonal *modal mass matrix, $\boldsymbol{M}$*, and a diagonal *modal stiffness matrix, $\boldsymbol{K}$*, given by

$$\boldsymbol{M} = \boldsymbol{\Phi}^T\mathbf{m}\boldsymbol{\Phi} = \text{diag}(M_1 M_2 \ldots M_N) \tag{13.28}$$

and

$$\boldsymbol{K} = \boldsymbol{\Phi}^T\mathbf{k}\boldsymbol{\Phi} = \text{diag}(K_1 K_2 \ldots K_N) \tag{13.29}$$

If the natural modes are normalized so that $M_r = 1$, they are said to form an *orthonormal* set of vectors. Then $\boldsymbol{M}$ becomes the unit matrix, that is,

$$\mathbf{\Phi}^T\mathbf{m}\mathbf{\Phi} = \mathbf{I} \tag{13.30}$$

and

$$\mathbf{\Phi}^T\mathbf{k}\mathbf{\Phi} = \mathbf{\Lambda} \tag{13.31}$$

where

$$\mathbf{\Lambda} = \text{diag}(\omega_1^2, \omega_2^2, \ldots, \omega_N^2) \tag{13.32}$$

Expansion Theorem

An *expansion theorem* similar to that stated for continuous systems in Eq. 10.54 can be stated for MDOF systems. The normal modes, $\boldsymbol{\phi}_r$, $r = 1, 2, \ldots, N$ form a mutually orthogonal set of N-dimensional vectors. We can show (See Problem 13.8) that the N modes also form a linearly independent set of N-dimensional vectors. Hence, an arbitrary vector $\mathbf{u}$ can be expressed as a linear combination of the normal modes by the equation

$$\mathbf{u} = \sum_{r=1}^{N} c_r \boldsymbol{\phi}_r \tag{13.33}$$

where the c_r's are determined by

$$c_r = \left(\frac{1}{M_r}\right) \boldsymbol{\phi}_r^T \mathbf{m}\mathbf{u} \tag{13.34}$$

Equation 13.33 forms the basis of what is known as the *normal mode method* or the *mode-superposition method,* which was employed in Example 12.6 and will be treated in more detail in Chapter 15.

Example 13.4

A uniform beam was modeled in Example 13.3 by three lumped masses on a massless beam. The model is symmetric about its central mass. The mutually orthogonal normal modes can be identified as two symmetric modes and one antisymmetric mode.

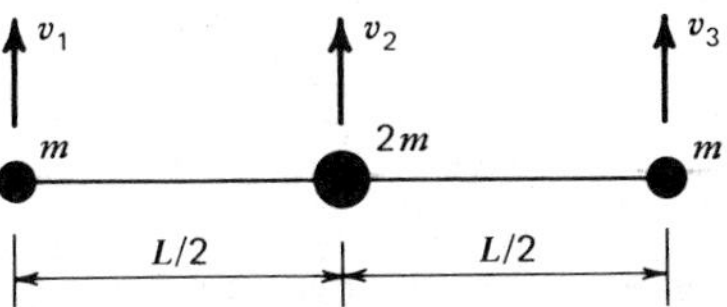

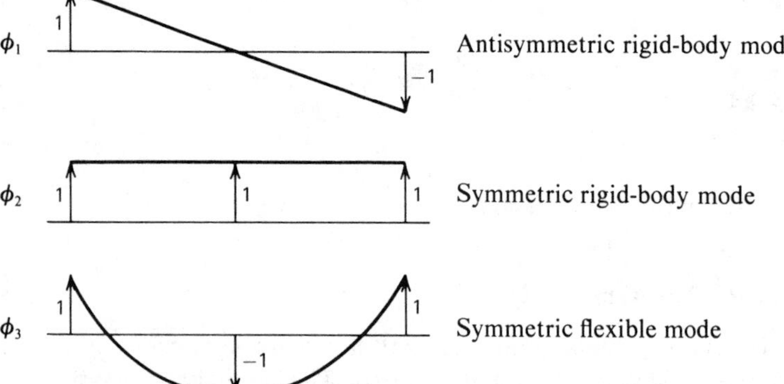

Show that the expansion of any arbitrary symmetric deflection $\mathbf{v}$ using Eq. 13.35 will involve only the symmetric modes of the structure.

Solution

We need to show that for an arbitrary symmetric vector of the form

$$\mathbf{v} = \begin{Bmatrix} a \\ b \\ a \end{Bmatrix} \tag{1}$$

the coefficient c_1 vanishes in the expansion

$$\mathbf{v} = \sum_{r=1}^{3} c_r \boldsymbol{\phi}_r \tag{2}$$

Therefore, evaluate c_1 from Eq. 13.34.

$$c_1 = \frac{\boldsymbol{\phi}_1^T \mathbf{m} \mathbf{v}}{\boldsymbol{\phi}_1^T \mathbf{m} \boldsymbol{\phi}_1} \tag{3}$$

But,

$$\begin{aligned} \boldsymbol{\phi}_1^T \mathbf{m} \mathbf{v} &= [1 \quad 0 \quad -1] \begin{bmatrix} m & 0 & 0 \\ 0 & 2m & 0 \\ 0 & 0 & m \end{bmatrix} \begin{Bmatrix} a \\ b \\ a \end{Bmatrix} \\ &= [1 \quad 0 \quad -1] \begin{Bmatrix} ma \\ 2mb \\ ma \end{Bmatrix} = 0 \end{aligned} \tag{4}$$

Therefore, $c_1 = 0$, and the expansion of any symmetric deflection involves only the symmetric modes of the system.

General Solution for Free Vibration

Returning to Eq. 13.3, we see that the free vibration of an undamped N-DOF system* vibrating in its rth natural mode is given by

$$\mathbf{u}_r = c_r \boldsymbol{\phi}_r \cos(\omega_r t - \alpha_r) \tag{13.35}$$

Thus, the *general solution* for free vibration may be written in the form

$$\mathbf{u} = \sum_{r=1}^{N} c_r \boldsymbol{\phi}_r \cos(\omega_r t - \alpha_r) \tag{13.36}$$

or, in the form that was employed in Example 12.2

$$\mathbf{u} = \sum_{r=1}^{N} \boldsymbol{\phi}_r [a_r \cos(\omega_r t) + b_r \sin(\omega_r t)] \tag{13.37}$$

The $2N$ coefficients (c_r, α_r) or (a_r, b_r) are determined by the initial conditions $\mathbf{u}(0)$ and $\dot{\mathbf{u}}(0)$. The coefficients a_r and b_r can be determined as in Eqs. 13.33 and 13.34. Thus,

$$\mathbf{u}(0) = \sum_{r=1}^{N} \boldsymbol{\phi}_r a_r \tag{13.38a}$$

$$\dot{\mathbf{u}}(0) = \sum_{r=1}^{N} \boldsymbol{\phi}_r \omega_r b_r \tag{13.38b}$$

Multiplying Eqs. 13.38 by $(\boldsymbol{\phi}_s^T \mathbf{m})$ and observing the orthogonality of the modes, we obtain

$$a_r = \frac{\boldsymbol{\phi}_r^T \mathbf{m} \mathbf{u}(0)}{M_r} \tag{13.39a}$$

$$b_r = \frac{\boldsymbol{\phi}_r^T \mathbf{m} \dot{\mathbf{u}}(0)}{M_r \omega_r} \tag{13.39b}$$

This provides a straightforward procedure for determining the coefficients in Eq. 13.37 and, hence, for determining the free vibration due to $\mathbf{u}(0)$ and $\dot{\mathbf{u}}(0)$.

Example 13.5

Using Eqs. 13.37 and 13.39, determine expressions for the free vibration response of the 2-DOF system of Examples 12.1 and 12.2 if the initial conditions are

$$\mathbf{u}(0) = \begin{Bmatrix} 0 \\ u_o \end{Bmatrix}, \qquad \dot{\mathbf{u}}(0) = \begin{Bmatrix} 0 \\ 0 \end{Bmatrix}$$

*The results here are expressed for systems having no rigid-body modes and no damping. More general cases of systems having rigid-body modes and damping will be treated in Chapter 15.

The mass matrix and mode shapes for the system are

$$\mathbf{m} = \begin{bmatrix} m & 0 \\ 0 & m \end{bmatrix}, \qquad \boldsymbol{\phi}_1 = \begin{Bmatrix} 1 \\ 1 \end{Bmatrix}, \qquad \boldsymbol{\phi}_2 = \begin{Bmatrix} 1 \\ -1 \end{Bmatrix}$$

Solution

a. Determine the modal masses M_r from Eq. 13.9.

$$M_r = \boldsymbol{\phi}_r^T \mathbf{m} \boldsymbol{\phi}_r \tag{1}$$

$$M_1 = [1 \quad 1] \begin{bmatrix} m & 0 \\ 0 & m \end{bmatrix} \begin{Bmatrix} 1 \\ 1 \end{Bmatrix} = 2m \tag{2}$$

$$M_2 = [1 \quad -1] \begin{bmatrix} m & 0 \\ 0 & m \end{bmatrix} \begin{Bmatrix} 1 \\ -1 \end{Bmatrix} = 2m$$

b. Determine the coefficients a_r and b_r from Eqs. 13.39.

$$a_r = \frac{\boldsymbol{\phi}_r^T \mathbf{m}\mathbf{u}(0)}{M_r} \qquad b_r = \frac{\boldsymbol{\phi}_r^T \mathbf{m}\dot{\mathbf{u}}(0)}{M_r \omega_r} \tag{3}$$

$$\mathbf{m}\mathbf{u}(0) = \begin{bmatrix} m & 0 \\ 0 & m \end{bmatrix} \begin{Bmatrix} 0 \\ u_o \end{Bmatrix} = \begin{Bmatrix} 0 \\ mu_o \end{Bmatrix} \qquad \mathbf{m}\dot{\mathbf{u}}(0) = \begin{bmatrix} m & 0 \\ 0 & m \end{bmatrix} \begin{Bmatrix} 0 \\ 0 \end{Bmatrix} = \begin{Bmatrix} 0 \\ 0 \end{Bmatrix} \tag{4}$$

Therefore,

$$a_1 = \frac{\boldsymbol{\phi}_1^T \mathbf{m}\mathbf{u}(0)}{M_1} = \left(\frac{1}{2m}\right) [1 \quad 1] \begin{Bmatrix} 0 \\ mu_o \end{Bmatrix} = \frac{u_o}{2} \qquad a_2 = \frac{\boldsymbol{\phi}_2^T \mathbf{m}\mathbf{u}(0)}{M_2} = \left(\frac{1}{2m}\right) [1 \quad -1] \begin{Bmatrix} 0 \\ mu_o \end{Bmatrix} = \frac{-u_o}{2} \tag{5}$$

and finally, from Eq. 13.37,

$$\mathbf{u} = \sum_{r=1}^{2} a_r \boldsymbol{\phi}_r \cos \omega_r t$$

$$\begin{Bmatrix} u_1 \\ u_2 \end{Bmatrix} = \frac{u_o}{2} \begin{Bmatrix} 1 \\ 1 \end{Bmatrix} \cos(\omega_1 t) - \frac{u_o}{2} \begin{Bmatrix} 1 \\ -1 \end{Bmatrix} \cos(\omega_2 t)$$

or

$$\boxed{\begin{aligned} u_1(t) &= \left(\frac{u_o}{2}\right) [\cos(\omega_1 t) - \cos(\omega_2 t)] \\ u_2(t) &= \left(\frac{u_o}{2}\right) [\cos(\omega_1 t) + \cos(\omega_2 t)] \end{aligned}}$$

which is the same result determined by a different procedure in Example 12.2.

Although the procedure of Example 12.2 works well for a 2-DOF problem, the method employed in this example is systematic and can be easily applied in a computer solution for the free vibration response of any size system.

13.2 Rayleigh Method; Rayleigh-Ritz Method

The Rayleigh quotient for continuous systems was defined in Eq. 10.56. The *Rayleigh quotient* for MDOF systems is defined by

$$\omega_R^2 \equiv R(\mathbf{U}) = \frac{\mathbf{U}^T\mathbf{k}\mathbf{U}}{\mathbf{U}^T\mathbf{m}\mathbf{U}} \tag{13.40}$$

Let $\mathbf{U}$ be expanded in a series of the orthonormal modes ϕ_r, that is, $M_r = 1$.

$$\mathbf{U} = \sum_{r=1}^{N} c_r \phi_r \tag{13.41}$$

Then, due to the orthogonality properties of the modes

$$R(\mathbf{U}) = \frac{\omega_1^2 c_1^2 + \omega_2^2 c_2^2 + \ldots + \omega_N^2 c_N^2}{c_1^2 + c_2^2 + \ldots + c_N^2} \tag{13.42}$$

Using Eq. 13.42 we can prove that

$$\omega_1^2 \leq R(\mathbf{U}) \leq \omega_N^2 \tag{13.43}$$

If $\omega_1 \neq 0$, we can write Eq. 13.42 in the form

$$R(\mathbf{U}) = \omega_1^2 \left[\frac{1 + (c_2/c_1)^2(\omega_2/\omega_1)^2 + \ldots + (c_N/c_1)^2(\omega_N/\omega_1)^2}{1 + (c_2/c_1)^2 + \ldots + (c_N/c_1)^2} \right] \tag{13.44}$$

Since $\omega_1 \leq \omega_2 \ldots \leq \omega_N$, each term in the numerator is greater than, or equal to, the corresponding term in the denominator. Hence,

$$R(\mathbf{U}) \geq \omega_1^2 \tag{13.45}$$

A similar procedure can be employed to show that $R(\mathbf{U}) \leq \omega_N^2$, and hence Eq. 13.43 is proved.

The *Rayleigh method* was introduced in Sec. 10.3 as a procedure for approximating a continuous system by a SDOF system and using energy conservation to calculate an approximate fundamental frequency, which was denoted ω_R. A similar procedure can also be used to reduce an N-DOF system to a SDOF system which approximates the fundamental mode of the system. Let

$$\mathbf{u}(t) = \mathbf{U}\cos(\omega_R t) = \psi \hat{U} \cos(\omega_R t) \tag{13.46}$$

Then,

$$\dot{\mathbf{u}} = -\omega_R \mathbf{U}\sin(\omega_R t) = -\omega_R \boldsymbol{\psi}\hat{U}\sin(\omega_R t) \tag{13.47}$$

Since

$$V = \tfrac{1}{2}\mathbf{u}^T\mathbf{k}\mathbf{u}, \qquad T = \tfrac{1}{2}\dot{\mathbf{u}}^T\mathbf{m}\dot{\mathbf{u}} \tag{13.48}$$

then

$$V_{\max} = \tfrac{1}{2}\hat{k}\hat{U}^2, \qquad T_{\max} = \tfrac{1}{2}\,\omega_R^2\hat{m}\hat{U}^2 \tag{13.49}$$

where

$$\hat{k} = \boldsymbol{\psi}^T\mathbf{k}\boldsymbol{\psi}, \qquad \hat{m} = \boldsymbol{\psi}^T\mathbf{m}\boldsymbol{\psi} \tag{13.50}$$

For energy conservation to hold, $V_{\max} = T_{\max}$, which gives

$$R(\mathbf{U}) = \omega_R^2 = \frac{\hat{k}}{\hat{m}} \tag{13.51}$$

From Eqs. 13.43 and 13.51, then,

$$\omega_1^2 \le \omega_R^2 \le \omega_N^2 \tag{13.52}$$

An assumed mode shape $\boldsymbol{\psi}$ which closely approximates the shape of the true fundamental mode $\boldsymbol{\phi}_1$ will give a close upper bound to ω_1.

The *Rayleigh-Ritz method* for MDOF systems permits approximate values of the frequencies of $\hat{N}$ modes ($\hat{N} < N$) to be computed. The displacement is assumed to be harmonic in time

$$\mathbf{u}(t) = \mathbf{U}\cos(\omega t - \alpha) \tag{13.53}$$

The shape vector $\mathbf{U}$ is assumed to be given by the series expansion

$$\mathbf{U} = \sum_{i=1}^{\hat{N}} \hat{U}_i \boldsymbol{\psi}_i = \boldsymbol{\Psi}\hat{\mathbf{U}} \tag{13.54}$$

where

$$\boldsymbol{\Psi} = [\boldsymbol{\psi}_1\,\boldsymbol{\psi}_2 \cdots \boldsymbol{\psi}_{\hat{N}}] \tag{13.55}$$

The $\boldsymbol{\psi}_i$'s are preselected linearly independent assumed-mode vectors. Setting $T_{\max} = V_{\max}$ leads to the Rayleigh quotient

$$R(\mathbf{U}) \equiv \hat{\omega}^2 = \frac{\displaystyle\sum_{i=1}^{\hat{N}}\sum_{j=1}^{\hat{N}} \hat{U}_i\hat{U}_j\hat{k}_{ij}}{\displaystyle\sum_{i=1}^{\hat{N}}\sum_{j=1}^{\hat{N}} \hat{U}_i\hat{U}_j\hat{m}_{ij}} = \frac{\hat{\mathbf{U}}^T\hat{\mathbf{k}}\hat{\mathbf{U}}}{\hat{\mathbf{U}}^T\hat{\mathbf{m}}\hat{\mathbf{U}}} \tag{13.56}$$

where

$$\hat{\mathbf{k}} = \mathbf{\Psi}^T\mathbf{k}\mathbf{\Psi}, \qquad \hat{\mathbf{m}} = \mathbf{\Psi}^T\mathbf{m}\mathbf{\Psi} \tag{13.57}$$

Comparing Eq. 13.51 with Eq. 13.56, we see that a definite frequency ω_R is established by the former, whereas the value of $\hat{\omega}^2$ in Eq. 13.56 depends on the values of $\hat{U}_i$, $i = 1, 2, \ldots, \hat{N}$.

Ritz proposed that the coefficients $\hat{U}_i$ be chosen to make $R(\mathbf{U})$ stationary. That is, let

$$\frac{\partial R(\mathbf{U})}{\partial \hat{U}_i} = 0, \qquad i = 1, 2, \ldots, \hat{N} \tag{13.58}$$

Let

$$R(\mathbf{U}) = \frac{N(\mathbf{U})}{D(\mathbf{U})} \tag{13.59}$$

Then, Eq. 13.58 gives

$$N(\mathbf{U})\frac{\partial D(\mathbf{U})}{\partial \hat{U}_i} - D(\mathbf{U})\frac{\partial N(\mathbf{U})}{\partial \hat{U}_i} = 0 \tag{13.60}$$

But, since $\hat{k}_{ij} = \hat{k}_{ji}$,

$$\frac{\partial N(\mathbf{U})}{\partial \hat{U}_i} = 2\sum_{j=1}^{\hat{N}} \hat{k}_{ij}\hat{U}_j \tag{13.61}$$

and similarly for the derivative of $D(\mathbf{U})$. Combining Eqs. 13.56, 13.60, and 13.61, we get

$$\sum_{j=1}^{\hat{N}} (\hat{k}_{ij} - \hat{\omega}^2\hat{m}_{ij})\hat{U}_j = 0, \qquad i = 1, 2, \ldots, \hat{N} \tag{13.62}$$

or

$$(\hat{\mathbf{k}} - \hat{\omega}^2\hat{\mathbf{m}})\hat{\mathbf{U}} = \mathbf{0} \tag{13.63}$$

This eigenvalue problem leads to a set of $\hat{N}$ approximate frequencies $\hat{\omega}_r$ and corresponding modes $\hat{\mathbf{U}}_r$. The relationship of the $\hat{N}$ approximate frequencies to the N exact frequencies is shown later in this section.

The Rayleigh-Ritz procedure described above can be considered to be a special case of applying the assumed-modes method to free vibration of a MDOF system. The assumed-modes method can be used to reduce an N-DOF system to an $\hat{N}$-DOF system by assuming

$$\mathbf{u}(t) = \sum_{i=1}^{\hat{N}} \psi_i\hat{u}_i(t) = \mathbf{\Psi}\hat{\mathbf{u}}(t) \tag{13.64}$$

This approximation can be substituted into expressions for V and T in Eqs. 13.48, and the result substituted into Lagrange's equation, Eq. 11.6, to give

$$\hat{\mathbf{m}}\ddot{\hat{\mathbf{u}}} + \hat{\mathbf{k}}\hat{\mathbf{u}} = \mathbf{0} \tag{13.65}$$

for undamped free vibration, or, for forced vibration of a viscous-damped system

$$\hat{\mathbf{m}}\ddot{\hat{\mathbf{u}}} + \hat{\mathbf{c}}\dot{\hat{\mathbf{u}}} + \hat{\mathbf{k}}\hat{\mathbf{u}} = \hat{\mathbf{p}}(t) \tag{13.66}$$

where

$$\hat{\mathbf{c}} = \mathbf{\Psi}^T\mathbf{c}\mathbf{\Psi}, \qquad \hat{\mathbf{p}}(t) = \mathbf{\Psi}^T\mathbf{p}(t) \tag{13.67}$$

Several specific procedures for selecting the assumed-mode vectors ψ_i will be discussed in Sec. 16.6.

A very interesting and useful property of eigenvalues (natural frequencies) is the *eigenvalue separation property*[13.1]. Let the original N-DOF eigenvalue problem be stated as

$$(\mathbf{k} - \lambda\mathbf{m})\mathbf{U} = \mathbf{0} \tag{13.68}$$

where $\lambda = \omega^2$, and let

$$(\mathbf{k}^{(m)} - \lambda^{(m)}\mathbf{m}^{(m)})\mathbf{U}^{(m)} = \mathbf{0}, \qquad m = 0, 1, \ldots, (N-1) \tag{13.69}$$

be the mth constrained eigenvalue problem, where $\mathbf{k}^{(m)}$ and $\mathbf{m}^{(m)}$ are obtained by deleting the last m rows and columns of $\mathbf{k}$ and $\mathbf{m}$, respectively. $\mathbf{k}^{(0)} = \mathbf{k}$ and

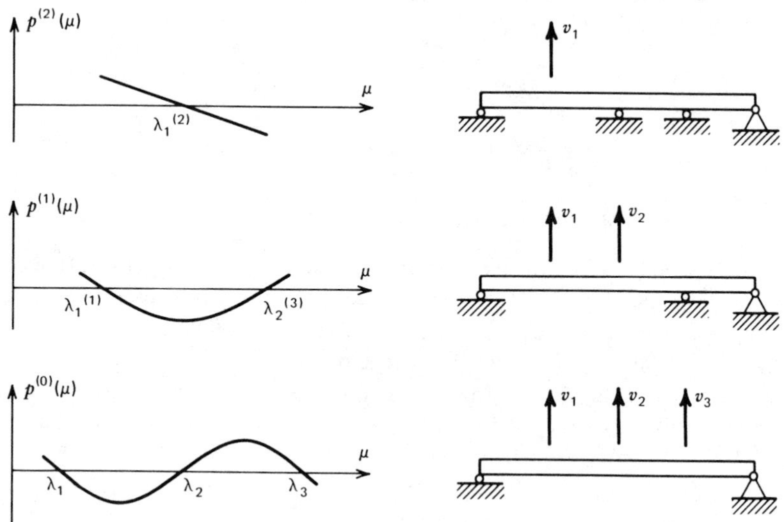

Figure 13.2. Illustration of eigenvalue separation property.

$\mathbf{m}^{(0)} = \mathbf{m}$. Then, the *eigenvalue separation theorem* states that

$$\lambda_1^{(m)} \leq \lambda_1^{(m+1)} \leq \lambda_2^{(m)} \leq \lambda_2^{(m+1)} \leq \ldots \leq \lambda_{(N-m)}^{(m)}$$
$$\text{for } m = 0, 1, 2, \ldots, (N-2) \quad (13.70)$$

That is, the eigenvalues of the $(m + 1)$st problem separate the eigenvalues of the mth problem as illustrated by the sketch in Fig. 13.2, where

$$p^{(m)}(\mu) = \det(\mathbf{k}^{(m)} - \mu\mathbf{m}^{(m)}) \quad (13.71)$$

The eigenvalue separation theorem of Eq. 13.70 can be employed directly to show the convergence properties of frequencies obtained by the Rayleigh-Ritz method. The result is shown in Table 13.1. Thus, each of the N eigenvalues produced by a Rayleigh-Ritz approximation to an N-DOF system is an upper bound to the corresponding exact eigenvalue, and the eigenvalues approach the exact values from above as the number of degrees of freedom, $\hat{N}$, increases.

*Table 13.1. **Convergence Properties of Rayleigh-Ritz Frequencies***

DOF = $\hat{N}$ =	1	2	3	...	N
"Constraints" = m =	$N-1$	N − 2	N − 3	...	0
1st eigenvalue	$\lambda_1^{(N-1)} \geq$	$\lambda_1^{(N-2)} \geq$	$\lambda_1^{(N-3)} \geq$	...	$\searrow \lambda_1$
2nd eigenvalue		$\lambda_2^{(N-2)} \geq$	$\lambda_2^{(N-3)} \geq$	...	$\searrow \lambda_2$
3rd eigenvalue			$\lambda_3^{(N-3)} \geq$	...	$\searrow \lambda_3$
.					.
.					.
.					.
*N*th eigenvalue					$\searrow \lambda_N$

Reference

13.1 K-J. Bathé and E. L. Wilson, *Numerical Methods in Finite Element Analysis*, Prentice-Hall, Englewood Cliffs, NJ (1976).

Problem Set 13.1

13.1 This problem illustrates the fact that the symmetry of **k** and **m** depend on the coordinates employed. For the 2-DOF system below:

(a) Derive the equations of motion by using Newton's laws and using as coordinates u_1 and u_2, the absolute motion of the two masses.

(b) Repeat (a) but using as coordinates the absolute motion, u_1, of mass

m_1 and the displacement of mass m_2 relative to m_1, which may be called u_r.

(c) Using the coordinates of part (b), use Lagrange's equations to derive the equations of motion. What conclusions can you draw?

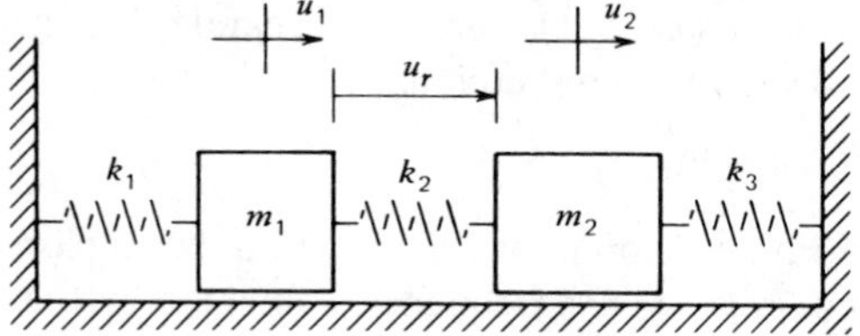

Figure P13.1

13.2 (a) For the system below, determine the characteristic polynomial

$$p(\omega^2) = \det(\mathbf{k} - \omega^2\mathbf{m})$$

(b) Plot $p(\omega^2)$ from $\omega^2 = 0$ to $\omega^2 = 3.2$ using increments $\Delta\omega^2 = 0.2$.
(c) Solve for ω_1^2, ω_2^2, and ω_3^2.

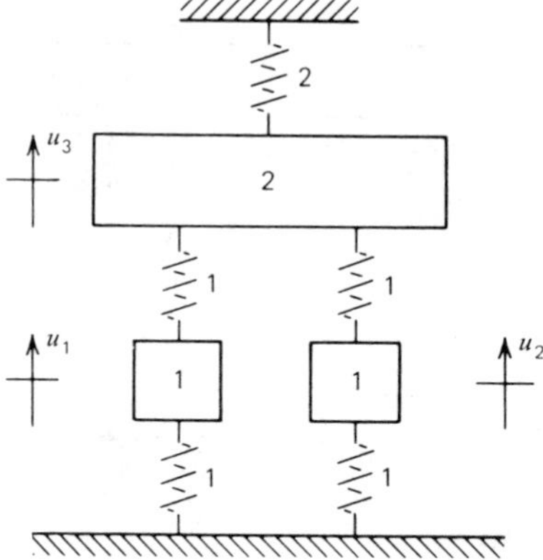

Figure P13.2

13.3 (a) Solve for modes ϕ_1 and ϕ_3 of the system in Example 13.1 using the method of Eq. 13.14.
(b) Scale modes ϕ_1 and ϕ_3 so that $M_1 = M_3 = 1$.
(c) Show that ϕ_1 and ϕ_3 are orthogonal with respect to the mass matrix.

13.4 (a) Solve for modes ϕ_2 and ϕ_3 of the system in Example 13.2 using the method of Eq. 13.14.
(b) Scale modes ϕ_2 and ϕ_3 so that $M_2 = M_3 = \rho AL$.
(c) Show that ϕ_2 is orthogonal (with respect to mass) to both ϕ_1 and ϕ_3.

13.5 (a) The system in Fig. P13.5 has a repeated frequency $\omega_2 = \omega_3$. Determine the three frequencies.

(b) Using the method of Eq. 13.23, determine ϕ_2 and ϕ_3.
(c) Are the modes you have obtained orthogonal with respect to the mass matrix?

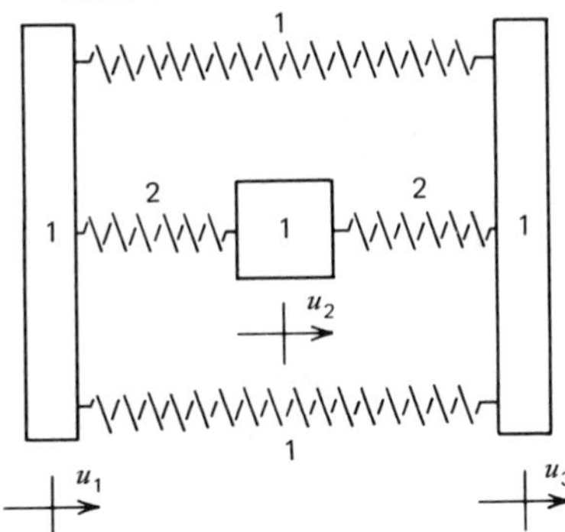

Figure P13.5

13.6 Repeat Problem 12.5 using Eqs. 13.37 and 13.39 to determine the free-vibration response.

13.7 For a system with rigid-body modes, the free-vibration response of Eq. 13.37 can be replaced by

$$\mathbf{u} = \sum_{r=1}^{N_r} \phi_r(a_r + b_r t) + \sum_{r=N_r+1}^{N} \phi_r[a_r \cos(\omega_r t) + b_r \sin(\omega_r t)]$$

(a) If the modes $\phi_r(r = 1, 2, \ldots, N)$ form a mutually orthogonal set of modes, determine expressions for a_r and b_r for $r = 1, 2, \ldots, N_r$.
(b) The modes ϕ_1, ϕ_2, and ϕ_3 of Example 13.4 form a mutually orthogonal set of modes for the system of Example 13.3. Determine $\mathbf{v}(t)$ if

$$\mathbf{v}(0) = \begin{Bmatrix} 0 \\ 0 \\ 1 \end{Bmatrix}, \qquad \dot{\mathbf{v}}(0) = \begin{Bmatrix} 0 \\ 0 \\ 0 \end{Bmatrix}$$

for this system.

13.8 Prove that, since the normal modes ϕ_r are mutually orthogonal, they form a linearly independent set of N-dimensional vectors.

Problem Set 13.2

13.9 By restraining, respectively, mass 3 and then masses 2 and 3, show that the eigenvalue separation theorem of Eq. 13.70 holds true for the system in Example 13.1.

14 NUMERICAL EVALUATION OF MODES AND FREQUENCIES OF MDOF SYSTEMS

In Chapter 12 you learned how to solve for the modes and frequencies of a 2-DOF system by determining the roots of the characteristic (polynomial) equation. In Chapter 13 many properties of modes and frequencies of MDOF systems were discussed. This chapter introduces you to the variety of procedures which are used in obtaining numerical solutions to large eigenproblems, that is, to the determination of modes and frequencies of systems having tens, hundreds, or even thousands of degrees of freedom. A detailed presentation of these methods, however, is beyond the scope of this text.

Upon completion of this chapter you should:

- Be aware of the variety of numerical methods for solving the structural dynamics eigenproblem and know the principal factors to consider in choosing a method.
- Be able to apply vector iteration to determine at least two modes and frequencies of an N-DOF system ($N \leq 5$).
- Be able to solve an eigenproblem by using the ISMIS computer program, and interpret the results obtained in the computer output produced by ISMIS.

14.1 Introduction to Methods for Solving Algebraic Eigenproblems

MDOF models of structures may have from ten to ten thousand (or maybe more) degrees-of-freedom, and dynamic analysis of these structures generally involves determining some, or all, of the natural frequencies (eigenvalues) and natural modes (eigenvectors) by solving the equation

$$(\mathbf{k} - \lambda_i \mathbf{m})\boldsymbol{\phi}_i = \mathbf{0} \tag{14.1}$$

where $\lambda_i \equiv \omega_i^2$. The polynomial root-finding method which was employed in Chapter 12 for determining the natural frequencies of 2-DOF systems is only one of a number of methods for solving the algebraic eigenproblem, Eq. 14.1. Other methods may be far superior for certain situations. Most large structural dynamics computer programs provide several methods so that the analyst can choose the one most appropriate for a given problem. The purpose of this section is merely to make you aware of the variety of algebraic eigensolvers, and to suggest to you some of the important factors you need to consider in selecting an eigensolver. More advanced texts[14.1] and recent literature[14.2,14.3] provide details on eigensolvers which are particularly appropriate for structural dynamics eigenproblems.

The fundamental properties of natural frequencies and natural modes which were discussed in Chapter 13 provide the basis for a variety of eigensolution techniques. These techniques fall into three broad categories, depending on the fundamental property that is most prominently used in the solution procedure.

Methods that directly employ the property stated in Eq. 14.1 are called *vector iteration* methods (or power methods). *Matrix transformation* methods constitute the second group. They employ the orthogonality properties

$$\boldsymbol{\Phi}^T\mathbf{k}\boldsymbol{\Phi} = \boldsymbol{\Lambda} \tag{14.2a}$$

and

$$\boldsymbol{\Phi}^T\mathbf{m}\boldsymbol{\Phi} = \mathbf{I} \tag{14.2b}$$

where $\boldsymbol{\Phi} = [\phi_1\ \phi_2 \ldots \phi_N]$ and $\boldsymbol{\Lambda} = \text{diag}(\lambda_i)$, $i = 1, 2, \ldots, N$. *Polynomial root-finding* is the basis of a third group of eigensolution techniques. The equation most prominently used is

$$\det(\mathbf{k}-\lambda_i\mathbf{m}) = 0 \tag{14.3}$$

or

$$\det(\mathbf{k}^{(r)}-\lambda_i^{(r)}\mathbf{m}^{(r)}) = 0 \tag{14.4}$$

where $\mathbf{k}^{(r)}$ and $\mathbf{m}^{(r)}$ were defined in Sec. 13.2. Finally, many practical eigensolvers, such as the HQRI method and the subspace iteration method[14.1], employ techniques based on a combination of the above fundamental properties.

In each of the above three groups there are several specific methods. For example, among the vector iteration methods are: direct iteration, inverse iteration (method of Vianello and Stodola), and inverse iteration with spectrum shift. Matrix transformation methods include: Jacobi's method, Lanczos' method, Givens' method, Householder's method, and the QR method. Polynomial root-finding is involved in determinant search methods and Stürm sequence methods.

With the above methods and more to choose from, a person wishing to use the computer to solve structural dynamics eigenproblems is obliged to consider which methods are most suited to which classes of problems. The factors that most directly influence the choice of eigensolver are: (1) the order, N, and bandwidths, M_k and M_m, of **k** and **m**, and (2) the number of eigenvalues and eigenvectors required. Bathé and Wilson[14.1] conclude the following:

1. The Householder-QR-Inverse Iteration (HQRI) solution is most efficient when all eigenvalues and eigenvectors of a matrix are required and the matrix has a large bandwidth or is full.
2. The determinant search technique is very effectively used to calculate the lowest eigenvalues and corresponding eigenvectors of systems with small bandwidth.
3. The subspace iteration solution is very effective in the calculation of the lowest eigenvalues and corresponding eigenvectors of systems with large bandwidth and which are too large for the high-speed storage of the computer.

A tridiagonal reduction method, FEER[14.2,14.3], has also been found to be very effective in solving problems of the class described in (3) above.

Vector iteration can be used as a "stand-alone" eigensolver or can be combined with other procedures, for example, in the HQRI method named above. When one or two modes of a small system ($N \leq 5$) are required, a vector iteration solution can be carried out by using a hand calculator. In the next section, therefore, you will be introduced to vector iteration techniques.

A review of eigensolvers for vibration analysis is presented in Reference 14.10.

14.2 Vector Iteration Methods

The most effective procedure for hand computation of a few eigenvalues and eigenvectors of a small ($N \leq 10$) system is the *method of inverse iteration.* Vianello (1898) used it to solve for buckling of a strut, and Stodola (1904) used it to solve for the vibration of a rotating shaft. We begin by considering Eq. 14.1, which we write in the form

$$\mathbf{k}\boldsymbol{\phi} = \lambda\mathbf{m}\boldsymbol{\phi} \tag{14.5}$$

We will assume that the fundamental frequency is distinct and that the eigenvalues, λ_i, are ordered according to

$$\lambda_1 < \lambda_2 \leq \lambda_3 \leq \cdots \leq \lambda_N \tag{14.6}$$

When the left-hand side of Eq. 14.5 is equal to the right-hand side, an eigenvalue λ_i and corresponding eigenvector $\boldsymbol{\phi}_i$ have been obtained. An iterative procedure based on Eq. 14.5 can be set up by writing Eq. 14.5 in the form

$$\mathbf{k}\mathbf{v}_{(s+1)} = \mathbf{m}\mathbf{u}_{(s)} \tag{14.7}$$

or

$$\mathbf{v}_{(s+1)} = \mathbf{D}\mathbf{u}_{(s)} \tag{14.8}$$

where **D** is the *dynamical matrix* given by

$$\mathbf{D} = \mathbf{k}^{-1}\mathbf{m} \tag{14.9}$$

Then, $\mathbf{u}_{(s+1)}$ is defined by

$$\mathbf{u}_{(s+1)} = \lambda_{(s+1)}\mathbf{v}_{(s+1)} \tag{14.10}$$

where $\lambda_{(s+1)}$ is an appropriately chosen scaling factor. For example, $\lambda_{(s+1)}$ may be chosen to make the largest element of $\mathbf{u}_{(s+1)}$ equal to $+1$, or $\lambda_{(s+1)}$ can be chosen to make $\mathbf{u}^T_{(s+1)}\mathbf{m}\mathbf{u}_{(s+1)} = 1$.

Note that Eq. 14.8 can be interpreted as solving for the $(s + 1)$st deflection shape produced by the inertia forces associated with the (s)th deflection shape. The procedure is called "inverse iteration" because of the use of the inverse of **k** in Eq. 14.8.

We will first look at an example of the above procedure (Example 14.1). Then we will prove that the procedure always produces convergence to the fundamental mode if $\lambda_1 < \lambda_2$ as assumed in Eq. 14.6.

Example 14.1

Use the inverse iteration procedure to solve for the fundamental mode and frequency of the 3-DOF system shown below.

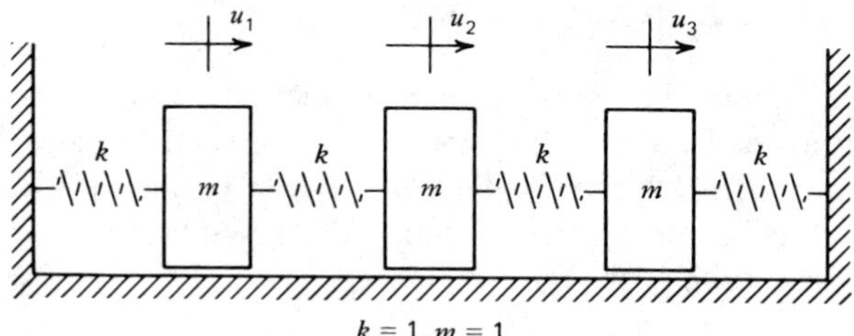

For this system

$$\mathbf{k} = \begin{bmatrix} 2 & -1 & 0 \\ -1 & 2 & -1 \\ 0 & -1 & 2 \end{bmatrix}, \qquad \mathbf{m} = \begin{bmatrix} 1 & 0 & 0 \\ 0 & 1 & 0 \\ 0 & 0 & 1 \end{bmatrix}$$

$$\lambda_1 = 2 - \sqrt{2} = 0.5858, \qquad \lambda_2 = 2.0, \quad \lambda_3 = 2 + \sqrt{2} = 3.412$$

Also, $\mathbf{k}^{-1}$ can be shown to be

$$\mathbf{k}^{-1} = \frac{1}{4}\begin{bmatrix} 3 & 2 & 1 \\ 2 & 4 & 2 \\ 1 & 2 & 3 \end{bmatrix} = \begin{bmatrix} 0.75 & 0.50 & 0.25 \\ 0.50 & 1.00 & 0.50 \\ 0.25 & 0.50 & 0.75 \end{bmatrix}$$

Solution

a. Form the dynamical matrix.

$$\mathbf{D} = \mathbf{k}^{-1}\mathbf{m} = \begin{bmatrix} 0.75 & 0.50 & 0.25 \\ 0.50 & 1.00 & 0.50 \\ 0.25 & 0.50 & 0.75 \end{bmatrix} \tag{1}$$

b. Choose a starting vector $\mathbf{u}_{(0)}$. Let

$$\mathbf{u}_{(0)} = \begin{Bmatrix} 1 \\ 1 \\ 1 \end{Bmatrix} \tag{2}$$

c. Use Eq. 14.8 to obtain $\mathbf{v}_{(s+1)}$, $s = 0, 1, 2, \ldots$, and scale $\mathbf{v}_{(s+1)}$ to make the largest element of $\mathbf{u}_{(s+1)}$ equal to $+1$. The results of two iteration cycles are tabulated below.

$u_{(0)}$	$v_{(1)}$	$u_{(1)}$	$v_{(2)}$	$u_{(2)}$
1.000	1.500	0.750	1.250	0.714
1.000	2.000	1.000	1.750	1.000
1.000	1.500	0.750	1.250	0.714
		$\lambda_{(1)} = 0.500$		$\lambda_{(2)} = 0.571$

Note that the eigenvalue estimates are approaching the true value, $\lambda_1 = 0.5858$, from below. A better estimate of the eigenvalue can be obtained by computing the Rayleigh quotient corresponding to each shape $\mathbf{v}_{(s+1)}$.

$$\lambda_{(s+1)} = \frac{\mathbf{v}_{(s+1)}^T \mathbf{k} \mathbf{v}_{(s+1)}}{\mathbf{v}_{(s+1)}^T \mathbf{m} \mathbf{v}_{(s+1)}} = \frac{\mathbf{v}_{(s+1)}^T \mathbf{m} \mathbf{u}_{(s)}}{\mathbf{v}_{(s+1)}^T \mathbf{m} \mathbf{v}_{(s+1)}} \tag{3}$$

The denominator of the above expression can also be used in the scaling equation

$$\mathbf{u}_{(s+1)} = \frac{\mathbf{v}_{(s+1)}}{(\mathbf{v}_{(s+1)}^T \mathbf{m} \mathbf{v}_{(s+1)})^{1/2}} \tag{4}$$

The table below summarizes two iteration cycles using Eqs. 3 and 4.

$u_{(0)}$	$v_{(1)}$	$u_{(1)}$	$v_{(2)}$	$u_{(2)}$
1.000	1.500	0.5145	0.8575	0.5024
1.000	2.000	0.6860	1.2010	0.7037
1.000	1.500	0.5145	0.8575	0.5024
		$\lambda_{(1)} = \frac{5.000}{8.500} = 0.588$		$\lambda_{(2)} = \frac{1.706}{2.913} = 0.586$

Although calculation of the Rayleigh quotient involves more calculation than the scaling used in (c) above, you can see that even after one iteration cycle the Rayleigh quotient gives a very good approximation to the exact value of λ_1. After two cycles of iteration the fundamental frequency and mode shape are approximated by

$$\lambda_1 \doteq 0.586 \qquad \boldsymbol{\phi}_1 \doteq \begin{Bmatrix} 0.5024 \\ 0.7037 \\ 0.5024 \end{Bmatrix}$$

Reference 14.1 presents a useful computational algorithm for carrying out a computer implementation of inverse iteration including Rayleigh quotient calculations.

We will now prove that the inverse iteration procedure described above produces convergence to the fundamental mode of a system. Recall (Sec. 13.3) that an arbitrary vector can be written in terms of an eigenvector expansion. Therefore, we can write $\mathbf{u}_{(0)}$ as

$$\mathbf{u}_{(0)} = \sum_{i=1}^{N} c_i \boldsymbol{\phi}_i \tag{14.11}$$

where $\boldsymbol{\phi}_i$, $i = 1, 2, \ldots, N$, are the exact mode shapes.

From Eqs. 14.1 and 14.9,

$$\mathbf{D}\boldsymbol{\phi}_i = \frac{1}{\lambda_i} \boldsymbol{\phi}_i \tag{14.12}$$

Then,

$$\mathbf{D}\mathbf{u}_{(0)} = \sum_{i=1}^{N} c_i \mathbf{D}\boldsymbol{\phi}_i = \sum_{i=1}^{N} c_i \left(\frac{1}{\lambda_i}\right) \boldsymbol{\phi}_i \tag{14.13}$$

As before, assume $\lambda_1 < \lambda_2$. Each cycle of iteration using Eqs. 14.8 and 14.10 involves a solution of Eq. 14.8 and a scaling operation, Eq. 14.10. Since the scaling does not affect the convergence, we can omit the scaling step and apply Eq. 14.8 s times to get

$$\begin{aligned} \mathbf{D}^s\mathbf{u}_{(0)} &= \sum_{i=1}^{N} c_i \left(\frac{1}{\lambda_i}\right)^s \boldsymbol{\phi}_i \\ &= \left(\frac{1}{\lambda_1}\right)^s \sum_{i=1}^{N} c_i \left(\frac{\lambda_1}{\lambda_i}\right)^s \boldsymbol{\phi}_i \end{aligned} \tag{14.14}$$

Since $\lambda_1 < \lambda_i$, $(\lambda_1/\lambda_i)^s \to 0$ for $i > 1$. Therefore,

$$\mathbf{D}^s\mathbf{u}_{(0)} \to \left(\frac{1}{\lambda_1}\right)^s c_1\boldsymbol{\phi}_1 \tag{14.15a}$$

Then

$$\mathbf{D}^{s+1}\mathbf{u}_{(0)} \to \left(\frac{1}{\lambda_1}\right)^{s+1} c_1\boldsymbol{\phi}_1 \tag{14.15b}$$

so the convergence is to a vector proportional to $\boldsymbol{\phi}_1$. The convergence rate depends on the (λ_1/λ_i) ratios, particularly (λ_1/λ_2), and on the c_i's, that is, on the make-up of the starting vector.

Equations 14.15 indicate that if $c_1 \neq 0$ in the starting vector $\mathbf{u}_{(0)}$, inverse iteration will always converge to the fundamental mode, $\boldsymbol{\phi}_1$. This suggests that to produce convergence to $\boldsymbol{\phi}_2$ it will be necessary to remove the $\boldsymbol{\phi}_1$ component from $\mathbf{u}_{(0)}$. But, since roundoff error can serve to reintroduce $\boldsymbol{\phi}_1$ into later iterates, it will be necessary to remove the $\boldsymbol{\phi}_1$ component from each $\mathbf{u}_{(s)}$ used in iterating for $\boldsymbol{\phi}_2$. To determine $\boldsymbol{\phi}_3$ by inverse iteration we must remove both $\boldsymbol{\phi}_1$ and $\boldsymbol{\phi}_2$ from each $\mathbf{u}_{(s)}$ used in the iteration procedure, and so on. *Gram-Schmidt orthogonalization* and *matrix deflation* are the names applied to procedures used for carrying out this removal of previously determined eigenvectors.[14.4]

Gram-Schmidt orthogonalization can be described as follows. Let

$$\hat{\mathbf{u}}_{(s)} = \mathbf{u}_{(s)} - \alpha_1\boldsymbol{\phi}_1 \tag{14.16}$$

and choose α_1 so that $\hat{\mathbf{u}}_{(s)}$ is orthogonal to $\boldsymbol{\phi}_1$, that is,

$$\boldsymbol{\phi}_1^T\mathbf{m}\hat{\mathbf{u}}_{(s)} = 0 \tag{14.17}$$

Then,

$$\alpha_1 = \frac{\boldsymbol{\phi}_1^T\mathbf{m}\mathbf{u}_{(s)}}{\boldsymbol{\phi}_1^T\mathbf{m}\boldsymbol{\phi}_1}$$

or

$$\hat{\mathbf{u}}_{(s)} = \mathbf{S}_1\mathbf{u}_{(s)} \tag{14.18}$$

where

$$\mathbf{S}_1 = \mathbf{I} - \frac{\boldsymbol{\phi}_1\boldsymbol{\phi}_1^T\mathbf{m}}{\boldsymbol{\phi}_1^T\mathbf{m}\boldsymbol{\phi}_1} \tag{14.19}$$

To produce convergence to $\boldsymbol{\phi}_2$, $\mathbf{u}_{(s)}$ in Eq. 14.8 is replaced by $\hat{\mathbf{u}}_{(s)}$, which is defined by Eqs. 14.18 and 14.19. Thus,

$$\mathbf{v}_{(s+1)} = \mathbf{D}_2\mathbf{u}_{(s)} \tag{14.20}$$

where

$$\mathbf{D}_2 = \mathbf{D}\mathbf{S}_1 = \mathbf{D} - \frac{\mathbf{D}\boldsymbol{\phi}_1\boldsymbol{\phi}_1^T\mathbf{m}}{\boldsymbol{\phi}_1^T\mathbf{m}\boldsymbol{\phi}_1} \tag{14.21}$$

Combining Eqs. 14.12 and 14.21 gives

$$\mathbf{D}_2 = \mathbf{D} - \frac{1}{\lambda_1}\left(\frac{\boldsymbol{\phi}_1\boldsymbol{\phi}_1^T\mathbf{m}}{\boldsymbol{\phi}_1^T\mathbf{m}\boldsymbol{\phi}_1}\right) \tag{14.22}$$

Procedures analogous to Eqs. 14.16 through 14.22 can be used to define $\mathbf{D}_3$, $\mathbf{D}_4$, and so forth for removing higher modes. The procedure is called *matrix deflation*[14.4].

For a hand computation of modes and frequencies higher than the fundamental, a *sweeping matrix* procedure[14.5] is preferable to the Gram-Schmidt orthogonalization procedure described above. This procedure is illustrated by Example 14.2.

Example 14.2

Use a sweeping matrix to solve iteratively for the second frequency and mode of the system in Example 14.1. Use as the mode shape $\boldsymbol{\phi}_1$ the following result from Example 14.1.

$$\boldsymbol{\phi}_1 = \begin{Bmatrix} 1.000 \\ 1.400 \\ 1.000 \end{Bmatrix}$$

Solution

The orthogonality condition, Eq. 14.17, must be enforced. Let

$$\hat{\mathbf{u}}_{(s)} = \begin{Bmatrix} \hat{u}_1 \\ \hat{u}_2 \\ \hat{u}_3 \end{Bmatrix} \tag{1}$$

Then Eq. 14.17 becomes

$$\boldsymbol{\phi}_1^T\mathbf{m}\hat{\mathbf{u}}_{(s)} = \lfloor 1.000 \quad 1.400 \quad 1.000 \rfloor \begin{bmatrix} 1 & 0 & 0 \\ 0 & 1 & 0 \\ 0 & 0 & 1 \end{bmatrix} \begin{Bmatrix} \hat{u}_1 \\ \hat{u}_2 \\ \hat{u}_3 \end{Bmatrix} \tag{2}$$

or

$$\hat{u}_1 + 1.400\hat{u}_2 + \hat{u}_3 = 0 \tag{3}$$

We can treat this as a constraint on $\hat{u}_1$ and write

$$\hat{u}_1 = -1.400\hat{u}_2 - \hat{u}_3 \tag{4}$$

So,

$$\begin{Bmatrix} \hat{u}_1 \\ \hat{u}_2 \\ \hat{u}_3 \end{Bmatrix} = \begin{bmatrix} -1.400 & -1.000 \\ 1 & 0 \\ 0 & 1 \end{bmatrix} \begin{Bmatrix} \hat{u}_2 \\ \hat{u}_3 \end{Bmatrix} \tag{5}$$

But, we can choose $\hat{u}_2$ and $\hat{u}_3$ arbitrarily, so we can let $\hat{u}_2 = u_2$, $\hat{u}_3 = u_3$ and write Eq. 5 in the form

$$\begin{Bmatrix} \hat{u}_1 \\ \hat{u}_2 \\ \hat{u}_3 \end{Bmatrix} = \begin{bmatrix} 0 & -1.400 & -1.000 \\ 0 & 1 & 0 \\ 0 & 0 & 1 \end{bmatrix} \begin{Bmatrix} u_1 \\ u_2 \\ u_3 \end{Bmatrix} \tag{6}$$

where u_1 is a dummy as far as Eq. 6 is concerned. Equation 6 is in the form of Eq. 14.18. We can form

$$\mathbf{D}_2 = \mathbf{D}\mathbf{S}_1 \tag{7}$$

where $\mathbf{S}_1$ is called a *sweeping matrix* because it sweeps ϕ_1 out of $\hat{\mathbf{u}}$. Using $\mathbf{D}$ from Example 14.1 and $\mathbf{S}_1$ from Eq. 6, we get

$$\mathbf{D}_2 = \begin{bmatrix} 0.75 & 0.50 & 0.25 \\ 0.50 & 1.00 & 0.50 \\ 0.25 & 0.50 & 0.75 \end{bmatrix} \begin{bmatrix} 0 & -1.400 & -1.000 \\ 0 & 1 & 0 \\ 0 & 0 & 1 \end{bmatrix} = \begin{bmatrix} 0 & -0.550 & -0.500 \\ 0 & 0.300 & 0 \\ 0 & 0.150 & 0.500 \end{bmatrix} \tag{8}$$

We next use the iteration equation

$$\mathbf{v}_{(s+1)} = \mathbf{D}_2\mathbf{u}_{(s)} \tag{9}$$

and scale $\mathbf{u}_{(s+1)}$ so that its largest active element is unity. The results of four iteration cycles are tabulated below.

$\mathbf{u}_{(0)}$	$\mathbf{v}_{(1)}$	$\mathbf{u}_{(1)}$	$\mathbf{v}_{(2)}$	$\mathbf{u}_{(2)}$
1.000	(−1.050)	(−1.615)	(−0.754)	(−1.325)
1.000	0.300	0.462	0.139	0.244
1.000	0.650	1.000	0.569	1.000
	$\lambda_{(1)} = 1.539$		$\lambda_{(2)} = 1.757$	

$\mathbf{u}_{(2)}$	$\mathbf{v}_{(3)}$	$\mathbf{u}_{(3)}$	$\mathbf{v}_{(4)}$	$\mathbf{u}_{(4)}$
(−1.325)	(−0.634)	(−1.181)	−0.575	−1.106
0.244	0.073	0.136	0.041	0.079
1.000	0.537	1.000	0.520	1.000
	$\lambda_{(3)} = 1.864$		$\lambda_{(4)} = 1.922$	

The exact eigenvalue is $\lambda_2 = 2.0$, to which the above is converging. The eigenvector, $\mathbf{u}_{(3)}$, has not yet converged to the exact (antisymmetric) eigenvector

$$\phi_2 = \begin{Bmatrix} -1.0 \\ 0.0 \\ 1.0 \end{Bmatrix}$$

The reason for the slow convergence observed is that a symmetric starting vector $\mathbf{u}_{(0)}$ was supplied whereas the true eigenvector is antisymmetric. The above procedure must therefore introduce the antisymmetry.

In the above solution the numbers enclosed in parentheses were not essential to obtaining the final estimate of frequency and mode shape.

Orthogonalization procedures such as those described above can be used to obtain four or five modes altogether, provided that a small convergence tolerance parameter is employed in order that the calculated eigenvectors closely approximate the true eigenvectors.

An effective way to produce convergence to an eigenvalue other than the fundamental is to employ *inverse iteration with spectrum shift*. Let μ be a number in the vicinity of the frequency of interest. For example, in Fig. 14.1, μ is closer to λ_2 than to the other eigenvalues. Then, using $\mu\mathbf{m}$, modify Eq. 14.1 to give the shifted eigenproblem

$$[(\mathbf{k} - \mu\mathbf{m}) - (\lambda_i - \mu)\mathbf{m}]\phi_i = 0 \tag{14.23}$$

Let

$$\hat{\mathbf{k}} = \mathbf{k} - \mu\mathbf{m}, \qquad \hat{\lambda}_i = \lambda_i - \mu \tag{14.24}$$

Then Eq. 14.23 becomes

$$(\hat{\mathbf{k}} - \hat{\lambda}_i\mathbf{m})\phi_i = 0 \tag{14.25}$$

Figure 14.1 shows the *spectrum* (i.e., plot on a frequency axis) of λ and of $\hat{\lambda}$. Note from the figure that if $\lambda_i < \mu$, $\hat{\lambda}_i$ will be negative. A proof similar to that in Eqs. 14.11 through 14.15 can be used to show that inverse iteration applied to the shifted eigenproblem, Eq. 14.25, produces convergence to the λ_i closest to the shift value μ.

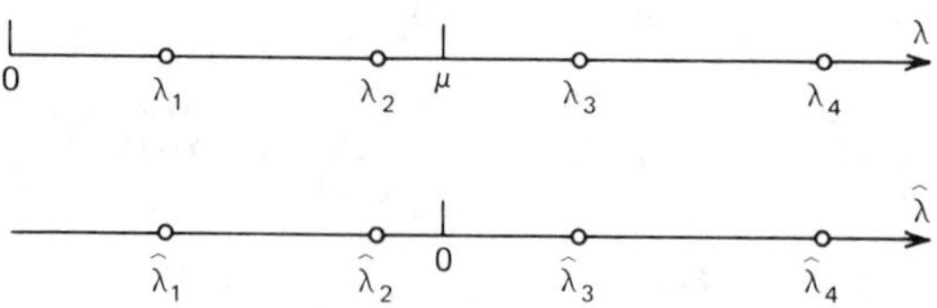

Figure 14.1. Eigenvalue spectra.

Inverse iteration with spectrum shift is a widely used procedure (e.g., in the HQRI method) for finding eigenvectors after the eigenvalues have been found by some other method. This method is also used to avoid the problem associated with the fact that $\mathbf{k}^{-1}$ does not exist for systems with rigid-body modes. A small shift value can be used to remove the singularity. Other methods for applying inverse iteration to systems with rigid-body modes are described in Reference 14.6 and other references.

14.3 Use of ISMIS to Solve for Modes and Frequencies of MDOF Systems

In Sec. 14.1 you learned of the existence of a large number of methods for solving for the modes and frequencies of MDOF systems. In Sec. 14.2 you learned about inverse iteration, inverse iteration with orthogonalization, and inverse iteration with spectrum shift. Although it is very worthwhile for you to have solved a few small ($N < 5$) problems by iteration using a hand calculator, you will also learn a great deal using a computer program such as ISMIS to do your number crunching while you spend your time in system modeling and in evaluating the physical meaning of the modes and frequencies obtained by the computer.

As noted in Sec. 14.1, one of the best numerical procedures for calculating the natural frequencies and mode shapes of systems having more than 2 degrees of freedom but less than a hundred or so (the exact number depends on the computer speed and high-speed storage capability, etc.) is the Householder-QR-Inverse Iteration (HQRI) procedure, which has been coded in the EIGEN command of the ISMIS computer program. The purpose of the present section is to acquaint you with the EIGEN command and allow you to use it in solving some lumped-parameter vibration problems. In Chapter 16 the finite element method will be used to generate the stiffness and mass matrices for use in the EIGEN command, and in Chapter 17 numerous examples will be presented to illustrate important features of numerical eigensolutions, for example, convergence. A detailed study of the HQRI method is beyond the scope of this text, but is treated in more advanced texts[14.1].

THE EIGEN command for solving Eq. 14.1 has the format

```
EIGEN,K,M,L,PT,N1
```

where

K = N×N stiffness matrix previously defined
M = N×N mass matrix previously defined, or previously defined 1×N row vector containing elements of a diagonal mass matrix
L = 1×N row vector of eigenvalues calculated by EIGEN

PT = (N1)×N matrix whose *rows* are the eigenvectors corresponding to the first N1 eigenvalues in L
N1 = number of eigen*vectors* calculated

The eigenvectors are normalized so that

$$\boldsymbol{\phi}_i^T \mathbf{m} \boldsymbol{\phi}_i = 1 \tag{14.26}$$

Two examples are presented here. Other examples appear in the ISMIS *User's Manual.* In the interest of clarity, the number of degrees of freedom is kept small, although the computer program is capable of handling much larger problems.

Example 14.3

Let the spring constant in Example 12.1 be $k = 10$ lb/in. and let the masses weigh 5 lb each ($g = 386$ in./sec²). Use EIGEN to determine the modes and frequencies of the system. Sketch the mode shapes.

Solution

$$(\mathbf{k} - \omega_i^2 \mathbf{m})\boldsymbol{\phi}_i = \mathbf{0}$$

$$m = \frac{W}{g} = \frac{5 \text{ lb}}{386 \text{ in./sec}^2} = 0.013 \text{ lb-sec}^2/\text{in.} \tag{1}$$

$$k = 10 \text{ lb/in.}$$

From Example 12.1,

$$\mathbf{k} = k\begin{bmatrix} 2 & -1 \\ -1 & 2 \end{bmatrix} = \begin{bmatrix} 20 & -10 \\ -10 & 20 \end{bmatrix} \text{ lb/in.}$$

$$\mathbf{m} = m\begin{bmatrix} 1 & 0 \\ 0 & 1 \end{bmatrix} = \begin{bmatrix} 0.013 & 0 \\ 0 & 0.013 \end{bmatrix} \text{ lb-sec}^2/\text{in.}$$

Let the ISMIS notation be

$\mathbf{k} \to$ K, $\mathbf{m} \to$ M, $\omega^2 \to$ W2, $\boldsymbol{\Phi}^T \to$ PT
EIGEN,K,M,W2,PT,2

Below is the output produced at a computer terminal during interactive execution of an ISMIS program to perform to above eigensolution. A colon preceding a line indicates a line of input. The remaining lines are generated by the computer program.

```
:START
:*
:EXAMPLE 14.3
:LOAD,K,2,2
       2 ROWS     2 COLUMNS
:20,-10,-10,20
```

```
:PRINT,K
  SCALED BY    1.0E+01
        1    2
   1  2.00-1.00
   2 -1.00 2.00
:LOAD,M,1,2
        1 ROWS      2 COLUMNS
:.013,.013
:PRINT,M
  SCALED BY    1.0E-02
        1    2
   1  1.30 1.30
:EIGEN,K,M,W2,PT,2
    NUMBER OF EIGENVECTORS DESIRED=          2
      ORIGINAL TRACE   = 3.07692308E+03
      SUM OF EIGENVALUES   = 3.07692308E+03
:PRINT,W2,2
  SCALED BY    1.0E+03
              1          2
   1    .76923077 2.30769231
:TRANS,PT,P
:PRINT,P,1
  SCALED BY    1.0E+00
           1         2
   1  6.20174-6.20174
   2  6.20174 6.20174
:STOP
```

From the computer output above we have

$$\omega_1^2 = 769.2 \text{ rad}^2/\text{s}^2$$
$$\omega_2^2 = 2308 \text{ rad}^2/\text{s}^2$$

and

$$\boldsymbol{\phi}_1 = \begin{Bmatrix} 6.20 \\ 6.20 \end{Bmatrix}, \qquad \boldsymbol{\phi}_2 = \begin{Bmatrix} -6.20 \\ 6.20 \end{Bmatrix}$$

The mode shapes are

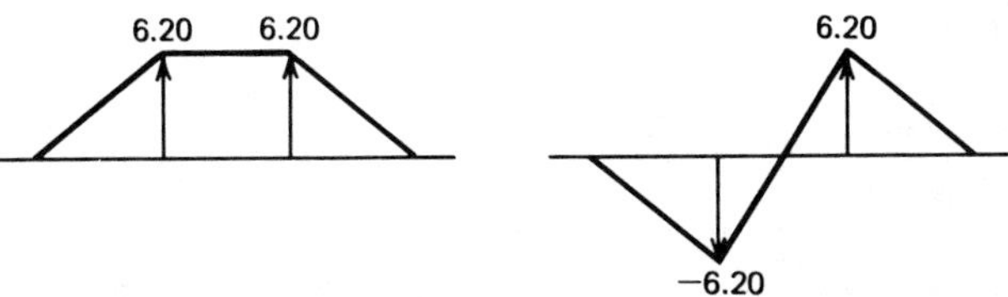

These are the same as obtained in Example 2.1 except that here they are scaled so that $\boldsymbol{\phi}_i^T \mathbf{m} \boldsymbol{\phi}_i = 1$.

When a structure must be designed to withstand specified dynamic loading conditions (e.g., gust loads and landing loads on an airplane, wind and wave loads on an offshore platform, etc.) one of the important steps in the design process is the determination of the natural frequencies and mode shapes of the structure. Frequently it becomes necessary to modify the structure in order to improve its dynamical behavior. Currently there is active research under way on the topic of computer automated design of structures, including design to meet specified dynamical requirements. In this text we will not go extensively into the subject of design, but Example 14.4 is intended to illustrate how the dynamical behavior (in this case the modes and frequencies) of a structure may be modified by changing the design parameters.

Example 14.4

A system has two masses and three springs as shown below. Solve the eigenvalue problem for the parameters given, tabulate the natural frequencies, sketch the mode shapes, and discuss the effect changing system parameters has on the modes and frequencies of the system.

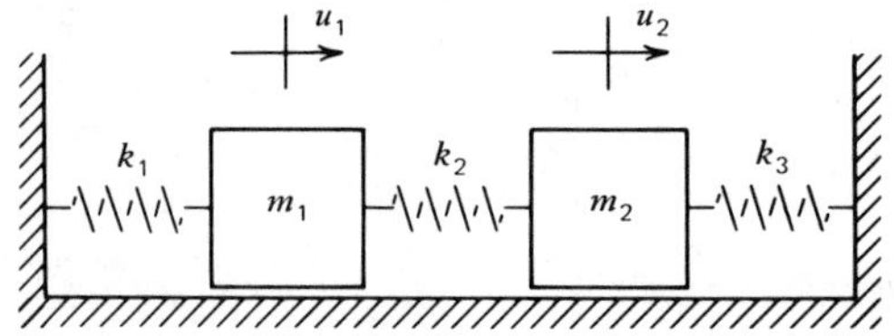

Case A. (symmetrical system)

$$k_1 = k_2 = k_3 = 1, \qquad m_1 = m_2 = 1$$

Case B. (symmetrical system, increased masses)

$$k_1 = k_2 = k_3 = 1, \qquad m_1 = m_2 = 2$$

Case C. (symmetrical system, increased coupling spring)

$$k_1 = k_3 = 1, \qquad k_2 = 2, \qquad m_1 = m_2 = 1$$

Case D. (unsymmetrical system)

$$k_1 = k_2 = k_3 = 1, \qquad m_1 = 1, \qquad m_2 = 2$$

Solution

By deriving the equations of motion of the system, the stiffness and mass matrices can be found to be

$$\mathbf{k} = \begin{bmatrix} (k_1 + k_2) & -k_2 \\ -k_2 & (k_2 + k_3) \end{bmatrix}, \qquad \mathbf{m} = \begin{bmatrix} m_1 & 0 \\ 0 & m_2 \end{bmatrix}$$

Thus,

$$\mathbf{k}_A = \begin{bmatrix} 2 & -1 \\ -1 & 2 \end{bmatrix}, \qquad \mathbf{m}_A = \begin{bmatrix} 1 & 0 \\ 0 & 1 \end{bmatrix}$$

$$\mathbf{k}_B = \mathbf{k}_A, \qquad \mathbf{m}_B = \begin{bmatrix} 2 & 0 \\ 0 & 2 \end{bmatrix}$$

$$\mathbf{k}_C = \begin{bmatrix} 3 & -2 \\ -2 & 3 \end{bmatrix}, \qquad \mathbf{m}_C = \mathbf{m}_A$$

$$\mathbf{k}_D = \mathbf{k}_A, \qquad \mathbf{m}_D = \begin{bmatrix} 1 & 0 \\ 0 & 2 \end{bmatrix}$$

The ISMIS program used to solve the eigenvalue problem

$$\mathbf{kU} = \omega^2\mathbf{mU}$$

follows.

```
:START
:*
:EXAMPLE 14.4
:LOAD,KA,2,2
      2 ROWS      2 COLUMNS
:2,-1,-1,2
:PRINT,KA
  SCALED BY    1.0E+00
       1    2
   1  2.00-1.00
   2 -1.00 2.00
:LOAD,MA,1,2
      1 ROWS      2 COLUMNS
:1,1
:PRINT,MA
  SCALED BY    1.0E+00
       1    2
   1  1.00 1.00
:DUPL,KA,KB
:LOAD,MB,1,2
      1 ROWS      2 COLUMNS
:2,2
:PRINT,MB
  SCALED BY    1.0E+00
       1    2
   1  2.00 2.00
:LOAD,KC,2,2
      2 ROWS      2 COLUMNS
:3,-2,-2,3
:PRINT,KC
  SCALED BY    1.0E+00
       1    2
   1  3.00-2.00
   2 -2.00 3.00
:DUPL,MA,MC
:DUPL,KA,KD
:LOAD,MD,1,2
      1 ROWS      2 COLUMNS
:1,2
:PRINT,MD
  SCALED BY    1.0E+00
       1    2
   1  1.00 2.00
:EIGEN,KA,MA,WA2,UAT,2
    NUMBER OF EIGENVECTORS DESIRED=         2
      ORIGINAL TRACE   = 4.00000000E+00
      SUM OF EIGENVALUES   = 4.00000000E+00
```

```
:PRINT,WA2,2
  SCALED BY    1.0E+00
              1          2
   1  1.00000000 3.00000000
:PRINT,UAT,1
  SCALED BY    1.0E-01
           1       2
   1  7.07107 7.07107
   2 -7.07107 7.07107
:EIGEN,KB,MB,WB2,UBT,2
    NUMBER OF EIGENVECTORS DESIRED=          2
      ORIGINAL TRACE  = 2.00000000E+00
      SUM OF EIGENVALUES   = 2.00000000E+00
:PRINT,WB2,2
  SCALED BY    1.0E+00
              1          2
   1    .50000000 1.50000000
:PRINT,UBT
  SCALED BY    1.0E-01
        1    2
   1  5.00 5.00
   2  5.00-5.00
:EIGEN,KC,MC,WC2,UCT,2
    NUMBER OF EIGENVECTORS DESIRED=          2
      ORIGINAL TRACE  = 6.00000000E+00
      SUM OF EIGENVALUES   = 6.00000000E+00
:PRINT,WC2,2
  SCALED BY    1.0E+00
              1          2
   1  1.00000000 5.00000000
:PRINT,UCT,1
  SCALED BY    1.0E-01
           1       2
   1  7.07107 7.07107
   2 -7.07107 7.07107
:EIGEN,KD,MD,WD2,UDT,2
    NUMBER OF EIGENVECTORS DESIRED=          2
      ORIGINAL TRACE  = 3.00000000E+00
      SUM OF EIGENVALUES   = 3.00000000E+00
:PRINT,WD2,2
  SCALED BY    1.0E+00
              1          2
   1    .63397460 2.36602540
:PRINT,UDT,1
  SCALED BY    1.0E-01
           1       2
   1  4.59701 6.27963
   2  8.88074-3.25058
:STOP
```

Eigenvalues:

Case	ω_1^2	ω_2^2
A	1.00	3.00
B	0.50	1.50
C	1.00	5.00
D	0.63	2.37

Eigenvectors:

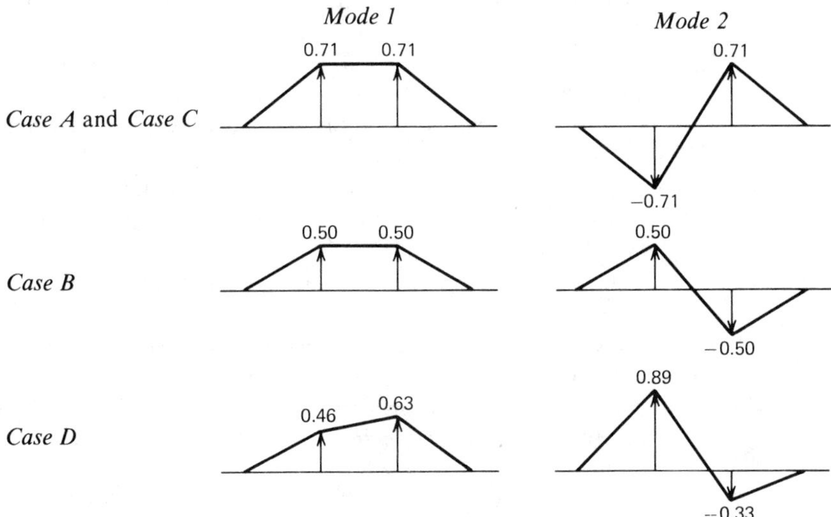

Conclusions:

a. Since cases A, B, and C are all for symmetrical systems, there is a symmetric mode and an antisymmetric mode for each. The symmetry property of modes is not present in case D because of the asymmetry of the mass distribution.

b. Increasing all masses proportionally simply reduces all eigenvalues in the same proportion, as seen by comparing case B with case A. The modes are unaffected except for overall amplitude (scaling).

c. An increase in the coupling spring constant in case C does not affect the symmetric mode (in which the spring remains undeformed) but increases the frequency of the antisymmetric mode in which the spring does deform.

d. The increase of mass m_2 only, case D, affects all frequencies and all mode shapes; the frequencies are reduced and the mode shapes are "skewed."

Design studies can be conducted as above by examining the effect of various design changes on the dynamical behavior of the system and by iterating until a suitable design is achieved. More sophisticated approaches to design (e.g., computerized optimization to determine minimum weight structures with specified natural frequency constraints) may also be taken (e.g., References 14.7–14.9).

References

14.1. K-J. Bathé and E. L. Wilson, *Numerical Methods in Finite Element Analysis,* Prentice-Hall, Englewood Cliffs, NJ (1976).

14.2. M. Newman and P. F. Flanagan, *Eigenvalue Extraction in NASTRAN by the Tridiagonal Reduction (FEER) Method-Real Eigenvalue Analysis,* NASA CR-2731, National Aeronautics and Space Administration, Washington, DC (1976).

14.3. A. Pipano, A. Raibstein, and A. Amnar, "Fast Modal Extraction in SAP Via the Tridiagonal Reduction Method," *First SAP Users' Conference and Computer Workship,* UCLA, Los Angeles, CA (1976).

14.4. L. Meirovitch, *Elements of Vibration Analysis,* McGraw-Hill, New York (1975).

14.5. W. C. Hurty and M. F. Rubinstein, *Dynamics of Structures,* Prentice-Hall, Englewood Cliffs, NJ (1964).

14.6. R. R. Craig, Jr. and M. C. C. Bampton, "On the Iterative Solution of Semidefinite Eigenvalue Problems," *Aero. J. of the Royal Aero. Soc.,* v. 75, 287–290 (1971).

14.7. M. J. Turner, "Design of Minimum Mass Structures with Specified Natural Frequencies," *AIAA Journal,* v. 5, 406–412 (1967).

14.8. W. J. Stroud, "Automated Structural Design with Aeroelastic Constaints: A Review and Assessment of the State of the Art," *Structural Optimization Symposium,* AMD-v. 7, ASME, New York (1974).

14.9. M. A. V. Rangacharyulu and G. T. S. Done; "A Survey of Structural Optimization Under Dynamic Constraints," *The Shock and Vibration Digest,* v. 11, n. 12, 15–25 (1979).

14.10. A. Jennings, "Eigenvalue Methods for Vibration Analysis," *The Shock and Vibration Digest,* v. 12, n. 2, 3–16 (1980).

Problem Set 14.2

14.1 (a) Use inverse iteration to solve for the fundamental frequency and mode shape of the system in Problem 13.2.

(b) Use a sweeping matrix, as in Example 14.2, to determine the second frequency and mode shape of this system.

14.2 (a) Use inverse iteration to solve for the fundamental frequency and mode shape of the three-story building below.

(b) Use a sweeping matrix, as in Example 13.2, to determine the second frequency and mode of the building.

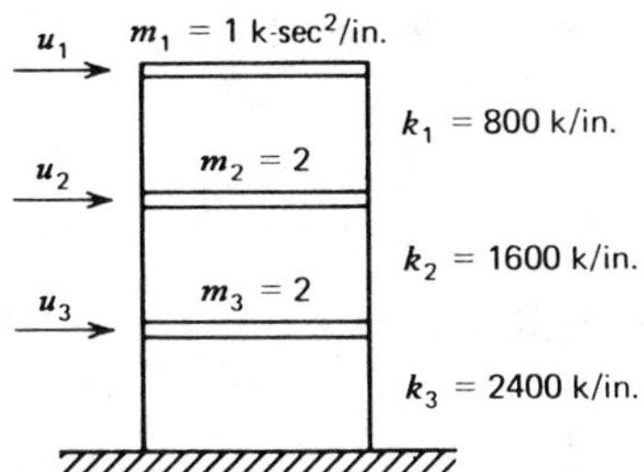

Figure P14.2

14.3 Perform a spectrum shift of $\mu = 0.5$, as in Eq. 14.23, on the system of Example 14.1 and use inverse iteration to compute the eigenvalue (frequency) and eigenvector (mode shape) of the shifted problem. How do these compare with the fundamental frequency and mode shape of the original system?

14.4 Perform a spectrum shift of $\mu = 2.2$, as in Eq. 14.23, on the system of Problem 13.2 and use inverse iteration to compute the fundamental frequency and mode of the shifted problem. To what frequency and mode of the original system do these results correspond?

14.5 By performing a spectrum shift on a system with rigid-body modes, it is possible to remove the singularity of **k** so that inverse iteration can be employed. Perform a spectrum shift of $\mu = -1$ on the system of Problem 13.5 and use inverse iteration to solve for the fundamental mode and frequency of the shifted problem.

14.6 The steps listed below form an efficient algorithm for carrying out inverse iteration with spectrum shift (e.g., see Reference 14.1). Construct a flowchart based on this algorithm.

1. Input **k**, **m**, μ.
2. Form $\hat{\mathbf{k}} = \mathbf{k} - \mu\mathbf{m}$.
3. Choose a starting vector $\mathbf{u}_{(0)}$.
4. Form $\mathbf{y}_{(0)} = \mathbf{m}\mathbf{u}_{(0)}$.

For $n = 0, 1, \ldots,$ evaluate the following:

5. Solve $\hat{\mathbf{k}}\mathbf{v}_{(n+1)} = \mathbf{y}_{(n)}$ for $\mathbf{v}_{(n+1)}$.
6. Form $\mathbf{w}_{(n+1)} = \mathbf{m}\mathbf{v}_{(n+1)}$.
7. Evaluate the Rayleigh quotient

$$\hat{\lambda}_{(n+1)} = \frac{\mathbf{v}_{(n+1)}^T\hat{\mathbf{k}}\mathbf{v}_{(n+1)}}{\mathbf{v}_{(n+1)}^T\mathbf{m}\mathbf{v}_{(n+1)}} = \frac{\mathbf{v}_{(n+1)}^T\hat{\mathbf{y}}_{(n)}}{\mathbf{v}_{(n+1)}^T\mathbf{w}_{(n+1)}}$$

8. Test for convergence.

$$|\hat{\lambda}_{(n+1)} - \hat{\lambda}_{(n)}| \le \epsilon?$$

9. If convergence has been achieved, scale the eigenvector so that $M = 1$.

$$\phi = \frac{\mathbf{v}_{(n+1)}}{[\mathbf{v}_{(n+1)}^T \mathbf{m} \mathbf{v}_{(n+1)}]^{1/2}} = \frac{\mathbf{v}_{(n+1)}}{[\mathbf{v}_{(n+1)}^T \mathbf{w}_{(n+1)}]^{1/2}}$$

10. If convergence has not been achieved, form

$$\mathbf{y}_{(n+1)} = \mathbf{m}\mathbf{u}_{(n+1)} = \frac{\mathbf{w}_{(n+1)}}{[\mathbf{v}_{(n+1)}^T \mathbf{w}_{(n+1)}]^{1/2}}$$

and continue iteration from step 5.

Problem Set 14.3

14.7 Use the ISMIS computer program (or another specified program) to solve for the modes and frequencies of the system in Problem 13.2.

14.8 Use the ISMIS computer program (or another specified program) to solve for the modes and frequencies of the system in Problem 13.5.

14.9 Use the ISMIS computer program (or another specified program) to solve for the modes and frequencies of the system in Problem 14.2.

14.10 (a) Resolve Problem 14.9 letting $m_1 = 2$ k-sec^2/in. Discuss the changes that occur in each of the three modes and frequencies.

(b) Resolve Problem 14.9 letting $k_3 = 2000$ k/in. Discuss the changes that occur in each of the three modes and frequencies as compared with those obtained in Problem 14.9.

15 DYNAMIC RESPONSE OF MDOF SYSTEMS: MODE-SUPERPOSITION METHOD

In Example 12.6 you were introduced to the subject of dynamic response of MDOF systems by considering the response of a 2-DOF system to harmonic excitation. You were also introduced to the very important mode-superposition method for solving for the dynamic response of MDOF systems. In this chapter you will make a more detailed study of the mode superposition method and will learn when and how to apply it. Chapter 18 will treat the dynamic response of systems for which the mode-superposition method is not appropriate.

Upon completion of this chapter you should be able to:

- Transform the equations of motion of an N-DOF system to principal coordinates.
- Determine the initial conditions in principal coordinates.
- Determine the displacement and internal stress response of an N-DOF system to harmonic or transient excitation using the mode-displacement method.
- Determine the displacement and internal stress response of an N-DOF system to harmonic or transient excitation using the mode-acceleration method if (a) the system has no rigid-body modes, or (b) the system has rigid-body modes.
- Generate complex frequency response plots for N-DOF systems having uncoupled modal damping, and describe the type of curves which can be expected if (a) the system has widely-spaced frequencies and light damping, or (b) the system has closely-spaced frequencies and light damping.
- Determine modal static displacements and describe the significance of the distribution of the input forces on the modal responses.

15.1 Introduction; Principal Coordinates

The equations of motion of a linear MDOF system were given in Eq. 11.30, repeated here.

$$\mathbf{m}\ddot{\mathbf{u}} + \mathbf{c}\dot{\mathbf{u}} + \mathbf{k}\mathbf{u} = \mathbf{p}(t) \tag{15.1}$$

In general, the coefficient matrices in Eq. 15.1, **m**, **c**, and **k**, may have nonzero coupling terms (e.g., $k_{ij} = k_{ji} \neq 0$), so that to solve Eq. 15.1 in its present form would require simultaneous solution of N equations in N unknowns. In this section we outline the *mode-superposition method,* or *normal-mode method,* by which such a set of coupled equations can be transformed into a set of uncoupled equations through use of the normal modes of the system. In Secs. 15.2 through 15.4, examples will be presented for undamped systems and for systems with a special form of viscous damping. Response of systems with more general viscous damping or with elastic forces that introduce nonlinearities will be considered in Chapter 18.

Equation 15.1 is the original set of coupled equations of motion for an N-DOF system, where $\mathbf{u}(t)$ may be physical or generalized coordinates. The response of the system to the excitation $\mathbf{p}(t)$ and to the initial conditions

$$\mathbf{u}(0) = \mathbf{u}_o, \qquad \dot{\mathbf{u}}(0) = \dot{\mathbf{u}}_o \tag{15.2}$$

is sought.

The first step in a mode-superposition solution is to obtain the natural frequencies and natural modes of the system. (We will assume here that all N modes are to be used. The important topic of truncation, that is, using fewer than N modes, will be discussed at length in Secs. 15.2 through 15.6.) Then the natural frequencies and modes satisfy

$$(\mathbf{k} - \omega_r^2\mathbf{m})\boldsymbol{\phi}_r = \mathbf{0} \tag{15.3}$$

giving $(\omega_r^2, \boldsymbol{\phi}_r)$, $r = 1, 2, \ldots, N$. The modes $\boldsymbol{\phi}_r$ are assumed to have been normalized by one of the schemes listed in Sec. 13.1, and the *modal mass,* M_r, and modal stiffness, K_r, calculated using

$$\begin{aligned} M_r &= \boldsymbol{\phi}_r^T\mathbf{m}\boldsymbol{\phi}_r \\ K_r &= \omega_r^2 M_r \end{aligned} \tag{15.4}$$

We will assume that if there are any repeated frequencies, the associated modes have been orthogonalized so that the orthogonality equations

$$\boldsymbol{\phi}_r^T\mathbf{m}\boldsymbol{\phi}_s = \boldsymbol{\phi}_r^T\mathbf{k}\boldsymbol{\phi}_s = 0 \tag{15.5}$$

are satisfied for all $r \neq s$. The modes are then collected to form the *modal matrix* $\boldsymbol{\Phi}$, that is,

$$\boldsymbol{\Phi} = [\boldsymbol{\phi}_1\ \boldsymbol{\phi}_2 \ldots \boldsymbol{\phi}_N] \tag{15.6}$$

The key step in the mode-superposition procedure is to introduce the coordinate transformation

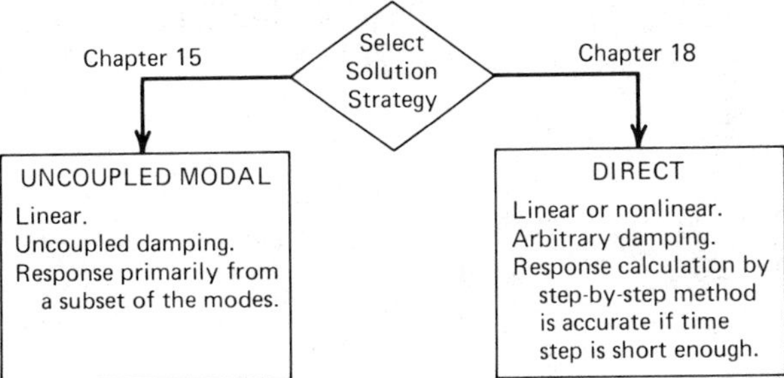

Figure 15.1. *Two basic solution strategies.*

$$\mathbf{u}(t) = \mathbf{\Phi}\boldsymbol{\eta}(t) = \sum_{r=1}^{N} \boldsymbol{\phi}_r \eta_r(t) \tag{15.7}$$

The coordinates $\eta_r(t)$ will be referred to as *principal coordinates.** Equation 15.7 is substituted into Eq. 15.1 and the resulting equation is multiplied by $\mathbf{\Phi}^T$ to give the equation of motion in principal coordinates, namely

$$\boldsymbol{M}\ddot{\boldsymbol{\eta}} + \boldsymbol{C}\dot{\boldsymbol{\eta}} + \boldsymbol{K}\boldsymbol{\eta} = \boldsymbol{P}(t) \tag{15.8}$$

where

$$\begin{aligned} \boldsymbol{M} &= \mathbf{\Phi}^T\mathbf{m}\mathbf{\Phi} = \text{modal mass matrix} \\ \boldsymbol{C} &= \mathbf{\Phi}^T\mathbf{c}\mathbf{\Phi} = \text{modal damping matrix} \\ \boldsymbol{K} &= \mathbf{\Phi}^T\mathbf{k}\mathbf{\Phi} = \text{modal stiffness matrix} \\ \boldsymbol{P}(t) &= \mathbf{\Phi}^T\mathbf{p}(t) = \text{modal force vector} \end{aligned} \tag{15.9}$$

Due to the orthogonality conditions of Eq. 15.5, $\boldsymbol{M}$ and $\boldsymbol{K}$ are diagonal matrices, so the equations of motion in modal coordinates, Eq. 15.8, are coupled only through nonzero off-diagonal coefficients in the modal damping matrix $\boldsymbol{C}$. For this reason we will consider mode-superposition solutions for undamped systems first in Secs. 15.2 and 15.3. Then, in Sec. 15.4, we will look at some special cases where the modal damping matrix is diagonal. In Chapter 18 we will return to the case of systems with damping coupling when we take up the topic of direct integration of the equations of motion. Figure 15.1 outlines the two basic strategies for integrating MDOF equations of motion to obtain dynamic response.

The total response $\boldsymbol{\eta}(t)$ can be obtained as a superposition of the response due to initial conditions alone and response due to the excitation alone. The

*Some authors refer to these as *natural coordinates* or as *normal coordinates*. Some authors[15.1,15.2] reserve the name normal coordinates to use when modes are normalized so that $M_r = 1$.

response due to initial conditions can be treated in a manner similar to that used in Sec. 13.1 (e.g., Eqs. 13.37 through 13.39). From Eq. 15.7,

$$\begin{aligned} \mathbf{u}(0) &= \mathbf{\Phi}\boldsymbol{\eta}(0) \\ \dot{\mathbf{u}}(0) &= \mathbf{\Phi}\dot{\boldsymbol{\eta}}(0) \end{aligned} \tag{15.10}$$

Multiplying these equations by $\mathbf{\Phi}^T\mathbf{m}$ we get

$$\begin{aligned} \mathbf{\Phi}^T\mathbf{m}\mathbf{u}(0) &= \boldsymbol{M}\boldsymbol{\eta}(0) \\ \mathbf{\Phi}^T\mathbf{m}\dot{\mathbf{u}}(0) &= \boldsymbol{M}\dot{\boldsymbol{\eta}}(0) \end{aligned} \tag{15.11}$$

Since $\boldsymbol{M}$ is diagonal, Eqs. 15.11 can be solved for the *modal initial conditions* giving

$$\left.\begin{aligned} \eta_r(0) &= \left(\frac{1}{M_r}\right)\boldsymbol{\phi}_r^T\mathbf{m}\mathbf{u}(0) \\ \\ \dot{\eta}_r(0) &= \left(\frac{1}{M_r}\right)\boldsymbol{\phi}_r^T\mathbf{m}\dot{\mathbf{u}}(0) \end{aligned}\right\} r = 1, 2, \ldots, N \tag{15.12}$$

An example of mode-superposition for free vibrations was given in Example 13.5.

15.2 Mode-Displacement Solution for Response of Undamped MDOF Systems

In Eq. 15.7 the coordinate transformation relating physical (or generalized) coordinates $\mathbf{u}$ and modal coordinates $\boldsymbol{\eta}$ includes all N system modes. However, in Fig 15.1 it was indicated that the mode-superposition procedure is most useful when the system response involves only a relatively small subset of the modes of the system. Hence, we will examine the possibility of truncation, and will determine the factors that should be considered in deciding upon the number of modes to be included in a mode-superposition solution in order to produce accurate results.

The displacement $\mathbf{u}(t)$ will be approximated by

$$\hat{u}(t) = \hat{\mathbf{\Phi}}\hat{\boldsymbol{\eta}}(t) = \sum_{r=1}^{\hat{N}} \boldsymbol{\phi}_r\eta_r(t) \tag{15.13}$$

where

$$\hat{\mathbf{\Phi}} = [\boldsymbol{\phi}_1\ \boldsymbol{\phi}_2 \ldots \boldsymbol{\phi}_{\hat{N}}] \tag{15.14}$$

Equation 15.13 is the coordinate transformation for mode-superposition with truncation of modes from N to $\hat{N} \ll N$ (e.g., $N = 1000$, $\hat{N} = 50$).

For an undamped system Eq. 15.8 reduces to

$$\hat{\mathcal{M}}\ddot{\eta} + \hat{\mathcal{K}}\eta = \hat{\mathcal{P}}(t) \tag{15.15}$$

Since the modal stiffness and mass matrices are diagonal, Eq. 15.15 can be written as $\hat{N}$ uncoupled equations

$$M_r\ddot{\eta}_r + K_r\eta_r = P_r(t), \qquad r = 1, 2, \ldots, \hat{N} \tag{15.16}$$

where M_r and K_r are given by Eqs. 15.4, and where

$$P_r = \boldsymbol{\phi}_r^T\mathbf{p} \tag{15.17}$$

The total response of the rth mode can be expressed as a superposition of response due to modal initial conditions, which are given by Eqs. 15.12, and modal response due to $P_r(t)$. The Duhamel integral can be used to represent symbolically the response. Thus,

$$\begin{aligned} \eta_r(t) = {} & \eta_r(0)\cos(\omega_r t) + \left(\frac{1}{\omega_r}\right)\dot{\eta}_r(0)\sin(\omega_r t) \\ & + \left(\frac{1}{M_r\omega_r}\right)\int_0^t P_r(\tau)\sin\omega_r(t-\tau)\,d\tau \end{aligned} \tag{15.18}$$

Any of the numerical techniques of Chapter 7 can, of course, be employed for evaluating the modal time histories, $\eta_r(t)$.

The process of computing modal responses and substituting these back into Eq. 15.13 to obtain the approximate system response will be called the *mode-displacement method.*

Consider the steady-state response of an undamped MDOF system with harmonic excitation given by

$$\mathbf{p}(t) = \mathbf{P}\cos\Omega t \tag{15.19}$$

From Eqs. 15.9d and 15.19,

$$\mathcal{P}(t) = (\boldsymbol{\phi}^T\mathbf{P})\cos\Omega t \tag{15.20a}$$

or

$$P_r(t) = F_r\cos\Omega t \tag{15.20b}$$

where

$$F_r = \boldsymbol{\phi}_r^T\mathbf{P} \tag{15.20c}$$

Substituting Eq. 15.20b into Eq. 15.16 and using Eqs. 4.4 and 4.9, we obtain the steady-state response

$$\eta_r(t) = \left(\frac{F_r}{K_r}\right)\left[\frac{1}{1-(\Omega/\omega_r)^2}\right]\cos\Omega t \tag{15.21}$$

Then, combining Eqs. 15.13 and 15.21, we obtain the following approximated steady-state response in the original coordinates

$$\hat{\mathbf{u}}(t) = \sum_{r=1}^{\hat{N}} \phi_r \left(\frac{F_r}{K_r}\right)\left[\frac{1}{1-(\Omega/\omega_r)^2}\right]\cos\Omega t \tag{15.22}$$

Now consider an example steady-state response solution.

Example 15.1

The four-story shear building below has the natural frequencies and modal matrix shown. Harmonic excitation $P_1 \cos \Omega t$ is applied at the top story. Consider steady-state response only in parts (c) through (e).

a. Determine the four modal masses M_r and modal stiffnesses K_r.

b. Determine the four modal forces $P_r(t)$.

c. Determine an expression for the steady-state responses $\eta_r(t)$.

d. Using Eq. 15.13 determine an expression for $\hat{u}_1(t)$.

e. Prepare a table showing the amplitude of $\hat{u}_1(t)$ using $N = 1$, $N = 2$, and $N = 3$ for excitation frequencies $\Omega = 0$, $\Omega = 0.5\omega_1$, and $\Omega = 1.3\omega_3$. Also, determine $u_1(t)$ using $N = 4$.

f. From the results tabulated in (e), what conclusions can you draw concerning truncation to one mode, two modes, or three modes?

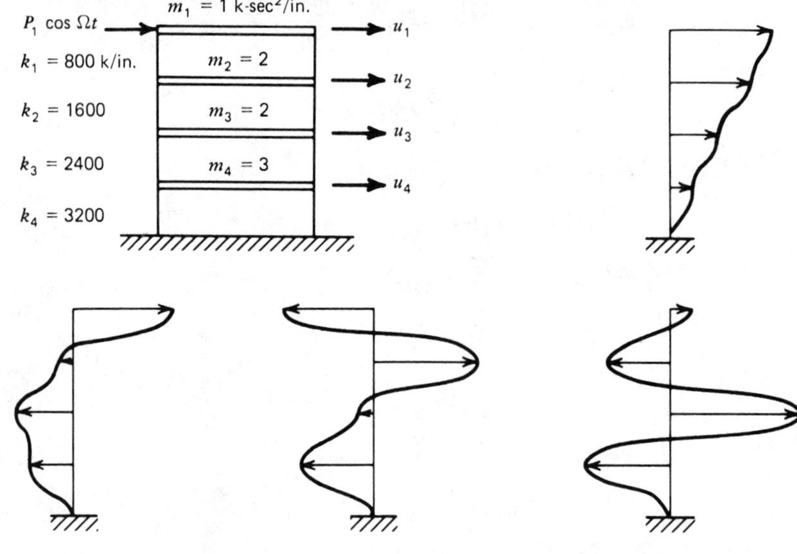

$$\mathbf{k} = 800 \begin{bmatrix} 1 & -1 & 0 & 0 \\ -1 & 3 & -2 & 0 \\ 0 & -2 & 5 & -3 \\ 0 & 0 & -3 & 7 \end{bmatrix}, \qquad \mathbf{m} = \begin{bmatrix} 1 & 0 & 0 & 0 \\ 0 & 2 & 0 & 0 \\ 0 & 0 & 2 & 0 \\ 0 & 0 & 0 & 3 \end{bmatrix}$$

$$\omega^2 = \begin{Bmatrix} 0.17672 \\ 0.87970 \\ 1.68746 \\ 3.12279 \end{Bmatrix} \times 10^3, \qquad \omega = \begin{Bmatrix} 13.294 \\ 29.660 \\ 41.079 \\ 55.882 \end{Bmatrix}$$

$$\mathbf{\Phi} = \begin{bmatrix} 1.00000 & 1.00000 & -0.90145 & 0.15436 \\ 0.77910 & -0.09963 & 1.00000 & -0.44817 \\ 0.49655 & -0.53989 & -0.15859 & 1.00000 \\ 0.23506 & -0.43761 & -0.70797 & -0.63688 \end{bmatrix}$$

Solution

a. Determine modal masses from Eq. 15.4a and modal stiffnesses from Eq. 15.4b.

$$M_r = \boldsymbol{\phi}_r^T \mathbf{m} \boldsymbol{\phi}_r, \qquad K_r = \omega_r^2 M_r \tag{1}$$

For example,

$$M_1 = \begin{Bmatrix} 1.00000 \\ 0.77910 \\ 0.49655 \\ 0.23506 \end{Bmatrix}^T \begin{bmatrix} 1 & 0 & 0 & 0 \\ 0 & 2 & 0 & 0 \\ 0 & 0 & 2 & 0 \\ 0 & 0 & 0 & 3 \end{bmatrix} \begin{Bmatrix} 1.00000 \\ 0.77910 \\ 0.49655 \\ 0.23506 \end{Bmatrix}$$

$$M_1 = 2.87288$$

$$K_1 = \omega_1^2 M_1$$

$$K_1 = 176.72(2.87288) = 507.695$$

The remaining masses and stiffnesses are calculated in a similar manner giving

$$\boxed{\begin{array}{ll} M_1 = 2.87288 & K_1 = 507.695 \\ M_2 = 2.17732 & K_2 = 1915.39 \\ M_3 = 4.36658 & K_3 = 7368.43 \\ M_4 = 3.64239 & K_4 = 11{,}374.4 \end{array}} \tag{2}$$

b. The modal forces are calculated from Eq. 15.17. Since

$$\mathbf{P} = \begin{Bmatrix} P_1 \\ 0 \\ 0 \\ 0 \end{Bmatrix} \tag{3}$$

and

$$F_r = \phi_r^T \mathbf{P} \tag{4}$$

then

$$\boxed{\begin{aligned} F_1 &= P_1 \\ F_2 &= P_1 \\ F_3 &= -0.90145P_1 \\ F_4 &= 0.15436P_1 \end{aligned}} \tag{5}$$

c. Since this is harmonic excitation, Eq. 15.21 gives

$$\boxed{\eta_r(t) = \frac{(F_r/K_r)\cos\Omega t}{(1 - r_r^2)}} \tag{6}$$

where

$$r_r = \frac{\Omega}{\omega_r} \tag{7}$$

d. The mode-displacement approximation $\hat{\mathbf{u}}(t)$ is given by Eq. 15.22. For $\hat{u}_1(t)$ we take the first component of each ϕ_r to get

$$\boxed{u_1(t) = \sum_{r=1}^{\hat{N}} \phi_{1r}\eta_r(t)} \tag{8}$$

e. We can write the expression for $u_1(t)$ without truncation, and then indicate the terms corresponding to $\hat{N} = 1$, $\hat{N} = 2$, and $\hat{N} = 3$. Thus,

$$\begin{aligned} u_1(t) &= \frac{(1.0)(P_1 \cos \Omega t)}{(507.695)[1 - (\Omega^2/176.72)]} \quad \Big] \; \hat{N} = 1 \\ &+ \frac{(1.0)(P_1 \cos \Omega t)}{(1915.39)[1 - (\Omega^2/879.70)]} \quad \Big] \; \hat{N} = 2 \\ &+ \frac{(-0.90145)(-0.90145P_1 \cos \Omega t)}{(7368.43)[1 - (\Omega^2/1687.46)]} \quad \Big] \; \hat{N} = 3 \\ &+ \frac{(0.15436)(0.15436\, P_1 \cos \Omega t)}{(11374.4)[1 - (\Omega^2/3122.79)]} \end{aligned} \tag{9}$$

For

$$\Omega = 0.5\omega_1, \qquad \Omega = 6.6468, \qquad \Omega^2 = 44.179$$

For

$$\Omega = 1.3\omega_3, \qquad \Omega = 53.402, \qquad \Omega^2 = 2851.80$$

Constant C in $u_1(t) = CP_1 \cos \Omega t$

	$\hat{N} = 1$	$\hat{N} = 2$	$\hat{N} = 3$	$\hat{N} = 4$
$\Omega = 0$	$1.970(10^{-3})$	$2.492(10^{-3})$	$2.602(10^{-3})$	$2.604(10^{-3})$
$\Omega = 0.5\omega_1$	$2.626(10^{-3})$	$3.176(10^{-3})$	$3.289(10^{-3})$	$3.291(10^{-3})$
$\Omega = 1.3\omega_3$	$-1.301(10^{-4})$	$-3.630(10^{-4})$	$-5.228(10^{-4})$	$-4.987(10^{-4})$

f. The following conclusions can be reached:

1. A one-mode solution is not accurate at any of the three frequencies.
2. A three-mode solution is quite accurate for $\Omega = 0$ or $\Omega = 0.5\omega_1$, but since $\Omega = 1.3\omega_3$ is almost equal to ω_4, an important contribution to $u_1(t)$ at this frequency is the mode 4 contribution. A truncated solution will be useless at this frequency.

Equation 9 of Example 15.1 illustrates the importance of the distribution and frequency of the input on the response, as is also seen in Eq. 15.22. The importance of the distribution of the input forces is also illustrated by the next example. By comparing Eq. 4.9 and 15.21 we see that the *modal static deflection*[15.3] defined by

$$D_r \equiv \frac{F_r}{K_r} = \frac{\boldsymbol{\phi}_r^T \mathbf{P}}{K_r} \tag{15.23}$$

plays the same role as the static displacement, U_o, of a SDOF system.

Example 15.2

For the four-story building of Example 15.1 compute F_r and D_r for the two load distributions shown.

$$\mathbf{P}_a = \begin{Bmatrix} 1 \\ 0 \\ 0 \\ 0 \end{Bmatrix}, \qquad \mathbf{P}_b = \begin{Bmatrix} 1 \\ 1 \\ 1 \\ 1 \end{Bmatrix}$$

Solution

a. Using the mode shapes from Example 15.1, we obtain

$$F_{1a} = \boldsymbol{\phi}_1^T \mathbf{P}_a = \begin{Bmatrix} 1.00000 \\ 0.77910 \\ 0.49655 \\ 0.23506 \end{Bmatrix}^T \begin{Bmatrix} 1 \\ 0 \\ 0 \\ 0 \end{Bmatrix} = 1.00000 \tag{1a}$$

$$F_{1b} = \boldsymbol{\phi}_1^T \mathbf{P}_b = \begin{Bmatrix} 1.00000 \\ 0.77910 \\ 0.49655 \\ 0.23506 \end{Bmatrix}^T \begin{Bmatrix} 1 \\ 1 \\ 1 \\ 1 \end{Bmatrix} = 2.51071 \tag{1b}$$

The other modal forces are determined in a similar manner.

$$\boxed{\begin{array}{llll} F_{1a} = & 1.00000, & F_{1b} = & 2.51071 \\ F_{2a} = & 1.00000, & F_{2b} = & -0.07713 \\ F_{3a} = & -0.90145, & F_{3b} = & -0.76801 \\ F_{4a} = & 0.15436, & F_{4b} = & 0.06931 \end{array}} \tag{2}$$

b. Using the results of (a) together with the generalized stiffnessess calculated in Example 15.1, we get

$$D_r = \frac{F_r}{K_r} \tag{3}$$

$$\boxed{\begin{array}{llll} D_{1a} = & 1.9697(10^{-3}), & D_{1b} = & 4.9453(10^{-3}) \\ D_{2a} = & 5.2209(10^{-4}), & D_{2b} = & -0.4027(10^{-4}) \\ D_{3a} = & -1.2234(10^{-4}), & D_{3b} = & -1.0423(10^{-4}) \\ D_{4a} = & 1.3571(10^{-5}), & D_{4b} = & 0.6094(10^{-5}) \end{array}} \tag{4}$$

Note that, although the total force applied in case (b) is four times that in case (a), the mode shape together with the force distribution determines the value of F_r, and hence the excitation force level of a particular mode. Also note that when modes are normalized so that the largest component is 1.0, the stiffness increases with mode number. This influences the modal static deflections D_r.

15.3 Mode-Acceleration Solution for Response of Undamped MDOF Systems

As noted in Example 15.1 a mode-displacement solution may fail to give an accurate solution, even when static loading is applied. Frequently the convergence is slow and many modes would be needed to give an accurate mode-displacement solution. This difficulty can be alleviated by use of the mode-acceleration method. Because of the improved convergence properties of this method, $\hat{N}$ can be reduced, and fewer natural frequencies and modes are required from the eigensolution.

Consider the undamped MDOF system.

$$\mathbf{m}\ddot{\mathbf{u}} + \mathbf{k}\mathbf{u} = \mathbf{p}(t). \tag{15.24}$$

The mode-displacement (approximate) solution $\hat{\mathbf{u}}$ is given by Eq. 15.13. The modes from $(\hat{N} + 1)$ to N are completely ignored.

The *mode-acceleration* solution is based on the following. Equation 15.24 is written

$$\mathbf{u} = \mathbf{k}^{-1}(\mathbf{p} - \mathbf{m}\ddot{\mathbf{u}}) \tag{15.25}$$

and the $\ddot{\mathbf{u}}$ term is approximated by the mode-displacement term $\ddot{\hat{\mathbf{u}}}$. Thus, the mode-acceleration solution $\tilde{\mathbf{u}}$ is given by*

$$\tilde{\mathbf{u}} = \mathbf{k}^{-1}(\mathbf{p} - \mathbf{m}\ddot{\hat{\mathbf{u}}}) \tag{15.26}$$

Combining Eqs. 15.26 and 15.13 we obtain

$$\tilde{\mathbf{u}} = \mathbf{k}^{-1}\mathbf{p} - \mathbf{k}^{-1}\sum_{r=1}^{\hat{N}} \mathbf{m}\boldsymbol{\phi}_r\ddot{\eta}_r \tag{15.27}$$

and, incorporating Eq. 15.3 we get

$$\tilde{\mathbf{u}} = \mathbf{k}^{-1}\mathbf{p} - \sum_{r=1}^{\hat{N}} \left(\frac{1}{\omega_r^2}\right) \boldsymbol{\phi}_r\ddot{\eta}_r \tag{15.28}$$

The first term in the above equation is the *pseudo static response,* while the second term gives the method its name, the *mode-acceleration method.* The presence of ω_r^2 in the denominator of this term improves the convergence of the method as compared to the mode-displacement method[15.4,15.5]. Williams[15.6] is credited with first suggesting the mode-acceleration method.

Example 15.3

For the four-story building of Example 15.1:

a. Use Eq. 15.28 to determine an expression for $\tilde{u}_1(t)$.

b. Prepare a table showing the amplitude of $\tilde{u}_1(t)$ using $\hat{N} = 1$, $\hat{N} = 2$, and $\hat{N} = 3$ for excitation frequencies $\Omega = 0$, $\Omega = 0.5\omega_1$, and $\Omega = 1.3\omega_3$.

c. By comparing the table prepared in (b) with the mode-displacement table of Example 15.1, state any conclusions you may reach about convergence of the two methods.

$$\mathbf{a} \equiv \mathbf{k}^{-1} = \begin{bmatrix} 2.60417 & 1.35417 & 0.72917 & 0.31250 \\ 1.35417 & 1.35417 & 0.72917 & 0.31250 \\ 0.72917 & 0.72917 & 0.72917 & 0.31250 \\ 0.31250 & 0.31250 & 0.31250 & 0.31250 \end{bmatrix} (10^{-3})$$

*Systems with rigid-body modes will be treated in Sec. 15.6.

Solution

a. As in part (e) of Example 15.1 we can write an expression for $u_1(t)$ without truncation and then indicate the terms corresponding to $\hat{N} = 1$, $\hat{N} = 2$, and $\hat{N} = 3$. Thus, from Eq. 15.28,

$$\tilde{u}_1(t) = a_{11}P_1 \cos \Omega t - \sum_{r=1}^{\hat{N}} \left(\frac{1}{\omega_r^2}\right) \phi_{1r}\ddot{\eta}_r \tag{1}$$

which, when combined with Eq. 15.21, gives

$$u_1(t) = a_{11}P_1 \cos \Omega t + \sum_{r=1}^{\hat{N}} \left(\frac{\Omega^2}{\omega_r^2}\right) \phi_{1r}\eta_r \tag{2}$$

Thus

$$\begin{aligned} u_1(t) &= 2.60417(10^{-3})(P_1 \cos \Omega t) \\ &+ \frac{(\Omega^2/176.72)(1.0)(P_1 \cos \Omega t)}{(507.695)[1 - (\Omega^2/176.72)]} \quad \Big] \; \hat{N} = 1 \\ &+ \frac{(\Omega^2/879.70)(1.0)(P_1 \cos \Omega t)}{(1915.39)[1 - (\Omega^2/879.70)]} \quad \Big] \; \hat{N} = 2 \\ &+ \frac{(\Omega^2/1687.46)(-0.90145)(-0.90145 P_1 \cos \Omega t)}{(7368.43)[1 - (\Omega^2/1687.46)]} \quad \Big] \; \hat{N} = 3 \\ &+ \frac{(\Omega^2/3122.79)(0.15436)(0.15436\, P_1 \cos \Omega t)}{(11374.4)[1 - (\Omega^2/3122.79)]} \end{aligned} \tag{3}$$

b. As in Example 15.1, the three frequencies of interest give

$\Omega^2 = 0, \qquad \Omega^2 = 44.197, \qquad \text{and} \qquad \Omega^2 = 2851.80$

***Constant* C *in* u_1(t) = CP$_1$ *cos* Ω t**

	$\hat{N} = 1$	$\hat{N} = 2$	$\hat{N} = 3$	$\hat{N} = 4$
$\Omega = 0$	$2.604(10^{-3})$	$2.604(10^{-3})$	$2.604(10^{-3})$	$2.604(10^{-3})$
$\Omega = 0.5\omega_1$	$3.261(10^{-3})$	$3.288(10^{-3})$	$3.291(10^{-3})$	$3.291(10^{-3})$
$\Omega = 1.3\omega_3$	$5.044(10^{-4})$	$-2.506(10^{-4})$	$-5.207(10^{-4})$	$-4.987(10^{-4})$

c. From the above table we can conclude that:

1. The exact static solution is produced at $\Omega = 0$ without any contribution from normal modes.
2. At the low frequency of $0.5\omega_1$, even a one-term solution is fairly accurate, and the solution is accurate to two places for $\hat{N} = 2$ as compared with $\hat{N} = 3$ for the mode-displacement solution.

3. Since the forcing frequency $\Omega = 1.3\omega_3$ lies between ω_3 and ω_4, a truncated mode-acceleration solution is not any better than a truncated mode-displacement solution—the fourth mode is needed in either case.

As noted in Example 15.3 the mode-acceleration method is particularly useful in improving the static or low-frequency response convergence. Whether the mode-displacement method or the mode-acceleration method is used, it is always important to include all modes whose natural frequencies are "in the vicinity of" any excitation frequency.

Equation 15.18 can be used to obtain a general expression for $\ddot{\eta}_r(t)$ to substitute into the mode-acceleration equation, Eq. 15.28. Thus,

$$\ddot{\eta}_r(t) = -\omega_r^2\eta_r(0)\cos\omega_r t - \omega_r\dot{\eta}_r(0)\sin\omega_r t + \frac{P_r(t)}{M_r} - \frac{\omega_r}{M_r}\int_0^t P_r(\tau)\sin\omega_r(t-\tau)\,d\tau \tag{15.29}$$

Integration by parts may be employed to convert Eq. 15.29 to the alternative form

$$\ddot{\eta}_r(t) = -\omega_r^2\eta_r(0)\cos\omega_r t - \omega_r\dot{\eta}_r(0)\sin\omega_r t + \frac{1}{M_r}\left\{P_r(0)\cos\omega_r t + \int_0^t \frac{d}{d\tau}[P_r(\tau)]\cos\omega_r(t-\tau)\,d\tau\right\} \tag{15.30}$$

Numerical procedures of Chapter 7 can also be employed to determine modal acceleration histories, for example, see Fig. 7.4.

In Sec. 15.2 and the present section it has generally been implied that $\hat{N} < N$ modes could be employed for mode superposition with the first $\hat{N}$ modes (i.e., starting from lowest frequency and counting the $\hat{N}$ lowest frequency modes out of the set of N modes). Engineering judgment is required in determining how many modes to include. Examples 15.1 through 15.3 are intended to point out the importance of frequency ratio and mode shape in determining how many modes are important.

15.4 Mode-Superposition Solutions for Response of Certain Viscous-Damped Systems

In this section we consider mode-superposition solutions for the response of systems with viscous damping distributed in such a way that the modal equations of motion are uncoupled. We will also use this opportunity to consider the nature of the response of such systems to harmonic excitation. Such response plays a very important role in the testing of structures to determine their dynamical properties (e.g., References 15.7 and 15.8).

Consider a MDOF system with viscous damping. Equation 15.8 gives the modal equation of motion. In this chapter we will assume that the damping matrix **c** is such that

$$\boldsymbol{\phi}_r^T \mathbf{c} \boldsymbol{\phi}_s = 0, \qquad r \neq s \tag{15.31}$$

that is, the damping matrix is uncoupled by the transformation to modal coordinates. (In Chapter 18 the conditions under which this is possible will be discussed.) Then, Eq. 15.8 becomes the set of uncoupled modal equations which may be written in the form

$$\ddot{\eta}_r + 2\zeta_r\omega_r\dot{\eta}_r + \omega_r^2\eta_r = \left(\frac{1}{M_r}\right) P_r(t), \qquad r = 1, 2, \ldots, N \tag{15.32}$$

where ζ_r is the modal damping factor defined by

$$\zeta_r = \frac{C_r}{2M_r\omega_r} = \left(\frac{1}{2M_r\omega_r}\right) \boldsymbol{\phi}_r^T \mathbf{c} \boldsymbol{\phi}_r \tag{15.33}$$

The solution of Eq. 15.32 can be written in the same form as the SDOF solution given in Eq. 6.7, that is,

$$\begin{aligned} \eta_r(t) &= \left(\frac{1}{M_r\omega_{dr}}\right) \int_0^t P_r(\tau) e^{-\zeta_r\omega_r(t-\tau)} \sin \omega_{dr}(t-\tau)\, d\tau \\ &\quad + \eta_r(0) e^{-\zeta_r\omega_r t} \cos \omega_{dr} t \\ &\quad + \left(\frac{1}{\omega_{dr}}\right) [\dot{\eta}_r(0) + \zeta_r\omega_r\eta_r(0)]\, e^{-\zeta_r\omega_r t} \sin \omega_{dr} t \end{aligned} \tag{15.34}$$

where

$$\omega_{dr} = \omega_r \sqrt{1 - \zeta_r^2} \tag{15.35}$$

Frequently, in the absence of more definitive information about damping, modal damping as in Eq. 15.32 is simply assumed to be valid and "reasonable" values of ζ_r are assumed.

If the solutions to the N modal equations, Eqs. 15.32, are substituted back into Eq. 15.7, the response $\mathbf{u}(t)$ is determined. If only $\hat{N}$ modes are retained in the solution and Eq. 15.13 is used, we have a *mode-displacement* solution that ignores completely the contribution of the modes from $(\hat{N} + 1)$ to N.

As in the case of undamped systems, a *mode-acceleration* solution can be obtained for these viscous-damped systems. This can be obtained by writing Eq. 15.1 in the form

$$\mathbf{u} = \mathbf{k}^{-1}[\mathbf{p}(t) - \mathbf{c}\dot{\mathbf{u}} - \mathbf{m}\ddot{\mathbf{u}}] \tag{15.36}$$

and approximating the velocity and acceleration terms with their mode-displacement approximations. Thus, the mode-acceleration approximation is given by

$$\tilde{\mathbf{u}}(t) = \mathbf{k}^{-1}\mathbf{p}(t) - \mathbf{k}^{-1}\mathbf{c}\sum_{r=1}^{\hat{N}} \boldsymbol{\phi}_r\dot{\eta}_r(t) - \mathbf{k}^{-1}\mathbf{m}\sum_{r=1}^{\hat{N}} \boldsymbol{\phi}_r\ddot{\eta}_r(t) \tag{15.37}$$

The last term can be simplified as was done for the undamped system in arriving at Eq. 15.28. The damping term can also be simplified (see Problem 15.5). The resulting *mode-acceleration* solution is

$$\tilde{\mathbf{u}}(t) = \mathbf{k}^{-1}\mathbf{p}(t) - \sum_{r=1}^{\hat{N}} \left(\frac{2\zeta_r}{\omega_r}\right) \boldsymbol{\phi}_r\dot{\eta}_r(t) - \sum_{r=1}^{\hat{N}} \left(\frac{1}{\omega_r^2}\right) \boldsymbol{\phi}_r\ddot{\eta}_r(t) \tag{15.38}$$

Reference 15.9 gives some interesting numerical convergence studies based on Eq. 15.38.

Consider now the mode-superposition solution for the steady-state response of a viscous-damped system subjected to harmonic excitation $\mathbf{p} = \mathbf{P}\cos\Omega t$. Then

$$\ddot{\eta}_r + 2\zeta_r\omega_r\dot{\eta}_r + \omega_r^2\eta_r = \left(\frac{1}{M_r}\right) F_r \cos\Omega t \tag{15.39}$$

As in the case of a SDOF system, we can solve Eq. 15.39 using complex frequency response techniques.

$$\ddot{\bar{\eta}}_r + 2\zeta_r\omega_r\dot{\bar{\eta}}_r + \omega_r^2\bar{\eta}_r = \omega_r^2\left(\frac{F_r}{K_r}\right) e^{i\Omega t} \tag{15.40}$$

Then the steady-state solution $\bar{\eta}_r$ can be written in the form

$$\bar{\eta}_r = \bar{H}_{\eta_r/F_r}(\Omega)F_r e^{i\Omega t} \tag{15.41}$$

where $\bar{H}_{\eta_r/F_r}(\Omega)$ is the *complex frequency response function for principal coordinates,* given by

$$\bar{H}_r(\Omega) \equiv \bar{H}_{\eta_r/F_r}(\Omega) = \frac{1/K_r}{(1 - r_r^2) + i(2\zeta_r r_r)} \tag{15.42}$$

As in Sec. 4.3, we can determine the magnitude and phase of $\bar{H}_r(\Omega)$, and then $\eta_r(t)$ becomes

$$\eta_r(t) = \frac{F_r/K_r}{\sqrt{(1 - r_r^2)^2 + (2\zeta_r r_r)^2}} \cos(\Omega t - \alpha_r) \tag{15.43a}$$

where

$$\tan \alpha_r = \frac{2\zeta_r r_r}{1 - r_r^2} \tag{15.43b}$$

The complex frequency response for physical (or generalized) coordinates **u** can be obtained by writing Eq. 15.7 in complex form as

$$\bar{\mathbf{u}}(t) = \mathbf{\Phi}\bar{\boldsymbol{\eta}}(t) = \sum_{r=1}^{N} \boldsymbol{\phi}_r \bar{\eta}_r(t) \tag{15.44}$$

Combining Eqs. 15.20c, 15.41, 15.42, and 15.43 we get

$$\bar{\mathbf{u}}(t) = \sum_{r=1}^{N} \left(\frac{\boldsymbol{\phi}_r \boldsymbol{\phi}_r^T \mathbf{P}}{K_r} \right) \left[\frac{1}{(1 - r_r^2) + i(2\zeta_r r_r)} \right] e^{i\Omega t} \tag{15.45}$$

The complex frequency response function, $\overline{H}_{ij}(\Omega)$, for physical coordinates gives the response at coordinate u_i due to unit harmonic excitation at p_j. Thus, from Eq. 15.45,

$$\overline{H}_{ij}(\Omega) \equiv \overline{H}_{u_i/p_j}(\Omega) = \sum_{r=1}^{N} \left(\frac{\phi_{ir}\phi_{jr}}{K_r} \right) \left[\frac{1}{(1 - r_r^2) + i(2\zeta_r r_r)} \right] \tag{15.46}$$

The steady-state response $\mathbf{u}(t)$ can also be obtained from Eq. 15.45. It is

$$\mathbf{u}(t) = \sum_{r=1}^{N} \left(\frac{\boldsymbol{\phi}_r \boldsymbol{\phi}_r^T \mathbf{P}}{K_r} \right) \left[\frac{1}{\sqrt{(1 - r_r^2)^2 + (2\zeta_r r_r)^2}} \right] \cos(\Omega t - \alpha_r) \tag{15.47}$$

where α_r is again given by Eq. 15.43b.

A plot of Eq. 15.46 on the complex plane is referred to as a *complex frequency response plot*. In this case, Ω is a parameter and $\mathcal{I}(\overline{H})$ is plotted versus $\mathcal{R}(\overline{H})$. On the other hand, it is frequently convenient to display $\mathcal{R}(\overline{H})$ versus $f = \Omega/2\pi$ and $\mathcal{I}(\overline{H})$ versus f. Expressions for these can be obtained from Eq. 15.46. Thus,

$$\mathcal{R}(\overline{H}_{ij}) = \sum_{r=1}^{N} \left(\frac{\phi_{ir}\phi_{jr}}{K_r} \right) \left[\frac{(1 - r_r^2)}{(1 - r_r^2)^2 + (2\zeta_r r_r)^2} \right] \tag{15.48a}$$

and

$$\mathcal{I}(\overline{H}_{ij}) = \sum_{r=1}^{N} \left(\frac{\phi_{ir}\phi_{jr}}{K_r} \right) \left[\frac{-2\zeta_r r_r}{(1 - r_r^2)^2 + (2\zeta_r r_r)^2} \right] \tag{15.48b}$$

Complex frequency response functions (sometimes called *transfer functions*) as given by Eq. 15.46 are frequently employed in determining the vibrational characteristics of a system experimentally. Two examples will be presented which show the nature of complex frequency response functions. Example 15.4 treats a 2-DOF system with light damping and widely separated natural frequencies, while Example 15.5 indicates the problems that arise when the two frequencies are very close.

Example 15.4

For the 2-DOF system below:

a. Determine f_1, f_2, ϕ_1, and ϕ_2.
b. Determine $\boldsymbol{M}$, $\boldsymbol{C}$, and $\boldsymbol{K}$.
c. Determine ζ_1 and ζ_2.
d. Determine $\overline{H}_{11}(\Omega)$ and $\overline{H}_{21}(\Omega)$.
e. Plot $\overline{H}_{11}$ and $\overline{H}_{21}$ in the complex plane for

$$f = 4 \text{ Hz} \quad \text{to} \quad f = 7 \text{ Hz}$$

f. Plot $\mathcal{R}(\overline{H}_{11})$ and $\mathcal{I}(\overline{H}_{11})$ versus f for $f = 4$ Hz to $f = 7$ Hz.

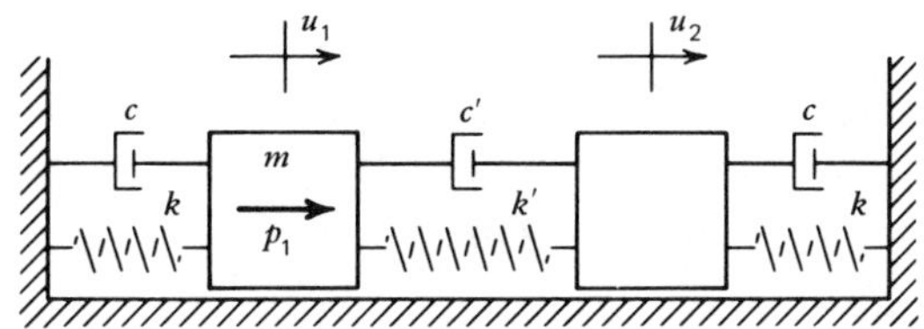

$$p_1 = P_1 \cos \Omega t$$
$$k = 987, \quad k' = 217, \quad m = 1$$
$$c = 0.6284, \quad c' = 0.0628$$

Solution

a. Determine the frequencies and mode shapes using the methods of Chapter 12.

The equation of motion is

$$\begin{bmatrix} m & 0 \\ 0 & m \end{bmatrix} \begin{Bmatrix} \ddot{u}_1 \\ \ddot{u}_2 \end{Bmatrix} + \begin{bmatrix} (c + c') & -c' \\ -c' & (c + c') \end{bmatrix} \begin{Bmatrix} \dot{u}_1 \\ \dot{u}_2 \end{Bmatrix} + \begin{bmatrix} (k + k') & -k' \\ -k' & (k + k') \end{bmatrix} \begin{Bmatrix} u_1 \\ u_2 \end{Bmatrix} = \begin{Bmatrix} p_1 \\ 0 \end{Bmatrix} \quad (1)$$

For undamped free vibration

$$\begin{bmatrix} m & 0 \\ 0 & m \end{bmatrix} \begin{Bmatrix} \ddot{u}_1 \\ \ddot{u}_2 \end{Bmatrix} + \begin{bmatrix} (k + k') & -k' \\ -k' & (k + k') \end{bmatrix} \begin{Bmatrix} u_1 \\ u_2 \end{Bmatrix} = \begin{Bmatrix} 0 \\ 0 \end{Bmatrix} \quad (2)$$

Let

$$\mathbf{u} = \boldsymbol{\phi} \cos \omega t \tag{3}$$

Then the algebraic eigenvalue equation can be written

$$\left[\begin{bmatrix} (k + k') & -k' \\ -k' & (k + k') \end{bmatrix} - \omega^2 \begin{bmatrix} m & 0 \\ 0 & m \end{bmatrix}\right] \begin{Bmatrix} \phi_1 \\ \phi_2 \end{Bmatrix} = \begin{Bmatrix} 0 \\ 0 \end{Bmatrix} \tag{4}$$

The frequencies and modes can be found to be

$$\omega_1^2 = \frac{k}{m}, \qquad \omega_2^2 = \frac{k + 2k'}{m} \tag{5}$$

$$\boxed{\boldsymbol{\phi}_1 = \begin{Bmatrix} 1 \\ 1 \end{Bmatrix}, \qquad \boldsymbol{\phi}_2 = \begin{Bmatrix} 1 \\ -1 \end{Bmatrix}} \tag{6}$$

For the specific values of k, k', and m above

$$\omega_1^2 = \frac{987}{1} = 987, \qquad \omega_2^2 = \frac{1421}{1} = 1421$$

$$\omega_1 = 31.42, \qquad \omega_2 = 37.70$$

$$\boxed{f_1 = \frac{\omega_1}{2\pi} = 5.00 \text{ Hz}, \qquad f_2 = \frac{\omega_2}{2\pi} = 6.00 \text{ Hz}} \tag{7}$$

b. Determine $\boldsymbol{M}$, $\boldsymbol{C}$, and $\boldsymbol{K}$.

$$\boldsymbol{\Phi} = \begin{bmatrix} 1 & 1 \\ 1 & -1 \end{bmatrix} \tag{8a}$$

$$\boldsymbol{M} = \boldsymbol{\Phi}^T \mathbf{m} \boldsymbol{\Phi} = \begin{bmatrix} 1 & 1 \\ 1 & -1 \end{bmatrix} \begin{bmatrix} 1 & 0 \\ 0 & 1 \end{bmatrix} \begin{bmatrix} 1 & 1 \\ 1 & -1 \end{bmatrix} = \begin{bmatrix} 2 & 0 \\ 0 & 2 \end{bmatrix} \tag{8b}$$

$$\boldsymbol{C} = \boldsymbol{\Phi}^T \mathbf{c} \boldsymbol{\Phi} = \begin{bmatrix} 1 & 1 \\ 1 & -1 \end{bmatrix} \begin{bmatrix} 0.6912 & -0.0628 \\ -0.0628 & 0.6912 \end{bmatrix} \begin{bmatrix} 1 & 1 \\ 1 & -1 \end{bmatrix}$$

$$= \begin{bmatrix} 1 & 1 \\ 1 & -1 \end{bmatrix} \begin{bmatrix} 0.6284 & 0.7540 \\ 0.6284 & -0.7540 \end{bmatrix} = \begin{bmatrix} 1.2568 & 0 \\ 0 & 1.5080 \end{bmatrix}$$

$$K_1 = \omega_1^2 M_1 = 987(2) = 1974$$

$$K_2 = \omega_2^2 M_2 = 1421(2) = 2842$$

Therefore

$$\boldsymbol{K} = \begin{bmatrix} 1974 & 0 \\ 0 & 2842 \end{bmatrix} \tag{8c}$$

c. Determine ζ_1 and ζ_2. Using Eq. 15.33,

$$\zeta_r = \frac{C_r}{2 M_r \omega_r} \tag{9}$$

$$\boxed{\begin{aligned}\zeta_1 &= \frac{1.2568}{2(2)(31.42)} = 0.0100\\ \zeta_2 &= \frac{1.5080}{2(2)(37.70)} = 0.0100\end{aligned}} \tag{10}$$

d. Determine $\overline{H}_{11}(\Omega)$ and $\overline{H}_{21}(\Omega)$. Using Eq. 15.46,

$$\overline{H}_{ij}(\Omega) = \sum_{r=1}^{2} \left(\frac{\phi_{ir}\phi_{jr}}{K_r}\right) \left[\frac{1}{1 - (\Omega/\omega_r)^2 + i(2\zeta_r\Omega/\omega_r)}\right]$$

$$\begin{aligned}\overline{H}_{11}(\Omega) &= \left[\frac{(1)(1)}{1974}\right] \left[\frac{1}{1 - (\Omega/31.42)^2 + i[2(0.01)\Omega/31.42]}\right]\\ &+ \left[\frac{(1)(1)}{2842}\right] \left[\frac{1}{1 - (\Omega/37.70)^2 + i[2(0.01)\Omega/37.70]}\right]\end{aligned} \tag{11}$$

$$\boxed{\begin{aligned}\overline{H}_{11} &= \frac{(5.066 \times 10^{-4})}{1 - (\Omega/31.42)^2 + i(0.02\Omega/31.42)}\\ &+ \frac{(3.519 \times 10^{-4})}{1 - (\Omega/37.70)^2 + i(0.02\Omega/37.70)}\end{aligned}} \tag{12a}$$

Similarly, from Eq. 11,

$$\begin{aligned}\overline{H}_{21}(\Omega) &= \left[\frac{(1)(1)}{1974}\right] \left[\frac{1}{1 - (\Omega/31.42)^2 + i[2(0.01)\Omega/31.42]}\right]\\ &+ \left[\frac{(-1)(1)}{2842}\right] \left[\frac{1}{1 - (\Omega/37.70)^2 + i[2(0.01)\Omega/37.70}\right]\end{aligned}$$

Then

$$\boxed{\begin{aligned}\overline{H}_{21} &= \frac{(5.066 \times 10^{-4})}{1 - (\Omega/31.42)^2 + i(0.02\Omega/31.42)}\\ &- \frac{(3.519 \times 10^{-4})}{1 - (\Omega/37.70)^2 + i(0.02\Omega/37.70)}\end{aligned}} \tag{12b}$$

e. Plot $\overline{H}_{11}$ and $\overline{H}_{21}$ in the complex plane.

We will need to plot $\mathcal{R}(\overline{H}_{ij})$ versus $\mathcal{I}(\overline{H}_{ij})$. Equations 15.48a and 15.48b give these.

$$\mathcal{R}(\overline{H}_{ij}) = \sum_{r=1}^{2} \left(\frac{\phi_{ir}\phi_{jr}}{K_r}\right) \left[\frac{1 - (\Omega/\omega_r)^2}{[1 - (\Omega/\omega_r)^2]^2 + (2\zeta_r\Omega/\omega_r)^2}\right] \tag{13a}$$

$$\mathcal{I}(\overline{H}_{ij}) = \sum_{r=1}^{2} \left(\frac{\phi_{ir}\phi_{jr}}{K_r}\right) \left[\frac{-2\zeta_r\Omega/\omega_r}{[1 - (\Omega/\omega_r)^2]^2 + (2\zeta_r\Omega/\omega_r)^2}\right] \tag{13b}$$

Numerical values needed in Eqs. 13 for $\overline{H}_{11}$ and $\overline{H}_{21}$ can be taken from Eqs. 12a and 12b and will not be repeated here. (*Note:* Different scales are used below.)

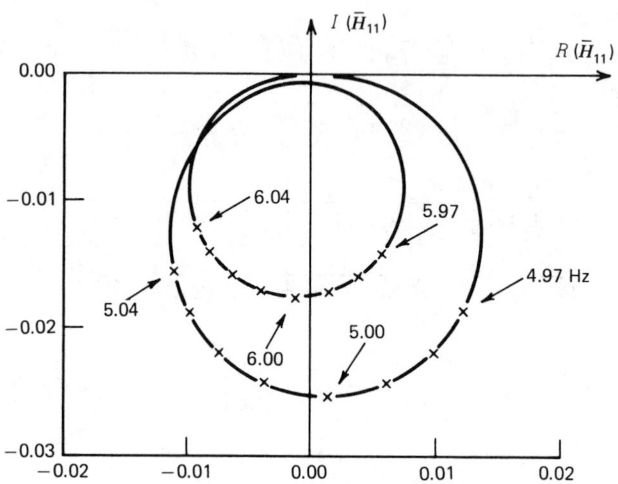

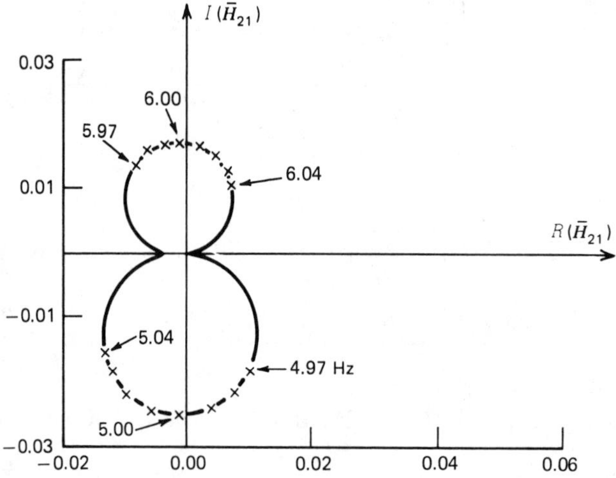

f. Equations 13 can also be used to plot $R(\overline{H}_{11})$ and $I(\overline{H}_{11})$ as functions of the excitation frequency $f = \Omega/2\pi$.

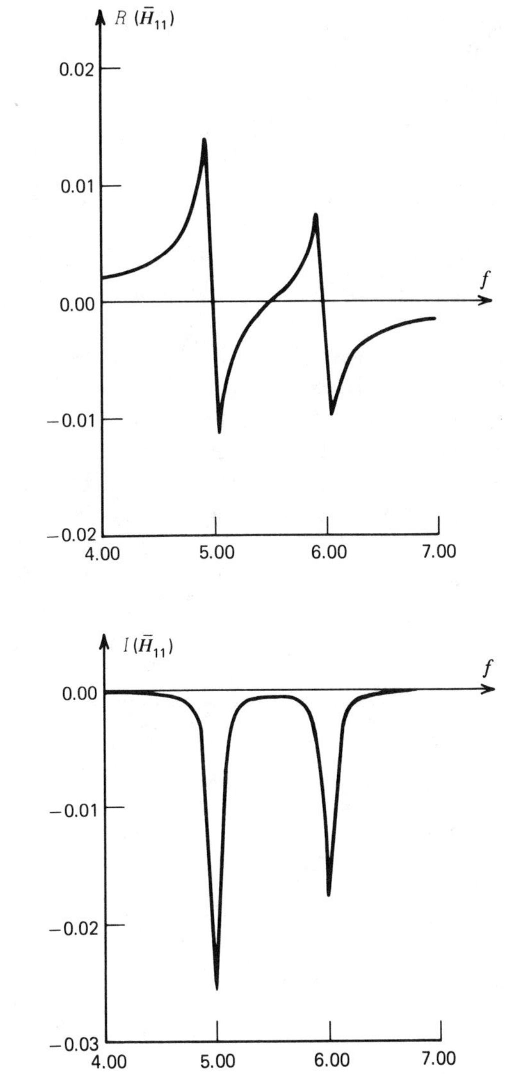

From the complex frequency response plot of $\overline{H}_{11}$ in part (e) of the above example it can be seen that:

a. The response resembles two distinct SDOF complex frequency response plots as seen in Fig. 4.9, that is, the response of this 2-DOF system resembles 2 SDOF independent systems.

b. Frequency spacing is greatest near the undamped natural frequencies $f_1 = 5.00$ Hz and $f_2 = 6.00$ Hz.

c. $\mathcal{R}(\overline{H}_{11}) = 0$ approximately at these natural frequencies.

d. $\mathcal{I}(\overline{H}_{11})$ has its maxima at these natural frequencies.

Conclusions (c) and (d) can also be deduced from the plots of $\mathcal{R}(\overline{H}_{11})$ and $\mathcal{I}(\overline{H}_{11})$, respectively.

The complex frequency response plot of $\overline{H}_{21}$ leads to similar conclusions, but it also exhibits the phase change due to the sign difference between modes 1 and 2.

Real structures of any complexity at all frequently have closely spaced frequencies. That is, a structure may have well over 100 natural frequencies between 0 and 50 Hz. In Sec. 13.1 techniques for treating systems with repeated frequencies were discussed and it was pointed out that repeated frequencies do not possess unique corresponding mode shapes. In Example 15.5 we consider a system with "closely spaced" frequencies, which indicates how response plots of such a system may differ from those of a system with more widely spaced frequencies as in Example 15.4[15.10,15.11].

Example 15.5

Repeat Example 15.4 using the following system parameters.

$$k = 987, \qquad k' = 10, \qquad c = 0.6284, \qquad c' = 0.0031$$

(*Note:* These parameters were chosen to give 1% damping and to give frequencies separated by 1%.)

Solution

a. The equation of motion has the same form as in Example 15.4, so the frequencies will be given by

$$\omega_1^2 = \frac{k}{m}, \qquad \omega_2^2 = \frac{k + 2k'}{m} \tag{1}$$

and the mode shapes will be the same as in Example 15.4.

$$\omega_1^2 = \frac{987}{1} = 987, \qquad \omega_2^2 = \frac{1007}{1} = 1007$$

$$\omega_1 = 31.42, \qquad \omega_2 = 31.73$$

$$\boxed{f_1 = \frac{\omega_1}{2\pi} = 5.00\text{ Hz}, \qquad f_2 = \frac{\omega_2}{2\pi} = 5.05\text{ Hz}} \tag{2}$$

b. From Example 15.4,

$$\boldsymbol{\Phi} = \begin{bmatrix} 1 & 1 \\ 1 & -1 \end{bmatrix}, \qquad \boldsymbol{M} = \begin{bmatrix} 2 & 0 \\ 0 & 2 \end{bmatrix} \tag{3a,b}$$

$$\boldsymbol{C} = \boldsymbol{\Phi}^T \mathbf{c} \boldsymbol{\Phi} = \begin{bmatrix} 1 & 1 \\ 1 & -1 \end{bmatrix} \begin{bmatrix} 0.6315 & -0.0031 \\ -0.0031 & 0.6315 \end{bmatrix} \begin{bmatrix} 1 & 1 \\ 1 & -1 \end{bmatrix}$$

$$= \begin{bmatrix} 1 & 1 \\ 1 & -1 \end{bmatrix} \begin{bmatrix} 0.6284 & 0.6346 \\ 0.6284 & -0.6346 \end{bmatrix} = \begin{bmatrix} 1.2568 & 0 \\ 0 & 1.2692 \end{bmatrix} \tag{3c}$$

$$K_1 = \omega_1^2 M_1 = 987(2) = 1974$$

$$K_2 = \omega_2^2 M_2 = 1007(2) = 2014$$

$$\boldsymbol{K} = \begin{bmatrix} 1974 & 0 \\ 0 & 2014 \end{bmatrix} \tag{3d}$$

c. Determine ζ_1 and ζ_2 using Eq. 15.33.

$$\zeta_r = \frac{C_r}{2M_r\omega_r} \tag{4}$$

$$\boxed{\begin{aligned} \zeta_1 &= \frac{1.2568}{2(2)(31.42)} = 0.0100 \\ \zeta_2 &= \frac{1.2692}{2(2)(31.73)} = 0.0100 \end{aligned}} \tag{5}$$

d. Using Eq. 11 of Example 15.4 together with numerical values from above, we get

$$\overline{H}_{11}(\Omega) = \left[\frac{(1)(1)}{1974}\right]\left[\frac{1}{1 - (\Omega/31.42)^2 + i[2(0.01)\Omega/31.42]}\right] + \left[\frac{(1)(1)}{2014}\right]\left[\frac{1}{1 - (\Omega/31.73)^2 + i[2(0.01)\Omega/31.73]}\right]$$

Then,

$$\boxed{\overline{H}_{11} = \frac{(5.066 \times 10^{-4})}{1 - (\Omega/31.42)^2 + i(0.02\Omega/31.42)} + \frac{(4.965 \times 10^{-4})}{1 - (\Omega/31.73)^2 + i(0.02\Omega/31.73)}} \tag{6a}$$

Similarly,

$$\boxed{\overline{H}_{21} = \frac{(5.066 \times 10^{-4})}{1 - (\Omega/31.42)^2 + i(0.02\Omega/31.42)} - \frac{(4.965 \times 10^{-4})}{1 - (\Omega/31.73)^2 + i(0.02\Omega/31.73)}} \tag{6b}$$

e. The plots of $\overline{H}_{11}$ and $\overline{H}_{21}$ are obtained from Eqs. 15.48 using numerical values from Eqs. 6a and 6b.

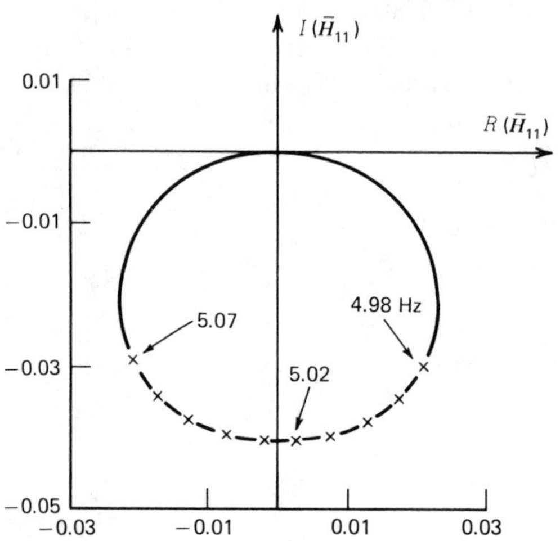

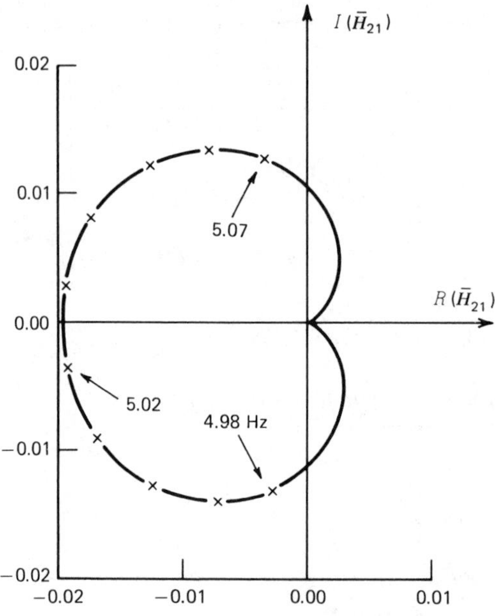

f. The real and imaginary plots are obtained as in Example 15.4. They are

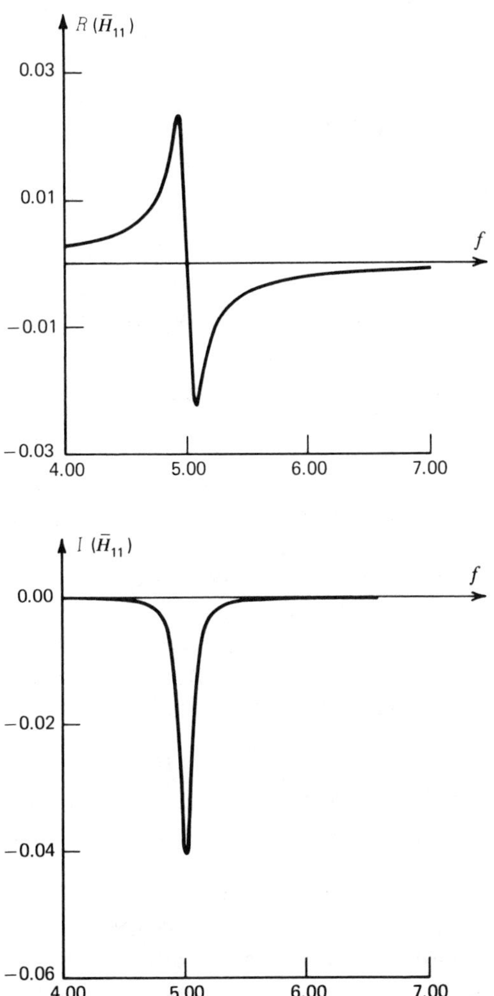

By comparing the frequency response plots of Examples 15.4 and 15.5 you can see that the complex frequency response plot of $\overline{H}_{11}$ of the system with closely spaced frequencies gives little indication that two modes are present. The same is true of the plot of $\overline{H}_{11}$ versus forcing frequency. On the other hand, the plot of $\overline{H}_{21}$ for the system with closely spaced modes does indicate a rapid phase change and this indicates the presence of two modes whose responses at

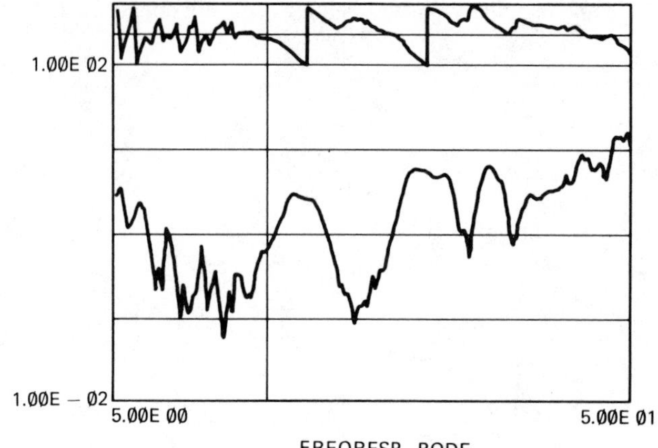

Figure 15.2. Inertance Bode plot for a complex structure.

this coordinate have opposite signs. Great difficulty is encountered in experimentally separating modes of systems with closely spaced frequencies or with heavy damping. Figure 15.2 is an *inertance* Bode plot, that is, Acceleration/Force as a function of frequency, of the response of a complex system with excitation at one point and response measured at another point.

15.5 Dynamic Stresses by Mode-Superposition

In Secs. 15.2 and 15.3 the basic techniques for determining the displacement history $\mathbf{u}(t)$ by using the mode-displacement method and the mode-acceleration method were described. In addition to determining the displacement history, a dynamic analysis usually includes the determination of stress histories (e.g., moment, shear, axial stress, etc.) or at least the determination of maximum values of stress at specified locations in the structure. The following are symbolic representations of stresses obtained by mode-superposition. For the mode-displacement method

$$\hat{\sigma}(t) = \sum_{r=1}^{\hat{N}} \mathbf{s}_r \eta_r(t) \tag{15.49}$$

where $\mathbf{s}_r$ is the contribution to the stress vector σ due to a unit displacement of the rth mode, that is, for $\eta_r = 1$. For the mode-acceleration method for an undamped system the displacement approximation in Eq. 15.28 leads to the stress approximation

$$\tilde{\sigma}(t) = \sigma_{\text{pseudostatic}} - \sum_{r=1}^{\hat{N}} \left(\frac{1}{\omega_r^2} \right) \mathbf{s}_r \ddot{\eta}_r(t) \tag{15.50}$$

The mode-acceleration method is particularly beneficial in speeding convergence of the internal stresses.

Example 15.6

For the shear building of Example 15.1 write an expression for the story shears, σ_i, corresponding to mode r.

Solution

The story stress are determined by the relative displacements between floors. Thus,

$$\begin{aligned} \sigma_1 &= k_1(u_1 - u_2) \\ \sigma_2 &= k_2(u_2 - u_3) \\ \sigma_3 &= k_3(u_3 - u_4) \\ \sigma_4 &= k_4 u_4 \end{aligned} \tag{1}$$

$$\begin{Bmatrix} \sigma_1 \\ \sigma_2 \\ \sigma_3 \\ \sigma_4 \end{Bmatrix} = \begin{bmatrix} k_1 & -k_1 & 0 & 0 \\ 0 & k_2 & -k_2 & 0 \\ 0 & 0 & k_3 & -k_3 \\ 0 & 0 & 0 & k_4 \end{bmatrix} \begin{Bmatrix} u_1 \\ u_2 \\ u_3 \\ u_4 \end{Bmatrix} \tag{2}$$

Thus, when the displacement shape is $\boldsymbol{\phi}_r$, Eq. 2 leads to the following expression for $\mathbf{s}_r$.

$$\boxed{\begin{Bmatrix} s_1 \\ s_2 \\ s_3 \\ s_4 \end{Bmatrix}_r = \begin{bmatrix} k_1 & -k_1 & 0 & 0 \\ 0 & k_2 & -k_2 & 0 \\ 0 & 0 & k_3 & -k_3 \\ 0 & 0 & 0 & k_4 \end{bmatrix} \begin{Bmatrix} \phi_1 \\ \phi_2 \\ \phi_3 \\ \phi_4 \end{Bmatrix}_r} \tag{3}$$

Using the numerical values of $\mathbf{k}$ and $\boldsymbol{\phi}$ from Example 15.1, we can use Eq. 3 to evaluate the modal shears.

$$\begin{aligned} \mathbf{s}_1 &= \begin{Bmatrix} 176.72 \\ 452.08 \\ 627.58 \\ 752.19 \end{Bmatrix}, \quad \mathbf{s}_2 = \begin{Bmatrix} 879.70 \\ 704.42 \\ -245.47 \\ -1400.35 \end{Bmatrix} \\ \mathbf{s}_3 &= \begin{Bmatrix} -1521.16 \\ 1853.74 \\ 1318.51 \\ -2265.50 \end{Bmatrix}, \quad \mathbf{s}_4 = \begin{Bmatrix} 482.02 \\ -2317.07 \\ 3928.51 \\ -2038.02 \end{Bmatrix} \end{aligned} \tag{4}$$

Note that, while the modes ϕ_r given in Example 15.1 and used in the above example to determine the modal shears $\mathbf{s}_r$ are scaled such that the maximum displacement in each mode is $+1$, the magnitudes of the modal shears tend to increase with increasing mode number. It is for this reason that convergence of stresses is slower than convergence of displacements, and hence the mode-acceleration method proves to be especially beneficial when stresses are to be computed[15.4].

15.6 Mode-Superposition for Undamped Systems with Rigid-Body Modes

Mode-superposition may be employed to determine the response of systems with rigid-body modes. In this section we consider the application of the mode-displacement method and the mode-acceleration method to undamped systems with rigid-body modes.

The basic equation of mode-superposition, Eq. 15.7, can be written in the form

$$\mathbf{u}(t) = \mathbf{u}_R(t) + \mathbf{u}_E(t) = \mathbf{\Phi}_R\eta_R(t) + \mathbf{\Phi}_E\eta_E(t) \tag{15.51}$$

where $\mathbf{\Phi}_R$ and $\mathbf{\Phi}_E$ are modal matrices containing the N_R rigid-body modes and the N_E elastic modes, respectively, $(N = N_R + N_E)$. Substituting Eq. 15.51 into the equation of motion, Eq. 15.1, with $\mathbf{c} = 0$, we get

$$\boldsymbol{M}_R\ddot{\eta}_R = \mathbf{\Phi}_R^T\mathbf{p} \tag{15.52a}$$

and

$$\boldsymbol{M}_E\ddot{\eta}_E + \boldsymbol{K}_E\eta_E = \mathbf{\Phi}_E^T\mathbf{p} \tag{15.52b}$$

From Eq. 15.52a, the rigid-body coordinates are obtained by double integration, that is,

$$\begin{aligned}\eta_r(t) = &\int_0^t \int_0^\tau \left(\frac{1}{M_r}\right) \phi_r^T\mathbf{p}(\xi)\,d\xi\,d\tau \\ &+ t\dot{\eta}_r(0) + \eta_r(0) \\ &\text{for } r = 1, 2, \ldots, N_R\end{aligned} \tag{15.53}$$

For the elastic coordinates, η_r is given by Eq. 15.18 as before.

Mode-Displacement Method

To determine the system displacements by the mode-displacement method, Eq. 15.51 is employed with the number of elastic modes truncated to $\hat{N}_E$. Thus,

$$\hat{\mathbf{u}}(t) = \mathbf{\Phi}_R \eta_R(t) + \hat{\mathbf{\Phi}}_E \hat{\eta}_E(t) \tag{15.54}$$

where $\hat{\mathbf{\Phi}}_E$ contains the $\hat{N}_E$ elastic modes. Equations 15.53 and 15.18 are used to determine the rigid-body and elastic modal displacement histories.

To determine stresses by the mode-displacement method, Eq. 15.49 is modified to account for the fact that rigid-body displacements do not give rise to internal stresses. Hence,

$$\hat{\sigma}(t) = \hat{\mathbf{S}}_E \hat{\eta}_E(t) = \sum_{r=1}^{\hat{N}_E} \mathbf{s}_r \eta_t(t) \tag{15.55}$$

where the columns of $\hat{\mathbf{S}}_E$ are the internal stress vectors corresponding to unit values of each of the $\hat{N}_E$ retained elastic modes.

Mode-Acceleration Method

Since the stiffness matrix for a system with rigid-body modes is singular and therefore cannot be inverted, the mode-acceleration method cannot be employed in the straightforward manner indicated by Eq. 15.38.

Let us begin by solving Eq. 15.52b for η_E and substituting this into Eq. 15.51. We get

$$\mathbf{u} = \mathbf{\Phi}_R \eta_R + (\mathbf{\Phi}_E \boldsymbol{K}_E^{-1} \mathbf{\Phi}_E^T)\mathbf{p} - (\mathbf{\Phi}_E \boldsymbol{K}_E^{-1} \boldsymbol{M}_E)\ddot{\eta}_E \tag{15.56}$$

The rigid-body term is determined just as for the mode-displacement equation, that is, all rigid-body modes are retained. To cast Eq. 15.56 into mode-acceleration form similar to Eq. 15.28 we can observe that truncating the last term in Eq. 15.56 leads to the same expression as in Eq. 15.28. Thus, let us rewrite Eq. 15.56 in the form

$$\mathbf{u} = \mathbf{\Phi}_R \eta_R + \mathbf{a}_E \mathbf{p} - \hat{\mathbf{\Phi}}_E \hat{\boldsymbol{K}}_E^{-1} \hat{\boldsymbol{M}}_E \ddot{\hat{\eta}}_E \tag{15.57}$$

where the only term leading to difficulty is the pseudostatic deflection term, which we have written $\mathbf{a}_E \mathbf{p}_E$. One expression for the *elastic flexibility matrix* $\mathbf{a}_E$ is

$$\mathbf{a}_E = \mathbf{\Phi}_E \boldsymbol{K}_E^{-1} \mathbf{\Phi}_E^T \tag{15.58}$$

which is found in Eq. 15.56. However, we wish to determine $\mathbf{a}_E$ without solving for all the elastic modes, as we would have to do if we used Eq. 15.58.

First, substitute Eq. 15.51 into the equation of motion, noting that $\mathbf{k}\mathbf{u}_R = \mathbf{0}$. Then,

$$\mathbf{m}\ddot{\mathbf{u}}_E + \mathbf{k}\mathbf{u}_E = \mathbf{p}_E \tag{15.59a}$$

where

$$\mathbf{p}_E = \mathbf{p} - \mathbf{m}\ddot{\mathbf{u}}_R \tag{15.59b}$$

Thus, in determining the elastic displacements, we use a self-equilibrated force system of applied forces and rigid-body inertia forces. Since, from Eqs. 15.51 and 15.52a,

$$\ddot{\mathbf{u}}_R = \mathbf{\Phi}_R \ddot{\eta}_R = \mathbf{\Phi}_R \boldsymbol{M}_R^{-1} \mathbf{\Phi}_R^T \mathbf{p} \tag{15.60}$$

we can write the elastic force vector $\mathbf{p}_E$ as

$$\mathbf{p}_E = \boldsymbol{R}\mathbf{p} \tag{15.61a}$$

where

$$\boldsymbol{R} = \boldsymbol{I} - \mathbf{m}\mathbf{\Phi}_R \boldsymbol{M}_R^{-1} \mathbf{\Phi}_R^T \tag{15.61b}$$

Let $\mathbf{a}$ be the flexibility matrix of the system relative to some statically determinate constraints, with zeros in the rows and columns corresponding to the constraints. Then, let

$$\mathbf{w} = \mathbf{a}\mathbf{p}_E \tag{15.62}$$

be the elastic displacement of the system relative to the imposed statically determinate constraints. But, the deflection $\mathbf{w}$ may contain some component of the rigid-body modes, so a new vector $\mathbf{w}_E$ is created which is orthogonal to the rigid-body modes. Let

$$\mathbf{w}_E = \mathbf{w} - \mathbf{\Phi}_R \mathbf{c}_R \tag{15.63}$$

where $\mathbf{c}_R$ is such that $\mathbf{\Phi}_R^T \mathbf{m} \mathbf{w}_E = 0$. Then,

$$\mathbf{\Phi}_R^T \mathbf{m}(\mathbf{w} - \mathbf{\Phi}_R \mathbf{c}_R) = 0 \tag{15.64}$$

or

$$\mathbf{c}_R = \boldsymbol{M}_R^{-1} \mathbf{\Phi}_R^T \mathbf{m}\mathbf{w} \tag{15.65}$$

Then

$$\mathbf{w}_E = (\mathbf{I} - \mathbf{\Phi}_R \boldsymbol{M}_R^{-1} \mathbf{\Phi}_R^T \mathbf{m})\mathbf{w} = \boldsymbol{R}^T \mathbf{w} \tag{15.66}$$

In conclusion, we get

$$\mathbf{w}_E = \mathbf{a}_E \mathbf{p} \tag{15.67}$$

where $\mathbf{a}_E$ has the same meaning as in Eq. 15.58, but is now given by

$$\mathbf{a}_E = \boldsymbol{R}^T \mathbf{a} \boldsymbol{R} \tag{15.68}$$

The mode-acceleration method for systems having rigid-body modes thus involves using Eq. 15.57 with $\mathbf{a}_E$ defined by Eq. 15.68.

The stresses within a system having rigid-body modes can be determined by the mode-acceleration expression

$$\tilde{\sigma}(t) = \sigma_{\text{pseudostatic}} - \sum_{r=1}^{\hat{N}_E} \left(\frac{1}{\omega_r^2} \right) \mathbf{s}_r \ddot{\eta}_r(t) \tag{15.69}$$

where $\sigma_{\text{pseudostatic}}$ is the vector of internal stresses resulting from the self-equilibrated elastic force vector $\mathbf{p}_E$ given by Eq. 15.59b. Determination of stresses by using Eq. 15.69 is considerably simpler than determining displacements through the use of Eqs. 15.57 and 15.68. Example 15.7 illustrates the use of the mode-displacement method and the mode-acceleration method to determine internal stresses.

Example 15.7

Use the mode-displacement and mode-acceleration methods to determine expressions for the maximum force in each of the two springs due to application of a step force $p_3(t) = p_o$, $t \geq 0$. Compare the convergence of the two methods. The system is at rest at $t = 0$.

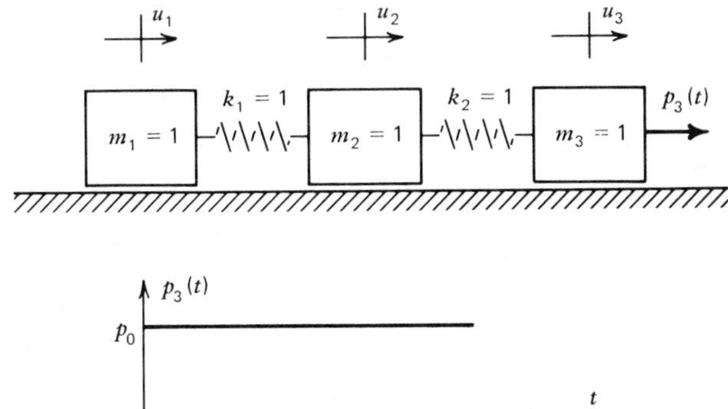

Solution

a. Solve for the modes and frequencies.

$$\begin{bmatrix} 1 & 0 & 0 \\ 0 & 1 & 0 \\ 0 & 0 & 1 \end{bmatrix} \begin{Bmatrix} \ddot{u}_1 \\ \ddot{u}_2 \\ \ddot{u}_3 \end{Bmatrix} + \begin{bmatrix} 1 & -1 & 0 \\ -1 & 2 & -1 \\ 0 & -1 & 1 \end{bmatrix} \begin{Bmatrix} u_1 \\ u_2 \\ u_3 \end{Bmatrix} = \begin{Bmatrix} 0 \\ 0 \\ 0 \end{Bmatrix} \tag{1}$$

Let $\mathbf{u} = \boldsymbol{\phi} \cos \omega t$. Then, the algebraic eigenvalue equation is (2)

$$\begin{bmatrix} (1 - \omega^2) & -1 & 0 \\ -1 & (2 - \omega^2) & -1 \\ 0 & -1 & (1 - \omega^2) \end{bmatrix} \begin{Bmatrix} \phi_1 \\ \phi_2 \\ \phi_3 \end{Bmatrix} = \begin{Bmatrix} 0 \\ 0 \\ 0 \end{Bmatrix} \tag{3}$$

Setting the determinant of coefficients of Eq. 3 to zero, we get the characteristic equation

$$\omega^2(\omega^4 - 4\omega^2 + 3) = 0 \tag{4}$$

whose roots are

$$\omega_1^2 = 0, \qquad \omega_2^2 = 1, \qquad \omega_3^2 = 3 \tag{5}$$

The corresponding modes are

$$\boldsymbol{\phi}_1 = \begin{Bmatrix} 1 \\ 1 \\ 1 \end{Bmatrix}, \qquad \boldsymbol{\phi}_2 = \begin{Bmatrix} 1 \\ 0 \\ -1 \end{Bmatrix}, \qquad \boldsymbol{\phi}_3 = \begin{Bmatrix} 1 \\ -2 \\ 1 \end{Bmatrix} \tag{6}$$

The generalized masses and stiffness are given by

$$M_r = \boldsymbol{\phi}_r^T \mathbf{m} \boldsymbol{\phi}_r = \boldsymbol{\phi}_r^T \boldsymbol{\phi}_r \tag{7}$$

and

$$K_r = \omega_r^2 M_r \tag{8}$$

so

$$\begin{aligned} M_1 &= 3, \qquad M_2 = 2, \qquad M_2 = 6 \\ K_1 &= 0, \qquad K_2 = 2, \qquad K_3 = 18 \end{aligned} \tag{9}$$

b. Since the system is initially at rest,

$$\eta_r(0) = \dot{\eta}_r(0) = 0, \qquad r = 1, 2, 3 \tag{10}$$

c. The generalized forces are given by

$$P_r(t) = \boldsymbol{\phi}_r^T \mathbf{p}(t) \tag{11}$$

so

$$P_1 = p_3(t) = p_o, \qquad P_2 = -p_3(t) = -p_o, \qquad P_3(t) = p_3(t) = p_o \tag{12}$$

d. The equations of motion in modal coordinates are thus

$$\left.\begin{aligned} 3\ddot{\eta}_1 &= p_o \\ 2\ddot{\eta}_2 + 2\eta_2 &= -p_o \\ 6\ddot{\eta}_3 + 18\eta_3 &= p_o \end{aligned}\right\} \; t \geq 0 \tag{13}$$

Then,

$$\begin{aligned} \eta_1 &= \frac{p_o t^2}{6} \\ \eta_2 &= \left(\frac{-p_o}{2}\right)(1 - \cos \omega_2 t) \\ \eta_3 &= \left(\frac{p_o}{18}\right)(1 - \cos \omega_3 t) \end{aligned} \tag{14}$$

where the latter two can be obtained from Eq. 5.8. From Eqs. 13 and 14,

$$\ddot{\eta}_2 = -\left(\frac{p_o}{2}\right)\cos\omega_2 t$$
$$\ddot{\eta}_3 = \left(\frac{p_o}{6}\right)\cos\omega_3 t \tag{15}$$

e. Determine the "modal stresses," that is, the spring forces due to the elastic modes $\boldsymbol{\phi}_2$ and $\boldsymbol{\phi}_3$.

$$\mathbf{s}_r = \begin{Bmatrix} s_1 \\ s_2 \end{Bmatrix}_r = \begin{bmatrix} -1 & 1 & 0 \\ 0 & -1 & 1 \end{bmatrix} \begin{Bmatrix} \phi_1 \\ \phi_2 \\ \phi_3 \end{Bmatrix}_r \tag{16}$$

Therefore,

$$\mathbf{s}_2 = \begin{Bmatrix} -1 \\ -1 \end{Bmatrix}, \qquad \mathbf{s}_3 = \begin{Bmatrix} -3 \\ 3 \end{Bmatrix} \tag{17}$$

f. The mode-displacement approximation to the spring forces ("internal stresses") is given by Eq. 15.55, that is,

$$\hat{\sigma}(t) = \mathbf{s}_2\eta_2(t) \tag{18a}$$

if one elastic mode is employed, or

$$\hat{\sigma} \equiv \sigma(t) = \mathbf{s}_2\eta_2(t) + \mathbf{s}_3\eta_3(t) \tag{18b}$$

if both elastic modes are included. Then,

$$\boxed{\begin{aligned} \hat{\sigma} = \begin{Bmatrix} \hat{\sigma}_1 \\ \hat{\sigma}_2 \end{Bmatrix} &= \begin{Bmatrix} -1 \\ -1 \end{Bmatrix}\left(\frac{-p_o}{2}\right)(1 - \cos\omega_2 t) \\ &= \left(\frac{p_o}{2}\right)\begin{Bmatrix} 1 \\ 1 \end{Bmatrix}(1 - \cos\omega_2 t) \end{aligned}} \tag{19a}$$

$$\boxed{\begin{aligned} \sigma = \begin{Bmatrix} \sigma_1 \\ \sigma_2 \end{Bmatrix} &= \left(\frac{p_o}{2}\right)\begin{Bmatrix} 1 \\ 1 \end{Bmatrix}(1 - \cos\omega_2 t) \\ &+ \left(\frac{p_o}{6}\right)\begin{Bmatrix} -1 \\ 1 \end{Bmatrix}(1 - \cos\omega_3 t) \end{aligned}} \tag{19b}$$

g. The mode-acceleration solution for internal stresses is based on Eq. 15.69. The self-equilibrated loading $\mathbf{p}_E$ is shown on the figure below.

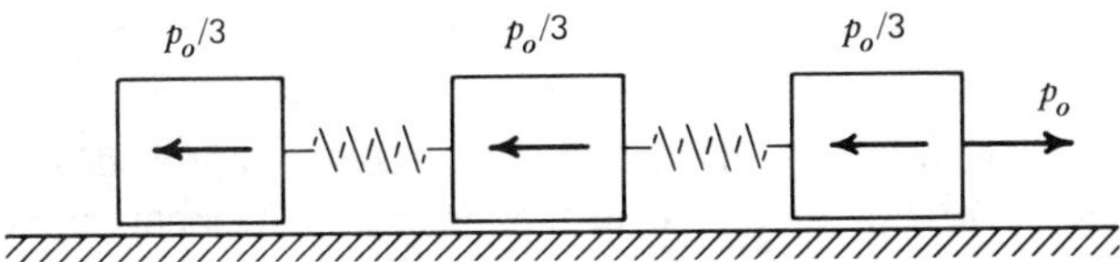

Therefore, by taking freebody diagrams, we get

$$\begin{Bmatrix} \sigma_1 \\ \sigma_2 \end{Bmatrix}_{\text{pseudostatic}} = \begin{Bmatrix} \dfrac{p_o}{3} \\ \dfrac{2p_o}{3} \end{Bmatrix} \tag{20}$$

Then, the mode-acceleration solution based on one elastic mode is

$$\tilde{\sigma} = \sigma_{\text{pseudostatic}} - \left(\frac{1}{\omega_2^2}\right) \mathbf{s}_2 \ddot{\eta}_2 \tag{21a}$$

and based on both elastic modes is

$$\tilde{\sigma} \equiv \sigma = \sigma_{\text{pseudostatic}} - \left(\frac{1}{\omega_2^2}\right) \mathbf{s}_2 \ddot{\eta}_2 - \left(\frac{1}{\omega_3^2}\right) \mathbf{s}_3 \ddot{\eta}_3 \tag{21b}$$

Finally,

$$\boxed{\tilde{\sigma} = \begin{Bmatrix} \tilde{\sigma}_1 \\ \tilde{\sigma}_2 \end{Bmatrix} = \left(\frac{p_o}{3}\right) \begin{Bmatrix} 1 \\ 2 \end{Bmatrix} - \left(\frac{p_o}{2}\right) \begin{Bmatrix} 1 \\ 1 \end{Bmatrix} \cos \omega_2 t}$$

and

$$\boxed{\begin{aligned} \tilde{\sigma} \equiv \sigma &= \begin{Bmatrix} \sigma_1 \\ \sigma_2 \end{Bmatrix} \\ &= \left(\frac{p_o}{3}\right) \begin{Bmatrix} 1 \\ 2 \end{Bmatrix} - \left(\frac{p_o}{2}\right) \begin{Bmatrix} 1 \\ 1 \end{Bmatrix} \cos \omega_2 t + \left(\frac{p_o}{6}\right) \begin{Bmatrix} 1 \\ -1 \end{Bmatrix} \cos \omega_3 t \end{aligned}} \tag{22b}$$

As should be expected, the two-mode solutions 19b and 22b are identical, since there are only two elastic modes.

Comparison of Maximum Spring Forces Computed by Mode-Displacement (M-D) Method and Mode-Acceleration (M-A) Method

	M-D *1 mode*	*M-A* *1 mode*	*Exact*[a]
σ_1/p_o	1.000000	0.833333	0.999933
σ_2/p_o	1.000000	1.166667	1.333241

[a]"Exact" values computed by evaluating σ_1 and σ_2 from Eq. 19b at 1° intervals to $t = 100\pi$.

The above example is really too small to indicate improved "convergence" of the mode-acceleration method over the mode-displacement method. It does, however, illustrate the procedures used in evaluating internal stresses by the two methods. A more extensive application of the two methods to a system with rigid-body modes is found in Reference 15.5.

References

15.1 S. P. Timoshenko, D. H. Young, and W. Weaver, Jr., *Vibration Problems in Engineering,* 4th Ed., Wiley, New York (1974).

15.2 L. Meirovitch, *Elements of Vibration Analysis,* McGraw-Hill, New York (1975).

15.3 J. M. Biggs, *Introduction to Structural Dynamics,* McGraw-Hill, New York (1964).

15.4 W. T. Thomson, *Theory of Vibration with Applications,* Prentice-Hall, Englewood Cliffs, NJ (1972).

15.5 R. L. Bisplinghoff, H. Ashley, and R. L. Halfman, *Aeroelasticity,* Addison-Wesley, Reading, MA (1955).

15.6 D. Williams, *Dynamic Loads in Aeroplanes Under Given Impulsive Loads with Particular Reference to Landing and Gust Loads on a Large Flying Boat,* Great Britain RAE Reports SME 3309 and 3316 (1945).

15.7 R. E. D. Bishop and G. M. L. Gladwell, "An Investigation Into the Theory of Resonance Testing," *Phil. Trans.,* v. 225, 241–280 (1963).

15.8 A. L. Klosterman, *On the Experimental Determination and Use of Modal Representations of Dynamic Characteristics,* Ph.D. Dissertation, U. of Cincinnati, Cincinnati, OH (1971).

15.9 R. E. Cornwell, *On the Applicability of the Mode-Acceleration Method to Structural Engineering Problems,* M.S. Thesis, The University of Texas at Austin, Austin, TX (1979).

15.10 J. W. Pendered, "Theoretical Investigation Into the Effects of Close Natural Frequencies," *J. Mech. Engr. Sci.,* v. 7, 372–379 (1965).

15.11 C. C. Kennedy and C. D. P. Pancu, "Use of Vectors in Vibration Measurement and Analysis," *J. Aero. Sci.,* v. 14, 603–625 (1947).

Problem Set 15.2

15.1 The three-story building of Problem 14.2 has the following stiffness and mass matrices, natural frequencies, and mode shapes.

$$\mathbf{k} = \begin{bmatrix} 800 & -800 & 0 \\ -800 & 2400 & -1600 \\ 0 & -1600 & 4000 \end{bmatrix} \text{ k/in.}$$

$$\mathbf{m} = \begin{bmatrix} 1 & 0 & 0 \\ 0 & 2 & 0 \\ 0 & 0 & 2 \end{bmatrix} \text{ k-sec}^2\text{/in.}$$

$$\omega_1^2 = 251.1, \qquad \omega_2^2 = 1200.0, \qquad \omega_3^2 = 2548.9 \text{ rad}^2/\text{s}^2$$

$$\mathbf{\Phi} = \begin{bmatrix} 1.00000 & 1.00000 & 0.31386 \\ 0.68614 & -0.50000 & -0.68614 \\ 0.31386 & -0.50000 & 1.00000 \end{bmatrix}$$

(a) Determine $\boldsymbol{M}$ and $\boldsymbol{K}$.
(b) If

$$\mathbf{p}(t) = \begin{Bmatrix} 100 \\ 100 \\ 100 \end{Bmatrix} \cos \Omega t \text{ kips}$$

determine F_r ($r = 1, 2, 3$).
(c) Determine expressions for the steady-state responses $\eta_r(t)$.
(d) Determine the response $u_1(t)$ by using the mode-superposition equation, 15.7, and clearly indicate the contribution from each mode as in Example 15.1.
(e) Form a table, as in Example 15.1, using $\Omega = 0$, $\Omega = 0.5\omega_1$, and $\Omega = \frac{1}{2}(\omega_1 + \omega_2)$.

15.2 For the three-story building of Problems 14.2 and 15.1:
(a) Determine F_r and D_r ($r = 1, 2, 3$) for the force distribution

$$\mathbf{p}(t) = \begin{Bmatrix} P_1 \\ 0 \\ 0 \end{Bmatrix} \cos \Omega t$$

(b) If the base of the structure in Fig. P14.2 has harmonic motion $z(t) = Z \cos \Omega t$, determine the effective force $\mathbf{p}_{\text{eff}}(t)$. Using this effective force vector, determine the corresponding F_r and D_r expressions.

15.3 The horizontal stabilizer of a light aircraft is modeled as a 3-DOF lumped mass system as shown below. The stiffness and mass matrices and the natural frequencies and mode shapes are given. If the airplane hits a sudden gust which produces a step force

$$\mathbf{p}(t) = \begin{Bmatrix} 500 \\ 100 \\ 100 \end{Bmatrix} f(t) \text{ lb.}$$

where $f(t)$ is the unit step force as shown below:
(a) Determine expressions for the modal responses $\eta_r(t)$ assuming $\mathbf{v}(0) = \dot{\mathbf{v}}(0) = \mathbf{0}$.
(b) Determine expressions for the response $v_1(t)$ using Eq. 15.7. Clearly indicate the contribution from each mode.

$$\mathbf{k} = \begin{bmatrix} 0.0656 & -0.1538 & 0.1220 \\ -0.1538 & 0.4797 & -0.5843 \\ 0.1220 & -0.5843 & 1.2593 \end{bmatrix} (10^5) \text{ lb/in.}$$

$$\mathbf{m} = \begin{bmatrix} 4.0 & 0 & 0 \\ 0 & 6.0 & 0 \\ 0 & 0 & 8.0 \end{bmatrix} \left(\frac{1}{386}\right) \text{lb-sec}^2/\text{in.}$$

$$\omega_1^2 = 5.99(10^4), \qquad \omega_2^2 = 1.33(10^6), \qquad \omega_3^2 = 8.40(10^6)$$

$$\phi_1 = \begin{Bmatrix} 8.31 \\ 4.08 \\ 1.10 \end{Bmatrix}, \qquad \phi_2 = \begin{Bmatrix} -4.96 \\ 5.36 \\ 3.80 \end{Bmatrix}, \qquad \phi_3 = \begin{Bmatrix} 1.70 \\ -4.35 \\ 5.71 \end{Bmatrix}$$

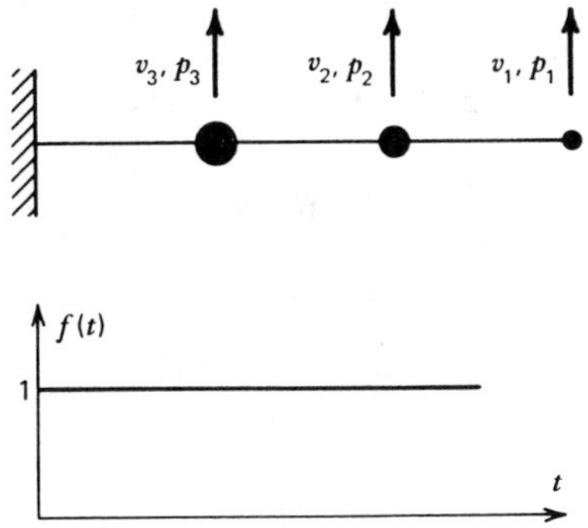

v_3, p_3 v_2, p_2 v_1, p_1

$f(t)$

1

t

Figure P15.3

Problem Set 15.3

15.4 The three-story building of Problem 14.2 and 15.1 has the following flexibility matrix

$$\mathbf{a} \equiv \mathbf{k}^{-1} = \begin{bmatrix} 2.29167 & 1.04167 & 0.41667 \\ 1.04167 & 1.04167 & 0.41667 \\ 0.41667 & 0.41667 & 0.41667 \end{bmatrix} (10^{-3}) \text{ in./k}$$

(a) Using the expressions for $\eta_r(t)$ from Problem 15.1(c) and using Eq. 15.28, determine the mode-acceleration expression for $u_1(t)$ using all three modes, and indicate clearly the portion of the solution corresponding to $\hat{N} = 1$ and $\hat{N} = 2$ as in Example 15.3.

(b) Form a table, as in Example 15.1, using $\Omega = 0$, $\Omega = 0.5\omega_1$, and $\Omega = \frac{1}{2}(\omega_1 + \omega_2)$.

Problem Set 15.4

The topic of response of multi-DOF systems to harmonic excitation is closely related to the topic of modal testing (recall Chapter 1). Response plots such as those presented in Section 15.4 are essential to the understanding of system behavior. Response equations such as Eqs. 15.46 through 15.48, while essential, must be supplemented by graphical representations. Readers are strongly encouraged to employ digital simulation, with graphics, or to employ Fourier analysis of experimental data to explore this important topic.

15.5 Show that the damping term in Eq. 15.37 can be expressed in the form given in Eq. 15.38 if the damping has the form given in Eqs. 15.31 through 15.33.

15.6 (a) For the system in Example 15.4 change the stiffness coupling coefficient to $k' = 40.25$ and the damping coupling coefficient to $c' = 0.0125$ and repeat all the steps in Example 15.4.

(b) Compare your results with those of Examples 15.4 and 15.5. Would you classify the frequencies in this problem to be closely spaced? Explain your answer.

(c) Could the natural frequencies be identified accurately from peaks in the plot of $I(\overline{H}_{11})$ versus f? Could the natural frequencies be identified accurately from peaks in the plot of $I(\overline{H}_{21})$ versus f?

15.7 The 3-DOF system below is a modification of the system in Problem 13.2. Assume that the dashpots provide damping such that $\zeta_1 = \zeta_2 = \zeta_3 = 0.01$.

(a) Determine expressions for $\overline{H}_{11}$ and $\overline{H}_{31}$.

(b) Plot $\overline{H}_{11}$ and $\overline{H}_{31}$ on the complex plane.

(c) Plot $I(\overline{H}_{11})$ and $I(\overline{H}_{31})$ versus f.

(d) Describe in a brief paragraph the important features of system response that you can observe in the plots of (b) and (c).

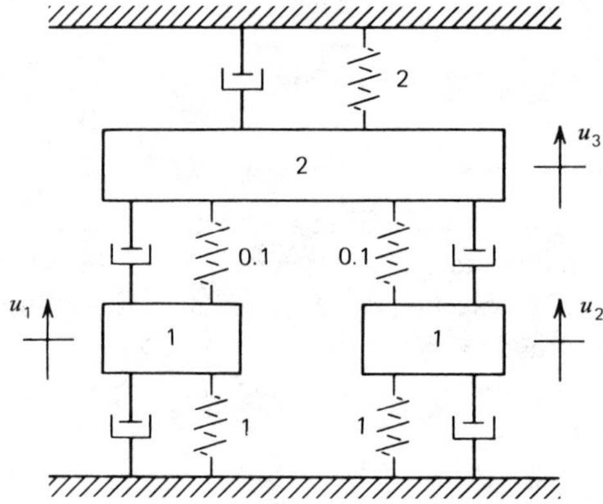

Figure P15.7

Problem Set 15.5

15.8 (a) Using the modal story shears $\mathbf{s}_r$ of Example 15.6 and the modal steady-state responses $\eta_r(t)$ from Example 15.1, determine expressions for the dynamic shears by the mode-displacement method, Eq. 15.49.

(b) Using the story shears again, determine expressions for the dynamic shears by the mode-acceleration method, Eq. 15.50.

(c) Let the bottom story shear, $\sigma_4(t)$, be written in the form

$$\sigma_4(t) = CP_1 \cos \Omega t$$

Construct tables like the ones in Examples 15.1 and 15.3 which show the convergence of the amplitude of $\sigma_4(t)$ when the expressions developed in (a) and (b) above are evaluated at $\Omega = 0$, $\Omega = 0.5\omega_1$, and $\Omega = 1.3\omega_3$.

Problem Set 15.6

15.9 Axial vibration of a free-free uniform bar was modeled in Problem 12.11 as a 3-DOF lumped-mass system. The system is shown below with a force $P_1(t)$ applied to the left end.

(a) Using the results of Problem 12.11, determine the modal axial stresses in the two "springs" due to the elastic modes, that is determine

$$\mathbf{s}_2 = \begin{Bmatrix} s_1 \\ s_2 \end{Bmatrix}_2 \quad \text{and} \quad \mathbf{s}_3 = \begin{Bmatrix} s_1 \\ s_2 \end{Bmatrix}_3$$

where s_1 is the axial stress in the rod joining masses m_1 and m_2 and s_2 is the axial stress in the rod joining masses m_2 and m_3.

(b) Let

$$\sigma(t) = \begin{Bmatrix} \sigma_1(t) \\ \sigma_2(t) \end{Bmatrix}$$

be the dynamic axial stresses in the two rods. Use the mode-displacement method, Eq. 15.55, to compute the steady-state stress vector $\sigma(t)$. Retain both elastic modes, and indicate the contribution from each elastic mode.

(c) Use the mode-acceleration method, Eq. (15.69), to compute the steady-state stress vector $\sigma(t)$. Retain both elastic modes and indicate the contribution of each.

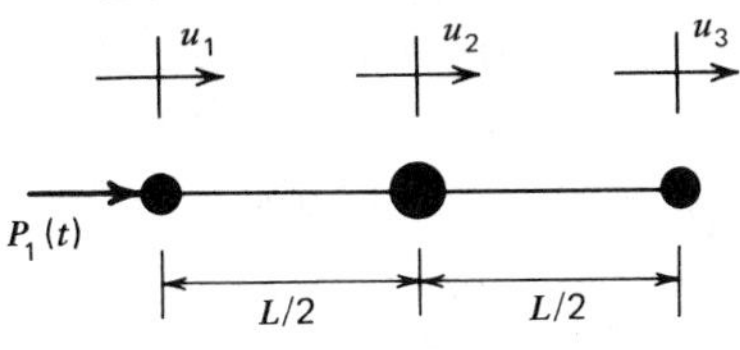

Figure P15.9

16 FINITE ELEMENT MODELING OF STRUCTURES

In Sec. 11.4 you were introduced to the assumed-modes method for generating a finite-DOF model of a continuous system. The procedures described there may be referred to as the global assumed-modes method because the assumed-modes, ψ_i, described deflection patterns throughout the entire structure. A powerful approximation method, which is a version of the assumed-modes method, is the finite element method. This chapter provides an introduction to the use of finite element models in structural dynamics analyses, and in Chapter 17 finite element models will be employed to study free vibration and forced response of structures.

Upon completion of Chapter 16 you should be able to:

- Derive expressions for shape functions for an axial element or a Bernoulli-Euler bending element.
- Generate the element stiffness matrix, mass matrix, or load vector referred to element axes for axial, bending, and torsion elements.
- Derive transformation matrices for truss and frame elements and transform element matrices to the global reference frame.
- Assemble system stiffness and mass matrices and load vector by the "direct stiffness" method and enforce boundary conditions.
- Reduce the number of degrees-of-freedom of a lumped-mass model by using static condensation.
- Generate a Ritz transformation matrix corresponding to a stated set of displacement constraint equations.

16.1 Introduction to the Finite Element Method

In Sec. 11.4 finite-DOF approximations to continuous systems were created using the *global assumed-modes method,* that is, by approximating the displacement function for the continuous system by an expression of the form

$$v(x, t) = \sum_{i=1}^{N} \psi_i(x) v_i(t) \tag{16.1}$$

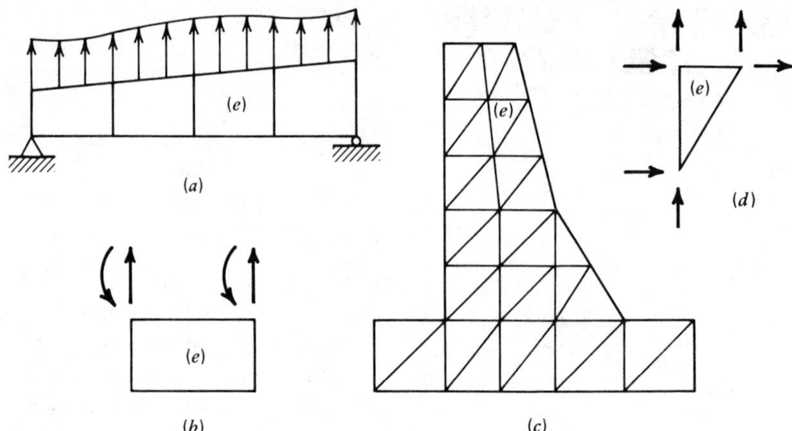

Figure 16.1. Structures represented by finite elements.

where each $\psi_i(x)$ describes a deflected shape of the entire structure. As noted in Sec. 11.4, an N-DOF mathematical model of the structure results, with generalized stiffness and mass coefficients and generalized forces being obtained from integrals involving the ψ_i's and their derivatives. The advent of the digital computer facilitated evaluation of these integrals, but it failed to remedy other serious drawbacks of the global assumed modes method, namely (1) it is extremely difficult for an analyst to choose a set of ψ_i's for a structure having complex geometry, (2) the equations that result from applying the global assumed modes procedure are usually highly coupled, and this requires more computer time and memory than is required if the coefficient matrices are sparsely populated, and (3) there is little carry-over from one problem to the next, that is, for each new geometry a new set of ψ_i's must be selected. The finite element method, to be introduced in this chapter, overcomes these difficulties. It is beyond the scope of this text, however, to present a thorough discussion of the finite element method in structural dynamics.

Finite element modeling of a structure may be considered to be an application of the assumed-mode method wherein the ψ_i's represent deflection shapes over a portion (finite element) of the structure, with the elements being assembled to form the structural system. Figures 16.1*a* and 16.1*b* show a tapered beam and a typical beam element of uniform cross section. Figures 16.1*c* and 16.1*d* show a gravity dam modeled by triangular elements. The elements are joined together at nodes, or joints,* and displacement compatibility

*The term node is generally used in finite element literature, but the term joint will be used in this text because of the alternate meaning that node has in structural dynamics and vibrations.

is enforced at these joints. Although the finite element method has a very wide range of applicability, only aspects of finite element theory applicable to structural dynamics will be treated here. To simplify the presentation, emphasis will be placed on one-dimensional elements such as the beam element of Fig. 16.1*b*.

In Sec. 16.2 stiffness and mass matrices and force vectors are derived for several finite element types. Section 16.3 treats the transformation of these element matrices so that all forces and displacements are referred to a global reference frame. Finally, in Sec. 16.4 the assembling of elements to form a structural system is described. Sections 16.5 through 16.7 discuss the enforcement of boundary conditions and other forms of constraints, including constraints imposed to reduce the number of active degrees-of-freedom in a system.

16.2 Element Stiffness and Mass Matrices and Element Force Vector

In this section we develop stiffness and mass matrices and load vectors for uniform one-dimensional elements subjected to axial deformation, bending, and torsion, or to a combination of these.

Axial Motion

Consider a uniform element of length L, mass density ρ, elastic modulus E, and cross-sectional area A. Choose a reference frame as shown in Fig. 16.2. This is referred to as an *element reference frame* because of the alignment of one axis along the element. The simplest approximation of axial displacement within the element employs the displacement at the two ends and is given by

$$u(x, t) = \psi_i(x)u_1(t) + \psi_2(x)u_2(t) \tag{16.2}$$

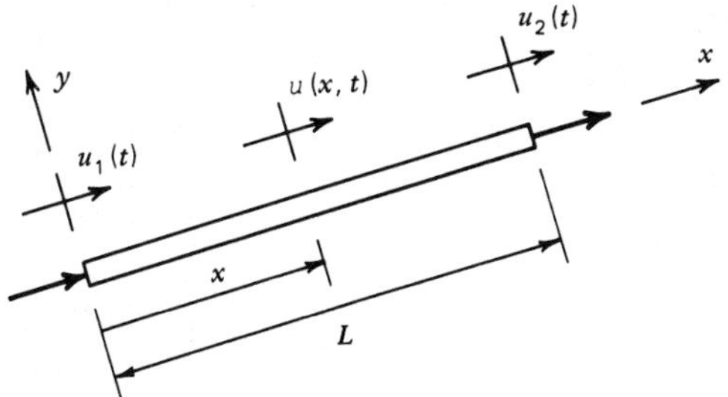

Figure 16.2. *A uniform element undergoing axial deformation.*

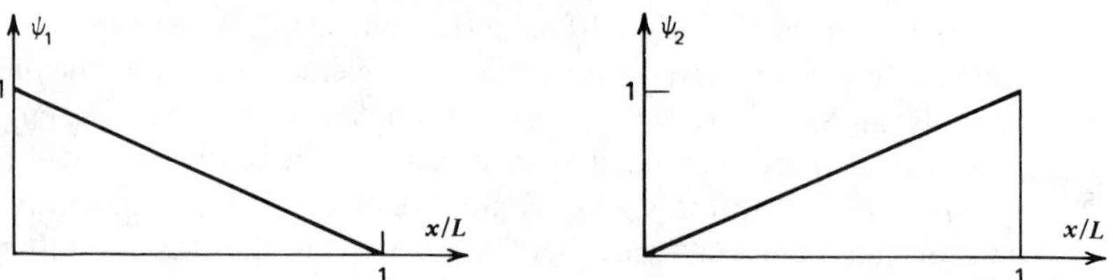

Figure 16.3. Shape functions for an axial element.

Since $u(0, t) = u_1(t)$ and $u(L, t) = u_2(t)$, the *shape functions* (assumed-modes) ψ_1 and ψ_2 must satisfy the boundary conditions

$$\begin{aligned} \psi_1(0) &= 1, \qquad \psi_1(L) = 0 \\ \psi_2(0) &= 0, \qquad \psi_2(L) = 1 \end{aligned} \tag{16.3}$$

The shape functions can be derived by considering axial deformation under static loads which would produce the boundary conditions of Eqs. 16.3. Specializing Eq. 9.7 to static deformation gives

$$(AEu')' = 0 \tag{16.4}$$

for the uniform element AE = constant, so

$$u(x) = c_1 + c_2\left(\frac{x}{L}\right) \tag{16.5}$$

The linear term is normalized to (x/L) so that c_1 and c_2 will both carry the dimensions of $u(x)$. From Eqs. 16.3 and 16.5, we obtain

$$\psi_1(x) = 1 - \frac{x}{L}, \qquad \psi_2(x) = \frac{x}{L} \tag{16.6}$$

Expressions for the stiffness coefficients k_{ij}, the mass coefficients m_{ij}, and the generalized forces p_i for axial motion are given by Eqs. 11.12, 11.16, and 11.20, respectively. Thus,

$$k_{ij} = \int_0^L EA\psi_i'\psi_j'\, dx \tag{16.7a}$$

$$m_{ij} = \int_0^L \rho A\psi_i\psi_j\, dx \tag{16.7b}$$

$$p_i = \int_0^L p(x, t)\psi_i\, dx \tag{16.7c}$$

Substituting Eqs. 16.6 into 16.7a and 16.7b we obtain the following stiffness and mass matrices for a uniform element.

$$\mathbf{k} = \left(\frac{AE}{L}\right)\begin{bmatrix} 1 & -1 \\ -1 & 1 \end{bmatrix} \tag{16.8a}$$

$$\mathbf{m} = \left(\frac{\rho AL}{6}\right)\begin{bmatrix} 2 & 1 \\ 1 & 2 \end{bmatrix} \tag{16.8b}$$

Use of Eq. 16.7c is illustrated in Example 16.1.

Transverse Motion—Bernoulli-Euler Theory

Consider a uniform beam element of length L, mass density ρ, elastic modulus E, cross-sectional area A, and moment of inertia I. Let the displacement coordinates for transverse motion be the end displacements and slopes numbered as shown in Fig. 16.4. Let

$$v(x, t) = \sum_{i=1}^{4} \psi_i(t) v_i(t) \tag{16.9}$$

where the functions $\psi_i(x)$ satisfy the boundary conditions

$$\begin{aligned}
\psi_1(0) &= 1, \qquad \psi_1'(0) = \psi_1(L) = \psi_1'(L) = 0 \\
\psi_2'(0) &= 1, \qquad \psi_2(0) = \psi_2(L) = \psi_2'(L) = 0 \\
\psi_3(L) &= 1, \qquad \psi_3(0) = \psi_3'(0) = \psi_3'(L) = 0 \\
\psi_4'(L) &= 1, \qquad \psi_4(0) = \psi_4'(0) = \psi_4(L) = 0
\end{aligned} \tag{16.10}$$

Appropriate shape functions can easily be derived by considering the beam element in Fig. 16.4 to be loaded statically by shears and moments to produce the various static deflection shapes that satisfy Eqs. 16.10. Thus, for a beam loaded only at its ends the equilibrium equation is

$$(EIv'')'' = 0 \tag{16.11}$$

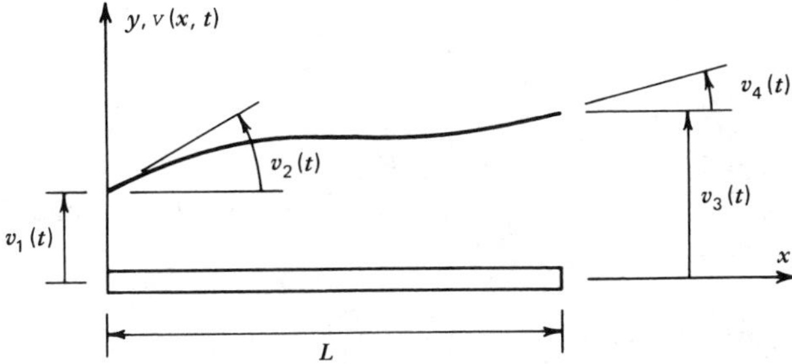

Figure 16.4. A uniform element undergoing transverse deflection.

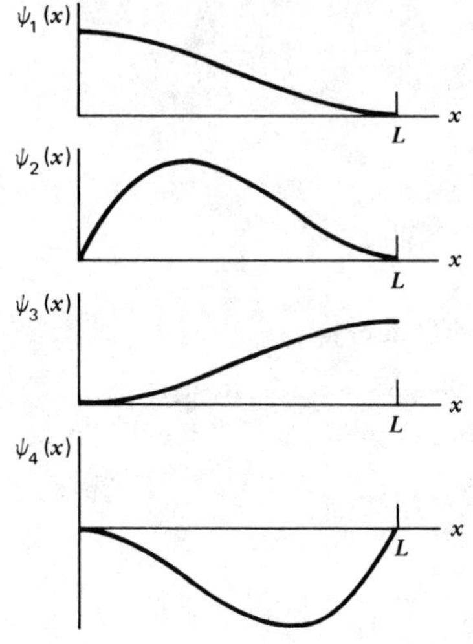

Figure 16.5. *Shape functions for transverse deformation of a beam element.*

The general solution to Eq. 16.9 for a uniform beam is a cubic polynomial

$$v(x) = c_1 + c_2\left(\frac{x}{L}\right) + c_3\left(\frac{x}{L}\right)^2 + c_4\left(\frac{x}{L}\right)^3 \qquad (16.12)$$

Note that (x/L) is again used instead of x so that all c_i's have the same dimensions. Substituting the four sets of boundary conditions of Eqs. 16.10 into Eq. 16.12 we obtain the following shape functions.

$$\begin{aligned} \psi_1 &= 1 - 3\left(\frac{x}{L}\right)^2 + 2\left(\frac{x}{L}\right)^3 \\ \psi_2 &= x - 2L\left(\frac{x}{L}\right)^2 + L\left(\frac{x}{L}\right)^3 \\ \psi_3 &= 3\left(\frac{x}{L}\right)^2 - 2\left(\frac{x}{L}\right)^3 \\ \psi_4 &= -L\left(\frac{x}{L}\right)^2 + L\left(\frac{x}{L}\right)^3 \end{aligned} \qquad (16.13)$$

These shape functions are illustrated in Fig. 16.5.

Equations 11.25 through 11.27 are expressions for k_{ij}, m_{ij}, and p_i for Bernoulli-Euler beams. Thus,

$$k_{ij} = \int_0^L EI\psi_i''\psi_j''\,dx \qquad (16.14a)$$

$$m_{ij} = \int_0^L \rho A\psi_i\psi_j\,dx \qquad (16.14b)$$

$$p_i = \int_0^L p(x, t)\psi_i \, dx \tag{16.14c}$$

Substituting Eqs. 16.13 into Eqs. 16.14a and 16.14b we obtain

$$\mathbf{k} = \left(\frac{EI}{L^3}\right)\begin{bmatrix} 12 & 6L & -12 & 6L \\ & 4L^2 & -6L & 2L^2 \\ & & 12 & -6L \\ \text{symm.} & & & 4L^2 \end{bmatrix} \tag{16.15a}$$

$$\mathbf{m} = \left(\frac{\rho AL}{420}\right)\begin{bmatrix} 156 & 22L & 54 & -13L \\ & 4L^2 & 13L & -3L^2 \\ & & 156 & -22L \\ \text{symm.} & & & 4L^2 \end{bmatrix} \tag{16.15b}$$

Example 16.1

Determine the generalized load vector **p** for a beam element subjected to a uniform transverse load $f(t)$.

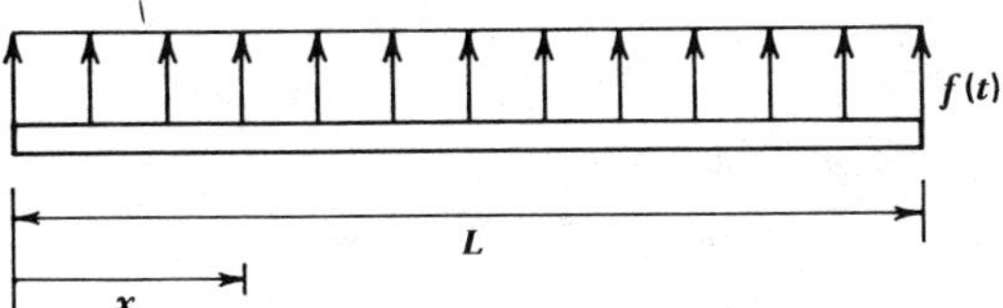

Solution

The beam element displacement coordinates referred to element axes are shown in Fig. 16.4 and the shape functions are given in Eqs. 16.13. Thus,

$$\begin{aligned}
\psi_1(x) &= 1 - 3\left(\frac{x}{L}\right)^2 + 2\left(\frac{x}{L}\right)^3 \\
\psi_2(x) &= x - 2L\left(\frac{x}{L}\right)^2 + L\left(\frac{x}{L}\right)^3 \\
\psi_3(x) &= 3\left(\frac{x}{L}\right)^2 - 2\left(\frac{x}{L}\right)^3 \\
\psi_4(x) &= -L\left(\frac{x}{L}\right)^2 + L\left(\frac{x}{L}\right)^3
\end{aligned} \tag{1}$$

The equation for the generalized force components is Eq. 16.14c, that is,

$$p_i(t) = \int_0^L p(x, t)\psi_i(x)\, dx \tag{2}$$

Combining Eqs. 1 and 2, we get

$$\begin{aligned} p_1(t) &= f(t) \int_0^L \psi_1(x)\, dx \\ &= f(t) \int_0^L \left[1 - 3\left(\frac{x}{L}\right)^2 + 2\left(\frac{x}{L}\right)^3\right] dx \\ &= fL\left[x - \left(\frac{x}{L}\right)^3 + \left(\frac{2}{4}\right)\left(\frac{x}{L}\right)^4\right]_0^L \end{aligned} \tag{3}$$

$$\boxed{p_1(t) = \frac{fL}{2}} \tag{4}$$

In a similar fashion,

$$\boxed{\begin{aligned} p_2(t) &= \tfrac{1}{12} fL^2 \\ p_3(t) &= \tfrac{1}{2} fL \\ p_4(t) &= -\tfrac{1}{12} fL^2 \end{aligned}}$$

Torsion

In Sec. 9.3 the equation of motion and boundary conditions for torsional deformation of a circular rod were obtained. We wish now to derive expressions for stiffness and mass coefficients and generalized forces (moments) for a uniform torsion element of length L as shown in Fig. 16.6. Let the local x-axis be the centroidal axis. Let I_p be the polar moment of inertia about the centroidal axis and let GJ be the torsional stiffness.* Then the strain energy and kinetic energy for pure torsion are given by

$$V = \frac{1}{2} \int_0^L GJ(\theta')^2\, dx \tag{16.16a}$$

*$J = I_p$ for a circular cross section. For other cross-sectional shapes the reader may consult texts on advanced strength of materials or elasticity for expressions for J. For noncircular members the axis of twist and the centroidal axis frequently do not coincide. This leads to a coupling of bending and torsion. An example of this, vibration of an airplane wing, has been discussed by numerous authors, for example, References 16.1, 16.2, and 16.3. Hence, care should be exercised in any structural dynamics problems where member torsion is involved.

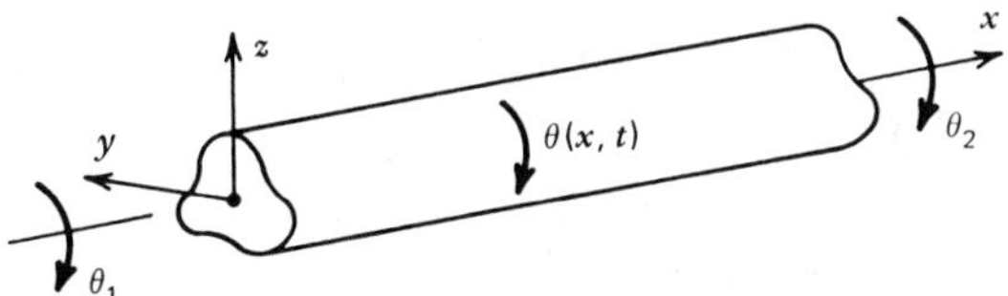

Figure 16.6. A uniform element undergoing torsional deformation.

$$T = \frac{1}{2}\int_0^L \rho I_p(\dot{\theta})^2\,dx \tag{16.16b}$$

The rotation along the element is to be given by the assumed-modes form

$$\theta(x,t) = \psi_1(x)\theta_1(t) + \psi_2(x)\theta_2(t) \tag{16.17}$$

where the appropriate boundary conditions for the shape functions are given by Eqs. 16.3. For a torsion member loaded statically by end torques, the equilibrium equation, obtained from Eq. 9.36, is

$$(GJ\theta')' = 0 \tag{16.18}$$

Since this equation has the same form as the equilibrium equation for axial deformation, Eq. 16.4, and since the torsion shape functions must satisfy the same boundary conditions as the axial shape functions, the shape functions for the torsion element and axial element are the same, that is, ψ_1 and ψ_2 are given by Eqs. 16.6.

Using the procedures employed in Chapter 11 we can combine Eqs. 16.16 and 16.17 to obtain the expressions

$$\begin{aligned} k_{ij} &= \int_0^L GJ\psi_i'\psi_j'\,dx \\ m_{ij} &= \int_0^L \rho I_p\psi_i\psi_j\,dx \\ p_i &= \int_0^L t(x,t)\psi_i\,dx \end{aligned} \tag{16.19}$$

where $t(x, t)$ is a distributed torque per unit length. Inserting the shape functions of Eqs. 16.6 into Eqs. 16.19, we obtain for a uniform torsion element

$$\mathbf{k} = \left(\frac{GJ}{L}\right)\begin{bmatrix} 1 & -1 \\ -1 & 1 \end{bmatrix} \tag{16.20a}$$

$$\mathbf{m} = \left(\frac{\rho I_p L}{6}\right)\begin{bmatrix} 2 & 1 \\ 1 & 2 \end{bmatrix} \tag{16.20b}$$

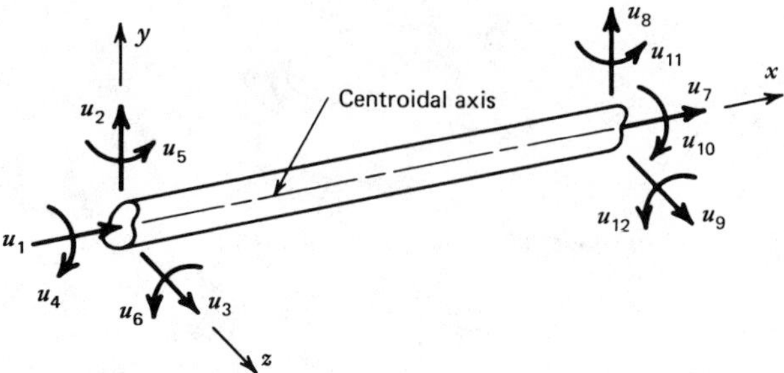

Figure 16.7. Notation for a three-dimensional frame element.

Three-Dimensional Frame Element

The stiffness and mass matrices for a slender three-dimensional frame element can be obtained from the axial, bending, and torsion elements discussed above. In particular, Bernoulli-Euler beam theory was used in deriving the bending element. Prezemeniecki[16.4] presents derivations for the stiffness and mass matrices of a three-dimensional frame element including shear deformation and rotatory inertia effects.

Figure 16.7 shows the local reference frame and the displacement coordinates for a uniform three-dimensional frame element. The x-axis is along the line of centroids of the cross sections and the y and z axes are principal axes in the cross section. I_y and I_z are moments of inertia of the cross section and I_p is the polar moment of inertia about the x-axis.

The stiffness and mass coefficients associated with u_1 and u_7 are from Eqs. 16.8; those associated with u_2, u_6, u_8, and u_{12} are based on Eqs. 16.15 for bending in the xy-plane, while u_9, u_{11}, u_3, and u_5 are for bending in the xz-plane. The order of these bending degrees of freedom preserves the sense of the rotations shown in Fig. 16.4. Finally, the coefficients associated with u_4 and u_{10} are obtained from Eqs. 16.20.

Other Elements

Stiffness and mass matrices for other finite elements, including two-dimensional and three-dimensional elements, may be found in reference texts and journal articles on finite elements, for example, Reference 16.4.

Lumped Mass Matrix for Beam Elements

The mass matrices derived above are referred to as *consistent mass matrices* because the same shape functions are used to derive the coefficients of the mass

$$
\mathbf{k} = \left[\begin{array}{cccccccccccc}
\frac{EA}{L} & & & & & & \frac{-EA}{L} & & & & & \\
 & \frac{12EI_z}{L^3} & & & & \frac{6EI_z}{L^2} & & \frac{-12EI_z}{L^3} & & & & \frac{6EI_z}{L^2} \\
 & & \frac{12EI_y}{L^3} & & \frac{-6EI_y}{L^2} & & & & \frac{-12EI_y}{L^3} & & \frac{-6EI_y}{L^2} & \\
 & & & \frac{GJ}{L} & & & & & & -\frac{GJ}{L} & & \\
 & & \frac{-6EI_y}{L^2} & & \frac{4EI_y}{L} & & & & \frac{6EI_y}{L^2} & & \frac{2EI_y}{L} & \\
 & \frac{6EI_z}{L^2} & & & & \frac{4EI_z}{L} & & \frac{-6EI_z}{L^2} & & & & \frac{2EI_z}{L} \\
-\frac{EA}{L} & & & & & & \frac{EA}{L} & & & & & \\
 & \frac{-12EI_z}{L^3} & & & & \frac{-6EI_z}{L^2} & & \frac{12EI_z}{L^3} & & & & \frac{-6EI_z}{L^2} \\
 & & \frac{-12EI_y}{L^3} & & \frac{6EI_y}{L^2} & & & & \frac{12EI_y}{L^3} & & \frac{6EI_y}{L^2} & \\
 & & & -\frac{GJ}{L} & & & & & & \frac{GJ}{L} & & \\
 & & \frac{-6EI_y}{L^2} & & \frac{2EI_y}{L} & & & & \frac{6EI_y}{L^2} & & \frac{4EI_y}{L} & \\
 & \frac{6EI_z}{L^2} & & & & \frac{2EI_z}{L} & & \frac{-6EI_z}{L^2} & & & & \frac{4EI_z}{L}
\end{array}\right]
\begin{array}{c} 1 \\ 2 \\ 3 \\ 4 \\ 5 \\ 6 \\ 7 \\ 8 \\ 9 \\ 10 \\ 11 \\ 12 \end{array}
\tag{16.21a}
$$

Columns: 1 2 3 4 5 6 7 8 9 10 11 12

$$
\mathbf{m} = \frac{\rho A L}{420}
\begin{bmatrix}
140 & & & & & & 70 & & & & & \\
 & 156 & & & & 22L & & 54 & & & & -13L \\
 & & 156 & & -22L & & & & 54 & & 13L & \\
 & & & \frac{140 I_p}{A} & & & & & & \frac{70 I_p}{A} & & \\
 & & -22L & & 4L^2 & & & & -13L & & -3L^2 & \\
 & 22L & & & & 4L^2 & & 13L & & & & -3L^2 \\
70 & & & & & & 140 & & & & & \\
 & 54 & & & & 13L & & 156 & & & & -22L \\
 & & 54 & & -13L & & & & 156 & & 22L & \\
 & & & \frac{70 I_p}{A} & & & & & & \frac{140 I_p}{A} & & \\
 & & 13L & & -3L^2 & & & & 22L & & 4L^2 & \\
 & -13L & & & & -3L^2 & & -22L & & & & 4L^2
\end{bmatrix}
\tag{16.21b}
$$

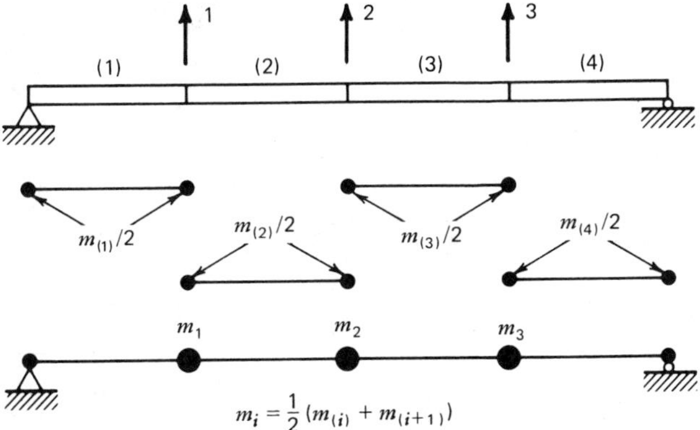

Figure 16.8. Lumped mass model of a beam.

matrix as are used for the stiffness matrix. This leads, for example, to coupling between translational coordinates and rotations in the mass matrix for a beam, Eq. 16.15b. A simpler model of the inertia properties of a beam is the *lumped mass* model. Figure 16.8 shows a beam divided into finite elements with the mass of elements lumped at the element ends and reassembled to form a lumped mass model of the beam. For elements of equal length the lumped mass matrix for the beam of Fig. 16.8 would be

$$\mathbf{m} = \frac{\rho AL}{4}\begin{bmatrix} 1 & 0 & 0 \\ 0 & 1 & 0 \\ 0 & 0 & 1 \end{bmatrix} \tag{16.22}$$

Archer[16.5], Leckie and Lindberg[16.6], and Tong, Pian, and Bucciarelli[16.7] have compared eigensolutions based on lumped mass and consistent mass formulations. We will return to this topic in Sec. 16.6 when we discuss static condensation.

Thus far, expressions have been obtained for element stiffness and mass matrices and for element load vectors. Some procedures for handling damping in structures will be discussed in Chapter 18.

16.3 Transformation of Element Matrices

In Sec. 16.1 it was indicated that the finite element method consists of locally approximating displacements and then tying elements together through displacement compatibility conditions at the joints. This approach is called the *displacement method* because it is the displacements of an element which are directly approximated. Texts on the finite element method also discuss the *force method,* wherein stresses are approximated directly, and *mixed methods.*

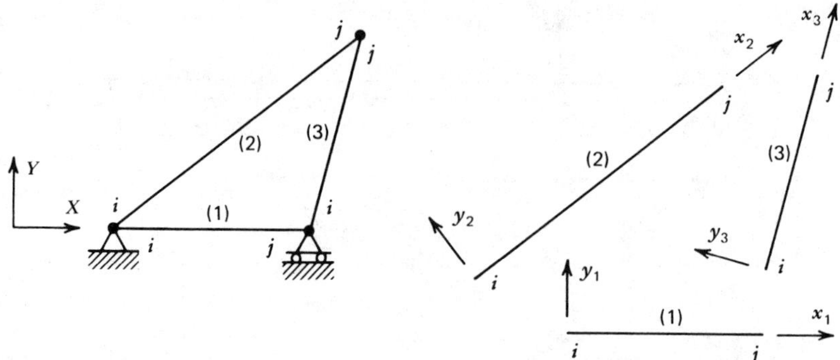

Figure 16.9. Global and element reference frames for a planar truss.

We will pursue only the displacement method in this text because it is the one most commonly employed in structural dynamics analyses.

In Sec. 16.2 stiffness and mass matrices were derived for elements with the displacement coordinates referred to an element reference frame, xyz. Truss and frame structures frequently have members that are not aligned with a common set of axes, or *global reference frame, XYZ,* as in the case of the simple planar truss in Fig. 16.9. The ends of elements are labeled i and j and the local axis sign convention places the origin at the i-end with the x-axis along the member from i toward j.

In this section we consider the transformation of element matrices (displacement and force vectors, stiffness and mass matrices) to the global reference so that compatibility equations and assembly of system matrices can be treated more directly. We begin with a plane truss element.

Plane Truss Transformations

Figure 16.10a shows the *e*lement displacement *c*oordinates referred to *e*lement axes (ECE), and Fig. 16.10b shows the *e*lement displacement *c*oordinates

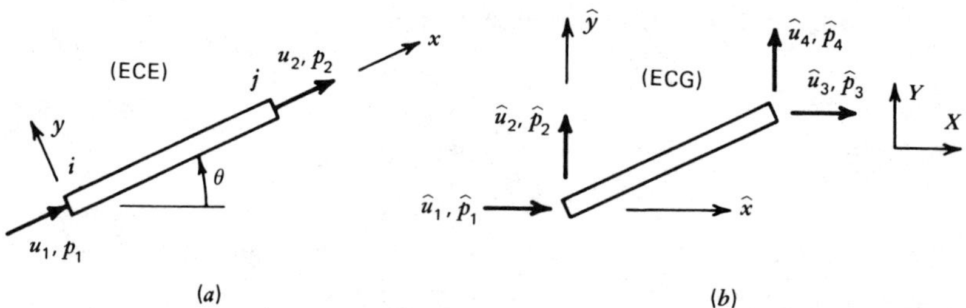

Figure 16.10 Element displacements and forces referred to element axes and global axes. (a) *Element reference.* (b) *Global reference.*

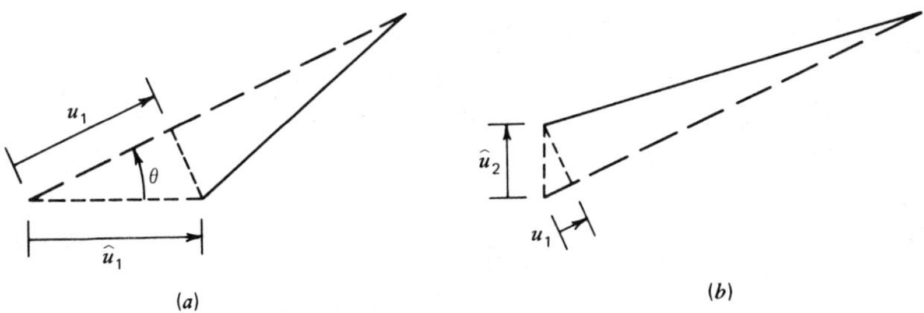

Figure 16.11. *Displacement transformation for a plane truss element.* (a) x̂ *displacement at* "i." (b) ŷ-*displacement at* "i."

referred to global axes (ECG) for a planar truss. In the former case only the axial displacements u_1 and u_2 are needed. When the truss element is assembled as one member of a truss, however, it is the displacements along the global directions $\hat{x}(=X)$ and $\hat{y}(=Y)$ which are employed. For the element, then, $\hat{u}_1$ and $\hat{u}_2$ are the displacements in the $\hat{x}$ and $\hat{y}$ directions at end i, while $\hat{u}_3$ and $\hat{u}_4$ are the corresponding displacements at end j of the element.

Figure 16.11 shows the change of configuration of a truss element due to a displacement $\hat{u}_1$ and also due to $\hat{u}_2$. We need to determine the axial component, u_1, of the displacement due to $\hat{u}_1$ and $\hat{u}_2$. From Fig. 16.11*a*

$$u_1 = \hat{u}_1 \cos\theta$$

and from Fig. 16.11*b*,

$$u_1 = \hat{u}_2 \sin\theta$$

Therefore, if both $\hat{u}_1$ and $\hat{u}_2$ displacement components are present

$$u_1 = \hat{u}_1 \cos\theta + \hat{u}_2 \sin\theta \tag{16.23a}$$

Similarly, for end j

$$u_2 = \hat{u}_3 \cos\theta + \hat{u}_4 \sin\theta \tag{16.23b}$$

These equations can be combined in a single matrix equation

$$\mathbf{u} = \mathbf{T}\hat{\mathbf{u}} \tag{16.24}$$

where

$$\mathbf{u}^{\mathrm{T}} = \lfloor u_1 u_2 \rfloor, \quad \hat{\mathbf{u}}^T = \lfloor \hat{u}_1 \hat{u}_2 \hat{u}_3 \hat{u}_4 \rfloor \tag{16.25}$$

and

$$\mathbf{T} = \begin{bmatrix} \cos\theta & \sin\theta & 0 & 0 \\ 0 & 0 & \cos\theta & \sin\theta \end{bmatrix} \tag{16.26}$$

Example 16.2

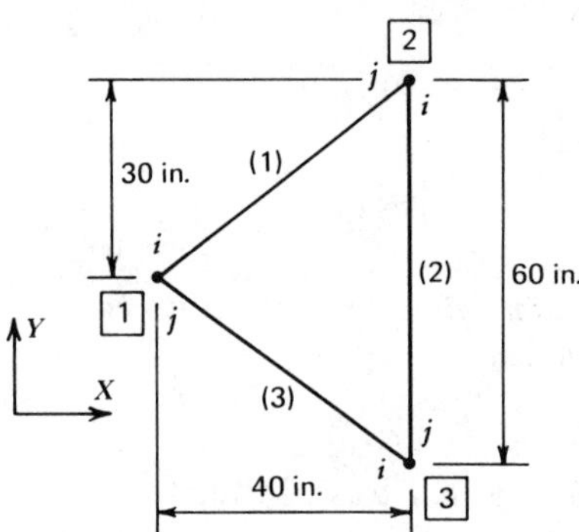

Three bars are connected as shown to form a plane truss. Set up the transformation matrix for element (1). Label the ends of the elements as indicated on the figure and use the global axes XY as shown.

Solution

$$L = \sqrt{(30)^2 + (40)^2} = 50 \text{ in.}$$

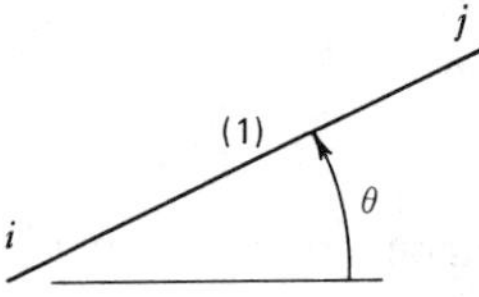

$$\cos\theta = \tfrac{4}{5}$$
$$\sin\theta = \tfrac{3}{5}$$

$$\boxed{\mathbf{T} = \frac{1}{5}\begin{bmatrix} 4 & 3 & 0 & 0 \\ 0 & 0 & 4 & 3 \end{bmatrix}}$$

A plane frame element has three displacement coordinates at each end. Figure 16.12 shows the notation for (a) element displacement coordinates referred to element axes, and (b) element displacement coordinates referred to global axes. The transformation from $\hat{\mathbf{u}}$-displacements to $\mathbf{u}$-displacements has the same form as Eq. 16.24, and, when written out, is

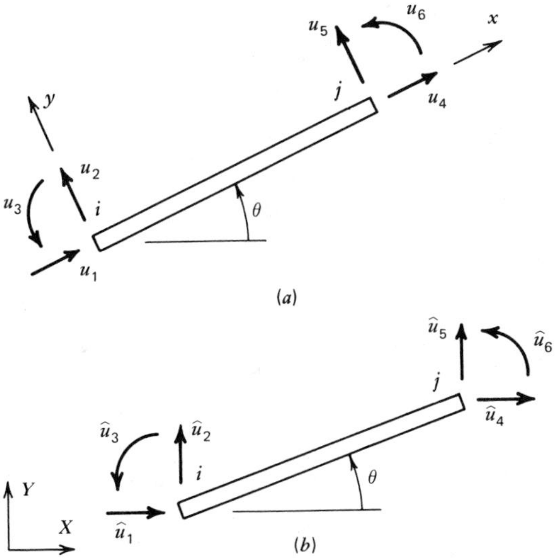

Figure 16.12. *Two sets of element displacement coordinates for a plane frame element.* (a) *Frame element displacements referred to element axes.* (b) *Frame element displacements referred to global axes*

$$\begin{Bmatrix} u_1 \\ u_2 \\ u_3 \\ u_4 \\ u_5 \\ u_6 \end{Bmatrix} = \begin{bmatrix} \cos\theta & \sin\theta & 0 & 0 & 0 & 0 \\ -\sin\theta & \cos\theta & 0 & 0 & 0 & 0 \\ 0 & 0 & 1 & 0 & 0 & 0 \\ 0 & 0 & 0 & \cos\theta & \sin\theta & 0 \\ 0 & 0 & 0 & -\sin\theta & \cos\theta & 0 \\ 0 & 0 & 0 & 0 & 0 & 1 \end{bmatrix} \begin{Bmatrix} \hat{u}_1 \\ \hat{u}_2 \\ \hat{u}_3 \\ \hat{u}_4 \\ \hat{u}_5 \\ \hat{u}_6 \end{Bmatrix} \tag{16.27}$$

Note that the rotations are the same in both coordinate systems, that is, $u_3 = \hat{u}_3$ and $u_6 = \hat{u}_6$. The expression for u_1 is the same as developed in Eq. 16.23a for the truss. Derivation of the expression for the transverse displacement u_5 is left as an exercise.

Figure 16.13 shows a three-dimensional truss element, which has three displacement coordinates at each end. The displacement transformation matrix that transforms $\hat{\mathbf{u}}$-displacements to $\mathbf{u}$-displacements can be written as a matrix of direction cosines. Thus,

$$\mathbf{T} = \begin{bmatrix} \mathbf{T}_c & 0 \\ 0 & \mathbf{T}_c \end{bmatrix} \tag{16.28}$$

where

$$\mathbf{T}_c = \begin{bmatrix} \cos(x\hat{x}) & \cos(x\hat{y}) & \cos(x\hat{z}) \\ \cos(y\hat{x}) & \cos(y\hat{y}) & \cos(y\hat{z}) \\ \cos(z\hat{x}) & \cos(z\hat{y}) & \cos(z\hat{z}) \end{bmatrix} \tag{16.29}$$

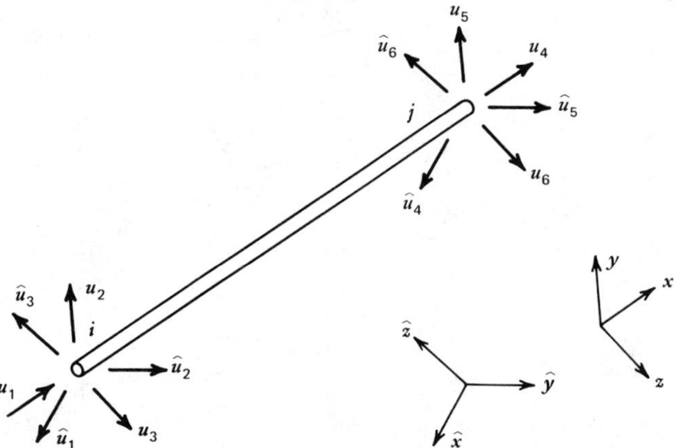

Figure 16.13. Displacement coordinates for a three-dimensional truss element.

To obtain the transformations for the force vector, the stiffness matrix, and the mass matrix it is convenient to employ virtual work and energy concepts. To transform the force vector **p** associated with the displacement vector **u** to a force vector $\hat{\mathbf{p}}$ associated with $\hat{\mathbf{u}}$, consider the virtual work. The scalar quantity δW should be the same for both representations, that is,

$$\delta W = \delta \mathbf{u}^T \mathbf{p} = \delta \hat{\mathbf{u}}^T \hat{\mathbf{p}} \tag{16.30}$$

Substituting Eq. 16.24 into the above we get

$$\delta W = \delta \mathbf{u}^T \mathbf{p} = (\delta \hat{\mathbf{u}}^T \mathbf{T}^T)\mathbf{p} = \delta \hat{\mathbf{u}}^T (\mathbf{T}^T \mathbf{p}) \tag{16.31}$$

By comparing the terms multiplying $\delta \hat{u}_1$, $\delta \hat{u}_2$, and so forth, in Eqs. 16.30 and 16.31 we conclude that

$$\hat{\mathbf{p}} = \mathbf{T}^T \mathbf{p} \tag{16.32}$$

since the $\delta \hat{u}_i$'s are independent.

To obtain the transformation equation for stiffness and mass matrices we can employ the strain energy and kinetic energy. From Eqs. 11.13 and 11.17, we have

$$V = \tfrac{1}{2}(\mathbf{u}^T \mathbf{k} \mathbf{u}) = \tfrac{1}{2}(\hat{\mathbf{u}}^T \hat{\mathbf{k}} \hat{\mathbf{u}}) \tag{16.33a}$$

and

$$T = \tfrac{1}{2}(\dot{\mathbf{u}}^T \mathbf{m} \dot{\mathbf{u}}) = \tfrac{1}{2}(\dot{\hat{\mathbf{u}}}^T \hat{\mathbf{m}} \dot{\hat{\mathbf{u}}}) \tag{16.33b}$$

Thus,

$$V = \tfrac{1}{2}(\hat{\mathbf{u}}^T \mathbf{T}^T)\mathbf{k}(\mathbf{T}\hat{\mathbf{u}}) = \tfrac{1}{2}\hat{\mathbf{u}}^T (\mathbf{T}^T \mathbf{k} \mathbf{T})\hat{\mathbf{u}} \tag{16.34}$$

or

$$\hat{\mathbf{k}} = \mathbf{T}^T\mathbf{k}\mathbf{T} \tag{16.35a}$$

Similarly, from the kinetic energy we obtain

$$\hat{\mathbf{m}} = \mathbf{T}^T\mathbf{m}\mathbf{T} \tag{16.35b}$$

For a plane truss element Eqs. 16.8a, 16.26, and 16.35a can be conveniently combined to give

$$\hat{\mathbf{k}} = k\begin{bmatrix} c^2 & cs & -c^2 & -cs \\ & s^2 & -cs & -s^2 \\ \text{symm.} & & c^2 & cs \\ & & & s^2 \end{bmatrix} \tag{16.36}$$

where $k = AE/L$, $c = \cos\theta$, and $s = \sin\theta$.

16.4 Assembly of System Matrices: The "Direct Stiffness" Method

Several processes are involved in arriving at a final set of equations of motion for a system based on finite element modeling. So far we have considered processes at the element level—generating element matrices and transforming them to global coordinates. There remains the assembly of system matrices, enforcing boundary conditions, and enforcing other constraints, if any (e.g., reducing the number of coordinates of the system). In many finite element codes these processes are carried out simultaneously, that is, boundary conditions and other constraints are imposed as the system matrices are assembled. For simplicity of presentation, we consider these separately.

In Sec. 16.3 the transformation matrix $\mathbf{T}$ was introduced to allow all element matrices to be referred to a common reference frame, the global frame XYZ. In this section the "direct stiffness" method is employed to assemble system matrices $\mathbf{M}$, $\mathbf{K}$, and $\mathbf{P}$ from the corresponding element matrices.

When several elements are combined to form a structure, the common set of *system displacement coordinates* will be designated by the vector $\mathbf{U}$. Figure 16.14 shows the six system displacements of an unrestrained three-bar truss. The joints are labeled 1, 2, and 3, and the system displacements are taken in the following order: U_1 = displacement in the X-direction at joint 1, U_2 = displacement in the Y-direction at joint 1, and so forth, as shown in Fig. 16.14. Notice that we have now introduced three distinct types of displacement coordinates: element displacements referred to element axes, element displacements referred to global axes, and system displacements. In this section we use the relationships between the latter two types in assembling the system matrices.

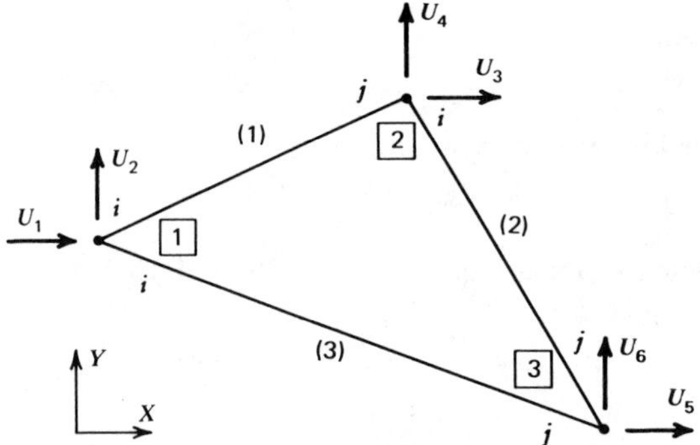

Figure 16.14. System displacement coordinates for a three-bar truss.

The "direct stiffness" method for assembling system matrices is based on the fact that work and energy are scalar quantities so that, for example, the total strain energy of a structure is the sum of the strain energy contributions of all of its elements. In addition, the element displacements referred to global axes can simply be identified with the appropriate system displacements, that is, the displacements $\hat{\mathbf{u}}_e$ of element e can be identified with the system displacements in $\mathbf{U}$. For example, consider element 3 of the three-bar truss in Fig. 16.14. The element displacements referred to global axes are noted on Fig. 16.15, where the orientation of element axes is defined by the i, j notation on Fig. 16.14.

By comparing Fig. 16.15 with Fig. 16.14 we note that the displacements of element 3 are related by

$$\begin{aligned} \hat{u}_1 &= U_1 \\ \hat{u}_2 &= U_2 \\ \hat{u}_3 &= U_5 \\ \hat{u}_4 &= U_6 \end{aligned} \tag{16.37}$$

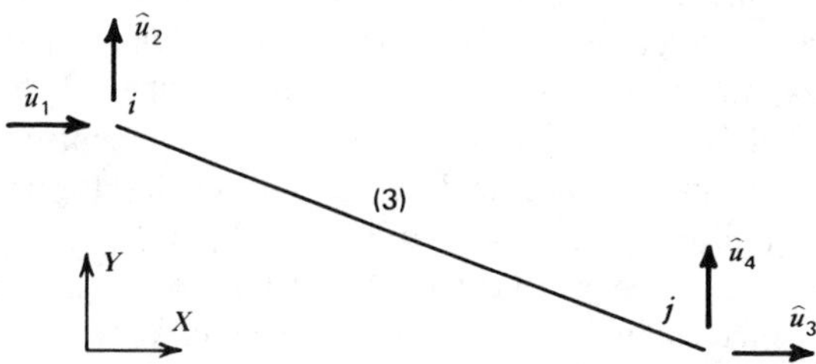

Figure 16.15. Element displacement coordinates referred to global axes (ECG).

The element coordinates referred to element axes can be related to system displacement coordinates through *locator* (or label) *matrices,* $\mathbf{L}_e$, such that

$$\hat{\mathbf{u}}_e = \mathbf{L}_e \mathbf{U} \tag{16.38}$$

For example, for element 3

$$\mathbf{L}_3 = \begin{bmatrix} 1 & 0 & 0 & 0 & 0 & 0 \\ 0 & 1 & 0 & 0 & 0 & 0 \\ 0 & 0 & 0 & 0 & 1 & 0 \\ 0 & 0 & 0 & 0 & 0 & 1 \end{bmatrix} \tag{16.39}$$

The number of rows of $\mathbf{L}_e$ is, of course, equal to the number of rows of $\hat{\mathbf{u}}_e$, and the number of columns of $\mathbf{L}_e$ is equal to the number of system degrees-of-freedom. Since $\mathbf{L}_e$ consists of only ones and zeros, it is more efficient in computer applications to store the locator information as a *locator vector* $\boldsymbol{l}_e$ which lists the system coordinates which correspond to the respective element coordinates in $\hat{\mathbf{u}}_e$, for example, for element 3

$$\boldsymbol{l}_3^T = \lfloor 1 \quad 2 \quad 5 \quad 6 \rfloor \tag{16.40}$$

We can arrive at the direct stiffness assembly procedure by expressing the strain energy of the structure as a sum of the element strain energies, that is,

$$V = \sum_{e=1}^{N_e} V_e = \sum_{e=1}^{N_e} \left(\frac{1}{2} \hat{\mathbf{u}}_e^T \hat{\mathbf{k}}_e \hat{\mathbf{u}}_e \right) = \sum_{e=1}^{N_e} \frac{1}{2} \mathbf{U}^T \mathbf{K}_e \mathbf{U} \tag{16.41}$$

where Eq. 16.38 has been used, and where

$$\mathbf{K}_e = \mathbf{L}_e^T \hat{\mathbf{k}}_e \mathbf{L}_e \tag{16.42}$$

The system strain energy can also be written as

$$V = \tfrac{1}{2} \mathbf{U}^T \mathbf{K} \mathbf{U} \tag{16.43}$$

where $\mathbf{K}$ is the *system stiffness matrix* given by

$$\mathbf{K} = \sum_{e=1}^{N_e} \mathbf{K}_e \tag{16.44}$$

Although the process of assembling the system stiffness matrix from the N_e element matrices $\hat{\mathbf{k}}_e$ appears to involve two steps given by Eqs. 16.42 and 16.44, the two steps can be combined in an efficient process called the *direct stiffness method,* which will be illustrated for the three-bar truss of Fig. 16.14.

First, we will show that Eq. 16.42 simply locates the elements of the $\hat{\mathbf{k}}_e$ matrix in the proper row and column locations corresponding to the system degrees-of-freedom of element *e*. Consider element 3, for which $\mathbf{L}_3$ is given in Eq. 16.38. Let

$$\hat{\mathbf{k}}_3 = \begin{bmatrix} \hat{k}_{11} & \hat{k}_{12} & \hat{k}_{13} & \hat{k}_{14} \\ \hat{k}_{21} & \hat{k}_{22} & \hat{k}_{23} & \hat{k}_{24} \\ \hat{k}_{31} & \hat{k}_{32} & \hat{k}_{33} & \hat{k}_{34} \\ \hat{k}_{41} & \hat{k}_{42} & \hat{k}_{43} & \hat{k}_{44} \end{bmatrix} \tag{16.45}$$

Then,

$$\hat{\mathbf{k}}_3\mathbf{L}_3 = \begin{bmatrix} \hat{k}_{11} & \hat{k}_{12} & 0 & 0 & \hat{k}_{13} & \hat{k}_{14} \\ \hat{k}_{21} & \hat{k}_{22} & 0 & 0 & \hat{k}_{23} & \hat{k}_{24} \\ \hat{k}_{31} & \hat{k}_{32} & 0 & 0 & \hat{k}_{33} & \hat{k}_{34} \\ \hat{k}_{41} & \hat{k}_{42} & 0 & 0 & \hat{k}_{43} & \hat{k}_{44} \end{bmatrix} \tag{16.46}$$

That is, postmultiplication by an $\mathbf{L}_e$ matrix simply places coefficients of $\hat{\mathbf{k}}_e$ into the proper columns of $\mathbf{K}_e$. Premultiplication by $\mathbf{L}_e^T$ expands the rows and places the coefficients in the proper row locations of $\mathbf{K}_e$. Finally, Eq. 16.44 simply takes coefficients and adds them into the proper locations of $\mathbf{K}$.

The postmultiplication by $\mathbf{L}_e$ and premultiplication by $\mathbf{L}_e^T$ involve many unnecessary multiplication operations, since the final result is simply a moving of elements of $\hat{\mathbf{k}}_e$ to proper row and column position for addition into $\mathbf{K}$. Bypassing these multiplications and just directly adding coefficients from $\hat{\mathbf{k}}_e$ into their proper locations in $\mathbf{K}$ with the aid of locator information from a locator vector $\boldsymbol{l}_e$, such as Eq. 16.40, is referred to as the *direct stiffness method* of assembling $\mathbf{K}$.* This process will be demonstrated in Example 16.3.

The system mass matrix can be assembled in the same fashion (i.e., by the direct "stiffness" method) as can be seen by considering the kinetic energy of the system.

$$\begin{aligned} T &= \frac{1}{2}\sum_{e=1}^{N_e} \dot{\hat{\mathbf{u}}}_e^T \hat{\mathbf{m}}_e \dot{\hat{\mathbf{u}}}_e \\ &= \frac{1}{2}\sum_{e=1}^{N_e} \dot{\mathbf{U}}^T \mathbf{M}_e \dot{\mathbf{U}} = \frac{1}{2}\dot{\mathbf{U}}^T \mathbf{M} \dot{\mathbf{U}} \end{aligned} \tag{16.47}$$

where

$$\mathbf{M}_e = \mathbf{L}_e^T \hat{\mathbf{m}}_e \mathbf{L}_e \tag{16.48a}$$

and

$$\mathbf{M} = \sum_{e=1}^{N_e} \mathbf{M}_e \tag{16.48b}$$

*The direct stiffness method has been formulated in terms of summing the contributions of each element to the total strain energy of the structure and hence to the stiffness matrix of the structure. A similar approach employing virtual work and strain energy is employed in Reference 16.4. An alternative discussion of the direct stiffness method, which is based on the summation of nodal forces to satisfy nodal equilibrium, is presented in Reference 16.8.

Example 16.3

Each of the truss members in Example 16.2 has the same cross-sectional area A and modulus of elasticity E. Using the lengths and the element and joint numbering shown in Example 16.2, form:

a. The three element stiffness matrices referred to global coordinates.
b. The system stiffness matrix.

Solution

The numbering of element coordinates and system coordinates is defined by the numbering of joints and elements and by the element ends (i, j) shown in Example 16.2.

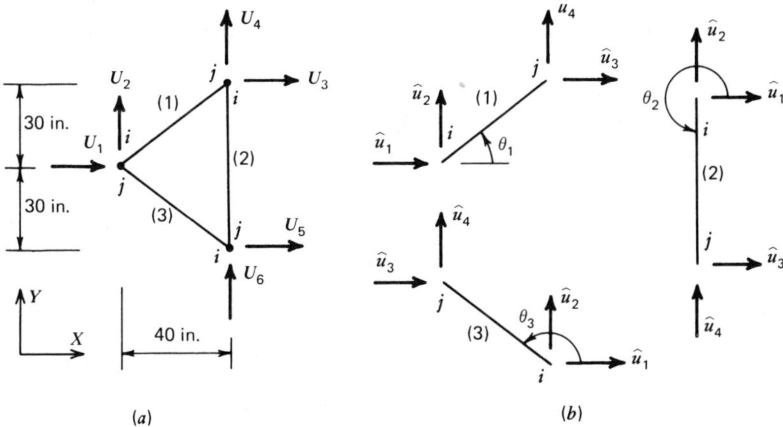

(a) (b)

a. Determine the element stiffness matrices $\hat{\mathbf{k}}_i$.
Since the elements are plane truss elements, Eq. 16.36 can be used for $\hat{\mathbf{k}}_i$.

$$\hat{\mathbf{k}}_i = \left(\frac{AE}{L}\right)_i \begin{bmatrix} c_i^2 & c_i s_i & -c_i^2 & -c_i s_i \\ & s_i^2 & -c_i s_i & -s_i^2 \\ & & c_i^2 & c_i s_i \\ \text{symm.} & & & s_i^2 \end{bmatrix} \tag{1}$$

From Fig. (a) above,

$$\begin{aligned} \cos\theta_1 &= \tfrac{4}{5}, & \sin\theta_1 &= \tfrac{3}{5} \\ \cos\theta_2 &= 0, & \sin\theta_2 &= -1 \\ \cos\theta_3 &= -\tfrac{4}{5}, & \sin\theta &= \tfrac{3}{5} \end{aligned} \tag{2}$$

and

$$L_1 = L_3 = 50 \text{ in.}, \qquad L_2 = 60 \text{ in.} \tag{3}$$

Thus,

$$\hat{\mathbf{k}}_1 = \left(\frac{AE}{50}\right)\left(\frac{1}{25}\right)\begin{array}{c}\begin{matrix}1 & 2 & 3 & 4\end{matrix}\\ \begin{bmatrix} 16 & 12 & -16 & -12 \\ 12 & 9 & -12 & -9 \\ -16 & -12 & 16 & 12 \\ -12 & -9 & 12 & 9 \end{bmatrix}\end{array}\begin{matrix}1\\2\\3\\4\end{matrix}$$

$$\hat{\mathbf{k}}_2 = \left(\frac{AE}{60}\right)\begin{array}{c}\begin{matrix}3 & 4 & 5 & 6\end{matrix}\\ \begin{bmatrix} 0 & 0 & 0 & 0 \\ 0 & 1 & 0 & -1 \\ 0 & 0 & 0 & 0 \\ 0 & -1 & 0 & 1 \end{bmatrix}\end{array}\begin{matrix}3\\4\\5\\6\end{matrix} \tag{4}$$

$$\hat{\mathbf{k}}_3 = \left(\frac{AE}{50}\right)\left(\frac{1}{25}\right)\begin{array}{c}\begin{matrix}5 & 6 & 1 & 2\end{matrix}\\ \begin{bmatrix} 16 & -12 & -16 & 12 \\ -12 & 9 & 12 & -9 \\ -16 & 12 & 16 & -12 \\ 12 & -9 & -12 & 9 \end{bmatrix}\end{array}\begin{matrix}5\\6\\1\\2\end{matrix}$$

b. Using the locator information adjacent to the rows and columns of the $\hat{\mathbf{k}}_i$ matrices above, assemble $\mathbf{K}$ by summing the elements of the $\hat{\mathbf{k}}_i$ matrices into the appropriate rows and columns of $\mathbf{K}$. Equation 16-44 is the basic equation for this direct stiffness assembly procedure.

$$\mathbf{K} = \left(\frac{AE}{25}\right)\begin{array}{c}\begin{matrix}1 & 2 & 3 & 4 & 5 & 6\end{matrix}\\ \begin{bmatrix} \left(\frac{16}{50}+\frac{16}{50}\right) & \left(\frac{12}{50}-\frac{12}{50}\right) & \frac{-16}{50} & \frac{-12}{50} & \frac{-16}{50} & \frac{12}{50} \\ \left(\frac{12}{50}-\frac{12}{50}\right) & \left(\frac{9}{50}+\frac{9}{50}\right) & \frac{-12}{50} & \frac{-9}{50} & \frac{12}{50} & \frac{-9}{50} \\ \frac{-16}{50} & \frac{-12}{50} & \frac{16}{50} & \frac{12}{50} & 0 & 0 \\ \frac{-12}{50} & \frac{-9}{50} & \frac{12}{50} & \left(\frac{9}{50}+\frac{25}{60}\right) & 0 & \frac{-25}{50} \\ \frac{-16}{50} & \frac{12}{50} & 0 & 0 & \frac{16}{50} & \frac{-12}{50} \\ \frac{12}{50} & \frac{-9}{50} & 0 & \frac{-25}{60} & \frac{-12}{50} & \left(\frac{9}{50}+\frac{25}{60}\right) \end{bmatrix}\end{array}\begin{matrix}1\\2\\3\\4\\5\\6\end{matrix} \tag{5}$$

The virtual work done by the loads can be used to obtain the system load vector $\mathbf{P}$ corresponding to the system displacements $\mathbf{U}$. Thus,

$$\delta W = \sum_{e=1}^{N_e} \delta\hat{\mathbf{u}}_e^T\hat{\mathbf{p}}_e = \delta\mathbf{U}^T\mathbf{P} \tag{16.49}$$

Combining Eq. 16.37 with Eq. 16.49, we get

$$\sum_{e=1}^{N_e} \delta\hat{\mathbf{u}}_e\hat{\mathbf{p}}_e = \delta\mathbf{U}^T\left(\sum_{e=1}^{N_e} \mathbf{L}_e^T\hat{\mathbf{p}}_e\right) \tag{16.50}$$

so, since the virtual displacements δU_i are independent,

$$\mathbf{P} = \sum_{e=1}^{N_e} \mathbf{L}_e^T\hat{\mathbf{p}}_e \tag{16.51}$$

Just as it was unnecessary to carry out the pre- and postmultiplications in Eq. 16.42, it is unnecessary to carry out the multiplications in Eq. 16.51. To assemble the system load vector **P** it is merely necessary to use the "locator" information to assign the components of each element load vector $\hat{\mathbf{p}}_e$ to the proper row of the system load vector **P**.

Example 16.4

A beam is divided into two elements. The transverse load on each element is a function of time only. Determine the system load vector, **P**.

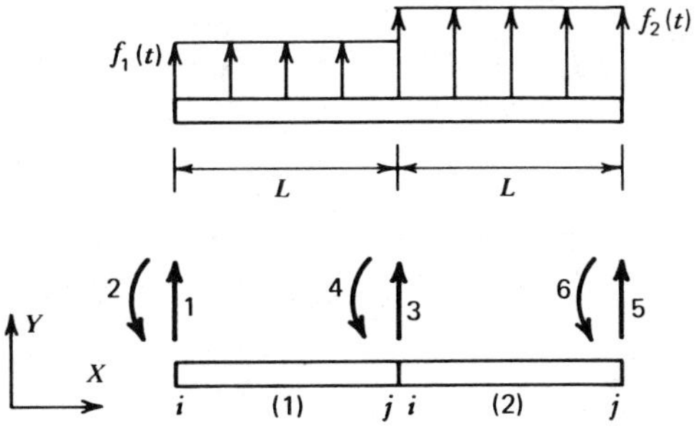

Solution

Since the elements are aligned with the global X-axis, $\hat{\mathbf{p}}_e = \mathbf{p}_e$. The generalized element load vector for a uniform load distribution on a beam element is given in Example 16.1. Thus,

$$\hat{\mathbf{p}}_e = \mathbf{p}_e = f_e L_e \begin{Bmatrix} \frac{1}{2} \\ \frac{L_e}{12} \\ \frac{1}{2} \\ \frac{-L_e}{12} \end{Bmatrix} \tag{1}$$

So,

$$\hat{\mathbf{p}}_1 = f_1 L \begin{Bmatrix} \frac{1}{2} \\ \frac{L}{12} \\ \frac{1}{2} \\ \frac{-L}{12} \end{Bmatrix} \begin{matrix} 1 \\ 2 \\ 3 \\ 4 \end{matrix}, \quad p_2 = f_2 L \begin{Bmatrix} \frac{1}{2} \\ \frac{L}{12} \\ \frac{1}{2} \\ \frac{-L}{12} \end{Bmatrix} \begin{matrix} 3 \\ 4 \\ 5 \\ 6 \end{matrix} \tag{2}$$

The "locator" information is shown alongside the element load vectors. This gives the system degree-of-freedom corresponding to the particular element degree-of-freedom. This information is used in adding the element contributions into the system load vector according to Eq. 16.51.

$$\mathbf{P}(t) = \begin{Bmatrix} \frac{f_1 L}{2} \\ \frac{f_1 L^2}{12} \\ \left(\frac{f_1 L}{2}\right) + \left(\frac{f_2 L}{2}\right) \\ -\left(\frac{f_1 L^2}{12}\right) + \left(\frac{f_2 L^2}{12}\right) \\ \frac{f_2 L}{2} \\ \frac{-f_2 L^2}{12} \end{Bmatrix} \begin{matrix} 1 \\ 2 \\ 3 \\ 4 \\ 5 \\ 6 \end{matrix} \tag{3}$$

16.5 Boundary Conditions

For dynamics problems stated in matrix form it is necessary to incorporate boundary conditions of two types: prescribed forces and prescribed displacements. Equation 16.51 provides for assembly of the system force vector from element force vectors. To this could be added any concentrated loads applied at the joints. We have previously considered (Sec. 4.4) structures having a base motion which imposes time-dependent displacements at some of the structural joints. In this section we will examine a procedure for enforcing boundary constraints on a finite element model of a structure.

In Sec. 16.4 the system matrices **K** and **M** were assembled as though all joints of the structure were unrestrained (e.g., see Fig. 16.14). This leads to a singular stiffness matrix **K**, and the system has rigid-body freedom. When a number of the joint displacements are prescribed to be zero, the most straightforward procedure for enforcing this condition is to partition the system into *a*ctive degrees-of-freedom and *c*onstrained degrees-of-freedom. Thus, for an undamped system the partitioned equation of motion is

$$\begin{bmatrix} \mathbf{M}_{aa} & \mathbf{M}_{ac} \\ \mathbf{M}_{ca} & \mathbf{M}_{cc} \end{bmatrix} \begin{Bmatrix} \ddot{\mathbf{U}}_a \\ \ddot{\mathbf{U}}_c \end{Bmatrix} + \begin{bmatrix} \mathbf{K}_{aa} & \mathbf{K}_{ac} \\ \mathbf{K}_{ca} & \mathbf{K}_{cc} \end{bmatrix} \begin{Bmatrix} \mathbf{U}_a \\ \mathbf{U}_c \end{Bmatrix} = \begin{Bmatrix} \mathbf{P}_a \\ \mathbf{P}_c \end{Bmatrix} \tag{16.52}$$

But, if $\mathbf{U}_c = \mathbf{0}$ due to constraint of the *c*-coordinates, Eq. 16.52 can be written as

$$\mathbf{M}_{aa}\ddot{\mathbf{U}}_a + \mathbf{K}_{aa}\mathbf{U}_a = \mathbf{P}_a \tag{16.53a}$$

$$\mathbf{P}_c = \mathbf{M}_{ca}\ddot{\mathbf{U}}_a + \mathbf{K}_{ca}\,\mathbf{U}_a \tag{16.53b}$$

Equation 16.53a must first be solved for the active displacement vector $\mathbf{U}_a(t)$. Then, reactions at the constraints could be obtained from Eq. 15.63b. However, since only $\mathbf{M}_{aa}$ and $\mathbf{K}_{aa}$ are required in the solution for the active displacement vector $\mathbf{U}_a$, the remaining portions of the **K** and **M** matrices need not be assembled from the corresponding element matrices. In this case, the reactions may be obtained from element forces after the latter have been calculated, rather than by use of Eq. 16.53b.

Example 16.5

A uniform propped cantilever beam is divided into two elements of length $L_1 = L_2 = L/2$. Enforce the system boundary conditions by assembling only the active matrices $\mathbf{M}_{aa}$ and $\mathbf{K}_{aa}$.

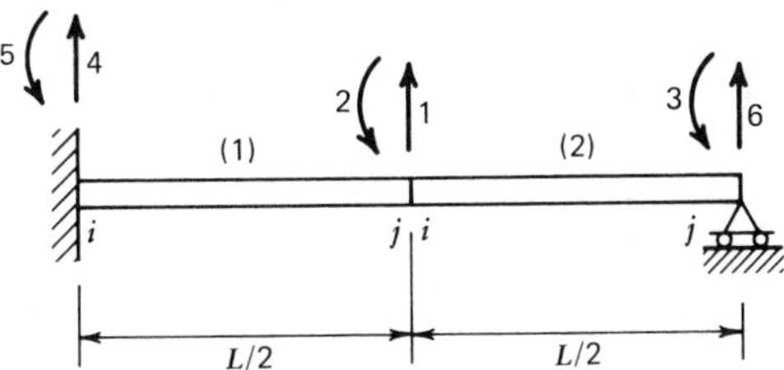

Solution

The system displacement coordinates are numbered such that the active coordinates are 1, 2, and 3, and the constraints are 4, 5, and 6. The element matrices are given by Eqs. 16.15.

$$\hat{\mathbf{k}}_e = \frac{EI}{L_e^3}\begin{bmatrix} 12 & 6L_e & -12 & 6L_e \\ 6L_e & 4L_e^2 & -6L_e & 2L_e^2 \\ -12 & -6L_e & 12 & -6L_e \\ 6L_e & 2L_e^2 & -6L_e & 4L_e^2 \end{bmatrix} \begin{matrix} 1 & 4 \\ 2 & 5 \\ 6 & 1 \\ 3 & 2 \end{matrix} \tag{1}$$

(column locators: 4, 5, 1, 2 / 1, 2, 6, 3)

$$\hat{\mathbf{m}}_e = \frac{\rho A L_e}{420}\begin{bmatrix} 156 & 22L_e & 54 & -13L_e \\ 22L_e & 4L_e^2 & 13L_e & -3L_e^2 \\ 54 & 13L_e & 156 & -22L_e \\ -13L_e & -3L_e^2 & -22L_e & 4L_e^2 \end{bmatrix} \begin{matrix} 1 & 4 \\ 2 & 5 \\ 6 & 1 \\ 3 & 2 \end{matrix} \tag{2}$$

(column locators: 4, 5, 1, 2 / 1, 2, 6, 3)

The element locator information is given alongside rows and columns of $\hat{\mathbf{k}}_e$ and $\hat{\mathbf{m}}_e$. This information enables $\mathbf{K}_{aa}$ and $\mathbf{M}_{aa}$ to be assembled.

$$\mathbf{K}_{aa} = \frac{EI}{(L/2)^3}\begin{bmatrix} (12+12) & (-3L+3L) & 3L \\ (-3L+3L) & (L^2+L^2) & \left(\frac{L^2}{2}\right) \\ 3L & \left(\frac{L^2}{2}\right) & L^2 \end{bmatrix} \begin{matrix} 1 \\ 2 \\ 3 \end{matrix} \tag{3}$$

(column locators: 1, 2, 3)

or

$$\boxed{\mathbf{K}_{aa} = \frac{4EI}{L^3}\begin{bmatrix} 48 & 0 & 6L \\ 0 & 4L^2 & L^2 \\ 6L & L^2 & 2L^2 \end{bmatrix}} \tag{4}$$

$$\mathbf{M}_{aa} = \frac{\rho A L}{840}\begin{bmatrix} (156+156) & (-11L+11L) & \left(\frac{-13L}{2}\right) \\ (-11L+11L) & (L^2+L^2) & \left(\frac{-3L^2}{4}\right) \\ \left(\frac{-13L}{2}\right) & \left(\frac{-3L^2}{4}\right) & L^2 \end{bmatrix} \begin{matrix} 1 \\ 2 \\ 3 \end{matrix} \tag{5}$$

(column locators: 1, 2, 3)

or

$$\boxed{\mathbf{M}_{aa} = \frac{\rho A L}{3360}\begin{bmatrix} 1248 & 0 & -26L \\ 0 & 8L^2 & -3L^2 \\ -26L & -3L^2 & 4L^2 \end{bmatrix}} \tag{6}$$

16.6 Constraints; Reduction of Degrees-of-Freedom

In the previous section we treated constraints imposed as boundary conditions on a structure. Frequently there arises a need for specifying relationships among system displacement coordinates. This transformation of coordinates may be written as

$$\mathbf{U} = \mathbf{T}\hat{\mathbf{U}} \tag{16.54}$$

where $\hat{N} \leq N$ and $\hat{\mathbf{U}}$ is a vector of generalized coordinates. Equation 16.54 may be referred to as a Ritz transformation; it is the finite-DOF analogue of Eq. 11.8.

Using energy equivalence, as we have done on many previous occasions, we obtain the equation of motion

$$\hat{\mathbf{M}}\ddot{\hat{\mathbf{U}}} + \hat{\mathbf{C}}\dot{\hat{\mathbf{U}}} + \hat{\mathbf{K}}\hat{\mathbf{U}} = \hat{\mathbf{P}} \tag{16.55}$$

where

$$\begin{aligned} \hat{\mathbf{M}} &= \mathbf{T}^T\mathbf{M}\mathbf{T}, \qquad \hat{\mathbf{C}} = \mathbf{T}^T\mathbf{C}\mathbf{T} \\ \hat{\mathbf{K}} &= \mathbf{T}^T\mathbf{K}\mathbf{T}, \qquad \text{and} \quad \hat{\mathbf{P}} = \mathbf{T}^T\mathbf{P} \end{aligned} \tag{16.56}$$

The transformation matrix $\mathbf{T}$ may arise as a result of the need for specifying a relationship among several system displacement coordinates. The equation of constraint may be written in matrix form as

$$\mathbf{R}\mathbf{U} \equiv [\mathbf{R}_{da}\ \mathbf{R}_{dd}] \begin{Bmatrix} \mathbf{U}_a \\ \mathbf{U}_d \end{Bmatrix} = \mathbf{0} \tag{16.57}$$

where $\mathbf{U}_d$ is the vector of N_d *d*ependent coordinates and $\mathbf{U}_a$ is the vector of independent, or *a*ctive, coordinates. Equation 16.57 may be solved for $\mathbf{U}_d$ giving

$$\mathbf{U}_d = -\mathbf{R}_{dd}^{-1}\mathbf{R}_{da}\mathbf{U}_a \tag{16.58}$$

Then, the original set of coordinates $\mathbf{U}$ can be related to the active coordinates by the equation

$$\mathbf{U} \equiv \begin{Bmatrix} \mathbf{U}_a \\ \mathbf{U}_d \end{Bmatrix} = \begin{bmatrix} \mathbf{I}_{aa} \\ \mathbf{T}_{da} \end{bmatrix} \mathbf{U}_a \equiv \mathbf{T}\mathbf{U}_a \tag{16.59}$$

where the active coordinates form $\hat{\mathbf{U}}$ and where

$$\mathbf{T}_{da} = -\mathbf{R}_{dd}^{-1}\mathbf{R}_{da} \tag{16.60}$$

A constraint equation such as Eq. 16.57 may arise from a situation where a rigid body is connected to an elastic structure. Example 16.6 treats this problem. Weaver[16.9] has discussed further aspects of the dynamics of elastically-connected rigid bodies.

Example 16.6

A rigid flat plate is connected to two joints of a planar truss as shown below. Write a constraint equation treating the translation and rotation coordinates of the plate as the active coordinates. Assume small rotations.

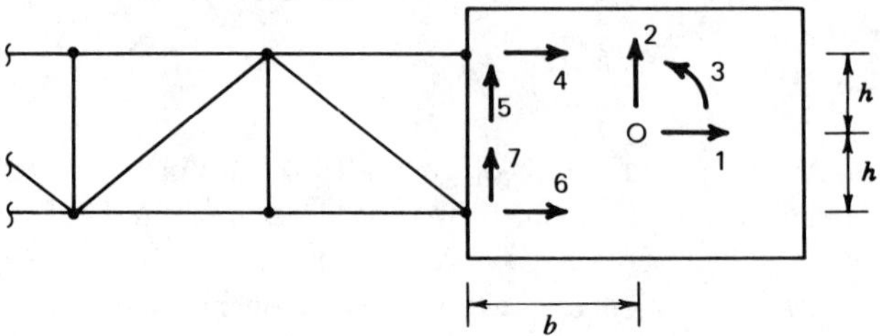

Solution

It is easiest to take each of the active displacements in turn and determine the resulting displacement of the dependent coordinates, $U_4, \ldots, U_7$.

For $U_1 \neq 0$:

$$U_4 = U_6 = U_1, \qquad U_5 = U_7 = 0 \tag{1}$$

For $U_2 \neq 0$:

$$U_5 = U_7 = U_2, \qquad U_4 = U_6 = 0 \tag{2}$$

For $U_3 \neq 0$:

$$\begin{aligned} U_4 &= -hU_3, & U_5 &= -bU_3 \\ U_6 &= hU_3, & U_7 &= -bU_3 \end{aligned} \tag{3}$$

Combining Eqs. 1 through 3, we can write the following constraint equation, which relates the dependent coordinates U_4, U_5, U_6, and U_7 to the active coordinates U_1, U_2, and U_3:

$$\left[\begin{array}{ccc|cccc} -1 & 0 & h & 1 & 0 & 0 & 0 \\ 0 & -1 & b & 0 & 1 & 0 & 0 \\ -1 & 0 & -h & 0 & 0 & 1 & 0 \\ 0 & -1 & b & 0 & 0 & 0 & 1 \end{array}\right] \begin{Bmatrix} U_1 \\ U_2 \\ U_3 \\ \hline U_4 \\ U_5 \\ U_6 \\ U_7 \end{Bmatrix} = \begin{Bmatrix} 0 \\ 0 \\ 0 \\ 0 \end{Bmatrix} \tag{4}$$

where the partitioning follows Eq. 16.57.

Static condensation is another example of the application of constraint equations to reduce out some coordinates. Consider an undamped system whose inertia is represented by a lumped mass matrix, as described in Sec. 16.2. The equation of motion may be written in partitioned-matrix form.

$$\begin{bmatrix} \mathbf{M}_{aa} & \mathbf{0} \\ \mathbf{0} & \mathbf{0} \end{bmatrix} \begin{Bmatrix} \ddot{\mathbf{U}}_a \\ \ddot{\mathbf{U}}_d \end{Bmatrix} + \begin{bmatrix} \mathbf{K}_{aa} & \mathbf{K}_{ad} \\ \mathbf{K}_{da} & \mathbf{K}_{dd} \end{bmatrix} \begin{Bmatrix} \mathbf{U}_a \\ \mathbf{U}_d \end{Bmatrix} = \begin{Bmatrix} \mathbf{P}_a \\ \mathbf{0} \end{Bmatrix} \tag{16.61}$$

The lower partition provides a static constraint equation

$$\mathbf{K}_{da}\mathbf{U}_a + \mathbf{K}_{dd}\mathbf{U}_d = \mathbf{0} \tag{16.62}$$

which is of the same form as Eq. 16.57. Then the transformation matrix is

$$\mathbf{T} \equiv \begin{bmatrix} \mathbf{I}_{aa} \\ \mathbf{T}_{da} \end{bmatrix} = \begin{bmatrix} \mathbf{I}_{aa} \\ -\mathbf{K}_{dd}^{-1}\mathbf{K}_{da} \end{bmatrix} \tag{16.63}$$

The transformation matrix of Eq. 16.63 can be used in conjunction with Eqs. 16.56 and 16.61 to produce the equation of motion for the active coordinates $\mathbf{U}_a$. The reduced mass and stiffness matrices are determined by

$$\hat{\mathbf{M}}_{aa} = \mathbf{T}^T\mathbf{M}\mathbf{T} = \begin{bmatrix} \mathbf{I}_{aa} & \mathbf{T}_{da}^T \end{bmatrix} \begin{bmatrix} \mathbf{M}_{aa} & \mathbf{0} \\ \mathbf{0} & \mathbf{0} \end{bmatrix} \begin{bmatrix} \mathbf{I}_{aa} \\ \mathbf{T}_{da} \end{bmatrix} = \mathbf{M}_{aa} \tag{16.64a}$$

$$\hat{\mathbf{K}}_{aa} = \mathbf{T}^T\mathbf{K}\mathbf{T} = \begin{bmatrix} \mathbf{I}_{aa} & \mathbf{T}_{da}^T \end{bmatrix} \begin{bmatrix} \mathbf{K}_{aa} & \mathbf{K}_{ad} \\ \mathbf{K}_{da} & \mathbf{K}_{dd} \end{bmatrix} \begin{bmatrix} \mathbf{I}_{aa} \\ \mathbf{T}_{da} \end{bmatrix} \tag{16.64b}$$

$$= \mathbf{K}_{aa} - \mathbf{K}_{da}^T\mathbf{K}_{dd}^{-1}\mathbf{K}_{da}$$

Thus, for a lumped mass model the problem size may be reduced by using only coordinates having associated inertia. The mass matrix is unchanged, while the stiffness matrix becomes the *reduced stiffness matrix* $\hat{\mathbf{K}}_{aa}$ defined by Eq. 16.64b.

Example 16.7

Create a 2-DOF lumped mass model of a uniform cantilever beam of length L.

Solution

One possible model is one having two elements of length $L/2$.

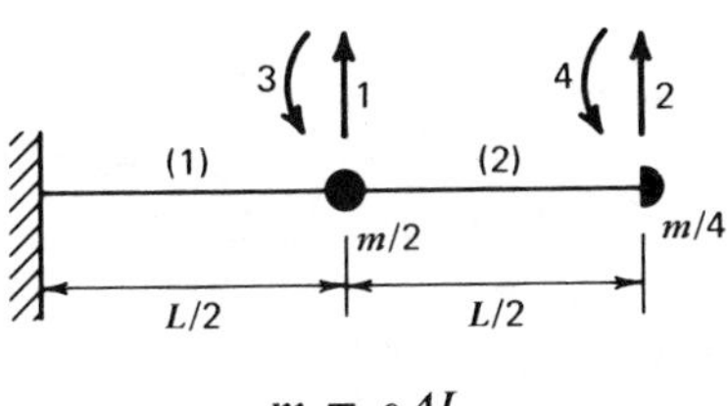

$m = \rho AL$

The element stiffness matrix is, from Eq. 16.15a,

$$\hat{\mathbf{k}}_e = \mathbf{k}_e = \frac{8EI}{L^3}\begin{bmatrix} 12 & 3L & -12 & 3L \\ 3L & L^2 & -3L & \dfrac{L^2}{2} \\ -12 & -3L & 12 & -3L \\ 3L & \dfrac{L^2}{2} & -3L & L^2 \end{bmatrix} \begin{matrix} \text{x} & 1 \\ \text{x} & 3 \\ 1 & 2 \\ 3 & 4 \end{matrix} \tag{1}$$

(column labels: 1, 3, 2, 4 over x, x, 1, 3)

The system equations of motion are assembled giving

$$\frac{\rho AL}{4}\left[\begin{array}{cc|cc} 2 & 0 & 0 & 0 \\ 0 & 1 & 0 & 0 \\ \hline 0 & 0 & 0 & 0 \\ 0 & 0 & 0 & 0 \end{array}\right]\left\{\begin{array}{c} \ddot{U}_1 \\ \ddot{U}_2 \\ \hline \ddot{U}_3 \\ \ddot{U}_4 \end{array}\right\} + \frac{8EI}{L^3}\left[\begin{array}{cc|cc} 24 & -12 & 0 & 3L \\ -12 & 12 & -3L & -3L \\ \hline 0 & -3L & 2L^2 & \dfrac{L^2}{2} \\ 3L & -3L & \dfrac{L^2}{2} & L^2 \end{array}\right]\left\{\begin{array}{c} U_1 \\ U_2 \\ \hline U_3 \\ U_4 \end{array}\right\} = \left\{\begin{array}{c} 0 \\ 0 \\ \hline 0 \\ 0 \end{array}\right\} \tag{2}$$

which is in the form of Eq. 16.61. Then

$$\mathbf{K}_{da} = \frac{8EI}{L^3}\begin{bmatrix} 0 & -3L \\ 3L & -3L \end{bmatrix}, \qquad \mathbf{K}_{dd} = \frac{8EI}{L}\begin{bmatrix} 2 & \frac{1}{2} \\ \frac{1}{2} & 1 \end{bmatrix}$$

$$\begin{aligned} \mathbf{K}_{dd}^{-1}\mathbf{K}_{da} &= \frac{L}{14EI}\begin{bmatrix} 1 & -\frac{1}{2} \\ -\frac{1}{2} & 2 \end{bmatrix}\left(\frac{8EI}{L^3}\right)\begin{bmatrix} 0 & -3L \\ 3L & -3L \end{bmatrix} \\ &= \frac{-6}{7L}\begin{bmatrix} 1 & 1 \\ -4 & 3 \end{bmatrix} \end{aligned} \tag{3}$$

From Eq. 16.64b,

$$\begin{aligned} \hat{\mathbf{K}}_{aa} &= \mathbf{K}_{aa} - \mathbf{K}_{da}^T\,\mathbf{K}_{dd}^{-1}\,\mathbf{K}_{da} \\ &= \frac{8EI}{L^3}\begin{bmatrix} 24 & -12 \\ -12 & 12 \end{bmatrix} - \frac{8EI}{L^3}\begin{bmatrix} 0 & 3L \\ -3L & -3L \end{bmatrix}\left(-\frac{6}{7L}\right)\begin{bmatrix} 1 & 1 \\ -4 & 3 \end{bmatrix} \end{aligned} \tag{4}$$

$$\hat{\mathbf{K}}_{aa} = \frac{8EI}{L^3}\begin{bmatrix} \left(24 - \dfrac{72}{7}\right) & \left(-12 + \dfrac{54}{7}\right) \\ \left(-12 + \dfrac{54}{7}\right) & \left(12 - \dfrac{72}{7}\right) \end{bmatrix}$$

or

$$\hat{\mathbf{K}}_{aa} = \frac{8EI}{7L^3}\begin{bmatrix} 96 & -30 \\ -30 & 12 \end{bmatrix} \tag{5}$$

From Eq. 16.64a,

$$\hat{\mathbf{M}}_{aa} = \mathbf{M}_{aa} = \frac{\rho AL}{4}\begin{bmatrix} 2 & 0 \\ 0 & 1 \end{bmatrix} \tag{6}$$

Thus, the system equation of motion in reduced coordinates is

$$\boxed{\frac{\rho AL}{4}\begin{bmatrix} 2 & 0 \\ 0 & 1 \end{bmatrix}\begin{Bmatrix} \ddot{U}_1 \\ \ddot{U}_2 \end{Bmatrix} + \frac{48EI}{7L^3}\begin{bmatrix} 16 & -5 \\ -5 & 2 \end{bmatrix}\begin{Bmatrix} U_1 \\ U_2 \end{Bmatrix} = \begin{Bmatrix} 0 \\ 0 \end{Bmatrix}} \tag{7}$$

A technique that has frequently been employed to reduce the number of degrees-of-freedom of a system is referred to as the *Guyan reduction method*[16.10]. Even though inertia is associated with all coordinates, a subset of the coordinates is arbitrarily selected as the set of active (or master) coordinates and the remaining coordinates are dependent (or slave) coordinates. Then, the Ritz transformation matrix of Eq. 16.54 is taken to be the same as that obtained for a lumped mass system in Eq. 16.63. The Ritz basis vectors, which are the columns of the Ritz transformation matrix **T**, are the displacement patterns associated with unit displacement of the respective *a*-coordinates while the *d*-coordinates are released. Anderson, Irons, and Zienkiewicz[16.11] discuss the application of this reduction procedure to vibration and buckling analyses and show the effect of the choice of coordinates and number of coordinates retained on the accuracy of the eigenvalues and eigenvectors obtained. References 16.12 and 16.13 discuss the use of substructuring as a procedure for selecting the degrees of freedom to be retained and those to be reduced out and discuss a "quadratic eigensolution" procedure for improving the frequencies and modes obtained by using Guyan reduction. An example Guyan reduction solution will be presented in Sec. 17.1. In Chapter 19 some particular forms of coordinate reduction associated with substructuring will be treated.

16.7 Systems with Rigid-Body Modes

Some algebraic eigensolvers, for example, iteration methods such as those discussed in Sec. 14.2, require the solution of the eigenvalue equation

$$\mathbf{K}^{-1}\mathbf{M}\boldsymbol{\phi} = \left(\frac{1}{\lambda}\right)\boldsymbol{\phi} \tag{16.65}$$

When a structure has rigid-body modes, the inversion of **K** is not possible, since **K** is singular. A procedure for eliminating the rigid-body freedoms can be expressed in the form of a transformation of coordinates that separates *r*igid-body coordinates from *e*lastic deformation coordinates. Let the original **K** matrix be written

$$\mathbf{K} = \begin{bmatrix} \mathbf{K}_{ee} & \mathbf{K}_{er} \\ \mathbf{K}_{re} & \mathbf{K}_{rr} \end{bmatrix} \tag{16.66}$$

A Ritz transformation having the form of Eq. 16.54, namely

$$\begin{Bmatrix} \mathbf{U}_e \\ \hline \mathbf{U}_r \end{Bmatrix} = \left[\begin{array}{c|c} \mathbf{I}_{ee} & (\mathbf{K}_{ee}^{-1}\mathbf{K}_{er}) \\ \hline \mathbf{0} & \mathbf{I}_{rr} \end{array}\right] \begin{Bmatrix} \hat{\mathbf{U}}_e \\ \hline \hat{\mathbf{U}}_r \end{Bmatrix} \tag{16.67}$$

together with Eq. 16.56c leads to the generalized stiffness matrix

$$\hat{\mathbf{K}} = \begin{bmatrix} \mathbf{K}_{ee} & \mathbf{0} \\ \mathbf{0} & \mathbf{0} \end{bmatrix} \tag{16.68}$$

The transformation given by Eq. 16.67 leads to a particularly simple form of $\hat{\mathbf{K}}$, which clearly shows that the coordinates $\mathbf{U}_r$ represent rigid-body displacements. Any linearly independent set of vectors could replace the $\mathbf{I}_{ee}$ matrix in Eq. 16.67, but other choices would not lead to $\hat{\mathbf{K}}_{ee} = \mathbf{K}_{ee}$ as in Eq. 16.68. Reference 16.14 discusses further numerical aspects of the rigid-body problem.

Example 16.8

A uniform bar is modeled by two elements of length $L/2$. Let $k = 2AE/L$, and let $U_r = U_3$.

a. Derive the transformation matrix of Eq. 16.67.

b. Sketch the shape functions which correspond to the vectors that are the Ritz basis vectors of this transformation.

c. Show that $\hat{\mathbf{K}}$ has the form of Eq. 16.68.

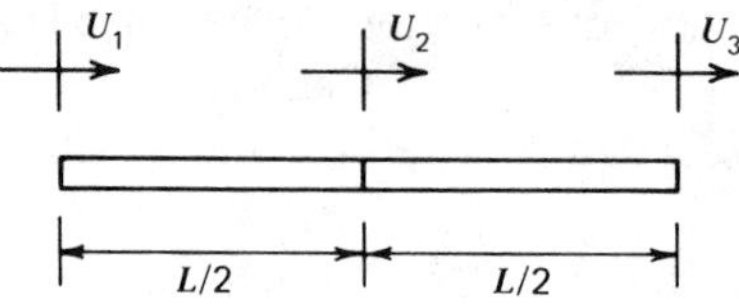

Solution

a. Set up the system stiffness matrix in partitioned form as in Eq. 16.66

$$\mathbf{K} = \begin{bmatrix} \mathbf{K}_{ee} & \mathbf{K}_{er} \\ \mathbf{K}_{re} & \mathbf{K}_{rr} \end{bmatrix} = k \left[\begin{array}{cc|c} 1 & -1 & 0 \\ -1 & 2 & -1 \\ \hline 0 & -1 & 1 \end{array}\right] \tag{1}$$

Therefore

$$\mathbf{K}_{ee}^{-1}\mathbf{K}_{er} = \left(\frac{1}{k}\right)\begin{bmatrix} 2 & 1 \\ 1 & 1 \end{bmatrix}(k)\begin{bmatrix} 0 \\ -1 \end{bmatrix} = \begin{bmatrix} -1 \\ -1 \end{bmatrix} \tag{2}$$

Then, from Eq. 16.67 and Eq. 2,

$$\mathbf{T} = \begin{bmatrix} \mathbf{I}_{ee} & (-\mathbf{K}_{ee}^{-1}\mathbf{K}_{er}) \\ \mathbf{0} & \mathbf{I}_{rr} \end{bmatrix} = \left[\begin{array}{cc|c} 1 & 0 & 1 \\ 0 & 1 & 1 \\ \hline 0 & 0 & 1 \end{array}\right] \tag{3}$$

b. Sketch the shape functions which correspond to the Ritz basis vectors, that is, the columns of **T**

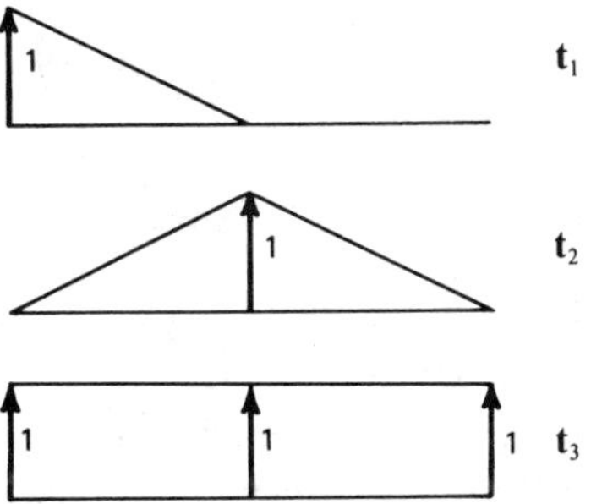

Note that column 3, that is, $\mathbf{t}_3$, corresponds to rigid-body translation.

c. Form $\hat{\mathbf{K}}$ using Eqs. 16.56c and 3

$$\hat{\mathbf{K}} = \mathbf{T}^T\mathbf{K}\mathbf{T} = \begin{bmatrix} 1 & 0 & 0 \\ 0 & 1 & 0 \\ 1 & 1 & 1 \end{bmatrix} (k) \begin{bmatrix} 1 & -1 & 0 \\ -1 & 2 & -1 \\ 0 & -1 & 1 \end{bmatrix} \begin{bmatrix} 1 & 0 & 1 \\ 0 & 1 & 1 \\ 0 & 0 & 1 \end{bmatrix}$$

$$\hat{\mathbf{K}} = k \begin{bmatrix} 1 & -1 & 0 \\ -1 & 2 & 0 \\ 0 & 0 & 0 \end{bmatrix} \tag{4}$$

$\hat{\mathbf{K}}$ does have the form indicated by Eq. 16.68.

References

16.1 Y. C. Fung, *An Introduction to the Theory of Aeroelasticity,* Wiley, New York (1955).

16.2 R. M. Rivello, *Theory and Analysis of Flight Structures,* McGraw-Hill, New York (1969).

16.3 W. C. Hurty and M. F. Rubinstein, *Dynamics of Structures,* Prentice-Hall, Englewood Cliffs, NJ (1964).

16.4 J. S. Przemieniecki, *Theory of Matrix Structural Analysis,* McGraw-Hill, New York (1968).

16.5 J. S. Archer, "Consistent Mass Matrix for Distributed Mass Systems," *J. Struct. Div., Proc. ASCE,* v. 89, 161–178 (1963).

16.6 F. A. Leckie and G. M. Lindberg, "The Effect of Lumped Parameters on Beam Frequencies," *Aeron. Quart.*, v. 14, 224–240 (1963).

16.7 P. Tong, T. H. H. Pian, and L. L. Bucciarelli, "Mode Shapes and Frequencies by Finite Element Method Using Consistent and Lumped Masses," *Comps. Struct.*, v. 1, 623–628 (1971).

16.8 W. McGuire and R. H. Gallagher, *Matrix Structural Analysis,* Wiley, New York (1979).

16.9 W. Weaver, Jr., "Dynamics of Elastically Connected Rigid Bodies," *Proc. 3rd Southeastern Conf. on T&AM,* Columbia, SC, 543–562 (1966).

16.10 R. J. Guyan, "Reduction of Stiffness and Matrices," *AIAA Journal,* v. 3, 380 (1965).

16.11 R. G. Anderson, B. M. Irons, and O. C. Zienkiewicz, "Vibration and Stability of Plates Using Finite Elements," *Int. J. Solids Struct.,* v. 4, 1031–1055 (1968).

16.12 C. P. Johnson et. al., "Quadratic Reduction for the Eigenproblem," Int. J. Num. Meth. in Engr., v. 15, 911–923 (1980).

16.13 C. P. Johnson, "Computational Aspects of a Quadratic Eigenproblem," *Proc. 7th Conference on Electronic Computation,* Washington University, St. Louis, MO, 418–431 (1979).

16.14 R. R. Craig, Jr. and M. C. C. Bampton, "On the Iterative Solution of Semidefinite Eigenvalue Problems," *Aero. J. of the Royal Aero. Soc.,* v. 75, 287–290 (1971).

Problem Set 16.2

16.1 Derive Eq. 16.13 from Eqs. 16.10 and 16.12.

16.2 Using Eqs. 16.13 and 16.14, verify the terms in k_{12} and m_{23} in Eqs. 16.15.

16.3 Stiffness coefficients can be defined in the following manner: "The stiffness coefficient k_{ij} is the force at degree-of-freedom i due to a unit displacement at degree-of-freedom j with all other degrees-of-freedom restrained." Using this definition, verify that k_{12} is the shear force at the left end when the beam is deflected in the shape $\psi_2(x)$.

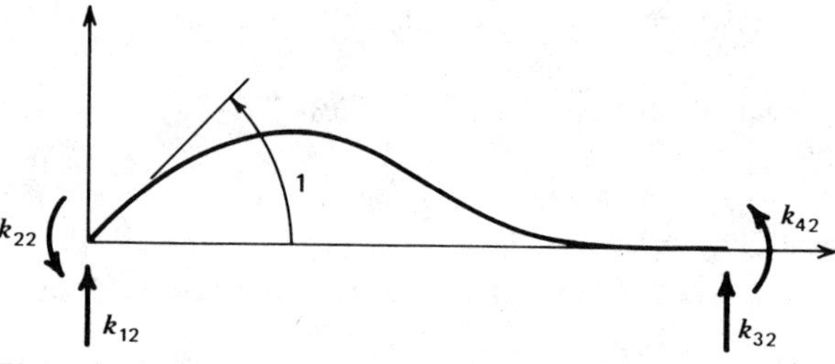

Figure P16.3

16.4 The forces p_i given by Eq. 16.14c are called consistent nodal loads, or fixed-end forces. Show that the force $p_2(t)$ determined in Example 16.1 is equal to the pseudostatic fixed-end moment at the left end of the beam below.

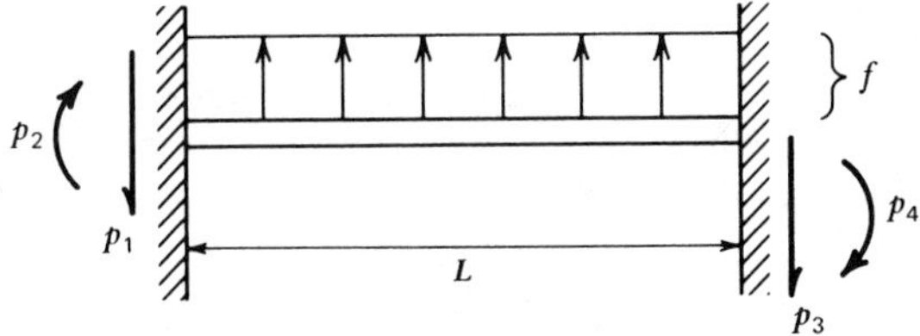

Figure P16.4

16.5 Determine the consistent nodal loads $p_1(t)$ through $p_4(t)$ for the triangular load distribution below.

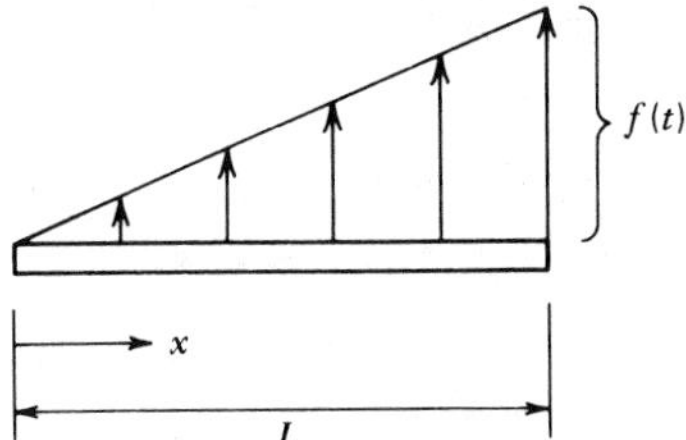

Figure P16.5

16.6 Determine the stiffness matrix **k** for the conical frustum torsion element below. Use Eqs. 16.19a and 16.6.

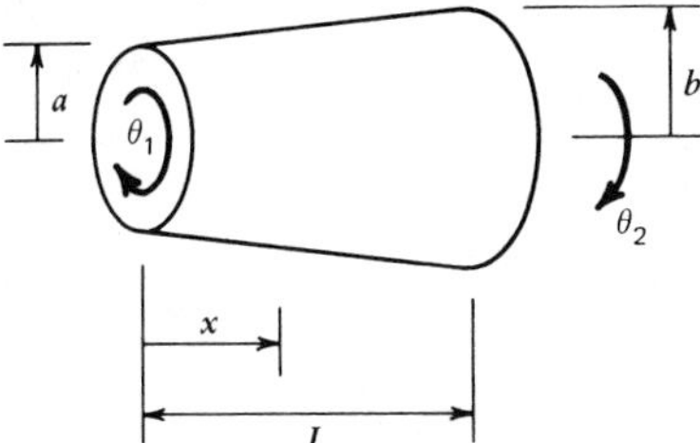

Figure P16.6

Problem Set 16.3

16.7 By performing displacements $\hat{u}_4$ and $\hat{u}_5$ on the plane frame element in Figure 16.12*b*, derive the expression for u_5 in Eq. 16.27.

16.8 By using a clearly drawn sketch, determine the expression for u_1 for the three-dimensional truss element in Fig. 16.13 in terms of the displacements $\hat{u}_1$, $\hat{u}_2$, and $\hat{u}_3$. (The answer should correspond to the first row of Eq. 16.28.)

16.9 Using the figure below determine the forces $\hat{p}_1$, $\hat{p}_2$, $\hat{p}_3$, and $\hat{p}_4$ which correspond to a displacement $\hat{u}_1$ as shown. (Note that the answer corresponds to the first column of Eq. 16.36, since $\hat{\mathbf{p}} = \hat{\mathbf{k}}\hat{\mathbf{u}}$.)

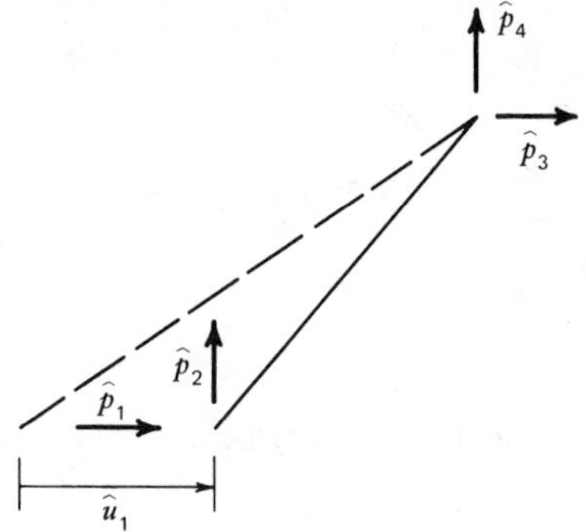

Figure P16.9

16.10 The force transformation in Eq. 16.32 was derived by using virtual work methods. It can also be derived by simple projection of force components. Using the figure below, verify Eq. 16.32 where **T** is given by Eq. 16.26 for the plane truss element.

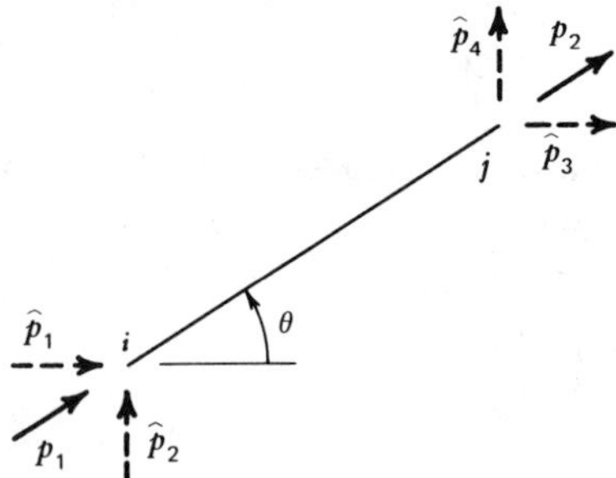

Figure P16.10

Problem Set 16.4

16.11 Each of the bars in the planar truss below has the same cross-sectional properties, AE.

(a) Determine the three element stiffness matrices $\hat{\mathbf{k}}_1$, $\hat{\mathbf{k}}_2$, and $\hat{\mathbf{k}}_3$ from Eq. 16.36.

(b) Assemble the 8×8 unconstrained system stiffness matrix using the direct stiffness method.

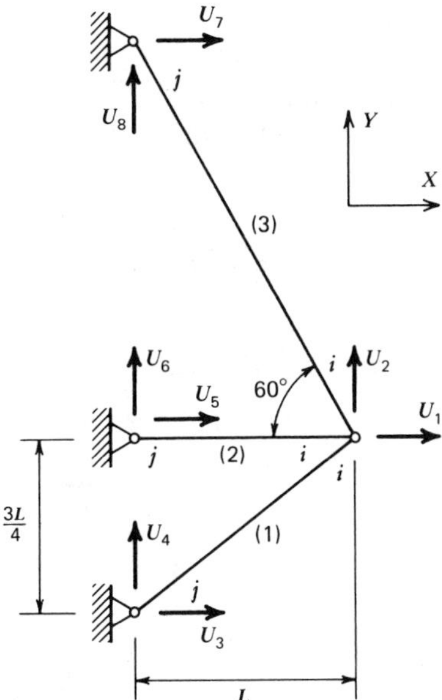

Figure P16.11

16.12 A crane boom consists of a uniform beam supported by a tie bar as shown below.

(a) Obtain the two element stiffness matrices $\hat{\mathbf{k}}_1$ and $\hat{\mathbf{k}}_2$.

(b) Assemble the 8 × 8 unconstrained system stiffness matrix.

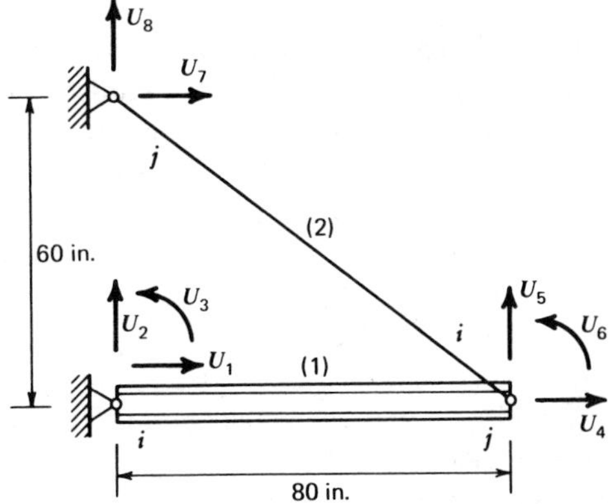

Figure P16.12

$$E = 30 \times 10^3 \text{ ksi}$$
$$I_1 = 10.8 \text{ in.}^4, \qquad A_1 = 4.1 \text{ in.}^2$$
$$A_2 = 0.8 \text{ in.}^2$$

16.13 For the plane frame below:

(a) Determine the element stiffness matrices $\hat{\mathbf{k}}_1 = \hat{\mathbf{k}}_3$ and $\hat{\mathbf{k}}_2$.

(b) Assemble the system equations of motion for the 3-DOF model which would result from neglecting axial deformation in all elements, that is, $U_5 = U_6 = 0$ and $U_4 = U_3$.

$$\begin{bmatrix} M_{11} & M_{12} & M_{13} \\ M_{21} & M_{22} & M_{23} \\ M_{31} & M_{32} & M_{33} \end{bmatrix} \begin{Bmatrix} \ddot{U}_1 \\ \ddot{U}_2 \\ \ddot{U}_3 \end{Bmatrix} + \begin{bmatrix} K_{11} & K_{12} & K_{13} \\ K_{21} & K_{22} & K_{23} \\ K_{31} & K_{32} & K_{33} \end{bmatrix} \begin{Bmatrix} U_1 \\ U_2 \\ U_3 \end{Bmatrix} = \begin{Bmatrix} P_1 \\ P_2 \\ P_3 \end{Bmatrix}$$

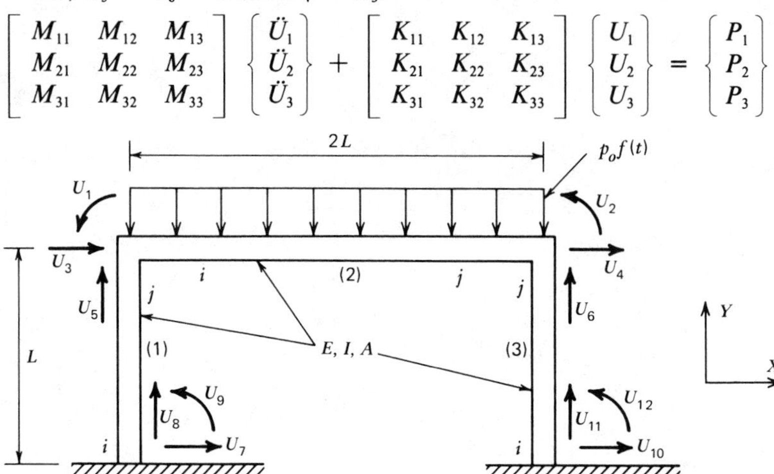

Figure P16.13

Problem Set 16.6

16.14 In Example 16.7 the rotations were reduced out yielding 2 × 2 stiffness and mass matrices. As noted in Eq. 16.64a the reduced mass matrix $\hat{\mathbf{M}}_{aa}$ is just the nonzero portion $\mathbf{M}_{aa}$ of the original mass matrix. For the beam in Fig. P16.14:

(a) Determine the 4 × 4 consistent mass matrix.

(b) Form the transformation matrix, $\mathbf{T}$, which will reduce out the rotations.

(c) Determine the 2 × 2 reduced mass matrix $\hat{\mathbf{M}}_{aa}$ by using the Guyan reduction method.

(d) Use Eq. 7 of Example 16.7 to determine the natural frequencies of the 2-DOF lumped mass model.

(e) Use the mass matrix determined in (c) above and the stiffness matrix $\hat{\mathbf{K}}_{aa}$ from Example 16.7 and determine the natural frequencies of the 2-DOF Guyan reduction model.

(f) Compare the results of (d) and (e) with exact frequencies of a uniform cantilever beam.

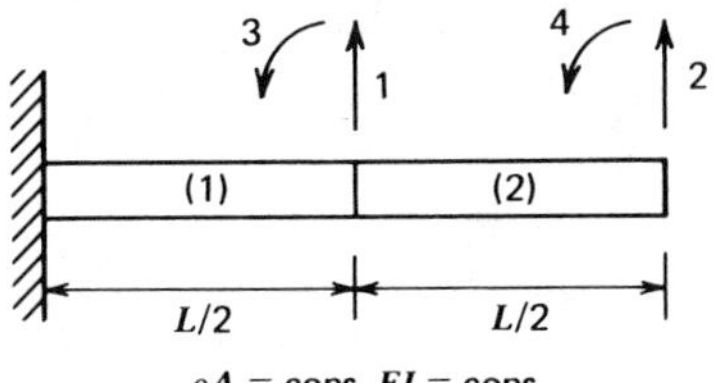

Figure P16.14

16.15 A large sign can be considered to be a uniform rigid plate attached rigidly to the top of a flexible column. Let the total mass matrix referred to the column degrees-of-freedom U_1, U_2, and U_3 be

$$\mathbf{M} = \mathbf{M}_c + \mathbf{M}_p$$

where

$$\mathbf{M}_p = \begin{bmatrix} M & 0 & 0 \\ 0 & M & 0 \\ 0 & 0 & I_G \end{bmatrix}$$

Use Eqs. 16.56, 16.59, and 16.60 to transfer the plate inertia terms from the center of gravity coordinates U_4, U_5, U_6 to the column coordinates U_1, U_2, U_3. Assume small rotations.

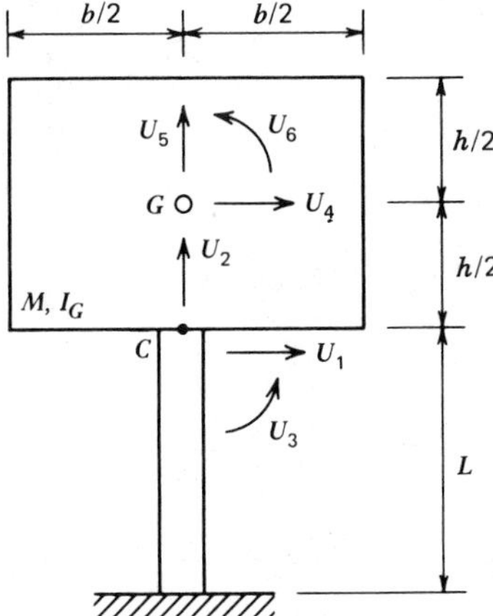

Figure P16.15

Problem Set 16.7

16.16 (a) Determine $\hat{\mathbf{M}}$ for the system in Example 16.8.

(b) Write the equation of motion in matrix form and solve for the modes and frequencies.

16.17 For the single uniform beam element below let U_1 and U_2 be *e*lastic deformation coordinates and let U_3 and U_4 be *r*igid body coordinates.

(a) Form **K** and **M** for this coordinate numbering.

(b) Form the transformation matrix of Eq. 16.67 for this system.

(c) Determine the mass matrix $\hat{\mathbf{M}}$ which corresponds to the $\hat{\mathbf{K}}$ of Eq. 16.68.

(d) Solve for the flexible modes of the system. Sketch the modes.

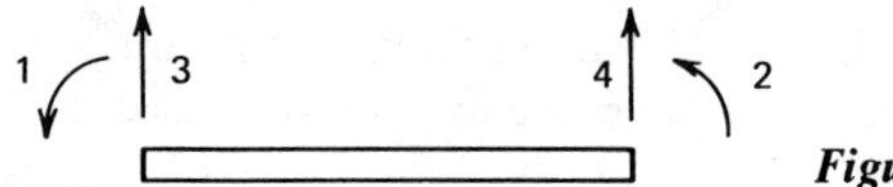

Figure P16.17

17 VIBRATION ANALYSIS EMPLOYING FINITE ELEMENT MODELS

The purpose of the present chapter is to present some solutions for modes and frequencies of simple structural members based on finite element representations of the members, to introduce the use of the ISMIS computer program for solving structural dynamics problems, and to present a complete structural dynamics response solution obtained by using the ISMIS program. Consistent mass and lumped mass finite element models are compared, and the reduction of coordinates is demonstrated.

Upon completion of this chapter you should be able to:

- Use the ISMIS computer program or some comparable finite element program to determine modes and frequencies of structures using consistent mass or lumped mass finite element models.
- Apply the Guyan reduction method to obtain reduced-DOF models.
- Describe the relationship of frequencies of simple structures based on consistent mass models and on lumped mass models with frequencies determined from continuum (i.e., partial differential equations) models.
- Use the ISMIS computer program or a comparable program to obtain the dynamic response of structures to specified excitation and initial conditions.

17.1 Finite Element Solutions for Natural Frequencies and Modes

In Example 11.9 a 2-DOF assumed-modes model of a uniform cantilever bar undergoing axial vibration was generated, and in Example 12.3 the natural frequencies and modes of this model were obtained. (Exercise 17.1 concerns a 2-DOF finite element model of the same cantilever bar undergoing axial vibration.) In the present section we concentrate on the transverse vibration of a uniform cantilever beam and use it to illustrate the following:

a. A "hand-crank" 2-DOF solution based on the finite element model.

b. The effect of increasing the number of degrees-of-freedom in a finite element model based on a consistent mass matrix.

c. The effect of reducing out freedoms by using the Guyan reduction method.

d. The comparison of lumped mass models with consistent mass models.

Consider the following single element, 2-DOF model of a uniform cantilever beam undergoing transverse vibration.

Example 17.1

a. Using the stiffness and mass matrices of a uniform beam as given in Eqs. 16.15, create a 2-DOF math model of a uniform cantilever beam.

b. Solve for the natural frequencies of this model and compare them with the "exact" frequencies given in Example 10.3.

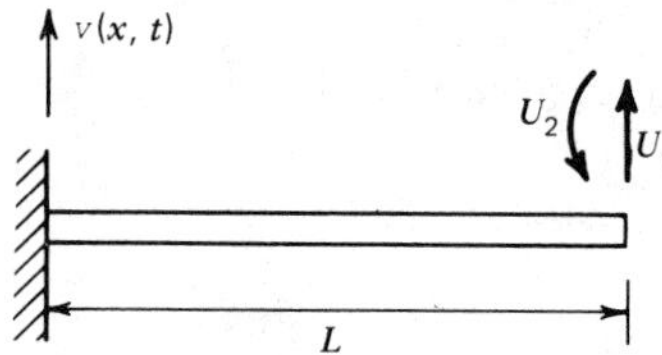

Solution

a. The stiffness and mass matrices in Eqs. 16.15 correspond to the element coordinate numbering

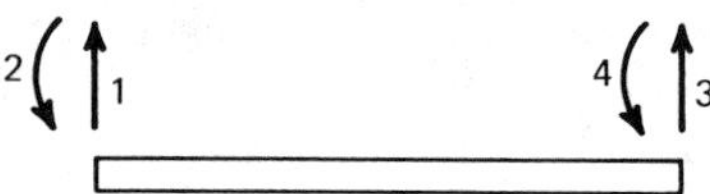

In creating "system" stiffness and mass matrices based on the two system degrees-of-freedom at the free end we can use the following locator information

Element	*System*
1	Constrained
2	Constrained
3	1
4	2

This results in the following system equations.

$$\frac{\rho AL}{420}\begin{bmatrix} 156 & -22L \\ -22L & 4L^2 \end{bmatrix}\begin{Bmatrix} \ddot{U}_1 \\ \ddot{U}_2 \end{Bmatrix} + \left(\frac{EI}{L^3}\right)\begin{bmatrix} 12 & -6L \\ -6L & 4L^2 \end{bmatrix}\begin{Bmatrix} U_1 \\ U_2 \end{Bmatrix} = \begin{Bmatrix} 0 \\ 0 \end{Bmatrix} \tag{1}$$

b. Equation 1 has the form

$$\mathbf{M\ddot{U}} + \mathbf{KU} = \mathbf{0} \tag{2}$$

Assume

$$\mathbf{U}(t) = \boldsymbol{\phi} \cos \omega t \tag{3}$$

and let

$$\mu = \frac{\omega^2 \rho AL^4}{420EI} \tag{4}$$

Then,

$$\left[\begin{bmatrix} 12 & -6L \\ -6L & 4L^2 \end{bmatrix} - \mu\begin{bmatrix} 156 & -22L \\ -22L & 4L^2 \end{bmatrix}\right]\begin{Bmatrix} \phi_1 \\ \phi_2 \end{Bmatrix} = \begin{Bmatrix} 0 \\ 0 \end{Bmatrix} \tag{5}$$

We set the determinant of the coefficients to zero, getting

$$(12 - 156\mu)(4L^2 - 4L^2\mu) - (-6L + 22L\mu)^2 = 0 \tag{6}$$

The roots of Eq. 6 are

$$\mu_1 = 2.97147 \times 10^{-2}$$
$$\mu_2 = 2.88457$$

Then,

$$\omega_1^2 = 12.4802\left(\frac{EI}{\rho AL^4}\right)$$

$$\omega_2^2 = 1211.52\left(\frac{EI}{\rho AL^4}\right)$$

$$\omega_1 = 3.533\left(\frac{EI}{\rho AL^4}\right)^{1/2} \qquad \omega_2 = 34.81\left(\frac{EI}{\rho AL^4}\right)^{1/2} \tag{7}$$

From Example 10.3,

$$\omega_{1_{\text{exact}}} = 3.516\left(\frac{EI}{\rho AL^4}\right)^{1/2} \qquad \omega_{2_{\text{exact}}} = 22.03\left(\frac{EI}{\rho AL^4}\right)^{1/2} \tag{8}$$

From Example 17.1, it can be seen that the single-element, 2-DOF model produces upper bounds to the first and second exact frequencies. The fundamental frequency of the 2-DOF model is quite accurate, while the second frequency is too high to be of any use. These results are in agreement with the information on Rayleigh-Ritz bounds summarized in Table 13.1, where the number of degrees of freedom of the continuum model is infinite, that is, $N = \infty$.

The ISMIS computer program will now be used to solve the 2-DOF problem of Example 17.1

Example 17.2

Use the ISMIS computer program to solve the problem stated in Example 17.1.

Solution

The ISMIS computer program has commands FRAMEL and FRAMMS that generate the 6×6 element stiffness and mass matrices for a plane frame element whose element coordinates are numbered as shown below. To obtain the constant that multiplies the factor $(EI/\rho AL^4)^{1/2}$ in Example 17.1, $E = \rho = I = A = L = 1$ was employed.

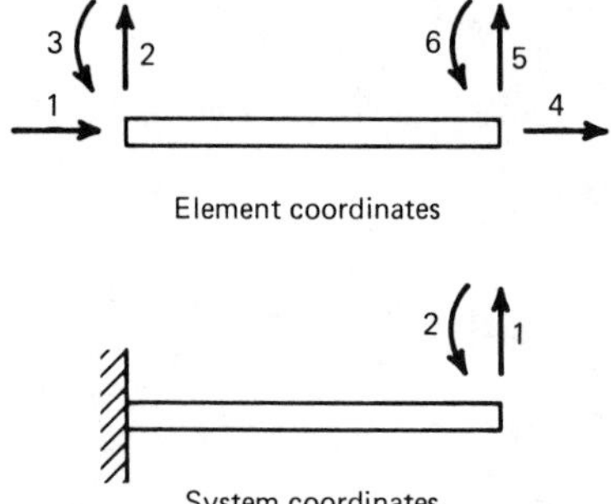

The ISMIS program listing is

```
START
*
**EXAMPLE 17.2**
FRAMEL,KE
1,0,1,1,1
PRINT,KE,1
FRAMMS,ME
1,0,1,1
PRINT,ME,1
```

```
RMVSM,KE,K,5,5,2,2
RMVSM,ME,M,5,5,2,2
EIGEN,K,M,VAL,VECT,2
TRANS,VECT,VEC
*
**OMEGA SQ.**
PRINT,VAL,2
SQREL,VAL
*
**OMEGA**
PRINT,VAL,2
*
**MODES (COLS.)**
PRINT,VEC,1
STOP
```

The resulting ISMIS output is

```
START
**EXAMPLE 17.2**
FRAMEL    KE
     6 ROWS      6 COLUMNS
 DX =  1.000E+00      A =  1.000E+00
 DY =  0.             E =  1.000E+00
  L =  1.000E+00      I =  1.000E+00
PRINT     KE
 SCALED BY    1.0E+01
          1       2       3       4       5       6
  1    .10000 0.00000 0.00000 -.10000 0.00000 0.00000
  2   0.00000 1.20000  .60000 0.00000-1.20000  .60000
  3   0.00000  .60000  .40000 0.00000 -.60000  .20000
  4   -.10000 0.00000 0.00000  .10000 0.00000 0.00000
  5   0.00000-1.20000 -.60000 0.00000 1.20000 -.60000
  6   0.00000  .60000  .20000 0.00000 -.60000  .40000
FRAMMS    ME
     6  ROWS      6 COLUMNS
 DX =  1.000E+00      DY =  0.
 RO =  1.000E+00      AR =  1.000E+00       L =  1.000E+00
PRINT     ME
 SCALED BY    1.0E-01
          1       2       3       4       5       6
  1   3.33333 0.00000 0.00000 1.66667 0.00000 0.00000
  2   0.00000 3.71429  .52381 0.00000 1.28571 -.30952
  3   0.00000  .52381  .09524 0.00000  .30952 -.07143
  4   1.66667 0.00000 0.00000 3.33333 0.00000 0.00000
  5   0.00000 1.28571  .30952 0.00000 3.71429 -.52381
  6   0.00000 -.30952 -.07143 0.00000 -.52381  .09524
RMVSM     KE        K
 ROW NUMBER    5,  COLUMN NUMBER    5
     2 ROWS     2 COLUMNS
RMVSM     ME        M
 ROW NUMBER    5,  COLUMN NUMBER    5
     2 ROWS     2 COLUMNS
EIGEN     K         M          VAL          VECT
   NUMBER OF EIGENVECTORS DESIRED=          2
     ORIGINAL TRACE  = 1.22400000E+03
     SUM OF EIGENVALUES   = 1.22400000E+03
```

```
TRANS      VECT       VEC
**OMEGA SQ.**
PRINT      VAL
 SCALED BY    1.0E+03
              1          2
  1    .01248019 1.21151981
SQREL      VAL
**OMEGA**
PRINT      VAL
 SCALED BY    1.0E+01
              1          2
  1    .35327315 3.48068931
**MODES (COLS.)**
PRINT      VEC
 SCALED BY    1.0E+01
          1        2
  1    .20195   .28145
  2    .27819 2.14537
STOP
```

Note that the frequencies printed under **OMEGA** agree with Eq. 8 of Example 17.1.

To illustrate the relationship between the number of degrees-of-freedom in a consistent mass finite element model and the accuracy of the frequencies obtained using the model, ISMIS programs similar to that in Example 17.2 were run, and the results tabulated in Table 17.1.

Table 17.1 illustrates the Rayleigh-Ritz bounds indicated previously in Table 13.1, that is, the accuracy of the frequencies deteriorates for the higher modes of a particular model, but the accuracy is increased by increasing the number of degrees-of-freedom of the model. Notice that the accuracy is quite good for a number of modes equal to the number of elements, but the frequencies for the remaining modes are poor.

Next, we use the Guyan reduction method described in Sec. 16.6 to reduce out the rotations in the above cantilever beam problem. Let the active coordinates be designated by "a" and the coordinates to be reduced out by "d" as in Sec. 16.6. Then, the partitioned mass and stiffness matrices are given by

$$\mathbf{M} = \begin{bmatrix} \mathbf{M}_{aa} & \mathbf{M}_{ad} \\ \mathbf{M}_{da} & \mathbf{M}_{dd} \end{bmatrix}, \qquad \mathbf{K} = \begin{bmatrix} \mathbf{K}_{aa} & \mathbf{K}_{ad} \\ \mathbf{K}_{da} & \mathbf{K}_{dd} \end{bmatrix} \tag{17.1}$$

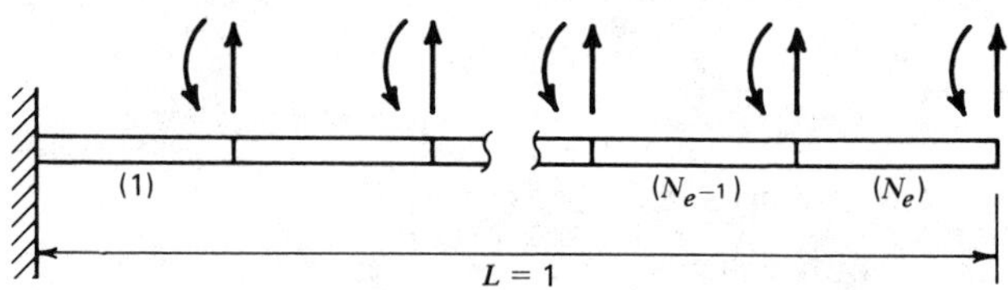

Figure 17.1. Finite element model of uniform cantilever beam.

*Table 17.1. **Comparison of Consistent Mass Finite Element Models of a Uniform Cantilever Beam with Continuum Model***

Mode No. \ N_e	*1*	*2*	*3*	*4*	*5*	*Exact (Reference 17.1)*
1	3.53273	3.51772	3.51637	3.51613	3.51606	3.51602
2	34.8069	22.2215	22.1069	22.0602	22.0455	22.0345
3		75.1571	62.4659	62.1749	61.9188	61.6972
4		218.138	140.671	122.657	122.320	120.902
5			264.743	228.137	203.020	199.860
6			527.796	366.390	337.273	298.556
7				580.849	493.264	416.991
8				953.051	715.341	555.165
9					1016.20	713.079
10					1494.88	890.732

Using the transformation matrix $\mathbf{T}$ defined by Eq. 16.63, we get the reduced stiffness matrix as given by Eq. 16.64b, repeated here

$$\hat{\mathbf{K}}_{aa} = \mathbf{K}_{aa} + \mathbf{K}_{ad}\mathbf{T}_{da} \tag{17.2}$$

where

$$\mathbf{T}_{da} = -\mathbf{K}_{dd}^{-1}\mathbf{K}_{da} \tag{17.3}$$

Since the consistent mass matrix has nonzero terms associated with the d-coordinates, we must use Eqs. 16.56a and 16.63 to compute $\hat{\mathbf{M}}_{aa}$. Thus,

$$\hat{\mathbf{M}}_{aa} = [\mathbf{I}_{aa}\ \mathbf{T}_{da}^T] \begin{bmatrix} \mathbf{M}_{aa} & \mathbf{M}_{ad} \\ \mathbf{M}_{da} & \mathbf{M}_{dd} \end{bmatrix} \begin{bmatrix} \mathbf{I}_{aa} \\ \mathbf{T}_{da} \end{bmatrix} \tag{17.4}$$

or

$$\hat{\mathbf{M}}_{aa} = \mathbf{M}_{aa} + \mathbf{M}_{ad}\mathbf{T}_{da} + (\mathbf{M}_{ad}\mathbf{T}_{da})^T + \mathbf{T}_{da}^T\mathbf{M}_{dd}\mathbf{T}_{da} \tag{17.5}$$

Example 17.3 illustrates how matrix manipulations can be combined with the formulation of element matrices and assembly of system matrices to carry out the Guyan reduction using the ISMIS program.

Example 17.3

Write an ISMIS program which will reduce out the rotational degrees of freedom for the two-element beam in the following figure.

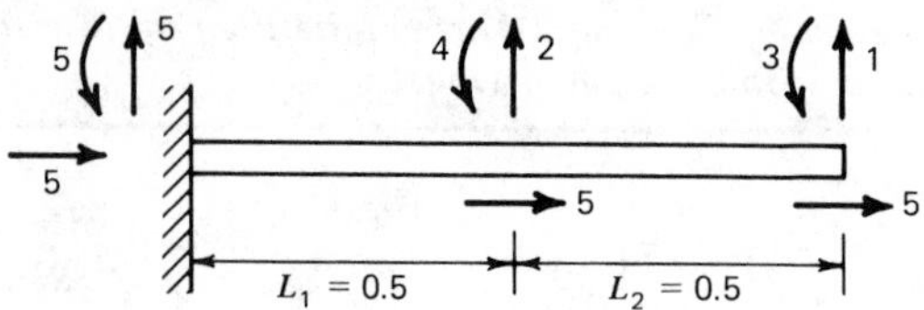

Solution

The plane frame element is used. All of the displacements which are constrained to be zero are designated by the constrained coordinate 5. The structure of the stiffness and mass matrices is thus

$$
\begin{array}{c} \\ 1 \\ 2 \\ 3 \\ 4 \\ 5 \end{array}
\begin{array}{c}
\begin{array}{ccccc} 1 & 2 & 3 & 4 & 5 \end{array} \\
\left[\begin{array}{c:c:c}
AA & AD & AC \\ \hdashline
DA & DD & DC \\ \hdashline
CA & CD & CC
\end{array}\right]
\end{array}
$$

```
START
*
** EXAMPLE 17.3 **
*
** GUYAN REDUCTION **
*
*. FORM ELEMENT MATRICES .*
FRAMEL,KE
.5,0,1,1,1
PRINT,KE,1
FRAMMS,ME
.5,0,1,1
PRINT,ME,1
*
*. ASSEMBLE SYSTEM MATRICES .*
FORM,KS,5,5
FORM,MS,5,5
ADDMAT,KS,KE
5,5,5,5,2,4
ADDMAT,MS,ME
5,5,5,5,2,4
ADDMAT,KS,KE
5,2,4,5,1,3
ADDMAT,MS,ME
5,2,4,5,1,3
*
*. REDUCE STIFFNESS MATRIX .*
RMVSM,KS,KAA,1,1,2,2
RMVSM,KS,KAD,1,3,2,2
RMVSM,KS,KDA,3,1,2,2
RMVSM,KS,KDD,3,3,2,2
SYMINV,KDD
MULT,KDD,KDA,TDA
SCALE,TDA,-1
TRANS,TDA,TAD
MULT,KAD,TDA,T
ADD,KAA,T
*
*. REDUCE MASS MATRIX .*
RMVSM,MS,MAA,1,1,2,2
```

```
RMVSM,MS,MAD,1,3,2,2
RMVSM,MS,MDD,3,3,2,2
MULT,MAD,TDA,T
ADD,MAA,T
TRANS,T,TT
ADD,MAA,TT
MULT,MDD,TDA,T
MULT,TAD,T,ADA
ADD,MAA,ADA
*
*. SOLVE EIGENPROBLEM .*
EIGEN,KAA,MAA,VAL,VECT,2
TRANS,VECT,VEC
*
*. OMEGA**2 .*
PRINT,VAL,2
SQREL,VAL
*
*. OMEGA .*
PRINT,VAL,2
*
*. MODES (COLUMNS) .*
PRINT,VEC,1
STOP
```

Table 17.2 gives the frequencies obtained by reducing out all rotations. Note that in each case the frequency is higher than the corresponding frequency in Table 17.1, illustrating again the Rayleigh-Ritz bound theorem. Note that the two underlined values in Table 17.2 are not smaller than the values immediately to their left. That is possible, since the values in column (N_e) of Table 17.2 were obtained by reducing out the rotations that were present in obtaining the corresponding column of Table 17.1, not by adding a degree of freedom to the model used for obtaining column (N_e–1) of Table 17.2.

Although the Guyan reduction method does reduce the size of the eigenproblem to be solved and does generally produce accurate frequencies for a

Table 17.2. ***Frequencies Obtained by Using Guyan Reduction to Reduce out Rotations of Uniform Cantilever Beams***

Mode No. \ N_e	*1*	*2*	*3*	*4*	*5*	*Exact (Reference 17.1)*
1	3.56753	3.52198	3.51699	3.51628	3.51611	3.51602
2		22.2790	22.2362	22.0946	22.0573	22.0345
3			62.6685	<u>62.9703</u>	62.2180	61.6972
4				123.545	<u>124.725</u>	120.902
5					205.277	199.860

good percentage of the modes retained, two cautions should be observed: (a) the accuracy of the frequencies obtained depends on which coordinates are retained and which are reduced out, as was shown by Anderson et. al.[17.2], and (b) whereas **K** and **M** may be banded, Guyan reduction destroys banding and may lead to a much more expensive eigensolution, even though the order of $\hat{\mathbf{K}}$ and $\hat{\mathbf{M}}$ is less than that of **K** and **M**.

Using the same cantilever beam as before, we now consider the accuracy of frequencies obtained by lumping the mass in the manner described in Sec. 16.2. The reduced stiffness matrix is again given by Eq. 17.2. The following frequencies are obtained.

Table 17.3. ***Frequencies of a Uniform Cantilever Beam Based on Lumped Mass Models***

N_e / *Mode No.*	*1*	*2*	*3*	*4*	*5*	*Exact (Reference 17.1)*
1	2.44949	3.15623	3.34568	3.41804	3.45266	3.51602
2		16.2580	18.8859	20.0904	20.7335	22.0345
3			47.0284	53.2017	55.9529	61.6972
4				92.7302	104.436	120.902
5					153.017	199.860

From Table 17.3 it can be seen that for the particular lumping procedure employed (i.e., half of the mass of an element distributed to each end of the element), the lumped mass procedure produces frequencies that converge quite slowly from below.

In a paper[17.3] in which he coined the name "consistent mass matrix," Archer compared frequencies of lumped mass and consistent mass models of uniform free-free and simply supported beams both including and excluding rotational freedoms. He concluded that "natural mode analysis and hence dynamic response analysis of beams with uniform stiffness and mass distribution is significantly improved if the mass matrix is constructed using equations corresponding to the Rayleigh-Ritz approach (i.e., the consistent mass approach) in place of the usual procedure of physical lumping of the structural mass at the coordinate points." In a paper published concurrently with Archer's paper, Leckie and Lindberg[17.4] presented error estimates for beam frequencies and showed that the number of elements required to achieve a desired accuracy depends on the boundary conditions if lumped mass procedures are employed, whereas this is not the case for consistent mass representations. Tong et. al.[17.5] established rates of convergence for mode shapes and frequencies obtained by using consistent mass and lumped mass finite element models. They showed that mass lumping is appropriate if the continuum model is represented by differential equations of second order, but that the consistent mass formulation should be used for higher order systems, for example, beams and plates.

Up to this point, only uniform finite elements have been considered, although Fig. 16.1 indicated that a tapered beam might be modeled as an assemblage of uniform elements. It is straightforward to use the shape functions presented in Sec. 16.2, together with expressions for k_{ij} and m_{ij}, for example, Eqs. 16.7, incorporating dimensions and physical properties that are functions of position within an element. Gallagher and Lee[17.6] compared eigensolutions for tapered beams represented by tapered elements and represented by uniform elements and showed that, for a given number of elements, much greater accuracy was achieved by using tapered elements.

17.2 Finite Element Solution for Dynamic Response by the Mode-Displacement Method

We now use the finite element method of Chapter 16 to generate a finite element model of a simple structure, and then use the mode-displacement method of Sec. 15.2 to compute its response to a given excitation. Although the ISMIS computer program is employed for the example presented, the primary purpose of the example is to illustrate the various steps in the solution of dynamic response problems, not to illustrate use of the ISMIS program itself. The following steps are used in the dynamic response solution:

1. Form element stiffness matrix.

2. Form element mass matrix.

3. Assemble system stiffness matrix and incorporate constraints.

4. Assemble system mass matrix and incorporate constraints.

5. Solve eigenproblem and obtain a vector of frequencies (ω_r) and mode shapes (ϕ_r). Print results.

6. Form excitation vector in physical coordinates.

7. Transform excitation vector to principal coordinates using only modes to be retained.

8. Generate time history of excitation using appropriate time step.

9. Assume modal damping factor(s) and use numerical integration to obtain time history of response in principal coordinates.

10. Transform response back to physical coordinates.

11. Print and/or plot response history.

Example 17.4

A uniform aluminum cantilever beam is at rest at $t = 0$ when a 10-lb concentrated mass is suddenly attached at the tip of the beam. The beam, its boundary conditions, and member properties are shown below. The beam is to be modeled by three 10-in. plane frame elements. The following information is to be calculated and printed (or plotted as required).

a. The natural frequencies of transverse vibration of the beam with tip mass.

b. The mode shapes corresponding to the three lowest frequencies.

c. The time history of the displacement at the tip of the beam. Use a one-mode representation and choose a time step that satisfies $\Delta t < 0.1\ T_1$, where T_1 is the fundamental period.

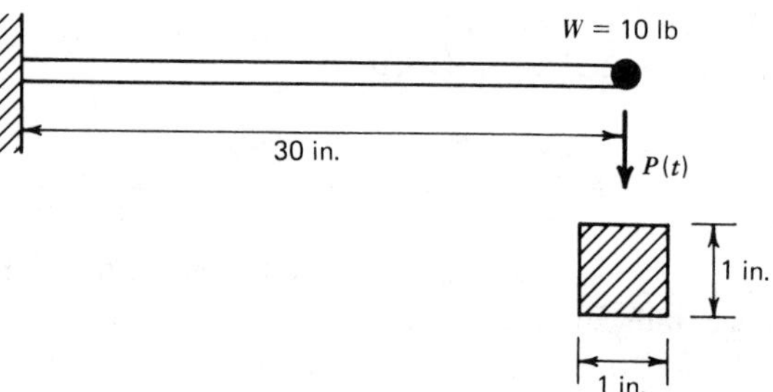

$$E = 10 \times 10^6 \text{ psi}$$
$$A = bh = 1 \text{ in.}^2$$
$$I = \frac{bh^3}{12} = 8.333 \times 10^{-2} \text{ in.}^4$$
$$\rho = \frac{\gamma}{g} = \frac{0.100}{386} = 2.591 \times 10^{-4} \text{ lb-sec}^2/\text{in.}^4$$
$$m = \frac{W}{g} = \frac{10}{386} = 0.02591 \text{ lb-sec}^2/\text{in.}$$

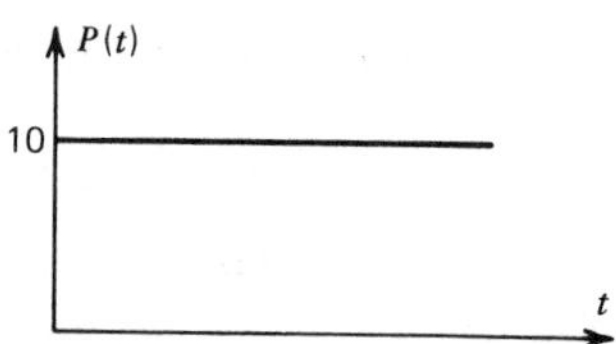

Solution

Let the system coordinates be as shown below.

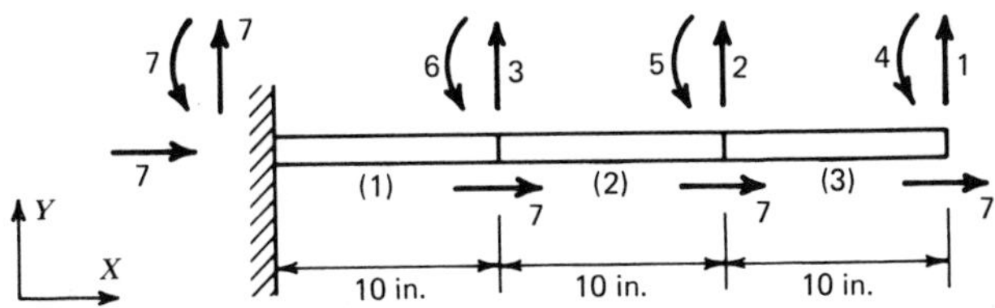

Listing of ISMIS input file:

```
START
*
** EXAMPLE 17.4 **
*
** DYNAMIC RESPONSE CALCULATION **
*
** MODE-DISPLACEMENT METHOD **
*
*. STEP 1. ELEMENT STIFFNESS .*
FRAMEL,KE
10.,0,10.E+6,8.333E-2,1
PRINT,KE,1
*
*. STEP 2. ELEMENT MASS .*
FRAMMS,ME
10.,0,2.591E-4,1
PRINT,ME,1
*
*. STEP 3. ASSEMBLE SYSTEM STIFFNESS .*
FORM,KS,7,7
ADDMAT,KS,KE
7,7,7,7,3,6
ADDMAT,KS,KE
7,3,6,7,2,5
ADDMAT,KS,KE
7,2,5,7,1,4
RMVSM,KS,K,1,1,6,6
PRINT,K,1
*
*. STEP 4. ASSEMBLE SYSTEM MASS .*
*
*.       INCORPORATE LUMPED MASS .*
FORM,MS,7,7
ADDMAT,MS,ME
7,7,7,7,3,6
ADDMAT,MS,ME
7,3,6,7,2,5
ADDMAT,MS,ME
7,2,5,7,1,4
LOAD,LM,1,1
.02591
PRINT,LM,1
ADDSM,MS,LM,1,1
RMVSM,MS,M,1,1,6,6
PRINT,M,1
*
```

```
*. STEP 5. SOLVE EIGENPROBLEM .*
EIGEN,K,M,VAL,VECT,6
TRANS,VECT,VEC
*
*OMEGA**2*
PRINT,VAL,2
SQREL,VAL
DUPL,VAL,HZ
SCALE,HZ,.1592
*
*OMEGA (RAD/SEC)*
PRINT,VAL,2
*
*FREQ (HZ)*
PRINT,HZ,2
*
*MODES (COLUMNS)*
PRINT,VEC,1
*
*. STEP 6. FORM EXCITATION VECTOR .*
LOAD,P,6,1
-10,0,0,0,0,0
PRINT,P,1
*
*. STEP 7. TRANSFORM EXCITATION VECTOR .*
RMVSM,VECT,U1T,1,1,1,6
MULT,U1T,P,F
*
*. STEP 8. GENERATE EXCITATION TIME HIST. .*
LOAD,FTIN,2,2
0,1,1,1
PRINT,FTIN,1
FUNGN,FTIN,FT,41,.005
*
*. STEP 9. PERFORM RESPONSE CALCS. .*
RMVSM,VAL,W1,1,1,1,1
RESPON,W1,F,FT,V,1,0,0,.005
*
*. STEP 10. TRANSFORM RESPONSE .*
RMVSM,U1T,U11,1,1,1,1
MSCALE,V,U11
*
*. STEP 11. PRINT AND PLOT RESULTS .*
PRINT,V,1
PLOT,V,1
***** TRANSIENT RESPONSE AT TIP OF BEAM *****
STOP
```

Listing of ISMIS output file:

```
START
** EXAMPLE 17.4 **
** DYNAMIC RESPONSE CALCULATION **
** MODE-DISPLACEMENT METHOD **
*. STEP 1. ELEMENT STIFFNESS .*
FRAMEL     KE
      6 ROWS      6 COLUMNS
 DX =  1.000E+01      A =  1.000E+00
 DY =  0.             E =  1.000E+07
  L =  1.000E+01      I =  8.333E-02
```

```
PRINT     KE
 SCALED BY    1.0E+06
          1         2         3         4         5         6
  1  1.00000 0.00000 0.00000-1.00000 0.00000 0.00000
  2  0.00000  .01000  .05000 0.00000 -.01000  .05000
  3  0.00000  .05000  .33332 0.00000 -.05000  .16666
  4 -1.00000 0.00000 0.00000 1.00000 0.00000 0.00000
  5  0.00000 -.01000 -.05000 0.00000  .01000 -.05000
  6  0.00000  .05000  .16666 0.00000 -.05000  .33332
*. STEP 2. ELEMENT MASS .*
FRAMMS    ME
     6  ROWS      6 COLUMNS
 DX =  1.000E+01      DY =  0.
 RO =  2.591E-04      AR =  1.000E+00       L =  1.000E+01
PRINT     ME
 SCALED BY    1.0E-03
          1         2         3         4         5         6
  1   .86367 0.00000 0.00000  .43183 0.00000 0.00000
  2  0.00000  .96237 1.35719 0.00000  .33313 -.80198
  3  0.00000 1.35719 2.46762 0.00000  .80198-1.85071
  4   .43183 0.00000 0.00000  .86367 0.00000 0.00000
  5  0.00000  .33313  .80198 0.00000  .96237-1.35719
  6  0.00000 -.80198-1.85071 0.00000-1.35719 2.46762
*. STEP 3. ASSEMBLE SYSTEM STIFFNESS .*
FORM      KS
     7 ROWS      7 COLUMNS
ADDMAT    KS          KE
   DEGREES OF FREEDOM =      7       7       7       7       3       6
ADDMAT    KS          KE
   DEGREES OF FREEDOM =      7       3       6       7       2       5
ADDMAT    KS          KE
   DEGREES OF FREEDOM =      7       2       5       7       1       4
RMVSM     KS          K
 ROW NUMBER    1,  COLUMN NUMBER   1
     6 ROWS      6 COLUMNS
PRINT     K
 SCALED BY    1.0E+05
          1         2         3         4         5         6
  1   .10000 -.10000 0.00000 -.49998 -.49998 0.00000
  2  -.10000  .19999 -.10000  .49998 0.00000 -.49998
  3  0.00000 -.10000  .19999 0.00000  .49998 0.00000
  4  -.49998  .49998 0.00000 3.33320 1.66660 0.00000
  5  -.49998 0.00000  .49998 1.66660 6.66640 1.66660
  6  0.00000 -.49998 0.00000 0.00000 1.66660 6.66640
*. STEP 4. ASSEMBLE SYSTEM MASS .*
*.        INCORPORATE LUMPED MASS .*
FORM      MS
     7 ROWS      7 COLUMNS
ADDMAT    MS          ME
   DEGREES OF FREEDOM =      7       7       7       7       3       6
ADDMAT    MS          ME
   DEGREES OF FREEDOM =      7       3       6       7       2       5
ADDMAT    MS          ME
   DEGREES OF FREEDOM =      7       2       5       7       1       4
LOAD      LM
     1 ROWS      1 COLUMNS
PRINT     LM
 SCALED BY    1.0E-02
          1
  1  2.59100
```

```
ADDSM      MS         LM
 ROW NUMBER   1,  COLUMN NUMBER   1
RMVSM      MS         M
 ROW NUMBER   1,  COLUMN NUMBER   1
    6 ROWS      6 COLUMNS
PRINT      M
 SCALED BY   1.0E-02
         1        2        3        4        5        6
  1  2.68724   .03331 0.00000 -.13572   .08020 0.00000
  2   .03331   .19247  .03331 -.08020  0.00000  .08020
  3  0.00000   .03331  .19247 0.00000  -.08020 0.00000
  4  -.13572  -.08020 0.00000  .24676  -.18507 0.00000
  5   .08020  0.00000 -.08020 -.18507   .49352 -.18507
  6  0.00000   .08020 0.00000 0.00000  -.18507  .49352
*. STEP 5. SOLVE EIGENPROBLEM .*
EIGEN      K          M          VAL        VECT
   NUMBER OF EIGENVECTORS DESIRED=        6
     ORIGINAL TRACE  = 9.23532264E+08
     SUM OF EIGENVALUES  = 9.23532264E+08
TRANS      VECT       VEC
*OMEGA**2*
PRINT      VAL
 SCALED BY   1.0E+08
            1           2           3           4           5           6
  1   .00003337   .00981368   .10378622   .56095007 2.06994600 6.49079329

SQREL      VAL
DUPL       VAL        HZ
SCALE      HZ
 SCALAR=  1.5920000E-01
*OMEGA (RAD/SEC)*
PRINT      VAL
 SCALED BY   1.0E+04
            1           2           3           4           5           6
  1   .00577673   .09906403   .32215869   .74896600 1.43873069 2.54770353

*FREQ (HZ)*
PRINT      HZ
 SCALED BY   1.0E+03
            1           2           3           4           5           6
  1   .00919655   .15770994   .51287664 1.19235388 2.29045926 4.05594402

*MODES (COLUMNS)*
PRINT      VEC
 SCALED BY   1.0E+01
         1        2        3        4        5        6
  1   .60031   .11347 -.06523  .05023   .04000  -.03948
  2   .31238-1.60078 1.17777  .45598  1.53259  -.47041
  3   .08961-1.11811-1.77354 -.92425  -.43371  -.76482
  4   .02985   .22509 -.39541  .76785  1.29914-2.21588
  5   .02665   .07322  .27246 -.71828  -.19961-1.83442
  6   .01677  -.14148  .02933  .60952-1.24188  -.88340
*. STEP 6. FORM EXCITATION VECTOR .*
LOAD       P
    6 ROWS      1 COLUMNS
```

```
PRINT      P
 SCALED BY    1.0E+01
         1
  1 -1.00000
  2  0.00000
  3  0.00000
  4  0.00000
  5  0.00000
  6  0.00000
*. STEP 7. TRANSFORM EXCITATION VECTOR .*
RMVSM      VECT         U1T
 ROW NUMBER    1,  COLUMN NUMBER    1
     1 ROWS      6 COLUMNS
MULT       U1T          P           F
*. STEP 8. GENERATE EXCITATION TIME HIST. .*
LOAD       FTIN
     2 ROWS      2 COLUMNS
PRINT      FTIN
 SCALED BY    1.0E+00
         1       2
  1  0.00000 1.00000
  2  1.00000 1.00000
FUNGN      FTIN         FT
   NUMBER OF VALUES DESIRED =      41
   X-INTERVAL DESIRED =  5.000E-03
*. STEP 9. PERFORM RESPONSE CALCS. .*
RMVSM      VAL          W1
 ROW NUMBER    1,  COLUMN NUMBER    1
     1 ROWS      1 COLUMNS
RESPON     W1           F           FT          V
   OUTPUT INTERVAL =        1
   DAMPING RATIO =       0.00000
   INTEGRATION INTERVAL -  5.000E-03
   COORDINATE      INITIAL DISP.      INITIAL VEL.
             1       0.                0.
*. STEP 10. TRANSFORM RESPONSE .*
RMVSM      U1T          U11
 ROW NUMBER    1,  COLUMN NUMBER    1
     1 ROWS      1 COLUMNS
MSCALE     V            U11
 SCALAR=  6.0030915E+00
*. STEP 11. PRINT AND PLOT RESULTS .*
PRINT      V
 SCALED BY    1.0E-01
         1       2       3       4       5       6       7       8
  1  0.00000 -.04443 -.17406 -.37822 -.64013 -.93821-1.24796-1.54388
         9      10      11      12      13      14      15      16
  1 -1.80162-1.99998-2.12263-2.15949-2.10751-1.97098-1.76113-1.49523
        17      18      19      20      21      22      23      24
  1 -1.19516 -.88560 -.59203 -.33860 -.14617 -.03057 -.00131 -.06080
        25      26      27      28      29      30      31      32
  1  -.20415 -.41955 -.68929 -.99118-1.30036-1.59140-1.84035-2.02674
        33      34      35      36      37      38      39      40
  1 -2.13522-2.15686-2.08989-1.93982-1.71899-1.44557-1.14207 -.83345
        41
  1  -.54512
PLOT       V
```

```
***** TRANSIENT RESPONSE AT TIP OF BEAM *****

-2.2000E-01                           -8.800E-02                            4.400E-02

       I                                   I                                   I
       I---------I---------I---------I---------I---------I---------I---------I
    1-I                                                        1          I
      I                                                        I          I
    2-I                                                       1I          I
      I                                                        I          I
    3-I                                                   1    I          I
      I                                                        I          I
    4-I                                             1          I          I
      I                                                        I          I
    5-I                                       1                I          I
      I                                                        I          I
    6-I                                 1                      I          I
      I                                                        I          I
    7-I                         1                              I          I
      I                                                        I          I
    8-I                  1                                     I          I
      I                                                        I          I
    9-I            1                                           I          I
      I                                                        I          I
   10-I     1                                                  I          I
      I                                                        I          I
   11-I 1                                                      I          I
      I                                                        I          I
   12-I1                                                       I          I
      I                                                        I          I
   13-I 1                                                      I          I
      I                                                        I          I
   14-I     1                                                  I          I
      I                                                        I          I
   15-I            1                                           I          I
      I                                                        I          I
   16-I                  1                                     I          I
      I                                                        I          I
   17-I                         1                              I          I
      I                                                        I          I
   18-I                                1                       I          I
      I                                                        I          I
   19-I                                       1                I          I
      I                                                        I          I
   20-I                                             1          I          I
      I---------I---------I---------I---------I---------I---------I---------I
   21-I                                                   1    I          I
      I                                                        I          I
   22-I                                                       1I          I
```

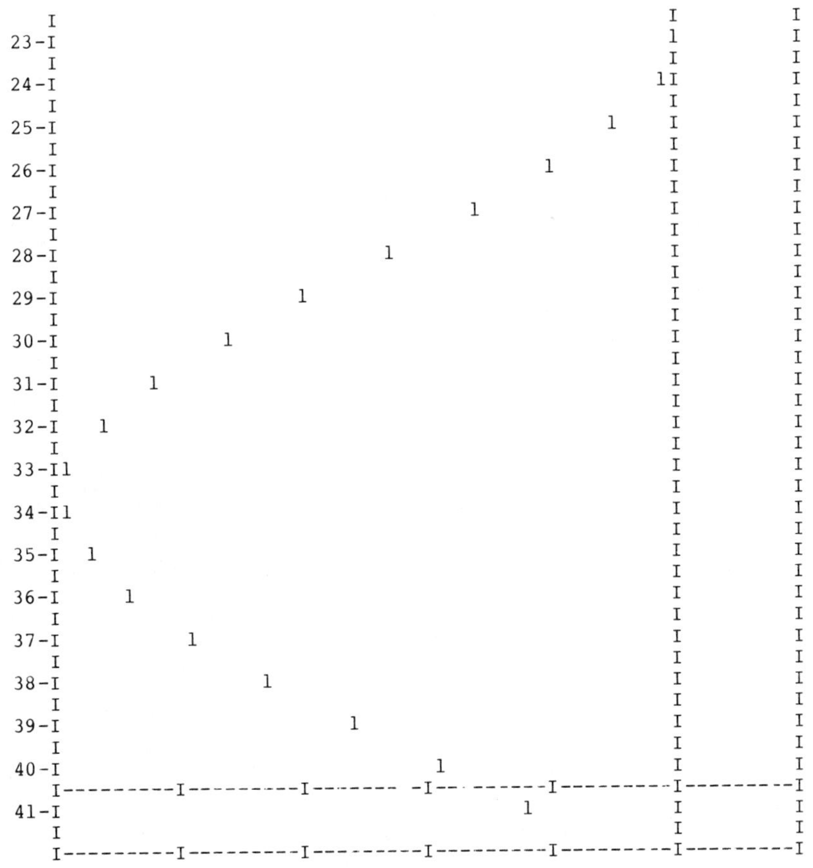

STOP

Note: A preliminary run through Step 5 established $f_1 = 9.2$ Hz. The time step $\Delta t = 0.005$, which is approximately $T_1/20$, was thus chosen for the subsequent response calculations. Since only one mode is retained in the response calculations, this time step is short enough to ensure stability of the linear acceleration method employed in the RESPON command (Chapter 18 discusses the topic of stability.) The plotted response has the $(1 - \cos \omega_n t)$ form that would be expected for a SDOF model subjected to a step input.

Since large quantities of output result from typical dynamic response analysis, it is important to select for printing or plotting output that is important. The knowledge of mode shapes and natural frequencies, which is available in a mode-superposition solution, can be valuable in interpreting output of response time histories.

References

17.1 T-C. Chang and R. R. Craig, Jr., "Normal Modes of Uniform Beams," *Proc. ASCE*, v. 95, n. EM4, 1025–1031 (1969).

17.2 R. G. Anderson, B. M. Irons, and O. C. Zienkiewicz, "Vibration and Stability of Plates Using Finite Elements," *Int. J. Solids Struct.*, v. 4, 1031–1055 (1968).

17.3 J. S. Archer, "Consistent Mass Matrix for Distributed Mass Systems," *J. Struct. Div., Proc. ASCE*, v. 89, 161–178 (1963).

17.4 F. A. Leckie and G. M. Lindberg, "The Effect of Lumped Parameters on Beam Frequencies," *Aeron. Quart.*, v. 14, 224–240 (1963).

17.5 P. Tong, T. H. H. Pian, and L. L. Bucciarelli, "Mode Shapes and Frequencies by Finite Element Method Using Consistent and Lumped Masses," *Comps. Struct.*, v. 1, 623–628 (1971).

17.6 R. Gallagher and B. Lee, "Matrix Dynamic and Instability Analysis with Non-uniform Elements," *Int. J. Num. Meth. Engr.*, v. 2, 265–276 (1970).

Problem Set 17.1

17.1 (a) Use the ISMIS computer program (or another specified program) to obtain the natural frequencies and mode shapes of the five-element uniform cantilever beam. Include both axial and transverse motion. Use $A = I = E = \rho = 1$. Use the consistent mass matrix.

(b) Compare the lowest three axial and lowest three transverse bending frequencies with exact values.

(c) Sketch the six mode shapes whose frequencies were tabulated in (b).

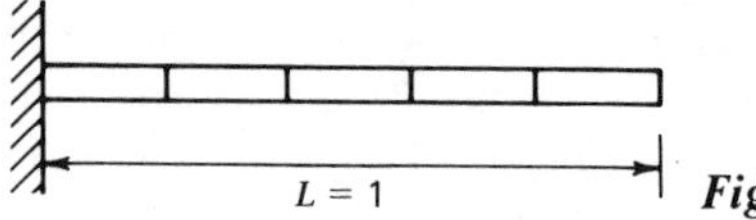

Figure P17.1

17.2 A space shuttle model (see Reference 19.13) was designed to enable preliminary tests to be conducted approximating some of the aspects of the dynamics of a coupled Orbiter-Booster configuration. The "Orbiter" consisted of an aluminum tube ($E = 10 \times 10^6$ psi) of 1-in. outside diameter and 0.035-

in. wall thickness 50.5 in. long. Equally spaced along the tube were six rigid masses which consisted of 0.50-in.-thick steel plates with outside diameter 6 in. and inside diameter 1 in. (to fit over the tube).

(a) Use the ISMIS computer program to compute the axial and transverse frequencies and mode shapes of this structure if it is in the "free-free" condition, that is, it has no boundary constraints.

(b) Sketch the mode shapes of the lowest three axial modes and lowest three transverse bending modes.

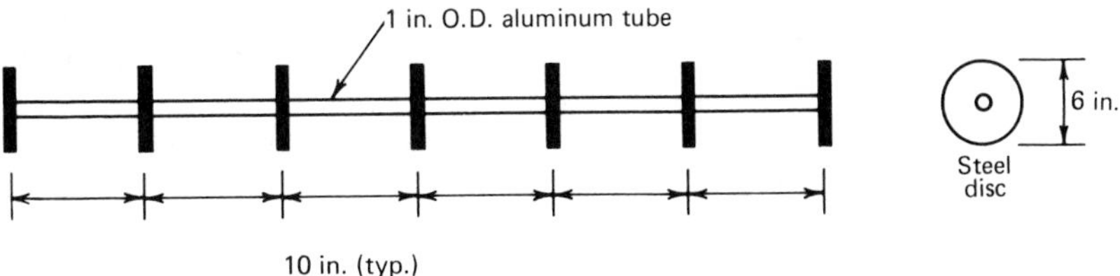

Figure P17.2

17.3 Repeat Problem 17.1 using Guyan reduction to reduce out all rotations.

17.4 Repeat Problem 17.3 replacing the original consistent mass matrix with a lumped mass matrix.

17.5 The basic design of a three-story structure is shown in Fig. P17.5.

(a) Determine the natural frequencies and mode shapes of this structure using ISMIS (or some other specified program). Sketch the mode shapes.

(b) If the stiffness of the bottom columns is reduced by 5%, what effect would this have on the frequencies and modes?

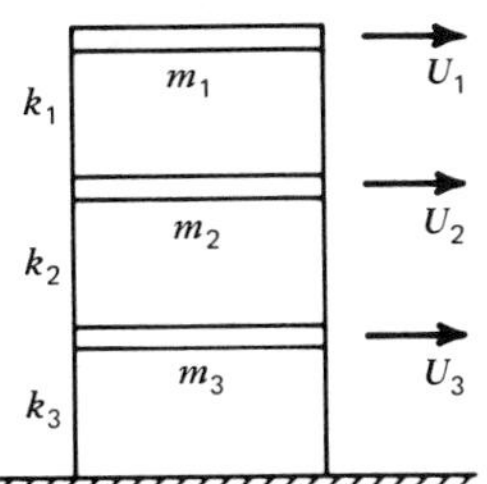

Figure P17.5

$m_1 = m_2 = 40$ k-sec²/ft, $\quad m_3 = 50$ k-sec²/ft

$k_1 = 30 \times 10^3$ k/ft, $\quad k_2 = 40 \times 10^3$ k/ft

$k_3 = 50 \times 10^3$ k/ft

17.6 A rigid platform is welded to steel pipe inclined columns as shown in Fig. P17.6. Due to the inclination of the columns, a simple SDOF model as in Problem 2.1 is not valid.

(a) Use the ISMIS computer program (or other specified program) to form a 3-DOF model of this structure using the system coordinates shown below. (*Note:* The method of Sec. 16.6 can be used to relate the column displacements to the three system displacements.) Neglect the inertia of the columns. Solve for the modes and frequencies of this model.

(b) Would it be valid to reduce this to a SDOF system by ignoring displacements 2 and 3?

(c) Would it be acceptable to ignore the mass moment of inertia, I_G, in the 3-DOF model?

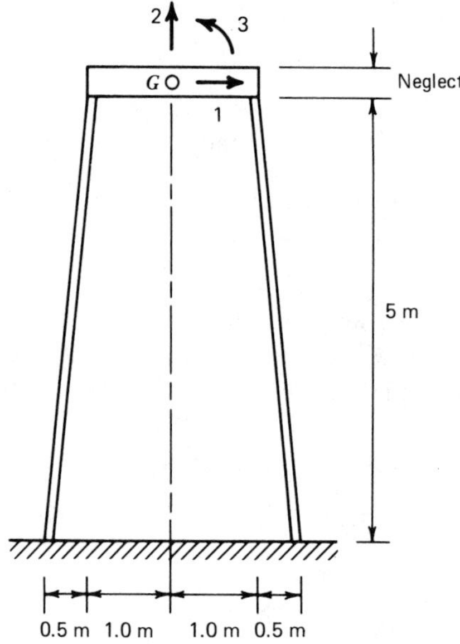

Figure P17.6

$m = 907$ kg

$I_G = 302$ kg·m^2

$E = 200$ GN/m^2

$A = 2000$ mm^2

$I = 300$ cm^4

Problem Set 17.2

17.7 The three-story shear building in Fig. P17.5 is subjected to wind loading that is assumed to have the form

$$\mathbf{P}(t) = \begin{Bmatrix} 0.4 \\ 0.7 \\ 1.0 \end{Bmatrix} f(t)\ \mathrm{k}$$

where $f(t)$ is shown below. Use the ISMIS computer program (or other specified program) to determine the dynamic response for $0 \leq t \leq 1.0$ sec if $\mathbf{U}(0) = \dot{\mathbf{U}}(0) = \mathbf{0}$.

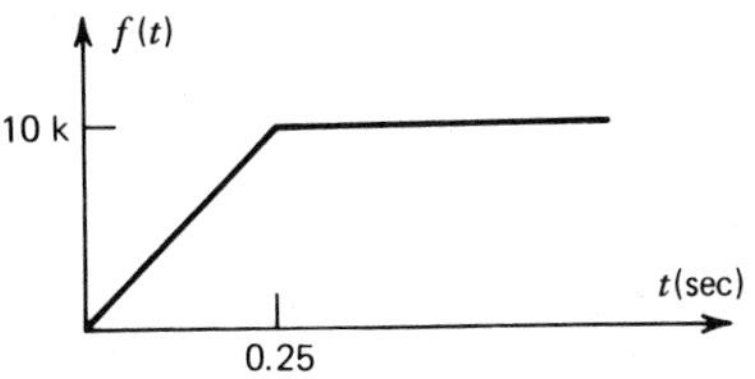

Figure P17.7

18 DIRECT INTEGRATION METHODS FOR DYNAMIC RESPONSE

In Chapter 15 mode-superposition methods were employed to obtain the dynamic response of linear MDOF systems by transforming the equations of motion to principal coordinates and solving the resulting set of uncoupled equations of motion. In Fig. 15.1 it was indicated that MDOF systems with nonlinearities or with coupled damping require direct integration of a set of coupled equations of motion. In this chapter we will first consider how to express the damping in systems whose response cannot properly be calculated by using mode-superposition. Then we will extend the incremental method of analysis of nonlinear systems, which was applied to SDOF systems in Chapter 7, to the MDOF case. Finally, we will examine the stability and accuracy properties of numerical integration algorithms when applied to SDOF systems, and will consider the implications when these algorithms are used for direct integration of the equations of motion of MDOF systems.

Upon completion of this chapter you should be able to:

- Generate a reasonable system damping matrix for an MDOF system.
- Set up the incremental equations for integrating the equations of motion of an MDOF system with material nonlinearity.
- Define the following terms: unconditionally stable, conditionally stable, unconditionally unstable, period error, amplitude error, numerical damping.
- Discuss the relationship between time step and the terms listed above.
- Name several step-by-step numerical integration methods, and discuss their suitability for direct integration of the equations of motion of MDOF systems.

18.1 Damping in MDOF Systems

In Sec. 15.4 the response of MDOF systems with a special form of viscous damping was considered. That form of damping satisfied the orthogonality equation

$$\boldsymbol{\phi}_r^T \mathbf{C} \boldsymbol{\phi}_s = 0, \qquad r \neq s \tag{18.1}$$

where **C** is the system viscous damping matrix in physical (or generalized) coordinates and where the $\boldsymbol{\phi}$'s are system modes. Then, modal damping factors ζ_r could be assumed on the basis of providing damping typical of the type of structure under consideration. This type of damping is referred to in the literature as *orthogonal, classical, modal,* or *proportional* damping.

There are a number of situations in which damping cannot properly be represented by the above model. One such instance is the modeling of a building and the surrounding soil. The damping level for the building on a rigid foundation is less than that for the soil, and when a combined finite element model is generated and soil-structure system modes are used to define the principal coordinates, the resulting damping matrix in principal coordinates contains off-diagonal coupling terms. Offshore structures with fluid damping and structures with concentrated energy absorbers provide other examples where nonproportional damping must be considered.

Several approaches have been suggested for solving for the response of systems with nonproportional damping. Hurty and Rubinstein[18.1] use complex modes in a procedure analogous to the mode-superposition method of Chapter 15. Warburton and Soni[18.2] and others have attempted to define equivalent diagonal damping matrices, or at least to establish criteria for when nonproportional damping can be approximated by proportional damping. Clough and Mojtahedi[18.3] compared various methods of solving for the response of systems with nonproportional damping and concluded that "the most efficient procedure is to express the response in terms of a truncated set of undamped modal coordinates and to integrate directly the resulting equations." Certainly this approach is straightforward, uses available analytical tools (eigensolvers that handle systems with real modes, direct integration algorithms, etc.), and avoids unnecessary approximations. It can be expected that research will continue on ways to treat damping both analytically and experimentally.

So far, we have considered two damping topics: (1) use of mode-superposition to solve for the response of linear MDOF systems with proportional damping (Chapter 15), and (2) response of systems with nonproportional damping. Now we take up a third topic, "How can damping be represented when a damping matrix in physical coordinates is required?" One situation where this arises is the soil-structure problem, where separate physical damping matrices for soil and structure are required. A physical damping matrix is also required for nonlinear problems, where a viscous damping matrix is introduced to account for the energy dissipated by all mechanisms other than material yielding.

In principle, element damping matrices could be derived by procedures analogous to those employed in Sec. 16.2 for deriving element stiffness and mass matrices. However, material damping properties are not defined well enough to permit this, and furthermore, much damping in structures results from joints and from nonstructural elements (e.g., partitions, etc.). Therefore, damping of a structure is usually defined directly at the system level rather than in terms of individual element properties.

One procedure for defining a system damping matrix is to employ a particular form of proportional damping called *Rayleigh damping,* defined by

$$\mathbf{C} = a_0\mathbf{M} + a_1\mathbf{K} \tag{18.2}$$

where, in the case of a nonlinear system, $\mathbf{K}$ could represent an initial tangent stiffness matrix, or $\mathbf{C}$ could be modified with each change in stiffness. The constants a_0 and a_1 can be chosen to produce specified modal damping factors for two given modes. Let $(\omega_r, \boldsymbol{\phi}_r)$ be the eigenpairs corresponding to

$$(\mathbf{K} - \omega_r^2\mathbf{M})\boldsymbol{\phi}_r = \mathbf{0}, \qquad r = 1, 2, \ldots, N \tag{18.3}$$

The orthogonality equations are

$$\boldsymbol{\phi}_r^T\mathbf{M}\boldsymbol{\phi}_s = M_r\delta_{rs} \tag{18.4a}$$

$$\boldsymbol{\phi}_r^T\mathbf{K}\boldsymbol{\phi}_s = \omega_r^2 M_r\delta_{rs} \tag{18.4b}$$

Then, for Rayleigh damping defined by Eq. 18.2,

$$\boldsymbol{\phi}_r^T\mathbf{C}\boldsymbol{\phi}_s = (a_0 + a_1\omega_r^2)M_r\delta_{rs} \tag{18.5}$$

Since, from Eq. 15.33,

$$C_r = \boldsymbol{\phi}_r^T\mathbf{C}\boldsymbol{\phi}_r = 2M_r\omega_r\zeta_r \tag{18.6}$$

the system damping factors are given by

$$\zeta_r = \frac{1}{2}\left(\frac{a_0}{\omega_r} + a_1\omega_r\right) \tag{18.7}$$

Thus, Rayleigh damping is easy to define by choosing ζ_r for two modes and solving for a_0 and a_1. The damping in the remaining modes is then determined by Eq. 18.7. The $a_0\mathbf{M}$ contribution to damping in Eq. 18.2 gives a contribution to ζ_r in Eq. 18.7 which is inversely proportional to ω_r. The $a_1\mathbf{K}$ term, on the other hand, leads to a contribution to ζ_r, which increases linearly with ω_r.

The disadvantage of Rayleigh damping is that it does not permit realistic damping to be defined for all the modes of interest. The method that follows permits a damping matrix to be generated which leads to proportional damping with specified damping factors for a given number of modes.

From Eq. 15.9b,

$$\boldsymbol{C} = \boldsymbol{\Phi}^T\mathbf{C}\boldsymbol{\Phi} = \text{diag}(2\zeta_r\omega_r M_r) \tag{18.8}$$

Then, the physical damping matrix $\mathcal{C}$ is given by

$$\mathbf{C} = \mathbf{\Phi}^{-T}\mathcal{C}\mathbf{\Phi}^{-1} \tag{18.9}$$

A convenient expression for $\mathbf{\Phi}^{-1}$ can be developed from the orthogonality property of the modes. Recall that

$$\mathcal{M} = \mathbf{\Phi}^T\mathbf{M}\mathbf{\Phi} \tag{18.10a}$$

Then,

$$\mathbf{I} = \mathcal{M}^{-1}\mathcal{M} = (\mathcal{M}^{-1}\mathbf{\Phi}^T\mathbf{M})\mathbf{\Phi} = \mathbf{\Phi}^{-1}\mathbf{\Phi} \tag{18.10b}$$

Therefore,

$$\mathbf{\Phi}^{-1} = \mathcal{M}^{-1}\mathbf{\Phi}^T\mathbf{M} \tag{18.10c}$$

Equations 18.9 and 18.10c can be combined to give

$$\mathbf{C} = (\mathbf{M}\mathbf{\Phi}\mathcal{M}^{-1})\mathcal{C}(\mathcal{M}^{-1}\mathbf{\Phi}^T\mathbf{M}) \tag{18.11}$$

Since $\mathcal{M}$ and $\mathcal{C}$ are diagonal, Eq. 18.11 can be written in the form

$$\mathbf{C} = \sum_{r=1}^{N}\left(\frac{2\zeta_r\omega_r}{M_r}\right)(\mathbf{M}\boldsymbol{\phi}_r)(\mathbf{M}\boldsymbol{\phi}_r)^T \tag{18.12}$$

Due to orthogonality of modes it can be seen that Eq. 18.12 gives

$$\boldsymbol{\phi}_s^T\mathbf{C}\boldsymbol{\phi}_s = 2\zeta_s\omega_s M_s \tag{18.13}$$

so that the modes for which a nonzero value of ζ_r is specified in Eq. 18.12 will have that damping present in **C**, while there will be no damping of those modes for which ζ_r is set to zero in Eq. 18.12.

If a limited number of the lower-frequency modes are considered to be important in the response calculations, a truncated form of Eq. 18.12 can be used as follows.

$$\mathbf{C} = \sum_{r=1}^{N_c}\left(\frac{2\zeta_r\omega_r}{M_r}\right)(\mathbf{M}\boldsymbol{\phi}_r)(\mathbf{M}\boldsymbol{\phi}_r)^T \tag{18.14}$$

This produces a damping matrix **C**, which yields no damping in the modes $(N_c + 1), (N_c + 2), \ldots, N$. It may, however, be desirable to provide damping in these higher modes. It is possible to modify Eq. 18.14 such that the modes $r = 1, 2, \ldots, N_c$ have specified damping ratios, and the modes $(N_c + 1)$, $(N_c + 2), \ldots, N$ have damping greater than that in mode N_c. This is possible by letting

$$\mathbf{C} = a_1\mathbf{K} + \sum_{r=1}^{N_c-1}\left(\frac{2\hat{\zeta}_r\omega_r}{M_r}\right)(\mathbf{M}\boldsymbol{\phi}_r)(\mathbf{M}\boldsymbol{\phi}_r)^T \tag{18.15}$$

where

$$a_1 = \frac{2\zeta_{N_c}}{\omega_{N_c}} \tag{18.16a}$$

$$\hat{\zeta}_r = \zeta_r - \zeta_{N_c}\left(\frac{\omega_r}{\omega_{N_c}}\right) \tag{18.16b}$$

Then

$$\zeta_s = \begin{cases} \text{specified value}, \; s = 1, 2, \ldots, N_c \\ \zeta_{N_c}\left(\dfrac{\omega_s}{\omega_{N_c}}\right), \qquad s = (N_c + 1), (N_c + 2), \ldots, N \end{cases} \tag{18.17}$$

Whenever direct integration requires use of a physical damping matrix, Rayleigh damping defined by Eq. 18.2 or generalized proportional damping defined by Eq. 18.12 or 18.15 may be employed to approximate the damping.

Example 18.1

a. Use Eq. 18.15 to define a physical damping matrix for the building of Example 15.1 corresponding to $\zeta_1 = \zeta_2 = 0.01$.

b. Determine the resulting damping ratios ζ_3 and ζ_4.

Solution

a. From Eq. 18.15

$$\mathbf{C} = a_1\mathbf{K} + \left(\frac{2\hat{\zeta}_1\omega_1}{M_1}\right)(\mathbf{M}\boldsymbol{\phi}_1)(\mathbf{M}\boldsymbol{\phi}_1)^T \tag{1}$$

where

$$a_1 = \frac{2\zeta_2}{\omega_2} \tag{2a}$$

$$\hat{\zeta}_1 = \zeta_1 - \zeta_2\left(\frac{\omega_1}{\omega_2}\right) \tag{2b}$$

Thus,

$$a_1 = \frac{2(0.01)}{29.660} = 6.7431 \times 10^{-4} \tag{3a}$$

$$\hat{\zeta}_1 = 0.01 - 0.01\left(\frac{13.294}{29.660}\right) = 5.5179 \times 10^{-3} \tag{3b}$$

$$\mathbf{M}\phi_1 = \begin{bmatrix} 1 & 0 & 0 & 0 \\ 0 & 2 & 0 & 0 \\ 0 & 0 & 2 & 0 \\ 0 & 0 & 0 & 3 \end{bmatrix} \begin{Bmatrix} 1.00000 \\ 0.77910 \\ 0.49655 \\ 0.23506 \end{Bmatrix} = \begin{Bmatrix} 1.00000 \\ 1.55820 \\ 0.99310 \\ 0.70518 \end{Bmatrix} \tag{4}$$

$$\mathbf{K} = 800 \begin{bmatrix} 1 & -1 & 0 & 0 \\ -1 & 3 & -2 & 0 \\ 0 & -2 & 5 & -3 \\ 0 & 0 & -3 & 7 \end{bmatrix} \tag{5}$$

The above may be substituted into Eq. 1 to give

$$\mathbf{C} = \begin{bmatrix} 0.59051 & -0.45988 & 0.05071 & 0.03601 \\ -0.45988 & 1.74233 & -0.99987 & 0.05611 \\ 0.05071 & -0.99987 & 2.74760 & -1.58258 \\ 0.03601 & 0.05611 & -1.58258 & 3.80153 \end{bmatrix} \tag{6}$$

b. From Eq. 18.17b,

$$\zeta_s = \zeta_2 \left(\frac{\omega_s}{\omega_2} \right), \ s = 3, 4 \tag{7}$$

$$\begin{aligned} \zeta_3 &= 0.01 \left(\frac{41.079}{29.660} \right) = 0.0138 \\ \zeta_4 &= 0.01 \left(\frac{55.882}{29.660} \right) = 0.0188 \end{aligned} \tag{8}$$

18.2 Nonlinear MDOF Systems

In Sec. 7.4 a step-by-step procedure for solving for the response of nonlinear SDOF systems was described. The topic of finite element analysis of transient nonlinear structural behavior is a subject of extensive current research interest (e.g., References 18.4, 18.5, and 18.6), and this brief section is not intended to do more than just introduce you to one approach that can be employed to solve some nonlinear MDOF structural dynamics problems.

The mode-superposition method, which was used for determining the response of linear MDOF systems with proportional damping, is not valid for determining nonlinear response. However, it may still be useful to determine system modes based on the initial mass and stiffness of a system and use these as a basis for a transformation from physical coordinates to a reduced set of principal coordinates even though, when nonlinearity develops or damping

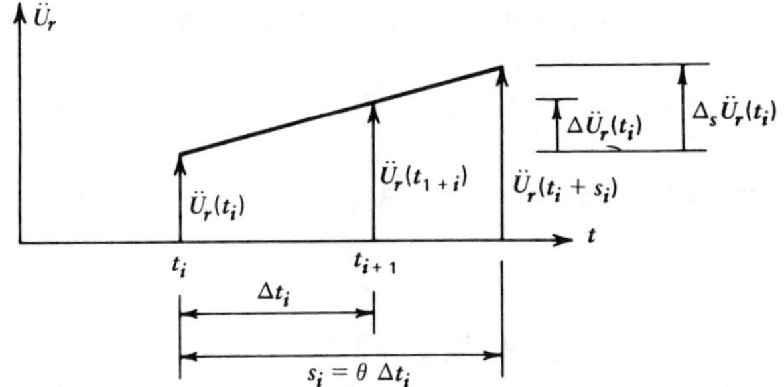

Figure 18.1. *Linear acceleration assumption of the Wilson-θ method.*

becomes coupled, the equations of motion in principal coordinates must be solved by step-by-step methods. The procedures discussed in the preceding section may be used to define a reasonable damping matrix for use in a nonlinear analysis.

Wilson et.al.[18.7] presented an incremental formulation for step-by-step integration of nonlinear equations of motion of MDOF systems. The procedure is analogous to that presented in Sec. 7.4 for nonlinear SDOF systems except that the Wilson-θ method, rather than the average acceleration method, is used for approximating the derivatives in the equations of motion so that step-by-step integration can be accomplished. (In Sec. 18.3 we will discuss why the Wilson-θ method and other methods might, for MDOF systems, be superior to the average acceleration method.)

The basic assumption of the *Wilson-θ method* is that each component $\ddot{U}_r$ of the acceleration vector $\ddot{\mathbf{U}}$ varies linearly with time over the *extended time step* $s_i = \theta\,\Delta t_i$ as indicated in Fig. 18.1. In Sec. 18.3 it will be shown that θ must satisfy $\theta \geq 1.37$. Usually the value $\theta = 1.4$ is used, although it has been shown(18.8) that the optimum value of θ is 1.420815. Using the above linear assumption it can be shown that

$$\Delta_s\dot{\mathbf{U}}(t_i) = s_i\ddot{\mathbf{U}}(t_i) + \frac{s_i}{2}\,\Delta_s\ddot{\mathbf{U}}(t_i) \tag{18.18a}$$

$$\Delta_s\mathbf{U}(t_i) = s_i\dot{\mathbf{U}}(t_i) + \frac{s_i^2}{2}\,\ddot{\mathbf{U}}(t_i) + \frac{s_i^2}{6}\,\Delta_s\ddot{\mathbf{U}}(t_i) \tag{18.18b}$$

where

$$\Delta_s\mathbf{U}(t_i) \equiv \mathbf{U}(t_i + s_i) - \mathbf{U}(t_i) \tag{18.19a}$$

$$\Delta_s\dot{\mathbf{U}}(t_i) \equiv \dot{\mathbf{U}}(t_i + s_i) - \dot{\mathbf{U}}(t_i) \tag{18.19b}$$

$$\Delta_s\ddot{\mathbf{U}}(t_i) \equiv \ddot{\mathbf{U}}(t_i + s_i) - \ddot{\mathbf{U}}(t_i) \tag{18.19c}$$

Equations 18.19 can be solved for $\Delta_s\ddot{\mathbf{U}}(t_i)$ and $\Delta_s\dot{\mathbf{U}}(t_i)$ in terms $\Delta_s\mathbf{U}(t_i)$ giving

$$\Delta_s\ddot{\mathbf{U}}(t_i) = \frac{6}{s_i^2}\Delta_s\mathbf{U}(t_i) - \frac{6}{s_i}\dot{\mathbf{U}}(t_i) - 3\ddot{\mathbf{U}}(t_i) \tag{18.20a}$$

$$\Delta_s\dot{\mathbf{U}}(t_i) = \frac{3}{s_i}\Delta_s\mathbf{U}(t_i) - 3\dot{\mathbf{U}}(t_i) - \frac{s_i}{2}\ddot{\mathbf{U}}(t_i) \tag{18.20b}$$

These are substituted into the incremental equation of equilibrium

$$\mathbf{M}\,\Delta_s\ddot{\mathbf{U}}(t_i) + \mathbf{C}(t_i)\,\Delta_s\dot{\mathbf{U}}(t_i) + \mathbf{K}(t_i)\,\Delta_s\mathbf{U}(t_i) = \hat{\mathbf{P}}(t_i + s_i) - \mathbf{P}(t_i) \tag{18.21}$$

where a "projected load" $\hat{\mathbf{P}}(t_i + s_i)$ given by

$$\hat{\mathbf{P}}(t_i + s_i) = \mathbf{P}(t_i) + \theta[\mathbf{P}(t_i + \Delta t_i) - \mathbf{P}(t_i)] \tag{18.22}$$

is needed since it is assumed that the loads are only given at t_i, $t_i + \Delta t_i$, and so forth. Equations 18.20 through 18.22 can be combined to give

$$\mathbf{K}^*(t_i)\,\Delta_s\mathbf{U}(t_i) = \mathbf{P}^*(t_i) \tag{18.23}$$

where

$$\mathbf{K}^*(t_i) = \mathbf{K}(t_i) + \frac{3}{s_i}\mathbf{C}(t_i) + \frac{6}{s_i^2}\mathbf{M} \tag{18.24a}$$

$$\mathbf{P}^*(t_i) = \theta[\mathbf{P}(t_i + \Delta t_i) - \mathbf{P}(t_i)] + \mathbf{M}\left[\frac{6}{s_i}\dot{\mathbf{U}}(t_i) + 3\ddot{\mathbf{U}}(t_i)\right] + \mathbf{C}(t_i)\left[3\dot{\mathbf{U}}(t_i) + \frac{s_i}{2}\ddot{\mathbf{U}}(t_i)\right] \tag{18.24b}$$

Equation 18.23 is then solved for $\Delta_s\mathbf{U}(t_i)$. Once this has been obtained, Eq. 18.20a gives the acceleration increment $\Delta_s\ddot{\mathbf{U}}(t_i)$. From Fig. 18.1 it can be seen that

$$\Delta\ddot{\mathbf{U}}(t_i) = \frac{1}{\theta}\Delta_s\ddot{\mathbf{U}}(t_i) \tag{18.25}$$

With this value of $\Delta\ddot{\mathbf{U}}(t_i)$, equations similar to Eq. 18.18 give

$$\dot{\mathbf{U}}(t_i + \Delta t_i) = \dot{\mathbf{U}}(t_i) + \Delta t_i\ddot{\mathbf{U}}(t_i) + \frac{\Delta t_i}{2}\Delta\ddot{\mathbf{U}}(t_i) \tag{18.26a}$$

$$\mathbf{U}(t_i + \Delta t_i) = \mathbf{U}(t_i) + \Delta t_i\dot{\mathbf{U}}(t_i) + \frac{\Delta t_i^2}{2}\ddot{\mathbf{U}}(t_i) + \frac{\Delta t_i^2}{6}\Delta\ddot{\mathbf{U}}(t_i) \tag{18.26b}$$

To ensure dynamic equilibrium at $t_i + \Delta t_i$, the acceleration $\ddot{\mathbf{U}}(t_i + \Delta t_i)$ is obtained from the equilibrium equation

$$\ddot{\mathbf{U}}(t_i + \Delta t_i) = \mathbf{M}^{-1}[\mathbf{P}(t_i + \Delta t_i) - \mathbf{F}_D(t_i + \Delta t_i) - \mathbf{F}_S(t_i + \Delta t_i)] \tag{18.26c}$$

rather than directly from the $\Delta\ddot{U}(t_i)$ of Eq. 18.25. Equations 18.26 give the initial conditions for the next time step.

18.3 Properties of Step-by-Step Numerical Integration Algorithms

In Chapter 7 the average acceleration algorithm was described, and in Sec. 18.2 the Wilson-θ method was employed. These are two examples of a large group of methods available for carrying out step-by-step numerical integration of linear or nonlinear equations of motion to obtain transient response of structures. Two very important properties of these numerical integrators, stability and accuracy, will now be discussed. We will first consider the use of numerical integration to solve linear SDOF problems and will then discuss the implications of using these methods for direct integration of the equations of motion of MDOF systems.

Operator Formulation of Step-by-Step Integration Algorithms

For the following discussions of stability and accuracy it will be useful to formulate the step-by-step integration algorithms in operator form[18.8,18.9]. We will use the average acceleration method of Sec. 7.2 as an example. The three relevant equations are the equation of motion at time $t_i + \Delta t_i$

$$M\ddot{U}_{i+1} + C\dot{U}_{i+1} + KU_{i+1} = P_{i+1} \tag{18.27}$$

and the kinematical approximations from Eqs. 7.10 and 7.11

$$\dot{U}_{i+1} = \dot{U}_i + \frac{\Delta t_i}{2}(\ddot{U}_i + \ddot{U}_{i+1}) \tag{18.28}$$

$$U_{i+1} = U_i + \Delta t_i\,\dot{U}_i + \frac{\Delta t_i^2}{4}(\ddot{U}_i + \ddot{U}_{i+1}) \tag{18.29}$$

These may conveniently be written in matrix form

$$\begin{bmatrix} K & C & M \\ 0 & 1 & \dfrac{-\Delta t_i}{2} \\ 1 & 0 & \dfrac{-\Delta t_i^2}{4} \end{bmatrix} \begin{Bmatrix} U_{i+1} \\ \dot{U}_{i+1} \\ \ddot{U}_{i+1} \end{Bmatrix} = \begin{bmatrix} 0 & 0 & 0 \\ 0 & 1 & \dfrac{\Delta t_i}{2} \\ 1 & \Delta t_i & \dfrac{\Delta t_i^2}{4} \end{bmatrix} \begin{Bmatrix} U_i \\ \dot{U}_i \\ \ddot{U}_i \end{Bmatrix} + \begin{Bmatrix} P_{i+1} \\ 0 \\ 0 \end{Bmatrix} \tag{18.30}$$

or

$$\mathbf{K}_1\mathbf{U}_{i+1} = \mathbf{K}_0\mathbf{U}_i + \mathbf{P}_{i+1} \tag{18.31}$$

Equation 18.30 can be solved by explicitly inverting $\mathbf{K}_1$, obtaining

$$\mathbf{L} \equiv \mathbf{K}_1^{-1} = \left(\frac{-1}{M + \frac{C\,\Delta t_i}{2} + \frac{K\,\Delta t_i^2}{4}} \right) \times \begin{bmatrix} \frac{-\Delta t_i^2}{4} & \frac{C\,\Delta t_i}{4} & -\left(M + \frac{C\,\Delta t_i}{2}\right) \\ \frac{-\Delta t_i}{2} & -\left(M + \frac{K\,\Delta t_i^2}{4}\right) & \frac{K\Delta t_i}{2} \\ -1 & C & -K \end{bmatrix} \tag{18.32}$$

which is also called the *load operator*. Then, the *recursion relation*

$$\mathbf{U}_{i+1} = \mathbf{A}\mathbf{U}_i + \mathbf{L}\mathbf{P}_{i+1} \tag{18.33}$$

can be formed, where

$$\mathbf{A} = \mathbf{K}_1^{-1}\mathbf{K}_0 = \left(\frac{1}{1 + \frac{\eta}{2} + \frac{\xi}{4}} \right) \begin{bmatrix} \left(1 + \frac{\eta}{2}\right) & \Delta t_i\left(1 + \frac{\eta}{4}\right) & \frac{\Delta t_i^2}{4} \\ \frac{-\xi}{2\,\Delta t_i} & \left(1 - \frac{\xi}{4}\right) & \frac{\Delta t_i}{2} \\ \frac{-\xi}{\Delta t_i^2} & \frac{-(\xi + \eta)}{\Delta t_i} & -\left(\frac{\eta}{2} + \frac{\xi}{4}\right) \end{bmatrix} \tag{18.34}$$

is the *amplification matrix*, and

$$\mathbf{F}_{i+1} \equiv \mathbf{L}\mathbf{P}_{i+1} = \left(\frac{1}{1 + \frac{\eta}{2} + \frac{\xi}{4}} \right) \begin{Bmatrix} \frac{\Delta t_i^2}{4} \\ \frac{\Delta t_i}{2} \\ 1 \end{Bmatrix} \left(\frac{P_{i+1}}{M} \right) \tag{18.35}$$

is the *load vector*, where

$$\begin{aligned} \xi &= \frac{K\,\Delta t_i^2}{M} = \omega_n^2\,\Delta t_i^2 \\ \eta &= \frac{C\,\Delta t_i}{M} = 2\zeta\omega_n\,\Delta t_i \end{aligned} \tag{18.36}$$

Example 18.2

Use the average acceleration method to obtain the free vibration response of an undamped SDOF system with

$$K = M = U(0) = 1, \qquad \dot{U}(0) = 0$$

Choose a reasonable time step and perform the integration for $0 \leq t \leq 2.0$ s.

Solution

$$\omega_n = \left(\frac{K}{M}\right)^{1/2} = 1 \tag{1}$$

$$T_n = \frac{2\pi}{\omega_n} = 2\pi \text{ s} \tag{2}$$

Use $\Delta t_i = 0.2$ s = constant, which will be shorter than the "rule-of-thumb" value of $T_n/10$.

$$\xi = \omega_n^2\, \Delta t^2 = 1(0.2)^2 = 0.04 \tag{3}$$

$$\eta = 2\zeta\omega_n\, \Delta t = 0 \tag{4}$$

Since this is free vibration, the recursion relation of Eq. 18.33 becomes

$$\mathbf{U}_{i+1} = \mathbf{A}\mathbf{U}_i \tag{5}$$

where $\mathbf{A}$ is given by Eq. 18.34.

$$\mathbf{A} = \frac{1}{1.01}\begin{bmatrix} 1.00 & 0.20 & 0.01 \\ -0.10 & 0.99 & 0.10 \\ -1.00 & -0.20 & -0.01 \end{bmatrix} \tag{6}$$

The equation of motion is

$$\ddot{U} + U = 0 \tag{7}$$

so the initial condition vector is

$$\mathbf{U}_0 \equiv \begin{Bmatrix} U(0) \\ \dot{U}(0) \\ \ddot{U}(0) \end{Bmatrix} = \begin{Bmatrix} 1 \\ 0 \\ -1 \end{Bmatrix} \tag{8}$$

Then, from Eq. 5,

$$\mathbf{U}(0.2) \equiv \mathbf{U}_1 = \mathbf{A}\mathbf{U}_0 = \frac{1}{1.01}\begin{Bmatrix} 0.99 \\ -0.20 \\ -0.99 \end{Bmatrix} = \begin{Bmatrix} 0.98020 \\ -0.19802 \\ -0.98020 \end{Bmatrix} \tag{9}$$

Continuing,

$$\mathbf{U}(0.4) \equiv \mathbf{U}_2 = \mathbf{A}\mathbf{U}_1 = \frac{1}{1.01}\begin{Bmatrix} 0.93079 \\ -0.39208 \\ -0.93079 \end{Bmatrix} = \begin{Bmatrix} 0.92158 \\ -0.38820 \\ 0.92158 \end{Bmatrix} \tag{10}$$

The table below summarizes the computations and compares the step-by-step solution with the exact solution, $U(t) = \cos(t)$.

Average Acceleration Solution for Free Vibration

i	t_i	$\ddot{U}_i$	$\dot{U}_i$	U_i	$\cos(t_i)$
0	0.0	−1.00000	0.00000	1.00000	1.00000
1	0.2	−0.98020	−0.19802	0.98020	0.98007
2	0.4	−0.92158	−0.38820	0.92158	0.92106
3	0.6	−0.82646	−0.56300	0.82646	0.82534
4	0.8	−0.69861	−0.71551	0.69861	0.69671
5	1.0	−0.54309	−0.83968	0.54309	0.54030
6	1.2	−0.36606	−0.93059	0.36606	0.36236
7	1.4	−0.17454	−0.98465	0.17454	0.16997
8	1.6	0.02390	−0.99971	−0.02390	−0.02920
9	1.8	0.22139	−0.97519	−0.22139	−0.22720
10	2.0	0.41011	−0.91204	−0.41011	−0.41615

Stability

As observed in Example 18.2 for free vibration

$$\mathbf{U}_{i+1} = \mathbf{A}\mathbf{U}_i \tag{18.37}$$

so

$$\mathbf{U}_1 = \mathbf{A}\mathbf{U}_0, \qquad \mathbf{U}_2 = \mathbf{A}\mathbf{U}_1 = \mathbf{A}^2\mathbf{U}_0 \tag{18.38}$$

and so forth, that is, each integration step corresponds to raising the power of the amplification matrix. The behavior of the integration operator can be characterized by its eigenvalues. Consider the eigenvalue problem

$$\mathbf{A}\phi_r = \lambda_r\phi_r \tag{18.39}$$

Since $\mathbf{A}$ is a 3×3 matrix whose elements depend in general on ξ, η and Δt_i (see Eq. 18.34), there will be three eigenvalues which depend on these quantities. The three eigenequations can be combined and written

$$\mathbf{A\Phi} = \mathbf{\Phi\Lambda} \tag{18.40}$$

Then

$$\mathbf{A} = \mathbf{\Phi\Lambda\Phi}^{-1} \tag{18.41}$$

Furthermore,

$$
\begin{aligned}
\mathbf{A}^2 &= \mathbf{A}\mathbf{A} = \mathbf{\Phi}\mathbf{\Lambda}^2\mathbf{\Phi}^{-1} \\
&\vdots \\
\mathbf{A}^s &= \mathbf{\Phi}\mathbf{\Lambda}^s\mathbf{\Phi}^{-1}
\end{aligned}
\tag{18.42}
$$

From Eq. 18.42 it can be seen that, if the magnitude of any of the three eigenvalues of **A** is greater than one, there will be an amplification of the solution with each time step. This leads to the following stability definitions.

Integration operators that cause the solution to grow without bound regardless of the time step are called *unconditionally unstable* operators. Operators that lead to bounded solutions if the time step satisfies $\Delta t_i \leq \Delta t_{cr}$ are termed *conditionally stable.* Operators that lead to bounded solutions regardless of the length of the time step are called *unconditionally stable* operators. Unconditionally stable operators are the most desirable, although conditionally stable operators may be used if the time step restriction is observed. The average acceleration operator is unconditionally stable. The Wilson-θ method is unconditionally stable only for $\theta \geq 1.37$. Further details on stability of operators may be found in References 18.8, 18.9, and 18.10.

The topic of stability of integration operators is closely related to topic of *numerical damping,* or *numerical dissipation.* The *spectral radius* of an integration operator is defined as

$$
\rho = \max |\lambda_r| \tag{18.43}
$$

If $\rho > 1$ the operator is unstable, because the solution will be amplified at each time step. On the other hand, if $\rho < 1$ there will be numerical damping of the solution. Figure 18.2 shows the relationship between time step and spectral radius for several operators, including a new operator introduced in Reference 18.11, and from this figure stability and numerical dissipation can be qualitatively inferred.

Accuracy

The accuracy of numerical integration algorithms can be characterized by two attributes: *amplitude accuracy,* and *period accuracy.* Figure 18.3 shows the exact free vibration response of an undamped SDOF system and a numerical solution computed with a short time step. It can be seen that the integration operator used produces an inaccurate solution even for a short time step. The amplitude decay results from numerical dissipation, and there is also period elongation. Figure 18.4 shows a corresponding response calculation based on

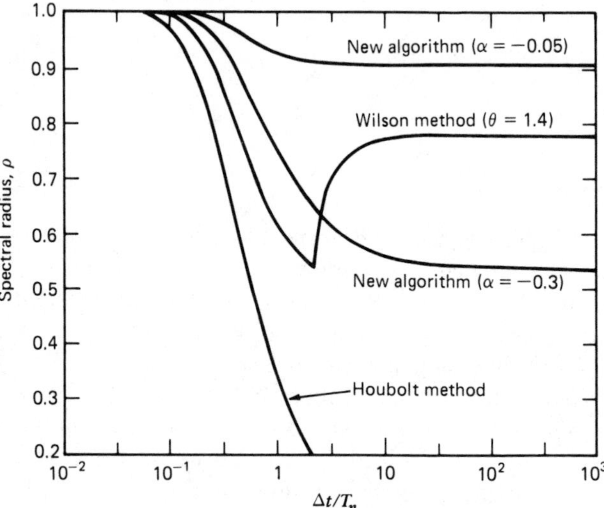

Figure 18.2. *Spectral radii versus* $\Delta t/T_n$. *(H. M. Hilber et al., "Improved Numerical Dissipation for Time Integration Algorithms in Structural Dynamics,"* Earthquake Engineering and Structural Dynamics, *1977. Reprinted with permission from Earthquake Engineering and Structural Dynamics, vol. 5, p. 239.)*

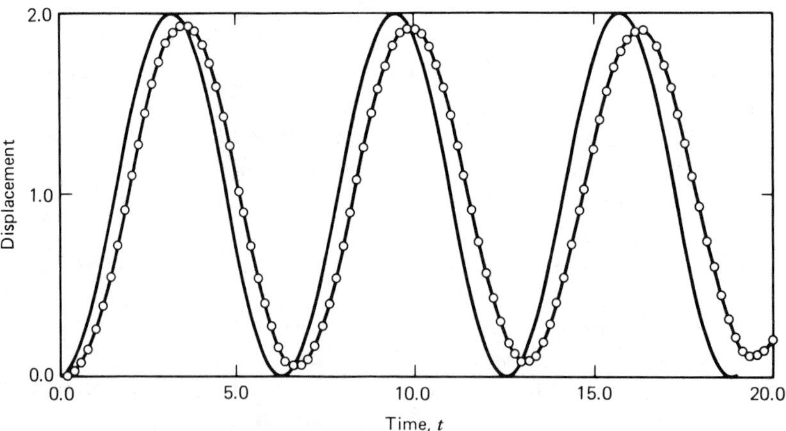

Figure 18.3. *Illustrations of amplitude and period error in numerical solution. (R. E. Nickell, "On the Stability of Approximation Operators in Problems of Structural Dynamics," Pergamon Press, Ltd., 1971. Reprinted with permission of Pergamon Press, Ltd.)*

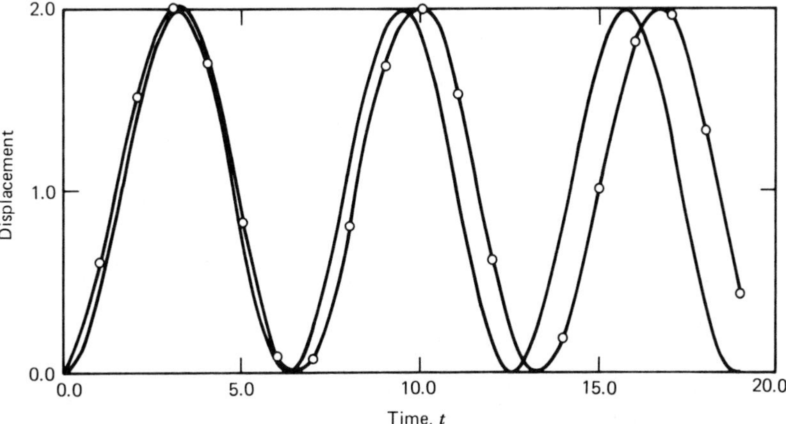

Figure 18.4. *Numerical integration using the average acceleration operator (Newmark-β operator with $\beta = \frac{1}{4}$). (R. E. Nickell, "On the Stability of Approximation Operators in Problems of Structural Dynamics," Pergamon Press, Ltd., 1971. Reprinted with permission of Pergamon Press, Ltd.)*

the average acceleration algorithm. In this case there is no numerical dissipation, but there is a period elongation. The period error can be reduced by decreasing the time step ratio $\Delta t/T_n$. Figures 18.5 and 18.6 present numerical damping factor and relative period error versus the time step ratio $\Delta t/T_n$ for several integration operators. References 18.8, 18.9, and 18.11 can be consulted for further details.

Direct Integration for MDOF Systems

From the preceding figures it is clear that the time step plays a key role in determining the stability and accuracy of numerical solutions of SDOF systems. Let us now consider the use of one of these numerical integration methods for direct integration of the equations of motion of a MDOF system. Although the equations being solved are not uncoupled equations, it is convenient to think of the response as being a superposition of the responses of the system in each of its N modes. The questions to resolve are: "What numerical integration algorithm should be used?" and "What time step should be used?" These questions are obviously related.

The first consideration in selecting a method is its stability. In most cases it is desirable to use a method that is unconditionally stable.* This leaves a

*Explicit methods, which are not unconditionally stable, are computationally simpler than any unconditionally stable methods and are preferred when the time step is limited already by other factors, as in nonlinear analysis.

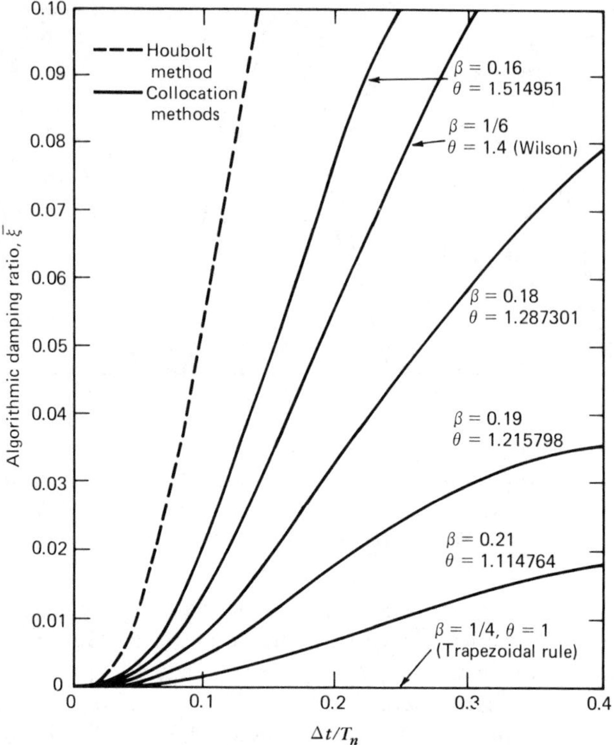

Figure 18.5. Numerical damping factors for several integration schemes. (H. M. Hilber and T. J. R. Hughes, "Collocation, Dissipation and 'Overshoot' for Time Integration Schemes in Structural Dynamics," Earthquake Engineering and Structural Dynamics, *1978. Reprinted with permission from Earthquake Engineering and Structural Dynamics, vol. 6, p. 108.)*

very wide selection, including the average acceleration method, the Wilson-θ method, and others. Next, it can be observed from Fig. 18.5 that the average acceleration method produces no amplitude error (i.e., no numerical dissipation) regardless of the time step, and from Fig. 18.6 it can be seen to have the lowest period error of the methods represented on the figure. These characteristics make the average acceleration method an excellent choice for integrating a SDOF system. However, for a MDOF system the requirements are more subtle. Recall from the discussion following Table 17.1 that in most finite-DOF models (e.g., assumed-modes or finite element models) the higher frequencies and mode shapes are inaccurate, indicating poor mathematical modeling of the higher modes. Hence, in direct integration it becomes desirable to use numerical dissipation to filter out the response of these higher modes, producing a result which is similar to that achieved by truncating the number of modes retained in a mode-superposition solution. Hence, there arises the goal of formulating a numerical integration algorithm which, in some sense, has optimal numerical dissipation properties. The Wilson-θ method does provide for the damping of higher modes, for example, modes for which $\Delta t/T_r \geq 1.0$, but other desirable methods have also been recently proposed[18.10,18.11].

The considerations of stability and accuracy discussed above apply strictly only to linear systems. The reader is urged to consult additional references on nonlinear dynamic analysis.

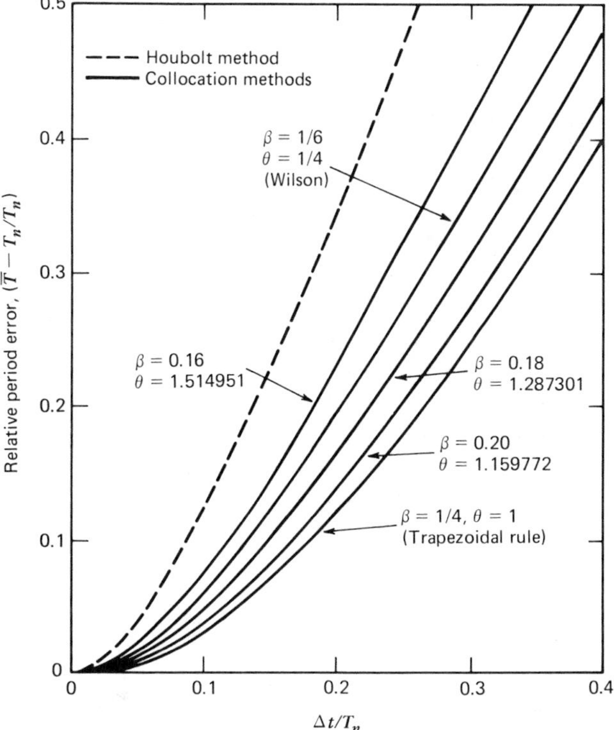

Figure 18.6. *Relative period error for several integration schemes. (H. M. Hilber and T. J. R. Hughes, "Collocation, Dissipation and 'Overshoot' for Time Integration Schemes in Structural Dynamics,"* Earthquake Engineering and Structural Dynamics, *1978. Reprinted with permission from Earthquake Engineering and Structural Dynamics, vol. 6, p. 108.)*

References

18.1 W. C. Hurty and M. F. Rubinstein, *Dynamics of Structures,* Prentice-Hall, Englewood Cliffs, NJ (1964).

18.2 G. B. Warburton and S. R. Soni, "Errors in Response Calculations for Non-Classically Damped Structures," *Earthquake Engr. and Str. Dyn.,* v. 5, 365–376 (1977).

18.3 R. W. Clough and Soheil Mojtahedi, "Earthquake Response Analysis Considering Non-proportional Damping," *Earthquake Engr. and Str. Dyn.,* v. 4, 489–496 (1976).

18.4 T. Belytschko, J. R. Osias, and P. V. Marcal (eds.), *Finite Element Analysis of Transient Nonlinear Structural Behavior,* AMD-v. 14, ASME, New York (1975).

18.5 J. T. Oden et.al. (eds.), *Computational Methods in Nonlinear Mechanics,* Texas Inst. for Comp. Mech., The Univ. of Texas at Austin, Austin, TX (1974).

18.6 K-J. Bathe, E. Ramm, and E. L. Wilson, "Finite Element Formulations for Large Deformation Dynamic Analysis," *Int. J. for Num. Meth. in Engr.,* v. 9, 353–386 (1975).

18.7 E. L. Wilson, I. Farhoomand, and K-J. Bathe, "Nonlinear Dynamic Analysis of Complex Structures," *Earthquake Engr. and Str. Dyn.,* v. 1, 241–252 (1973).

18.8 R. E. Nickell, "On the Stability of Approximation Operators in Problems of Structural Dynamics," *Int. J. Solids Struct.,* v. 7, 301–319 (1971).

18.9 K-J. Bathe and E. L. Wilson, *Numerical Methods in Finite Element Analysis,* Prentice-Hall, Englewood Cliffs, NJ (1976).

18.10 H. M. Hilber and T. J. R. Hughes, "Collocation, Dissipation and 'Overshoot' for Time Integration Schemes in Structural Dynamics," *Earthquake Engr. and Str. Dyn.,* v. 6, 99–177 (1978).

18.11 H. M. Hilber, T. J. R. Hughes, and R. L. Taylor, "Improved Numerical Dissipation for Time Integration Algorithms in Structural Dynamics," *Earthquake Engr. and Str. Dyn.,* v. 5, 283–292 (1977).

Problem Set 18.1

18.1 (a) Determine the Rayleigh damping coefficients a_0 and a_1 in Eq. 18.2 such that the 3-DOF system in Problems 14.2 and 15.1 has $\zeta_1 = \zeta_2 = 0.01$.

(b) What is the resulting value of ζ_3?

18.2 (a) Use Eq. 18.15 to determine a damping matrix **C** which would yield $\zeta_1 = \zeta_2 = 0.01$ for the 3-DOF system in Problems 14.2 and 15.1.

(b) What is the resulting value of ζ_3?

18.3 (a) Determine the Rayleigh damping coefficients a_0 and a_1 in Eq. 18.2 such that the transverse bending modes of the cantilever beam of Problem 17.1 would have $\zeta_1 = \zeta_2 = 0.01$.

(b) Evaluate the resulting damping factors ζ_3, ζ_4, and ζ_5.

18.4 Repeat Problem 18.3 using Eq. 18.15 rather than Rayleigh damping.

Problem Set 18.3

18.5 (a) In Eq. 18.32, the load operator for the average acceleration method is developed, and in Eq. 18.34 the amplification matrix for this method is determined. Following similar procedures, develop the load operator and the amplification matrix for the linear acceleration method, which is obtained by setting $\theta = 1$ in Fig. 18.1. (See Problem 7.5.)

(b) For what values of the time step Δt_i is the linear acceleration method stable?

18.6 Consider the undamped SDOF system of Example 18.2, that is, $K = M = U(0) = 1$, $\dot{U}(0) = 0$.

(a) Use the average acceleration method to solve for the response of this system for $0 \leq t \leq 2T_n$ using $\Delta t_i = (3/20)\,T_n$.

(b) Construct a table comparing your results from (a) with the exact values of U at the corresponding times. Do your results agree with Fig. 18.4?

19 COMPONENT MODE SYNTHESIS

In Chapter 16 you were introduced to finite element techniques for formulating MDOF models of structures for use in structural dynamics analyses. Stiffness and mass matrices were derived for finite elements and these were assembled to form system matrices. In Sec. 16.6 general techniques were introduced for reducing the order of system matrices by the use of assumed-modes and constraint equations. In the present chapter a class of reduction methods known as component mode synthesis, or substructure coupling for dynamic analysis, are introduced. These methods have been found to be very useful in solving very large structural dynamics problems, especially where the structure consists of several natural components, for example, a building and the surrounding soil, or the Space Shuttle Orbiter and its payloads.

Upon completion of this chapter you should be able to:

- Obtain component modes of the following types for a simple component: free-interface normal modes, fixed-interface normal modes, rigid-body modes, constraint modes, attachment modes, residual attachment modes, residual inertia relief attachment modes.
- Assemble the system matrices **K** and **M** using the generalized component coupling procedure of Sec. 19.3 and using any specified combination of component modes.

19.1 Introduction to Component Mode Synthesis

In previous chapters the use of finite element models to obtain solutions for the dynamic response of structures subjected to dynamic excitation has been described. Complete structures are frequently very complex, and major components are often designed and produced by different organizations, for example, the major components of the Space Shuttle. Thus, it may be difficult to assemble a finite element model of the entire structure in a timely manner. In addition, the finite element model of the entire structure might contain so many degrees of freedom that it would be infeasible to perform a dynamic analysis based on the finite element equations for the complete system. For these rea-

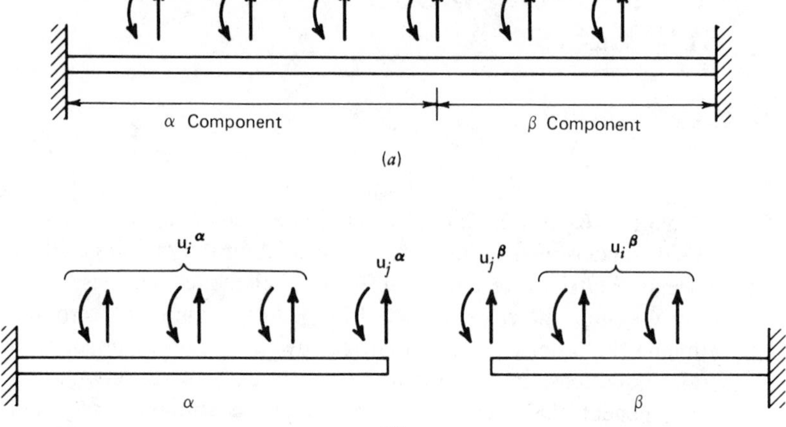

Figure 19.1. Components of a structure. (a) *Coupled structure.* (b) *Components with their physical coordinates.*

sons, methods have been developed which permit the structure to be subdivided into *components,* or *substructures,* with much of the analysis being done on the smaller components in order to develop an approximate mathematical model of of the full structural system. These methods have come to be called *methods of component mode synthesis,* or *methods of substructure coupling for dynamic analysis.* This chapter provides an introduction to methods of component mode synthesis. References 19.1 and 19.2 survey much of the literature currently available on this topic. Additional references may be found at the end of this chapter.

To simplify the analysis we will first consider undamped, free vibration of systems whose components do not possess rigid-body freedom. In Sec. 19.4 the applications will be extended to components that have rigid-body freedom.

Although any number of components could be coupled together to form a system, we will illustrate component mode synthesis methods by using only two components. Figure 19.1 shows a simple clamped-clamped beam which has been divided into two components, α and β. The set P of physical coordinates of the components may be divided into a set J of juncture coordinates, $\mathbf{u}_j$, and a set I of interior coordinates, $\mathbf{u}_i$. The juncture coordinates are those coordinates where components are joined together.

The equation of motion (undamped) for a component may be written in partitioned form

$$\begin{bmatrix} \mathbf{m}_{ii} & \mathbf{m}_{ij} \\ \mathbf{m}_{ji} & \mathbf{m}_{jj} \end{bmatrix} \begin{Bmatrix} \ddot{\mathbf{u}}_i \\ \ddot{\mathbf{u}}_j \end{Bmatrix} + \begin{bmatrix} \mathbf{k}_{ii} & \mathbf{k}_{ij} \\ \mathbf{k}_{ji} & \mathbf{k}_{jj} \end{bmatrix} \begin{Bmatrix} \mathbf{u}_i \\ \mathbf{u}_j \end{Bmatrix} = \begin{Bmatrix} \mathbf{f}_i \\ \mathbf{f}_j \end{Bmatrix} \tag{19.1}$$

The physical coordinates, $\mathbf{u}$, may be represented in terms of component generalized coordinates, $\mathbf{p}$, by the coordinate transformation

$$\mathbf{u} = \boldsymbol{\Psi}\mathbf{p} \tag{19.2}$$

where $\boldsymbol{\Psi}$ is a matrix of preselected component modes* of the following types: rigid-body modes, normal modes of free vibration, constraint modes, and attachment modes. In Sec. 19.2 we will describe these component modes, and in Sec. 19.3 we will discuss the procedure for coupling components together to form a system model based on component modes.

19.2 Component Modes for Constrained Components

Normal Modes

Component normal modes may be classified as fixed-interface normal modes, free-interface normal modes, or hybrid-interface normal modes depending on whether all, none or part of the juncture (interface) coordinates are restrained when the component normal modes are obtained using an eigenvalue problem of the form

$$(\mathbf{k} - \omega^2\mathbf{m})\boldsymbol{\phi} = \mathbf{0} \tag{19.3}$$

In addition, some synthesis methods[19.3,19.4] employ loaded-interface normal modes, wherein stiffness and mass coefficients are added to the matrices $\mathbf{k}_{jj}$ and $\mathbf{m}_{jj}$ of a component. It will be assumed that the modes are normalized with respect to $\mathbf{m}$, that is, that

$$\boldsymbol{\Phi}_n^T\mathbf{m}\boldsymbol{\Phi}_n = \mathbf{I}_{nn}, \qquad \boldsymbol{\Phi}_n^T\mathbf{k}\boldsymbol{\Phi}_n = \boldsymbol{\Lambda}_{nn} \equiv \operatorname{diag}(\omega_r^2) \tag{19.4}$$

where $\boldsymbol{\Phi}_n$ is a matrix whose columns are the component normal modes.† While the complete normal mode set will be identified by a subscript n, as in Eqs. 19.4, the normal mode set will usually be truncated to a set of normal modes which will be denoted by $\boldsymbol{\Phi}_k$ (for *k*ept modes).

Constraint Modes

Let the physical coordinates $\mathbf{u}$ be partitioned into a set C relative to which constraint modes are to be defined, and let V be the complement of C. A *con-*

*Note that Eq. 19.2 is actually an assumed-modes, or Ritz approximation, of the physical displacement vector $\mathbf{u}$.

†Note that $\boldsymbol{\Phi}$ will be used, rather than $\boldsymbol{\Psi}$, when assumed-modes are actually normal modes of free vibration.

straint mode is defined by statically imposing a unit displacement on one physical coordinate in the C set and zero displacement on the remaining coordinates of the C set. Thus, the set of constraint modes is defined by the equation

$$\begin{bmatrix} \mathbf{k}_{vv} & \mathbf{k}_{vc} \\ \mathbf{k}_{cv} & \mathbf{k}_{cc} \end{bmatrix} \begin{bmatrix} \boldsymbol{\Psi}_{vc} \\ \mathbf{I}_{cc} \end{bmatrix} = \begin{bmatrix} \mathbf{0}_{vc} \\ \mathbf{R}_{cc} \end{bmatrix} \tag{19.5}$$

where $\mathbf{R}_{cc}$ is the set of "reactions" at the C coordinates. From the top row partition

$$\boldsymbol{\Psi}_{vc} = -\mathbf{k}_{vv}^{-1}\mathbf{k}_{vc} \tag{19.6}$$

The constraint mode matrix is thus

$$\boldsymbol{\Psi}_c \equiv \begin{bmatrix} \boldsymbol{\Psi}_{vc} \\ \mathbf{I}_{cc} \end{bmatrix} = \begin{bmatrix} -\mathbf{k}_{vv}^{-1}\mathbf{k}_{vc} \\ \mathbf{I}_{cc} \end{bmatrix} \tag{19.7}$$

Attachment Modes

Let A be a subset of P relative to which attachment modes are to be defined. An *attachment mode* is defined as the static deflection of the component which results when a unit force is exerted on one coordinate of the A set, while the remaining coordinates in A are force free. Here we will treat only components that have no rigid-body freedom; in Sec. 19.4 components with rigid-body freedom will be treated. For a restrained component let W be the complement of A in P. Then, the attachment mode set $\boldsymbol{\Psi}_a$ is defined by

$$\begin{bmatrix} \mathbf{k}_{ww} & \mathbf{k}_{wa} \\ \mathbf{k}_{aw} & \mathbf{k}_{aa} \end{bmatrix} \begin{bmatrix} \boldsymbol{\Psi}_{wa} \\ \boldsymbol{\Psi}_{aa} \end{bmatrix} = \begin{bmatrix} \mathbf{0}_{wa} \\ \mathbf{I}_{aa} \end{bmatrix} \tag{19.8}$$

Let the flexibility matrix be designated by $\mathbf{g} \equiv \mathbf{k}^{-1}$. Then the attachment modes for a restrained component are just columns of the flexibility matrix.

$$\boldsymbol{\Psi}_a \equiv \begin{bmatrix} \boldsymbol{\Psi}_{wa} \\ \boldsymbol{\Psi}_{aa} \end{bmatrix} = \begin{bmatrix} \mathbf{g}_{wa} \\ \mathbf{g}_{aa} \end{bmatrix} \tag{19.9}$$

19.3 System Synthesis for Undamped Free Vibration

In this section we describe a generalized substructure coupling, or component modal synthesis, procedure[19.5] as applied to free vibration. Then we provide details on a specific method. The general procedure could be extended to damped structures or to forced vibration, but those topics are beyond the scope of this text.

Let the system be composed of two components, α and β, which have a common (generally redundant) interface. The physical displacements at the interface are constrained by

$$\mathbf{u}_j^\alpha = \mathbf{u}_j^\beta \tag{19.10}$$

and the interface forces are related by

$$\mathbf{f}_j^\alpha + \mathbf{f}_j^\beta = \mathbf{0} \tag{19.11}$$

The derivation of the system equation of motion will be based on Lagrange's equation of motion with undetermined multipliers (see Sec. 11.5). To that end, expressions for the system kinetic energy and potential energy are required. Then,

$$\begin{aligned} T &= \tfrac{1}{2}\dot{\mathbf{p}}^T\boldsymbol{\mu}\dot{\mathbf{p}} = \tfrac{1}{2}\dot{\mathbf{p}}^{\alpha T}\boldsymbol{\mu}^\alpha\dot{\mathbf{p}}^\alpha + \tfrac{1}{2}\dot{\mathbf{p}}^{\beta T}\boldsymbol{\mu}^\beta\dot{\mathbf{p}}^\beta \\ V &= \tfrac{1}{2}\mathbf{p}^T\boldsymbol{\kappa}\mathbf{p} = \tfrac{1}{2}\mathbf{p}^{\alpha T}\boldsymbol{\kappa}^\alpha\mathbf{p}^\alpha + \tfrac{1}{2}\mathbf{p}^{\beta T}\boldsymbol{\kappa}^\beta\mathbf{p}^\beta \end{aligned} \tag{19.12}$$

where

$$\boldsymbol{\mu}^\alpha = \boldsymbol{\Psi}^{\alpha T}\mathbf{m}^\alpha\boldsymbol{\Psi}^\alpha, \qquad \boldsymbol{\kappa}^\alpha = \boldsymbol{\Psi}^{\alpha T}\mathbf{k}^\alpha\boldsymbol{\Psi}^\alpha, \text{ etc.} \tag{19.13}$$

and

$$\mathbf{p} \equiv \begin{Bmatrix} \mathbf{p}^\alpha \\ \mathbf{p}^\beta \end{Bmatrix}, \qquad \boldsymbol{\mu} \equiv \begin{bmatrix} \boldsymbol{\mu}^\alpha & \mathbf{0} \\ \mathbf{0} & \boldsymbol{\mu}^\beta \end{bmatrix}, \qquad \boldsymbol{\kappa} \equiv \begin{bmatrix} \boldsymbol{\kappa}^\alpha & \mathbf{0} \\ \mathbf{0} & \boldsymbol{\kappa}^\beta \end{bmatrix} \tag{19.14}$$

Constraint equations such as Eq. 19.10 can be written in terms of generalized coordinates $\mathbf{p}$ and combined to form a matrix constraint equation of the form

$$\mathbf{Cp} = \mathbf{0} \tag{19.15}$$

The Lagrangian for the system may now be written

$$L = T - V + \boldsymbol{\sigma}^T\mathbf{Cp} \tag{19.16}$$

where $\boldsymbol{\sigma}$ is a vector of Lagrange multipliers. The system equations of motion can now be obtained by applying Lagrange's equation in the form

$$\frac{d}{dt}\left(\frac{\partial L}{\partial \dot{\zeta}_s}\right) - \frac{\partial L}{\partial \zeta_s} = Q_s \tag{19.17}$$

where ζ_s refers to either p_s or σ_s, and where Q_s is the "generalized force." For the free-vibration problem, forces are exerted only at the component interfaces, so

$$\begin{aligned} \delta W &= (\delta\mathbf{u}_j^\alpha)^T\mathbf{f}_j^\alpha + (\delta\mathbf{u}_j^\beta)^T\mathbf{f}_j^\beta \\ &= (\delta\mathbf{u}_j^\alpha)^T(\mathbf{f}_j^\alpha + \mathbf{f}_j^\beta) = 0 \end{aligned} \tag{19.18}$$

where Eqs. 19.10 and 19.11 have been used. Since $\delta W = 0$, $Q_s = 0$. When Eqs. 19.12 and 19.16 are substituted into Eq. 19.17 there results the system equation of motion

$$\boldsymbol{\mu}\ddot{\mathbf{p}} + \boldsymbol{\kappa}\mathbf{p} = \mathbf{C}^T\boldsymbol{\sigma} \tag{19.19}$$

together with the constraint equation, Eq. 19.15.

Reference 19.5 shows that practically all substructure coupling methods solve the coupled set of equations, Eqs. 19.15 and 19.19, by introducing a linear transformation of the form

$$\mathbf{p} = \mathbf{S}\mathbf{q} \tag{19.20}$$

Let $\mathbf{p}$ be rearranged if necessary and partitioned into dependent coordinates, $\mathbf{p}_d$, and linearly independent coordinates, $\mathbf{p}_l$, and let Eq. 19.20 be partitioned accordingly. Then,

$$[\mathbf{C}_{dd}\;\mathbf{C}_{dl}]\begin{Bmatrix}\mathbf{p}_d\\ \mathbf{p}_l\end{Bmatrix} = \mathbf{0} \tag{19.21}$$

where $\mathbf{C}_{dd}$ is a nonsingular square matrix, and the equation

$$\mathbf{p} \equiv \begin{Bmatrix}\mathbf{p}_d\\ \mathbf{p}_l\end{Bmatrix} = \begin{bmatrix}-\mathbf{C}_{dd}^{-1}\mathbf{C}_{dl}\\ \mathbf{I}_{ll}\end{bmatrix}\mathbf{p}_l \equiv \mathbf{S}\mathbf{q} \tag{19.22}$$

defines $\mathbf{S}$ and $\mathbf{q}$. From Eqs. 19.21 and 19.22 it is seen that $\mathbf{CS} = \mathbf{0}$.

Equations 19.19 and 19.20 can now be combined to give

$$\mathbf{M}\ddot{\mathbf{q}} + \mathbf{K}\mathbf{q} = \mathbf{S}^T\mathbf{C}^T\boldsymbol{\sigma} \tag{19.23}$$

where the system mass and stiffness matrices are

$$\mathbf{M} = \mathbf{S}^T\boldsymbol{\mu}\mathbf{S}, \qquad \mathbf{K} = \mathbf{S}^T\boldsymbol{\kappa}\mathbf{S} \tag{19.24}$$

Since $\mathbf{CS} = \mathbf{0}$, Eq. 19.23 reduces to the desired system equation of motion

$$\mathbf{M}\ddot{\mathbf{q}} + \mathbf{K}\mathbf{q} = \mathbf{0} \tag{19.25}$$

Example 19.1

The generalized coordinates employed in many component synthesis methods can be identified with the juncture freedoms and the interior freedoms. Let

$$\mathbf{p}^\alpha = \begin{Bmatrix}\mathbf{p}_i^\alpha\\ \mathbf{p}_j^\alpha\end{Bmatrix}, \qquad \mathbf{p}^\beta = \begin{Bmatrix}\mathbf{p}_i^\beta\\ \mathbf{p}_j^\beta\end{Bmatrix}$$

where specifically $\mathbf{p}_j^\alpha \equiv \mathbf{u}_j^\alpha$, $\mathbf{p}_j^\beta \equiv \mathbf{u}_j^\beta$.

Assume that Eq. 19.10 is the only constraint equation, let $\mathbf{p}_j^\beta$ be the set of dependent coordinates, and let

$$\mathbf{q} = \begin{Bmatrix} \mathbf{p}_i^\alpha \\ \mathbf{p}_j^\alpha \\ \mathbf{p}_i^\beta \end{Bmatrix}$$

a. Write the constraint equation, Eq. 19.10, in the form shown in Eq. 19.21 and determine the appropriate **C** matrix.

b. Determine the appropriate **S** matrix.

c. What is the physical interpretation of the Lagrange multiplier vector $\boldsymbol{\sigma}$ in Eq. 19.19 for this case?

Solution

a. The constraint equation is

$$\mathbf{p}_j^\alpha - \mathbf{p}_j^\beta = \mathbf{0} \tag{1}$$

The dependent coordinates are to be $\mathbf{p}_j^\beta$. Therefore, Eq. 1 can be written in the form of Eq. 19.21 as follows.

$$\left[-\mathbf{I} \;\vdots\; \mathbf{0} \quad \mathbf{I} \quad \mathbf{0}\right] \begin{Bmatrix} \mathbf{p}_j^\beta \\ \hline \mathbf{p}_i^\alpha \\ \mathbf{p}_j^\alpha \\ \mathbf{p}_i^\beta \end{Bmatrix} = \mathbf{0} \tag{2}$$

where the dimensions of the **I** and **0** matrices are determined by the rules of matrix compatibility. Thus,

$$\boxed{\mathbf{C}_{dd} = -\mathbf{I}, \qquad \mathbf{C}_{dl} = [\mathbf{0} \quad \mathbf{I} \quad \mathbf{0}]} \tag{3}$$

b. From the definition of **S** in Eq. 19.22 and from Eq. 3,

$$\boxed{\mathbf{S} = \begin{bmatrix} -\mathbf{C}_{dd}^{-1}\mathbf{C}_{dl} \\ \mathbf{I}_{ll} \end{bmatrix} = \begin{bmatrix} \mathbf{0} & \mathbf{I} & \mathbf{0} \\ \hline \mathbf{I} & \mathbf{0} & \mathbf{0} \\ \mathbf{0} & \mathbf{I} & \mathbf{0} \\ \mathbf{0} & \mathbf{0} & \mathbf{I} \end{bmatrix}} \tag{4}$$

Note that this gives the expected result, namely

$$\begin{Bmatrix} \mathbf{p}_j^\beta \\ \mathbf{p}_i^\alpha \\ \mathbf{p}_j^\alpha \\ \mathbf{p}_i^\beta \end{Bmatrix} = \begin{bmatrix} \mathbf{0} & \mathbf{I} & \mathbf{0} \\ \mathbf{I} & \mathbf{0} & \mathbf{0} \\ \mathbf{0} & \mathbf{I} & \mathbf{0} \\ \mathbf{0} & \mathbf{0} & \mathbf{I} \end{bmatrix} \begin{Bmatrix} \mathbf{p}_i^\alpha \\ \mathbf{p}_j^\alpha \\ \mathbf{p}_i^\beta \end{Bmatrix} \tag{5}$$

which has the form $\mathbf{p} = \mathbf{S}\mathbf{q}$.

c. For this part of the problem it will be more convenient to order $\mathbf{p}$ in the form indicated in Eq. 19.14a. Thus, Eqs. 2 and 5 can be reordered as follows.

$$[\mathbf{0} \quad \mathbf{I} \quad \mathbf{0} \quad -\mathbf{I}] \begin{Bmatrix} \mathbf{p}_i^\alpha \\ \mathbf{p}_j^\alpha \\ \mathbf{p}_i^\beta \\ \mathbf{p}_j^\beta \end{Bmatrix} = \mathbf{0} \tag{6}$$

and

$$\begin{Bmatrix} \mathbf{p}_i^\alpha \\ \mathbf{p}_j^\alpha \\ \mathbf{p}_i^\beta \\ \mathbf{p}_j^\beta \end{Bmatrix} = \begin{bmatrix} \mathbf{I} & \mathbf{0} & \mathbf{0} \\ \mathbf{0} & \mathbf{I} & \mathbf{0} \\ \mathbf{0} & \mathbf{0} & \mathbf{I} \\ \mathbf{0} & \mathbf{I} & \mathbf{0} \end{bmatrix} \begin{Bmatrix} \mathbf{p}_i^\alpha \\ \mathbf{p}_j^\alpha \\ \mathbf{p}_i^\beta \end{Bmatrix} \tag{7}$$

Combining Eqs. 19.14 and 6 with Eq. 19.19 we get

$$\begin{bmatrix} \mu_{ii}^\alpha & \mu_{ij}^\alpha & \mathbf{0} & \mathbf{0} \\ \mu_{ji}^\alpha & \mu_{jj}^\alpha & \mathbf{0} & \mathbf{0} \\ \mathbf{0} & \mathbf{0} & \mu_{ii}^\beta & \mu_{ij}^\beta \\ \mathbf{0} & \mathbf{0} & \mu_{ji}^\beta & \mu_{jj}^\beta \end{bmatrix} \begin{Bmatrix} \ddot{\mathbf{p}}_i^\alpha \\ \ddot{\mathbf{p}}_j^\alpha \\ \ddot{\mathbf{p}}_i^\beta \\ \ddot{\mathbf{p}}_j^\beta \end{Bmatrix} + \begin{bmatrix} \kappa_{ii}^\alpha & \kappa_{ij}^\alpha & \mathbf{0} & \mathbf{0} \\ \kappa_{ji}^\alpha & \kappa_{jj}^\alpha & \mathbf{0} & \mathbf{0} \\ \mathbf{0} & \mathbf{0} & \kappa_{ii}^\beta & \kappa_{ij}^\beta \\ \mathbf{0} & \mathbf{0} & \kappa_{ji}^\beta & \kappa_{jj}^\beta \end{bmatrix} \begin{Bmatrix} \mathbf{p}_i^\alpha \\ \mathbf{p}_j^\alpha \\ \mathbf{p}_i^\beta \\ \mathbf{p}_j^\beta \end{Bmatrix} = \begin{Bmatrix} \mathbf{0} \\ \sigma \\ \mathbf{0} \\ -\sigma \end{Bmatrix} \tag{8}$$

From the form of Eq. 8 it can be seen that σ is the generalized interface force on component α. As in Eq. 19.11, the generalized interface force on component β is equal and opposite, that is, it is $-\sigma$.

Equation 8 in Example 19.1 can be seen to be just a merging of the equations of motion of the two components, written in component generalized coordinates. These equations have been employed directly by previous authors, for example, see Reference 19.2. The use of the Lagrange multiplier technique for the present derivation, while more tedious than the direct derivation used by previous authors, permits constraints in addition to Eq. 19.10 to be incorporated in a systematic manner, as will be seen in one of the methods described subsequently in Example 19.5.

To summarize the generalized component mode synthesis procedure:

Step 1. Choose the component modes to be included in Ψ^α and Ψ^β. This defines $\mathbf{p}^\alpha$ and $\mathbf{p}^\beta$.

Step 2. Form μ^α, μ^β, κ^α, and κ^β.

Step 3. Establish which of the coordinates in $\mathbf{p}$ will be the dependent coordinates, $\mathbf{p}_d$; the remainder form $\mathbf{p}_l \equiv \mathbf{q}$.

Step 4. Write the constraint equations in the form of Eq. 19.15 and solve for $\mathbf{S}$ from Eq. 19.22.

Step 5. Determine $\mathbf{M}$ and $\mathbf{K}$ using Eqs. 19.24.

One of the most straightforward, and also one of the most accurate methods of component mode synthesis is the so-called Craig-Bampton method[19.6]. It differs only slightly from the Hurty Method[19.7], and the two give the same numerical results (e.g., system frequencies, etc.). In Ref. 19.15 Bajan and Feng also described the same modification of Hurty's original method. The *Craig-Bampton method* is described in Example 19.2.

Example 19.2

Let each of the two components in Fig. 19.1 be represented by a set of constraint modes defined for the interface coordinates plus a truncated set of fixed-interface normal modes. The figure below illustrates the two constraint modes for component α plus the lowest-frequency fixed-interface normal mode for component α.

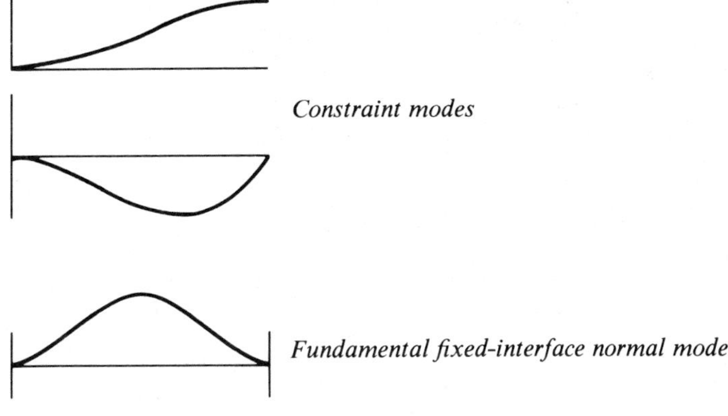

Thus, let

$$\mathbf{u} = \mathbf{\Phi}_k \mathbf{p}_k + \mathbf{\Psi}_c \mathbf{p}_c \tag{1}$$

for each component. Equation 1 can also be written in the partitioned form

$$\begin{Bmatrix} \mathbf{u}_i \\ \mathbf{u}_j \end{Bmatrix} = \begin{bmatrix} \mathbf{\Phi}_{ik} & \mathbf{\Psi}_{ic} \\ \mathbf{0} & \mathbf{I} \end{bmatrix} \begin{Bmatrix} \mathbf{p}_k \\ \mathbf{p}_c \end{Bmatrix} \tag{2}$$

since the normal modes are fixed-interface modes and since $\mathbf{p}_c \equiv \mathbf{u}_j$. Furthermore, from Eq. 19.6, $\mathbf{\Psi}_{ic}$ is given by

$$\mathbf{\Psi}_{ic} = -\mathbf{k}_{ii}^{-1} \mathbf{k}_{ij} \tag{3}$$

Assume that the fixed-interface normal modes have been normalized according to Eqs 19.4.

a. Determine μ^α and κ^α.

b. Let $\mathbf{p}_c^\beta$ be the dependent coordinate (as in Example 19.1). Determine **C** and **S** for this method.

c. Form **M** and **K** for this method.

Solution

a. From Eq. 19.13,

$$\boldsymbol{\mu}^{\alpha} = \boldsymbol{\Psi}^{\alpha^T} \mathbf{m}^{\alpha} \boldsymbol{\Psi}^{\alpha}, \qquad \boldsymbol{\kappa}^{\alpha} = \boldsymbol{\Psi}^{\alpha^T} \mathbf{k}^{\alpha} \boldsymbol{\Psi}^{\alpha} \tag{4}$$

Expanding $\mathbf{m}^{\alpha}$ and $\boldsymbol{\Psi}^{\alpha}$, we get

$$\boldsymbol{\mu}^{\alpha} = \begin{bmatrix} \boldsymbol{\mu}_{kk} & \boldsymbol{\mu}_{kc} \\ \boldsymbol{\mu}_{ck} & \boldsymbol{\mu}_{cc} \end{bmatrix}^{\alpha} = \begin{bmatrix} \boldsymbol{\Phi}_{ik}^{T} & \mathbf{0}^{T} \\ \boldsymbol{\Psi}_{ic}^{T} & \mathbf{I} \end{bmatrix}^{\alpha} \begin{bmatrix} \mathbf{m}_{ii} & \mathbf{m}_{ij} \\ \mathbf{m}_{ji} & \mathbf{m}_{jj} \end{bmatrix}^{\alpha} \begin{bmatrix} \boldsymbol{\Phi}_{ik} & \boldsymbol{\Psi}_{ic} \\ \mathbf{0} & \mathbf{I} \end{bmatrix}^{\alpha} \tag{5}$$

Carrying out the multiplications (and dropping the superscript α on all matrices) we get

$$\begin{aligned} \boldsymbol{\mu}_{kk} &= \mathbf{I}_{kk} \\ \boldsymbol{\mu}_{kc} &= \boldsymbol{\mu}_{ck}^{T} = \boldsymbol{\Phi}_{ik}^{T}(\mathbf{m}_{ii}\boldsymbol{\Psi}_{ic} + \mathbf{m}_{ij}) \\ \boldsymbol{\mu}_{cc} &= \boldsymbol{\Psi}_{ic}^{T}(\mathbf{m}_{ii}\boldsymbol{\Psi}_{ic} + \mathbf{m}_{ij}) + \mathbf{m}_{ji}\boldsymbol{\Psi}_{ic} + \mathbf{m}_{jj} \end{aligned} \tag{6}$$

Thus, in this model there is inertia coupling between the normal mode coordinates and the constraint mode coordinates.

In a similar manner we can determine $\boldsymbol{\kappa}^{\alpha}$. We get

$$\boldsymbol{\kappa}^{\alpha} = \begin{bmatrix} \boldsymbol{\kappa}_{kk} & \boldsymbol{\kappa}_{kc} \\ \boldsymbol{\kappa}_{ck} & \boldsymbol{\kappa}_{cc} \end{bmatrix}^{\alpha} \tag{7}$$

where

$$\begin{aligned} \boldsymbol{\kappa}_{kk} &= \boldsymbol{\Lambda}_{kk} \\ \boldsymbol{\kappa}_{kc}^{T} &= \boldsymbol{\kappa}_{ck} = \mathbf{0} \\ \boldsymbol{\kappa}_{cc} &= \mathbf{k}_{jj} - \mathbf{k}_{ji}\,\mathbf{k}_{ii}^{-1}\,\mathbf{k}_{ij} \end{aligned} \tag{8}$$

It can be seen from the above that the use of constraint modes leads to an uncoupled stiffness matrix in the component generalized coordinates. The matrix $\boldsymbol{\kappa}_{cc}$ is just the "reduced stiffness matrix" widely used in static substructure analysis (see Sec. 16.6).

b. Since $\mathbf{p}_c^{\alpha} = \mathbf{u}_j^{\alpha}$ and $\mathbf{p}_c^{\beta} = \mathbf{u}_j^{\beta}$, the equation of interface displacement compatibility, Eq. 19.10, can be written

$$[\mathbf{0} \;\; \mathbf{I} \;\; \mathbf{0} \;|\; -\mathbf{I}] \begin{Bmatrix} \mathbf{p}_k^{\alpha} \\ \mathbf{p}_c^{\alpha} \\ \mathbf{p}_k^{\beta} \\ \hline \mathbf{p}_c^{\beta} \end{Bmatrix} = \mathbf{0} \tag{9}$$

so

$$\boxed{\mathbf{C} = [\mathbf{0} \;\; \mathbf{I} \;\; \mathbf{0} \;|\; -\mathbf{I}]} \tag{10}$$

with $\mathbf{p}_c^{\beta}$ being the dependent coordinate. (The order of $\mathbf{p}_d$ and $\mathbf{p}_l$ is reversed from that in Eqs. 19.21 and 19.22.)

For this problem it is convenient to determine $\mathbf{S}$ by simply forming $\mathbf{p} = \mathbf{S}\,\mathbf{q}$. Thus,

$$\begin{Bmatrix} \mathbf{p}_k^\alpha \\ \mathbf{p}_c^\alpha \\ \mathbf{p}_k^\beta \\ \mathbf{p}_c^\beta \end{Bmatrix} = \begin{bmatrix} \mathbf{I} & \mathbf{0} & \mathbf{0} \\ \mathbf{0} & \mathbf{0} & \mathbf{I} \\ \mathbf{0} & \mathbf{I} & \mathbf{0} \\ \mathbf{0} & \mathbf{0} & \mathbf{I} \end{bmatrix} \begin{Bmatrix} \mathbf{p}_k^\alpha \\ \mathbf{p}_k^\beta \\ \mathbf{p}_c^\alpha \end{Bmatrix} \tag{11}$$

where we have chosen to group all the normal mode coordinates $\mathbf{p}_k$ prior to the juncture coordinates $\mathbf{p}_c^\alpha$.

c. Equations 19.24 determine $\mathbf{M}$ and $\mathbf{K}$.

$$\mathbf{M} = \mathbf{S}^T \boldsymbol{\mu} \mathbf{S}, \qquad \mathbf{K} = \mathbf{S}^T \boldsymbol{\mu} \mathbf{S} \tag{12}$$

$$M = \begin{bmatrix} \mathbf{I} & \mathbf{0} & \mathbf{0} & \mathbf{0} \\ \mathbf{0} & \mathbf{0} & \mathbf{I} & \mathbf{0} \\ \mathbf{0} & \mathbf{I} & \mathbf{0} & \mathbf{I} \end{bmatrix} \begin{bmatrix} \boldsymbol{\mu}_{kk}^\alpha & \boldsymbol{\mu}_{kc}^\alpha & \mathbf{0} & \mathbf{0} \\ \boldsymbol{\mu}_{ck}^\alpha & \boldsymbol{\mu}_{cc}^\alpha & \mathbf{0} & \mathbf{0} \\ \mathbf{0} & \mathbf{0} & \boldsymbol{\mu}_{kk}^\beta & \boldsymbol{\mu}_{kc}^\beta \\ \mathbf{0} & \mathbf{0} & \boldsymbol{\mu}_{ck}^\beta & \boldsymbol{\mu}_{cc}^\beta \end{bmatrix} \begin{bmatrix} \mathbf{I} & \mathbf{0} & \mathbf{0} \\ \mathbf{0} & \mathbf{0} & \mathbf{I} \\ \mathbf{0} & \mathbf{I} & \mathbf{0} \\ \mathbf{0} & \mathbf{0} & \mathbf{I} \end{bmatrix} \tag{13}$$

Then,

$$\mathbf{M} = \begin{bmatrix} \mathbf{M}_{kk}^\alpha & \mathbf{0} & \mathbf{M}_{kc}^\alpha \\ \mathbf{0} & \mathbf{M}_{kk}^\beta & \mathbf{M}_{kc}^\beta \\ \mathbf{M}_{ck}^\alpha & \mathbf{M}_{ck}^\beta & \mathbf{M}_{cc} \end{bmatrix} \tag{14}$$

where

$$\begin{aligned} \mathbf{M}_{kk}^\alpha &= \mathbf{I}_{kk}^\alpha \\ \mathbf{M}_{kk}^\beta &= \mathbf{I}_{kk}^\beta \\ \mathbf{M}_{kc}^\alpha &= (\mathbf{M}_{ck}^\alpha)^T = \boldsymbol{\mu}_{kc}^\alpha \\ \mathbf{M}_{kc}^\beta &= (\mathbf{M}_{ck}^\beta)^T = \boldsymbol{\mu}_{kc}^\beta \\ \mathbf{M}_{cc} &= \boldsymbol{\mu}_{cc}^\alpha + \boldsymbol{\mu}_{cc}^\beta \end{aligned} \tag{15}$$

The synthesized system stiffness matrix $\mathbf{K}$ is obtained in a similar manner giving

$$\mathbf{K} = \begin{bmatrix} \mathbf{K}_{kk}^\alpha & \mathbf{0} & \mathbf{0} \\ \mathbf{0} & \mathbf{K}_{kk}^\beta & \mathbf{0} \\ \mathbf{0} & \mathbf{0} & \mathbf{K}_{cc} \end{bmatrix} \tag{16}$$

where

$$\begin{aligned} \mathbf{K}_{kk}^\alpha &= \boldsymbol{\Lambda}_{kk}^\alpha, \qquad \mathbf{K}_{kk}^\beta = \boldsymbol{\Lambda}_{kk}^\beta \\ \mathbf{K}_{cc} &= \boldsymbol{\kappa}_{cc}^\alpha + \boldsymbol{\kappa}_{cc}^\beta \end{aligned} \tag{17}$$

Thus, the system equations of motion have only inertial coupling and, since $\boldsymbol{\Lambda}_{kk}^\alpha$ and $\boldsymbol{\Lambda}_{kk}^\beta$ are available from the component eigenproblems, only the following matrices are needed in order to assemble the system matrices: $\boldsymbol{\mu}_{kc}^\alpha$, $\boldsymbol{\mu}_{kc}^\beta$, $\boldsymbol{\mu}_{cc}^\alpha$, $\boldsymbol{\mu}_{cc}^\beta$, $\boldsymbol{\kappa}_{cc}^\alpha$, and $\boldsymbol{\kappa}_{cc}^\beta$.

Reference 19.6 presents the original derivation of the Craig-Bampton method and gives a numerical example. References 19.8 and 19.9 compare the results obtained by using this method with those obtained by using other component mode synthesis methods. Another method will be described in Sec. 19.5.

19.4 Component Modes for Unconstrained Components

In Sec. 19.2 component modes were defined for constrained components, that is, components having no rigid-body freedom. The only significant difference for unconstrained components lies in the addition of rigid-body modes and in the definition of attachment modes.

Normal Modes

Component normal modes are defined just as in Sec. 19.2 except that the set of normal modes may include one or more rigid body (zero frequency) modes.

Constraint Modes

Constraint modes are defined as in Sec. 19.2 except that the set C must be sufficient to prevent rigid-body motion so that $\mathbf{k}_{vv}$ will be nonsingular. Rigid-body modes, defined below, are a special case of constraint modes.

Rigid-Body Modes

Although rigid-body modes may be obtained in the process of solving the eigenproblem for component normal modes, they are also a special case of constraint modes. If a component has N_r rigid-body degrees of freedom, then an R set of coordinates may be used to restrain the component against rigid-body motion. The rigid-body modes corresponding to the R set are obtained by setting $c = r$ in Eq. 19.5 and noting that there is no reaction at the statically determinate constraint set R, that is, $\mathbf{R}_{rr} = \mathbf{0}$. Thus, if V is the complement of R in P, rigid-body modes are defined by

$$\boldsymbol{\Psi}_r \equiv \begin{bmatrix} \boldsymbol{\Psi}_{vr} \\ \mathbf{I}_{rr} \end{bmatrix} = \begin{bmatrix} -\mathbf{k}_{vv}^{-1}\mathbf{k}_{vr} \\ \mathbf{I}_{rr} \end{bmatrix} \tag{19.26}$$

Attachment Modes

Let P be divided into three sets: R, A, and W, where R is a statically determinate constraint set which provides restraint against rigid-body motion, and where the A set consists of physical coordinates where unit forces are to be applied to define attachment modes. Then the attachment modes relative to constraint set R are defined by

$$\left[\begin{array}{cc|c} \mathbf{k}_{ww} & \mathbf{k}_{wa} & \mathbf{k}_{wr} \\ \mathbf{k}_{aw} & \mathbf{k}_{aa} & \mathbf{k}_{ar} \\ \hline \mathbf{k}_{rw} & \mathbf{k}_{ra} & \mathbf{k}_{rr} \end{array}\right] \left[\begin{array}{c} \boldsymbol{\Psi}_{wa} \\ \boldsymbol{\Psi}_{aa} \\ \hline \mathbf{0}_{ra} \end{array}\right] = \left[\begin{array}{c} \mathbf{0}_{wa} \\ \mathbf{I}_{aa} \\ \hline \mathbf{R}_{ra} \end{array}\right] \tag{19.27}$$

As in Sec. 19.2, the attachment modes are essentially columns of a flexibility matrix. Thus,

$$\boldsymbol{\Psi}_a \equiv \begin{bmatrix} \boldsymbol{\Psi}_{wa} \\ \boldsymbol{\Psi}_{aa} \\ \mathbf{0}_{ra} \end{bmatrix} = \begin{bmatrix} \mathbf{g}_{wa} \\ \mathbf{g}_{aa} \\ \mathbf{0}_{ra} \end{bmatrix} \tag{19.28}$$

where $\mathbf{g}_{wa}$ and $\mathbf{g}_{aa}$ are from the inverse of the upper-left partition of $\mathbf{k}$ in Eq. 19.27. In the above formulation the R set may be any convenient set, exclusive of A, which will restrain the component against rigid-body motion.

Inertia Relief Attachment Modes

An alternative manner of defining attachment modes for a component with rigid-body freedom leads to so-called *inertia relief modes.* These were used by MacNeal[19.10] and by Rubin[19.11] in component mode synthesis studies. Inertia relief modes are obtained by applying to a body an equilibrated load system $\mathbf{f}_e$, which consists of the originally specified force vector $\mathbf{f}$ equilibrated by the rigid-body d'Alembert force vector $\mathbf{m}\ddot{\mathbf{u}}_r$, where $\mathbf{u}_r$ is the rigid-body motion due to $\mathbf{f}$. Let the rigid-body modes $\boldsymbol{\Psi}_r$ be orthonormalized so that

$$\boldsymbol{\Psi}_r^T \mathbf{m} \boldsymbol{\Psi}_r = \mathbf{I}_{rr} \tag{19.29}$$

Then

$$\mathbf{f}_e = \mathbf{f} - \mathbf{m}\ddot{\mathbf{u}}_r = \mathbf{P}\mathbf{f} \tag{19.30}$$

where

$$\mathbf{P} = \mathbf{I} - \mathbf{m}\boldsymbol{\Psi}_r\boldsymbol{\Psi}_r^T \tag{19.31}$$

This projection matrix premultiplies the force matrix on the right-hand side of Eq. 19.27 to define attachment modes $\hat{\boldsymbol{\Psi}}_a$ relative to the R constraints. Thus,

$$\begin{bmatrix} \mathbf{k}_{ww} & \mathbf{k}_{wa} & \mathbf{k}_{wr} \\ \mathbf{k}_{aw} & \mathbf{k}_{aa} & \mathbf{k}_{ar} \\ \mathbf{k}_{rw} & \mathbf{k}_{ra} & \mathbf{k}_{rr} \end{bmatrix} \begin{bmatrix} \hat{\boldsymbol{\Psi}}_{wa} \\ \hat{\boldsymbol{\Psi}}_{aa} \\ \mathbf{0}_{ra} \end{bmatrix} = \begin{bmatrix} \mathbf{P}_{ww} & \mathbf{P}_{wa} & \mathbf{P}_{wr} \\ \mathbf{P}_{aw} & \mathbf{P}_{aa} & \mathbf{P}_{ar} \\ \mathbf{P}_{rw} & \mathbf{P}_{ra} & \mathbf{P}_{rr} \end{bmatrix} \begin{bmatrix} \mathbf{0}_{wa} \\ \mathbf{I}_{aa} \\ \mathbf{0}_{ra} \end{bmatrix} \tag{19.32}$$

Since the loads are self-equilibrated, there are no reactions at the R constraints.

References 19.10 and 19.11 remove rigid-body modes from $\hat{\boldsymbol{\Psi}}_a$ by setting

$$\boldsymbol{\Psi}_a = \hat{\boldsymbol{\Psi}}_a + \boldsymbol{\Psi}_r\mathbf{C}_r \tag{19.33}$$

where $\mathbf{C}_r$ is chosen so that

$$\boldsymbol{\Psi}_r^T\mathbf{m}\boldsymbol{\Psi}_a = \mathbf{0} \tag{19.34}$$

This is satisfied if

$$\boldsymbol{\Psi}_a = \mathbf{P}^T\hat{\boldsymbol{\Psi}}_a = (\mathbf{P}^T\mathbf{G}\mathbf{P})\mathbf{F}_a \tag{19.35}$$

where $\mathbf{G}$ is the special flexibility matrix relative to R, as given by

$$\mathbf{G} = \begin{bmatrix} \mathbf{g}_{ww} & \mathbf{g}_{wa} & \mathbf{0} \\ \mathbf{g}_{aw} & \mathbf{g}_{aa} & \mathbf{0} \\ \mathbf{0} & \mathbf{0} & \mathbf{0} \end{bmatrix} \tag{19.36}$$

and

$$\mathbf{F}_a = \begin{bmatrix} \mathbf{0}_{wa} \\ \mathbf{I}_{aa} \\ \mathbf{0}_{ra} \end{bmatrix} \tag{19.37}$$

We will refer to the attachment modes defined by Eq. 19.35 as *inertia relief attachment modes.* In Eq. 19.35 the matrix

$$\mathbf{G}_e = \mathbf{P}^T\mathbf{G}\mathbf{P} \tag{19.38}$$

is the *elastic flexibility matrix.* It can also be shown that

$$\mathbf{G}_e = \boldsymbol{\Phi}_e\boldsymbol{\Lambda}_{ee}^{-1}\boldsymbol{\Phi}_e^T \tag{19.39}$$

where $\boldsymbol{\Phi}_e$ is the set of orthonormal elastic (i.e., flexible) normal modes.

Example 19.3

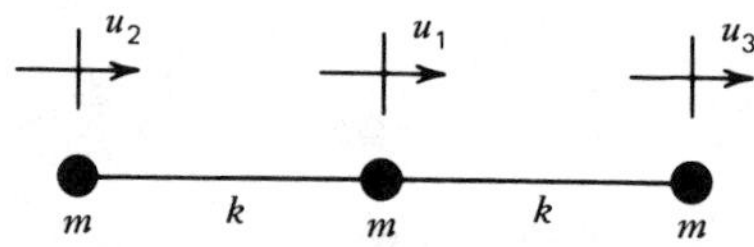

For axial motion of the spring-mass system shown above:

a. Determine the attachment mode as defined by Eqs. (19.27) and (19.28). Sketch the mode.

b. Determine the self-equilibrated elastic load vector $\mathbf{PF}_a$.

c. Determine the inertia relief attachment mode as defined by Eq. (19.35). Sketch the mode.

Let $w \to 1$, $a \to 2$, $r \to 3$. (The order on the figure above is chosen so that the attachment coordinate corresponds to the left-hand mass.) Then,

$$\mathbf{k} = k\begin{bmatrix} 2 & -1 & -1 \\ -1 & 1 & 0 \\ -1 & 0 & 1 \end{bmatrix}, \qquad \mathbf{m} = m\begin{bmatrix} 1 & 0 & 0 \\ 0 & 1 & 0 \\ 0 & 0 & 1 \end{bmatrix}$$

Solution

a. From Eq. 19.28,

$$\boldsymbol{\Psi}_a \equiv \begin{bmatrix} \boldsymbol{\Psi}_{wa} \\ \boldsymbol{\Psi}_{aa} \\ \mathbf{0}_{ra} \end{bmatrix} = \begin{bmatrix} \mathbf{g}_{wa} \\ \mathbf{g}_{aa} \\ \mathbf{0}_{ra} \end{bmatrix} \tag{1}$$

where

$$\begin{bmatrix} \mathbf{g}_{ww} & \mathbf{g}_{wa} \\ \mathbf{g}_{aw} & \mathbf{g}_{aa} \end{bmatrix} = \begin{bmatrix} 2k & -k \\ -k & k \end{bmatrix}^{-1} = \frac{1}{k}\begin{bmatrix} 1 & 1 \\ 1 & 2 \end{bmatrix} \tag{2}$$

Therefore,

$$\boldsymbol{\Psi}_a = \frac{1}{k}\begin{Bmatrix} 1 \\ 2 \\ 0 \end{Bmatrix} \tag{3}$$

$u_2 = \dfrac{2}{k}$ $\qquad u_1 = \dfrac{1}{k}$

b. From Eq. 19.31,

$$\mathbf{P} = \mathbf{I} - \mathbf{m}\boldsymbol{\Psi}_r\boldsymbol{\Psi}_r^T \tag{4}$$

$$\mathbf{F}_a = \begin{bmatrix} \mathbf{0}_{wa} \\ \mathbf{I}_{aa} \\ \mathbf{0}_{ra} \end{bmatrix} \tag{5}$$

There is one rigid-body mode, which could be computed from Eq. 19.26. However, in this case it is clear that all masses displace an equal amount. Thus, the normalized rigid-body mode is

$$\boldsymbol{\Psi}_r = \frac{1}{\sqrt{3m}}\begin{Bmatrix} 1 \\ 1 \\ 1 \end{Bmatrix} \tag{6}$$

$$\mathbf{P} = \begin{bmatrix} 1 & 0 & 0 \\ 0 & 1 & 0 \\ 0 & 0 & 1 \end{bmatrix} - \frac{1}{3m}\begin{bmatrix} m & 0 & 0 \\ 0 & m & 0 \\ 0 & 0 & m \end{bmatrix}\begin{Bmatrix} 1 \\ 1 \\ 1 \end{Bmatrix}\lfloor 1 \quad 1 \quad 1 \rfloor \tag{7}$$

$$\mathbf{P} = \frac{1}{3}\begin{bmatrix} 2 & -1 & -1 \\ -1 & 2 & -1 \\ -1 & -1 & 2 \end{bmatrix}$$

$$\mathbf{F}_a = \begin{Bmatrix} 0 \\ 1 \\ 0 \end{Bmatrix} \tag{8}$$

Therefore,

$$\boxed{\mathbf{PF}_a = \frac{1}{3}\begin{Bmatrix} -1 \\ 2 \\ -1 \end{Bmatrix}} \tag{9}$$

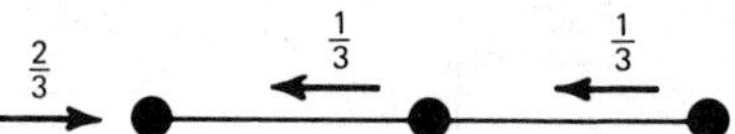

c. From Eq. 19.35 the inertia relief attachment mode is given by

$$\boldsymbol{\Psi}_a = \mathbf{G}_e\mathbf{F}_a = (\mathbf{P}^T\mathbf{GP})\mathbf{F}_a \tag{10}$$

From Eqs. 10, 19.36, 2, and 9,

$$\boldsymbol{\Psi}_a = \frac{1}{3}\begin{bmatrix} 2 & -1 & -1 \\ -1 & 2 & -1 \\ -1 & -1 & 2 \end{bmatrix}\left(\frac{1}{k}\right)\begin{bmatrix} 1 & 1 & 0 \\ 1 & 2 & 0 \\ 0 & 0 & 0 \end{bmatrix}\left(\frac{1}{3}\right)\begin{Bmatrix} -1 \\ 2 \\ -1 \end{Bmatrix} \tag{11}$$

so

$$\boxed{\boldsymbol{\Psi}_a = \frac{1}{9k}\begin{Bmatrix} -1 \\ 5 \\ -4 \end{Bmatrix}} \tag{12}$$

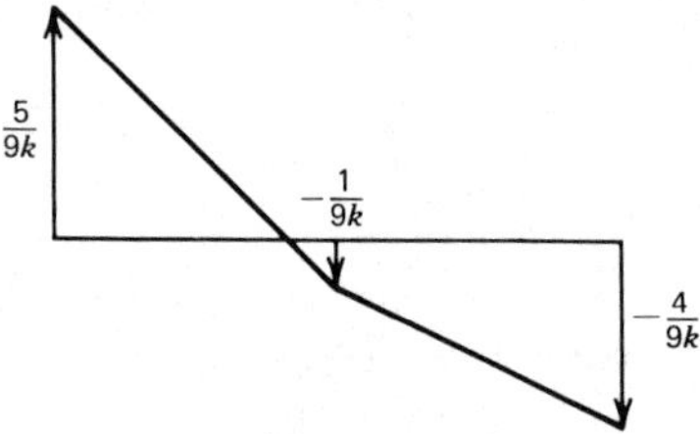

The reason for introducing inertia relief attachment modes by using Eq. 19.35, rather than just using attachment modes as defined by Eq. 19.28, is that this approach is consistent with the mode-acceleration approach of Sec. 15.6, where it was shown that convergence may be improved by use of the mode-acceleration method.

19.5 Residual Flexibility; Residual Component Modes

In Example 19.2 constraint modes and fixed-interface normal modes were employed as component modes. These mode sets are linearly independent since the constraint modes are defined by unit displacement of interface coordinates, while the component normal modes are based on a fully constrained interface. On the other hand, when attachment modes are employed in a modal synthesis method, the question of linear independence of modes arises. For example, if a complete set of free-interface modes were to be supplemented by attachment modes, the latter would be linearly dependent on the former.

Equation 19.39 provides a convenient method for making attachment modes and normal modes linearly independent. Let the elastic normal modes be separated into kept modes $\mathbf{\Phi}_k$ and deleted modes $\mathbf{\Phi}_d$. Then the elastic flexibility matrix is given by

$$\mathbf{G}_e = \mathbf{\Phi}_e\mathbf{\Lambda}_{ee}^{-1}\mathbf{\Phi}_e^T = \mathbf{\Phi}_k\mathbf{\Lambda}_{kk}^{-1}\mathbf{\Phi}_k^T + \mathbf{\Phi}_d\mathbf{\Lambda}_{dd}^{-1}\mathbf{\Phi}_d^T \tag{19.40}$$

The *residual flexibility matrix* is given by

$$\mathbf{G}_d = \mathbf{\Phi}_d\mathbf{\Lambda}_{dd}^{-1}\mathbf{\Phi}_d^T \tag{19.41}$$

This can be obtained without direct knowledge of $\mathbf{\Phi}_d$ and $\mathbf{\Lambda}_{dd}$ by combining Eqs. 19.38, 19.40, and 19.41 to give

$$\mathbf{G}_d = \mathbf{P}^T\mathbf{G}\mathbf{P} - \mathbf{\Phi}_k\mathbf{\Lambda}_{kk}^{-1}\mathbf{\Phi}_k^T \tag{19.42}$$

From Eqs. 19.35 and 19.42 a set of *residual inertia relief attachment modes,* $\mathbf{\Psi}_d$, is obtained.

$$\mathbf{\Psi}_d = \mathbf{G}_d\mathbf{F}_a \tag{19.43}$$

If the structure has no rigid-body modes, $\mathbf{P} = \mathbf{I}$, so Eq. 19.42 reduces to

$$\mathbf{G}_d = \mathbf{G} - \mathbf{\Phi}_k\mathbf{\Lambda}_{kk}^{-1}\mathbf{\Phi}_k^T \tag{19.44}$$

which is the *residual flexibility matrix,* and Eqs. 19.43 and 19.44 may then be combined to give the *residual attachment modes.*

Example 19.4

This example is a continuation of Example 19.3. It illustrates the use of Eqs. 19.42 and 19.43 to determine a residual inertia relief attachment mode for the three-mass system of Example 19.3.

a. Determine the normal modes of the three-mass sytem of Example 19.3.

b. Let $\mathbf{\Phi}_r = \phi_1$, $\mathbf{\Phi}_k = \phi_2$, and $\mathbf{\Phi}_d = \phi_3$. Determine the residual inertia relief attachment mode.

c. Give a physical explanation of your result in (b).

Solution

a. From Example 19.3 the algebraic eigenproblem can be written as

$$\begin{bmatrix} (2k - \omega^2 m) & -k & -k \\ -k & (k - \omega^2 m) & 0 \\ -k & 0 & (k - \omega^2 m) \end{bmatrix} \begin{Bmatrix} U_1 \\ U_2 \\ U_3 \end{Bmatrix} = \begin{Bmatrix} 0 \\ 0 \\ 0 \end{Bmatrix} \tag{1}$$

Then, the frequencies are given by

$$|\mathbf{k} - \omega^2 \mathbf{m}| = 0 \tag{2}$$

or

$$\begin{vmatrix} (2k - \omega^2 m) & -k & -k \\ -k & (k - \omega^2 m) & 0 \\ -k & 0 & (k - \omega^2 m) \end{vmatrix} = 0 \tag{3}$$

This gives the characteristic equation

$$\omega^2 m(\omega^2 m - k)(\omega^2 m - 3k) = 0 \tag{4}$$

whose roots are

$$\begin{aligned} \omega_1^2 &= 0 \\ \omega_2^2 &= \frac{k}{m} \\ \omega_3^2 &= \frac{3k}{m} \end{aligned} \tag{5}$$

Mode shapes can be obtained from the equations

$$\begin{aligned} (2k - \omega^2 m)U_1 - kU_2 - kU_3 &= 0 \\ -kU_1 + (k - \omega^2 m)U_2 &= 0 \end{aligned} \tag{6}$$

For mode 1, $\omega_1^2 = 0$, so

$$\begin{aligned} 2kU_1 - kU_2 - kU_3 &= 0 \\ -kU_1 + kU_2 &= 0 \end{aligned} \tag{7}$$

Then, as expected, the rigid-body mode is given by

$$\mathbf{U}_1 \sim \begin{Bmatrix} 1 \\ 1 \\ 1 \end{Bmatrix}$$

When normalized so that $\boldsymbol{\phi}_1^T \mathbf{m} \boldsymbol{\phi}_1 = 1$,

$$\boxed{\boldsymbol{\phi}_1 = \frac{1}{\sqrt{3m}} \begin{Bmatrix} 1 \\ 1 \\ 1 \end{Bmatrix}} \tag{8}$$

For mode 2, $\omega^2 m = k$, so Eq. 6 becomes

$$\begin{aligned} kU_1 - kU_2 - kU_3 &= 0 \\ -kU_1 &= 0 \end{aligned} \tag{9}$$

Thus, $U_1 = 0$ and $U_2 = -U_3$ or mode 2 has the form

$$\mathbf{U}_2 \sim \begin{Bmatrix} 0 \\ 1 \\ -1 \end{Bmatrix}$$

The normalized mode ϕ_2 is given by

$$\boxed{\phi_2 = \frac{1}{\sqrt{2m}} \begin{Bmatrix} 0 \\ 1 \\ -1 \end{Bmatrix}} \tag{10}$$

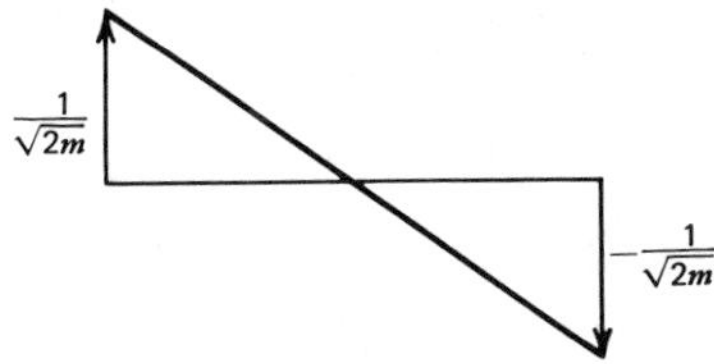

Finally, mode 3 can be shown to be

$$\mathbf{U}_3 \sim \begin{Bmatrix} -2 \\ 1 \\ 1 \end{Bmatrix}, \qquad \phi_3 = \frac{1}{\sqrt{6m}} \begin{Bmatrix} -2 \\ 1 \\ 1 \end{Bmatrix} \tag{11}$$

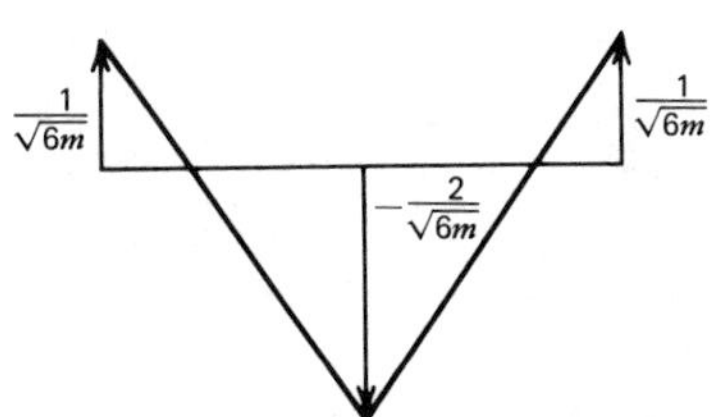

b. From Eqs. 19.42 and 19.43

$$\mathbf{\Psi}_d = (\mathbf{P}^T\mathbf{G}\mathbf{P})\mathbf{F}_a - (\mathbf{\Phi}_k\mathbf{\Lambda}_{kk}^{-1}\mathbf{\Phi}_k^T)\mathbf{F}_a \tag{12}$$

The first term in Eq. 12 was determined in Example 19.3. In the present problem

$$\mathbf{\Phi}_k = \phi_2 \quad \text{and} \quad \mathbf{\Lambda}_{kk} = \omega_2^2 = \frac{k}{m} \tag{13}$$

Hence,

$$\Phi_k \Lambda_{kk}^{-1} \Phi_k^T = \left(\frac{m}{k}\right) \phi_2 \phi_2^T$$

$$= \left(\frac{m}{k}\right)\left(\frac{1}{2m}\right) \begin{Bmatrix} 0 \\ 1 \\ -1 \end{Bmatrix} \lfloor 0 \quad 1 \quad -1 \rfloor \tag{14}$$

$$= \frac{1}{2k} \begin{bmatrix} 0 & 0 & 0 \\ 0 & 1 & -1 \\ 0 & -1 & 1 \end{bmatrix}$$

From Example 19.3,

$$\mathbf{F}_a = \begin{Bmatrix} 0 \\ 1 \\ 0 \end{Bmatrix} \tag{15}$$

Hence,

$$\psi_d = \frac{1}{9k} \begin{Bmatrix} -1 \\ 5 \\ -4 \end{Bmatrix} - \frac{1}{2k} \begin{bmatrix} 0 & 0 & 0 \\ 0 & 1 & -1 \\ 0 & -1 & 1 \end{bmatrix} \begin{Bmatrix} 0 \\ 1 \\ 0 \end{Bmatrix} \tag{16}$$

or

$$\boxed{\psi_d = \frac{1}{18k} \begin{Bmatrix} -2 \\ 1 \\ 1 \end{Bmatrix}} \tag{17}$$

c. As might have been expected, the residual attachment mode in the present case is exactly a multiple of ϕ_3, since it represents deformation of the three-mass system after removing rigid-body motion and after removing mode 2.

In determining the attachment mode, a unit force was placed at node 2 (left end). That applied force can be apportioned to the various modes as shown below.

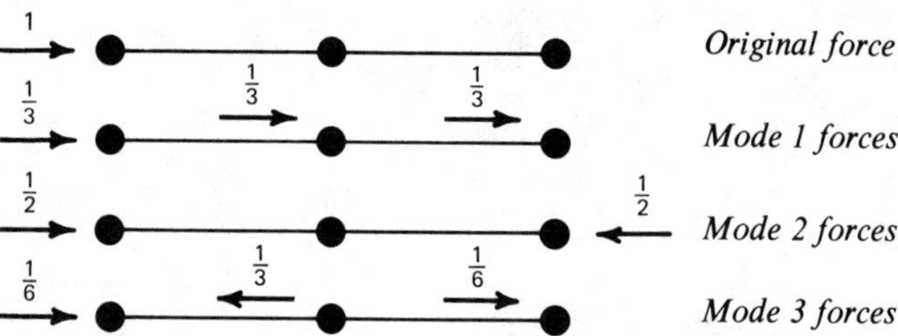

From Eqs. 14 and 15,

$$\Phi_k \Lambda_{kk}^{-1} \Phi_k^T \mathbf{F}_a = \frac{1}{2k} \begin{Bmatrix} 0 \\ 1 \\ -1 \end{Bmatrix}$$

This is exactly the deflection shape which would result from the application of the "Mode 2 forces" above. Also, ψ_d given by Eq. 17 would be exactly the deflection shape produced by the "Mode 3 forces."

From Example 19.4 it can be seen that residual inertia relief attachment modes provide the logical complement to a mode set consisting of rigid-body modes and kept free-interface normal modes.

Residual attachment modes have been employed in several component mode synthesis methods. One method, which was first presented in its present form by Craig and Chang[19.9], will be described below. Following a different procedure Rubin[19.11] arrived at a similar method. The method of Craig and Chang will be presented for constrained components α and β, although it can easily be extended to unrestrained components by using residual inertia relief attachment modes in place of simply residual attachment modes.

Let the displacement transformation equation, Eq. 19.2, be

$$\mathbf{u} = \begin{Bmatrix} \mathbf{u}_i \\ \mathbf{u}_j \end{Bmatrix} = \begin{bmatrix} \mathbf{\Phi}_{ik} & \mathbf{\Psi}_{id} \\ \mathbf{\Phi}_{jk} & \mathbf{\Psi}_{jd} \end{bmatrix} \begin{Bmatrix} \mathbf{p}_k \\ \mathbf{p}_d \end{Bmatrix} \tag{19.45}$$

where the normal mode matrix $\mathbf{\Phi}_k$ consists of free-interface normal modes and $\mathbf{\Psi}_d$ consists of residual attachment modes defined by Eqs. 19.41, 19.43, and 19.44 as

$$\mathbf{\Psi}_d = \mathbf{G}_d\mathbf{F}_a = (\mathbf{\Phi}_d\mathbf{\Lambda}_{dd}^{-1}\mathbf{\Phi}_d^T)\mathbf{F}_a \tag{19.46}$$

or

$$\mathbf{\Psi}_d = \mathbf{G}_d\mathbf{F}_a = (\mathbf{G} - \mathbf{\Phi}_k\mathbf{\Lambda}_{kk}^{-1}\mathbf{\Phi}_k)\mathbf{F}_a \tag{19.47}$$

In Eq. 19.45 the attachment modes are defined by placing unit forces at juncture coordinates, $\mathbf{u}_j$, so from Eq. 19.37

$$\mathbf{F}_a = \begin{bmatrix} \mathbf{0}_{ij} \\ \mathbf{I}_{jj} \end{bmatrix} \tag{19.48}$$

Hence, Eq. 19.46 can be written

$$\mathbf{\Psi}_d = \mathbf{\Phi}_d\mathbf{\Lambda}_{dd}^{-1}\mathbf{\Phi}_{jd}^T = \mathbf{\Phi}_d(\mathbf{\Lambda}_{dd}^{-1}\mathbf{\Phi}_{jd}^T) \tag{19.49}$$

The present method is based on the mode-acceleration method of Sec. 15.3, wherein residual attachment modes are included to account for the flexibility of the "deleted" modes. Let Eq. 19.45 be written in the simpler form

$$\mathbf{u} = [\mathbf{\Phi}_k \quad \mathbf{\Psi}_d] \begin{Bmatrix} \mathbf{p}_k \\ \mathbf{p}_d \end{Bmatrix} \tag{19.50}$$

The component equation of motion (undamped) in terms of physical coordinates is

$$\mathbf{m}\ddot{\mathbf{u}} + \mathbf{k}\mathbf{u} = \mathbf{f} \tag{19.51}$$

When this equation is transformed to component generalized coordinates as defined by Eq. 19.50, the following equations are obtained:

$$\boldsymbol{\mu}_{kk}\ddot{\mathbf{p}}_k + \boldsymbol{\kappa}_{kk}\mathbf{p}_k = \boldsymbol{\Phi}_k^T\mathbf{f} \tag{19.52a}$$

$$\boldsymbol{\mu}_{dd}\ddot{\mathbf{p}}_d + \boldsymbol{\kappa}_{dd}\mathbf{p}_d = \boldsymbol{\Psi}_d^T\mathbf{f} \tag{19.52b}$$

These equations are uncoupled because the modes in $\boldsymbol{\Psi}_d$ are linear combinations of the modes in $\boldsymbol{\Phi}_d$ (as can be seen in Eq. 19.49), which in turn are orthogonal to the modes in $\boldsymbol{\Phi}_k$.

The mode-acceleration concept enters by virtue of the fact that we will now choose to approximate the response of the coordinates $\mathbf{p}_d$ by the pseudostatic response given by ignoring $\ddot{\mathbf{p}}_d$ in Eq. 19.52b, that is,

$$\boldsymbol{\kappa}_{dd}\,\mathbf{p}_d = \boldsymbol{\Psi}_d^T\mathbf{f} \tag{19.53}$$

Since, for free vibration, $\mathbf{f}$ consists only of interface forces $\mathbf{f}_j$, and since

$$\boldsymbol{\kappa}_{dd} = \boldsymbol{\Psi}_d^T\mathbf{k}\boldsymbol{\Psi}_d \tag{19.54}$$

Equations 19.49, 19.53, and 19.54 can be combined to give

$$(\boldsymbol{\Phi}_{jd}\boldsymbol{\Lambda}_{dd}^{-1}\boldsymbol{\Phi}_{jd}^T)(\mathbf{p}_d - \mathbf{f}_j) = \mathbf{0} \tag{19.55}$$

Since the matrix product in parentheses is nonsingular,

$$\mathbf{p}_d = \mathbf{f}_j \tag{19.56}$$

is the pseudostatic approximation to the generalized coordinates $\mathbf{p}_d$ (which are the responses of the *residual* attachment modes).

Since components α and β satisfy Eq. 19.11 when they are coupled together, Eqs. 19.11 and 19.56 can be combined to give the constraint equation

$$\mathbf{p}_d^\alpha + \mathbf{p}_d^\beta = \mathbf{0} \tag{19.57}$$

Now we will apply the techniques of Sec. 19.3 to couple two components whose motion is approximated by Eq. 19.50 and which must satisfy both the displacement compatibility constraint, Eq. 19.10, and, independently, the constraint given by Eq. 19.57. The *Craig-Chang method*[19.5,19.9,19.12] is described in Example 19.5.

Example 19.5

Let each of the two components in Fig. 19.1 be represented by a set of free-interface normal modes and a set of residual attachment modes defined by forces applied to interface coordinates. The figures below illustrate the forces applied to component α to define its attachment modes and they also illustrate one free-interface normal mode.

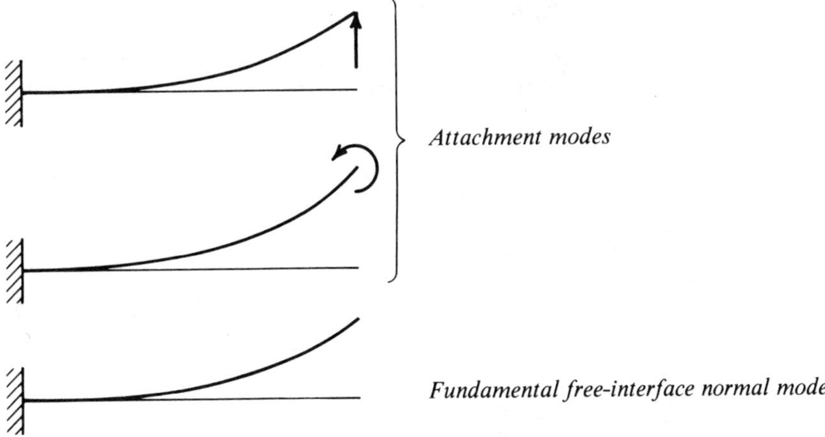

Let

$$\mathbf{u} = [\mathbf{\Phi}_k \quad \mathbf{\Psi}_d] \begin{Bmatrix} \mathbf{p}_k \\ \mathbf{p}_d \end{Bmatrix} = \begin{bmatrix} \mathbf{\Phi}_{ik} & \mathbf{\Psi}_{id} \\ \mathbf{\Phi}_{jk} & \mathbf{\Psi}_{jd} \end{bmatrix} \begin{Bmatrix} \mathbf{p}_k \\ \mathbf{p}_d \end{Bmatrix} \tag{1}$$

a. Determine μ^α and κ^α.

b. Let $\mathbf{p}_d^\alpha$ and $\mathbf{p}_d^\beta$ be the dependent coordinates. Determine **C** and **S** for this method.

c. Form **M** and **K** for this method.

Solution

a. From Eq. 19.13,

$$\mu^\alpha = \mathbf{\Psi}^{\alpha T}\mathbf{m}^\alpha\mathbf{\Psi}^\alpha, \qquad \kappa^\alpha = \mathbf{\Psi}^{\alpha T}\mathbf{k}^\alpha\mathbf{\Psi}^\alpha \tag{2}$$

$$\mu^\alpha = \begin{bmatrix} \mu_{kk} & \mu_{kd} \\ \mu_{dk} & \mu_{dd} \end{bmatrix}^\alpha = \begin{bmatrix} \mathbf{\Phi}_k^T \\ \mathbf{\Psi}_d^T \end{bmatrix}^\alpha \mathbf{m}^\alpha[\mathbf{\Phi}_k \quad \mathbf{\Psi}_d]^\alpha \tag{3}$$

Carrying out the multiplications (and dropping the superscript α on all matrices) and employing Eq. 19.49, we get

$$\begin{aligned} \mu_{kk} &= \mathbf{I}_{kk} \\ \mu_{kd} &= \mu_{dk}^T = \mathbf{0} \\ \mu_{dd} &= \mathbf{\Psi}_d^T\mathbf{m}\mathbf{\Psi}_d = \mathbf{\Phi}_{jd}\mathbf{\Lambda}_{dd}^{-1}\mathbf{\Lambda}_{dd}^{-1}\mathbf{\Phi}_{jd}^T \end{aligned} \tag{4}$$

In a similar manner we can determine κ^α.

$$\kappa^\alpha = \begin{bmatrix} \kappa_{kk} & \kappa_{kd} \\ \kappa_{dk} & \kappa_{dd} \end{bmatrix}^\alpha \tag{5}$$

where

$$\begin{aligned} \kappa_{kk} &= \mathbf{\Lambda}_{kk} \\ \kappa_{kd}^T &= \kappa_{dk} = \mathbf{0} \\ \kappa_{dd} &= \mathbf{\Psi}_d^T \mathbf{k} \mathbf{\Psi}_d = \mathbf{\Phi}_{jd} \mathbf{\Lambda}_{dd}^{-1} \mathbf{\Phi}_{jd}^T = \mathbf{\Psi}_{jd} \end{aligned} \tag{6}$$

Again, Eq. 19.49 is used in establishing the form of κ_{kd} and κ_{dd}. Comparing Eq. 6c with Eq. 19.41 we see that κ_{dd} is just the interface partition of the residual flexibility matrix $\mathbf{G}_d$. Thus, Eq. 19.42 or Eq. 19.44 could be employed in actually calculating κ_{dd}.

b. There are two constraint equations to be enforced, Eq. 19.10 and Eq. 19.57, repeated here

$$\mathbf{u}_j^\alpha - \mathbf{u}_j^\beta = \mathbf{0} \tag{7a}$$

$$\mathbf{p}_d^\alpha + \mathbf{p}_d^\beta = \mathbf{0} \tag{7b}$$

Let the coordinates be ordered as shown in Eq. 19.21, that is,

$$\mathbf{p} = \begin{Bmatrix} \mathbf{p}_d^\alpha \\ \mathbf{p}_d^\beta \\ \hline \mathbf{p}_k^\alpha \\ \mathbf{p}_k^\beta \end{Bmatrix}, \qquad \mathbf{q} = \begin{Bmatrix} \mathbf{p}_k^\alpha \\ \mathbf{p}_k^\beta \end{Bmatrix} \tag{8}$$

Combining Eqs. 1, 7, and 8, we get

$$\boxed{\mathbf{C} \equiv [\mathbf{C}_{dd} \quad \mathbf{C}_{dl}] = \left[\begin{array}{cc|cc} \mathbf{\Psi}_{jd}^\alpha & -\mathbf{\Psi}_{jd}^\beta & \mathbf{\Phi}_{jk}^\alpha & -\mathbf{\Phi}_{jk}^\beta \\ \mathbf{I} & \mathbf{I} & \mathbf{0} & \mathbf{0} \end{array}\right]} \tag{9}$$

From Eq. 19.22

$$\mathbf{S} = \begin{bmatrix} -\mathbf{C}_{dd}^{-1}\mathbf{C}_{dl} \\ \mathbf{I}_{ll} \end{bmatrix} \tag{10}$$

The format of $\mathbf{C}_{dd}$ in Eq. 9 permits its inversion, giving

$$\mathbf{C}_{dd}^{-1} = \begin{bmatrix} \mathbf{k}_1 & \mathbf{k}_1\mathbf{\Psi}_{jd}^\beta \\ -\mathbf{k}_1 & (\mathbf{I} - \mathbf{k}_1\mathbf{\Psi}_{jd}^\beta) \end{bmatrix} \tag{11}$$

where

$$\mathbf{k}_1 = (\mathbf{\Psi}_{jd}^\alpha + \mathbf{\Psi}_{jd}^\beta)^{-1} \tag{12}$$

Equations 9 through 11 can be combined to give

$$\boxed{\mathbf{S} = \begin{bmatrix} -\mathbf{k}_1\mathbf{\Phi}_{jk}^\alpha & \mathbf{k}_1\mathbf{\Phi}_{jk}^\beta \\ \mathbf{k}_1\mathbf{\Phi}_{jk}^\alpha & -\mathbf{k}_1\mathbf{\Phi}_{jk}^\beta \\ \mathbf{I} & \mathbf{0} \\ \mathbf{0} & \mathbf{I} \end{bmatrix}} \tag{13}$$

c. The system mass and stiffness matrices are given by

$$\mathbf{M} = \mathbf{S}^T\boldsymbol{\mu}\mathbf{S}, \qquad \mathbf{K} = \mathbf{S}^T\boldsymbol{\kappa}\mathbf{S} \tag{14}$$

Equations 3 through 6 can be used to rearrange μ and κ according to Eq. 8 to give

$$\mu = \begin{bmatrix} \mu_{dd}^{\alpha} & \mathbf{0} & \mathbf{0} & \mathbf{0} \\ \mathbf{0} & \mu_{dd}^{\beta} & \mathbf{0} & \mathbf{0} \\ \mathbf{0} & \mathbf{0} & \mathbf{I}_{kk}^{\alpha} & \mathbf{0} \\ \mathbf{0} & \mathbf{0} & \mathbf{0} & \mathbf{I}_{kk}^{\beta} \end{bmatrix}, \qquad \kappa = \begin{bmatrix} \kappa_{dd}^{\alpha} & \mathbf{0} & \mathbf{0} & \mathbf{0} \\ \mathbf{0} & \kappa_{dd}^{\beta} & \mathbf{0} & \mathbf{0} \\ \mathbf{0} & \mathbf{0} & \mathbf{\Lambda}_{kk}^{\alpha} & \mathbf{0} \\ \mathbf{0} & \mathbf{0} & \mathbf{0} & \mathbf{\Lambda}_{kk}^{\beta} \end{bmatrix} \tag{15}$$

Carrying out the multiplications of Eq. 14 using Eqs. 13 and 15, and noting from Eq. 6c that Eq. 12 can be written in the form

$$\mathbf{k}_1 = (\kappa_{dd}^{\alpha} + \kappa_{dd}^{\beta})^{-1} \tag{16}$$

we get

$$\mathbf{M} = \begin{bmatrix} \mathbf{M}_{\alpha\alpha} & \mathbf{M}_{\alpha\beta} \\ \mathbf{M}_{\beta\alpha} & \mathbf{M}_{\beta\beta} \end{bmatrix}, \qquad \mathbf{K} = \begin{bmatrix} \mathbf{K}_{\alpha\alpha} & \mathbf{K}_{\alpha\beta} \\ \mathbf{K}_{\beta\alpha} & \mathbf{K}_{\beta\beta} \end{bmatrix} \tag{17}$$

where

$$\begin{aligned} \mathbf{M}_{\alpha\alpha} &= \mathbf{I}_{kk}^{\alpha} + \mathbf{\Phi}_{jk}^{\alpha T}\mathbf{m}_1\mathbf{\Phi}_{jk}^{\alpha} \\ \mathbf{M}_{\alpha\beta} &= \mathbf{M}_{\beta\alpha}^{T} = -\mathbf{\Phi}_{jk}^{\alpha T}\mathbf{m}_1\mathbf{\Phi}_{jk}^{\beta} \\ \mathbf{M}_{\beta\beta} &= \mathbf{I}_{kk}^{\beta} + \mathbf{\Phi}_{jk}^{\beta T}\mathbf{m}_1\mathbf{\Phi}_{jk}^{\beta} \\ \mathbf{K}_{\alpha\alpha} &= \mathbf{\Lambda}_{kk}^{\alpha} + \mathbf{\Phi}_{jk}^{\alpha T}\mathbf{k}_1\mathbf{\Phi}_{jk}^{\alpha} \\ \mathbf{K}_{\alpha\beta} &= \mathbf{K}_{\beta\alpha}^{T} = -\mathbf{\Phi}_{jk}^{\alpha T}\mathbf{k}_1\mathbf{\Phi}_{jk}^{\beta} \\ \mathbf{K}_{\beta\beta} &= \mathbf{\Lambda}_{kk}^{\beta} + \mathbf{\Phi}_{jk}^{\beta T}\mathbf{k}_1\mathbf{\Phi}_{jk}^{\beta} \end{aligned} \tag{18}$$

where

$$\mathbf{m}_1 = \mathbf{k}_1(\mu_{dd}^{\alpha} + \mu_{dd}^{\beta})\mathbf{k}_1 \tag{19}$$

Component mode synthesis using the above method is not as straightforward as synthesis using the Craig-Bampton method of Example 19.2. However, since the final coordinates are just the normal mode coordinates $\mathbf{p}_k^{\alpha}$ and $\mathbf{p}_k^{\beta}$, and since residual attachment modes account for the flexibility of all modes, the method produces very good results. Figure 19.2 gives the results of some numerical convergence studies[19.9,19.12]. Other studies[19.12] have shown even more advantage of the Craig-Chang method when applied to structures having a large number of interface degrees of freedom.

Examples 19.2 and 19.5 have illustrated component mode synthesis when both components are represented by the same types of modes, for example, fixed-interface normal modes plus constraint modes in Example 19.2. It is equally possible to couple components having different representations. For example, a component represented by fixed-interface modes plus constraint modes could be coupled to a component represented by free-interface normal modes plus residual attachment modes. Such a representation has, in fact, been employed for studies of payload dynamics for Space Shuttle payloads.

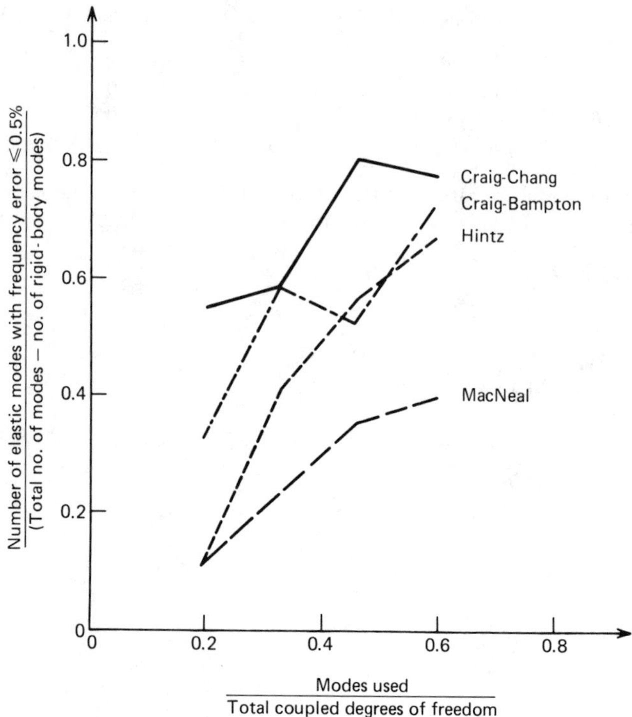

Figure 19.2. A comparison of the number of modes produced with frequency error ≦ 0.5% by various component mode synthesis methods.

Several authors have considered damping synthesis within the context of component mode synthesis methods (e.g., References 19.12, 19.13, and 19.14). In particular, Chang[19.12] has shown that the Craig-Chang method leads to system modal damping that is more accurate than that obtained by using other component mode synthesis methods. Further discussion of damping is, however, beyond the scope of this text.

References

19.1 R. R. Craig, Jr., "Methods of Component Mode Synthesis," *Shock and Vib. Digest,* Naval Research Lab., Washington, DC, v. 9, 3–10 (1977).

19.2 R. M. Hintz, "Analytical Methods in Component Modal Synthesis," *AIAA J.*, v. 13, 1007–1016 (1975).

19.3 W. A. Benfield and R. F. Hruda, "Vibration Analysis of Structures by Component Mode Substitution, *AIAA J.*, v. 9, 1255–1261 (1971).

19.4 S. Goldenberg and M. Shapiro, *A Study of Modal Coupling Procedures for the Space Shuttle,* NASA CR-112252, Grumman Aerospace Corp., Bethpage, NY (1972).

19.5 R. R. Craig, Jr. and C-J. Chang, "A Review of Substructure Coupling Methods for Dynamic Analysis," NASA CP-2001, National Aeronautics and Space Admin., Washington, DC, v. 2, 393–408 (1976).

19.6 R. R. Craig, Jr. and M. C. C. Bampton, "Coupling of Substructures for Dynamic Analysis," *AIAA J.*, v. 6, 1313–1319 (1968).

19.7 W. C. Hurty, "Dynamic Analysis of Structural Systems Using Component Modes," *AIAA J.*, v. 3, 678–685 (1965).

19.8 W. A. Benfield, C. S. Bodley, and G. Morosow, "Modal Synthesis Methods," *Symp. Substr. Testing and Synth.*, NASA Marshall Space Flight Center (1972).

19.9 R. R. Craig, Jr. and C-J. Chang, "On the Use of Attachment Modes in Substructure Coupling for Dynamic Analysis," Paper 77-405, *AIAA/ASME 18th Struct., Struct. Dyn., and Materials Conf.*, San Diego, CA (1977).

19.10 R. H. MacNeal, "A Hybrid Method of Component Mode Synthesis," *Comp. and Struct.*, v. 1, 581–601 (1971).

19.11 S. Rubin, "Improved Component-Mode Representation for Structural Dynamic Analysis," *AIAA J.*, v. 13, 995–1006 (1975).

19.12 C-J. Chang, *A General Procedure for Substructure Coupling in Dynamic Analysis*, Ph.D. Dissertation, The University of Texas at Austin, Austin, TX (1977).

19.13 D. D. Kana and S. Huzar, "Synthesis of Shuttle Vehicle Damping Using Substructure Test Results," *J. Spacecraft and Rockets*, v. 10, 790–797 (1973).

19.14 L. R. Klein and E. H. Dowell, "Analysis of Modal Damping by Component Modes Method Using Lagrange Multipliers," *J. Appl. Mech., Trans. ASME*, v. 41, 527, 528 (1974).

19.15 R. L. Bajan and C. C. Feng, "Free Vibration Analysis by the Modal Substitution Method," Paper 68-8-1, *Space Projections from the Rocky Mountain Region*, American Astronautical Society, Denver, CO (1968).

Problem Set 19.2

The problems in Chapter 19 employ simple axial components and are intended to clarify the concepts and methods presented. Since several of the problems are linked together, the instructor should consider assigning all problems which are related to one another, for example, 19.1, 19.2, 19.5. All of the problems in Problem Set 19.2 employ Fig. P19.1.

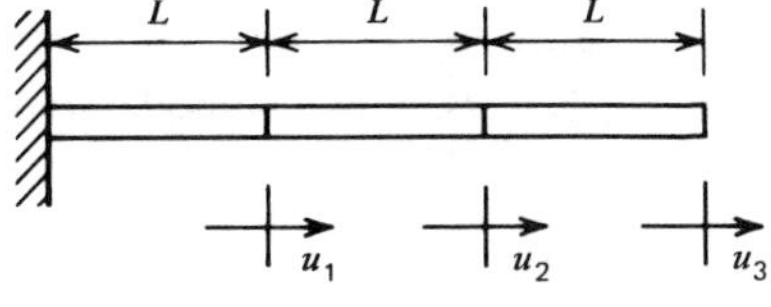

Figure P19.1

19.1 Use Eq. 19.7 to determine the constraint mode for the axial component in Fig. P19.1, where

$$\mathbf{u}_v = \begin{Bmatrix} u_1 \\ u_2 \end{Bmatrix}, \qquad \mathbf{u}_c = \{u_3\}$$

19.2 Letting $u_3 = 0$, determine the two fixed-interface normal modes for the axial bar in Fig. P19.1. Normalize the modes so that $M_1 = M_2 = 1$.

19.3 Use Eq. 19.9 to determine the attachment mode for the bar in Fig. P19.1, where

$$\mathbf{u}_w = \begin{Bmatrix} u_1 \\ u_2 \end{Bmatrix}, \qquad \mathbf{u}_a = \{u_3\}$$

19.4 (a) Determine the three free-interface normal modes for the bar in Fig. P19.1. Scale these three modes so that

$$M_1 = M_2 = M_3 = 1.$$

(b) Using the modal expansion theorem of Eq. 13.34, determine the coefficients c_r in the expansion of the attachment mode determined in Problem 19.3 in terms of the free-interface modes determined in (a) above.

Problem Set 19.3

19.5 In Example 19.2 constraint modes and a truncated set of fixed interface normal modes are used as component modes (see Eq. 2 of Example 19.2.). Repeat parts (a), (b), and (c) of Example 19.2 using attachment modes defined for interface coordinates plus a truncated set of free-interface modes. Assume that the modes form a linearly independent set. The attachment modes are defined for interface coordinates, so $\mathbf{u}_a \equiv \mathbf{u}_j$. The modal transformation is

$$\begin{Bmatrix} \mathbf{u}_i \\ \mathbf{u}_j \end{Bmatrix} = \begin{bmatrix} \mathbf{\Phi}_{ik} & \mathbf{\Psi}_{ia} \\ \mathbf{\Phi}_{jk} & \mathbf{\Psi}_{ja} \end{bmatrix} \begin{Bmatrix} \mathbf{p}_k \\ \mathbf{p}_a \end{Bmatrix}$$

19.6 For each of the two components below, use the component modes determined in Problems 19.1 and 19.2 to construct a model like the general mode-superposition model in Eq. 2 of Example 19.2.

(a) Form κ^α, κ^β, μ^α, μ^β keeping both normal modes for each component.
(b) Use Eqs. 14 and 15 of Example 19.2 to form **M**.
(c) Use Eqs. 16 and 17 of Example 19.2 to form **K**.
(d) Delete the coordinates corresponding to the second normal mode of each component and solve the resulting 3-DOF eigenproblem. Compare your results with exact frequencies, which can be obtained by the method of Sec. 10.1.

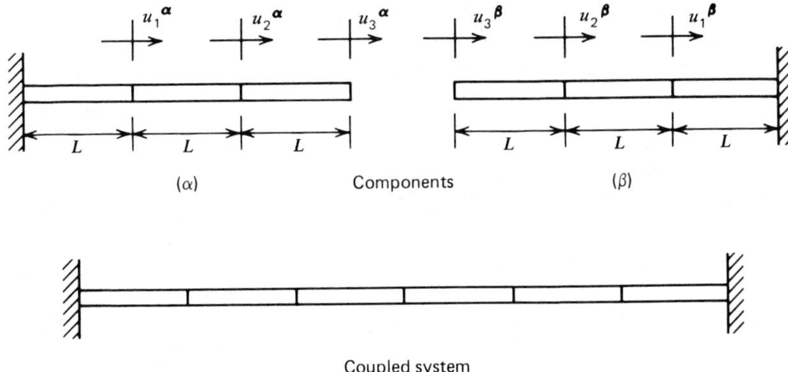

Figure P19.6

Problem Set 19.4

19.7 Repeat Example 19.3 using the bar below (i.e., using a consistent mass matrix). Note the different order of coordinates.

$$\mathbf{u}_w = \{u_1\}, \qquad \mathbf{u}_r = \{u_2\}, \qquad \mathbf{u}_a = \{u_3\}$$

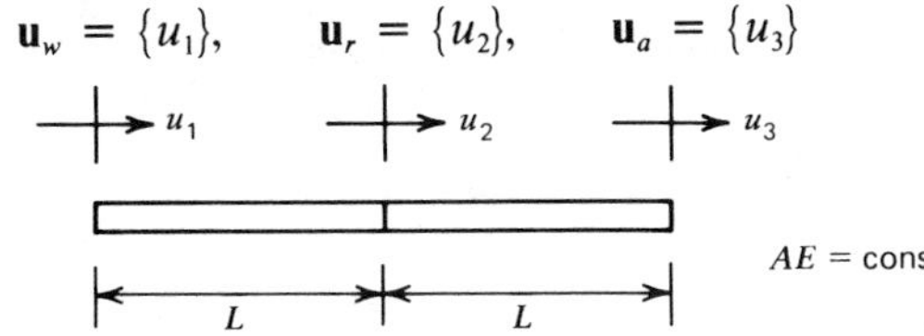

Figure P19.7

Problem Set 19.5

19.8 Determine the residual attachment mode using the attachment mode determined in Problem 19.3 and keeping the lowest-frequency free-interface mode determined in Problem 19.4.

19.9 Repeat Example 19.4 using the 3-DOF bar of Fig. P19.7, that is, using a consistent mass matrix.

20 INTRODUCTION TO EARTHQUAKE RESPONSE OF STRUCTURES

Although the structural dynamics analysis techniques presented so far in this book have applicability to a wide range of structural dynamics problems, one problem that merits special consideration is the analysis and design of structures to withstand earthquake excitation. Many of the ingredients of the analysis of earthquake response of structures which were covered in previous chapters are thus drawn together in the present chapter. However, the treatment here is only intended as a brief introduction to this important topic. Readers who require a more thorough coverage are encouraged to consult a reference on earthquake engineering[20.1-20.3] or a structural dynamics text with more extensive coverage of earthquake analysis and design of structures[20.4].

Upon completion of this chapter you should be able to:

- Define the three types of response spectra $-S_a$, S_v, and S_d in this chapter and describe how they are obtained.
- Use shock spectra to estimate the maximum displacement and maximum spring force for a lumped-mass SDOF system.
- Use shock spectra to estimate the maximum displacement and maximum base shear of a SDOF assumed-modes model of a column subjected to base acceleration.
- Use shock spectra to estimate the maximum displacement and maximum base shear of a lumped-mass MDOF model of a structure subjected to base acceleration.

20.1 Introduction

Although thousands of earthquakes occur each year, and although they are widely distributed over the earth's surface, comparatively few of these cause significant damage to property or loss of life. Those that are of interest to the structural engineer are those that can cause structural damage; these are called *strong-motion earthquakes*. The majority of earthquakes, both large and small, occur in two zones, or belts. One of these is the Circum-Pacific belt, which extends around the Pacific Ocean. The other is the Alpide belt, which extends from the Himalayan mountain range through Iran and Turkey to the Mediterranean Sea. Figure 20.1 illustrates the concentration of seismic activity in these two zones. In the continental United States the most active seismic zone is along the California coast and is associated with the San Andreas fault. The Richter scale provides a convenient means of classifying earthquakes according to size. Earthquakes of magnitude 5.0 or greater on the Richter scale generate ground motions that are severe enough to be damaging to structures. For example, the San Francisco earthquake of 1906 registered 8.2 on the Richter scale (Reference 20.1, p. 77).

Figure 20.2 shows a typical record of acceleration recorded during a strong-motion earthquake and velocity and displacement records obtained by integration of the acceleration record. Two fundamental difficulties encountered in earthquake response analysis are the random nature of the excitation[20.4,20.7,20.8] and the nonlinear nature of the response[20.9,20.10]. It is beyond the scope of this text to treat either of these topics in detail. Rather, earthquake engineers have found that a deterministic approach based on response spectra provides valuable insight into the response of structures to earthquake excitation[20.7,20.11-20.14], so this is the technique that will be presented here. Not only is a consideration of earthquake response important for the design of building structures, but the design of mechanical and electrical equipment to be housed in these buildings must take into account dynamic response to earthquake excitation as transmitted to the equipment through the structure of the building.

The response of a structure to earthquake excitation is a base motion problem (see Secs. 4.4 and 6.2). Section 20.2 treats the response of a SDOF system to earthquake type base motion, while Sec. 20.3 extends the analysis to MDOF systems.

20.2 Response of a SDOF System to Earthquake Excitation: Response Spectra

The earthquake response problem is essentially a base motion problem similar to those introduced in Secs. 4.4 and 6.2. Figure 20.3 shows a SDOF system subjected to ground motion. This is the simplest structural model that captures the base excitation. Since only linear response will be treated in this chapter, Fig. 20.3 represents a linear structure with spring constant k. It is assumed that the base is rigid and has a translational motion $z(t)$. The equations of motion for absolute motion $u(t)$ and relative motion $w(t)$ were found in Example 2.2 to be

$$m\ddot{u} + c\dot{u} + ku = c\dot{z} + kz \tag{20.1}$$

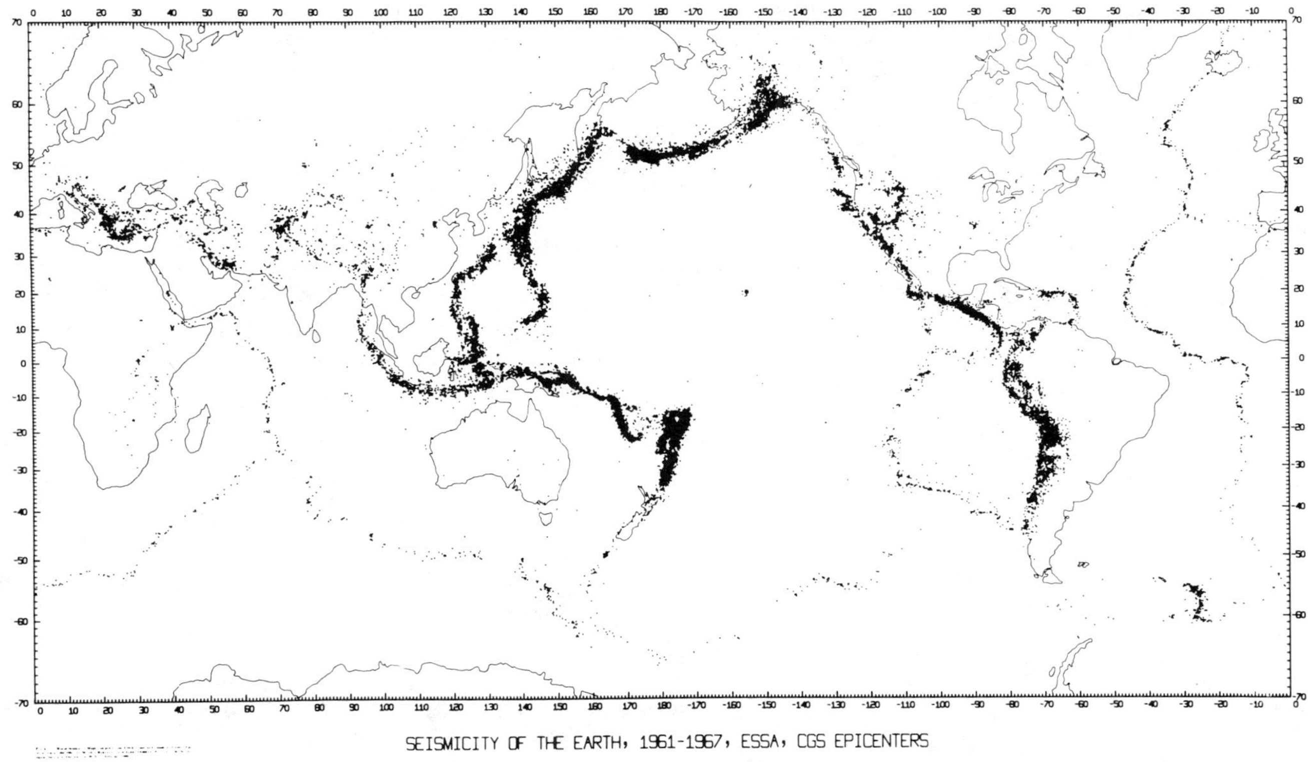

Figure 20.1. Earthquakes recorded during 1961 through 1967. *(M. Barazangi and J. Dorman, "Seismicity of the Earth 1961–67," Seismological Society of America, 1969.)*

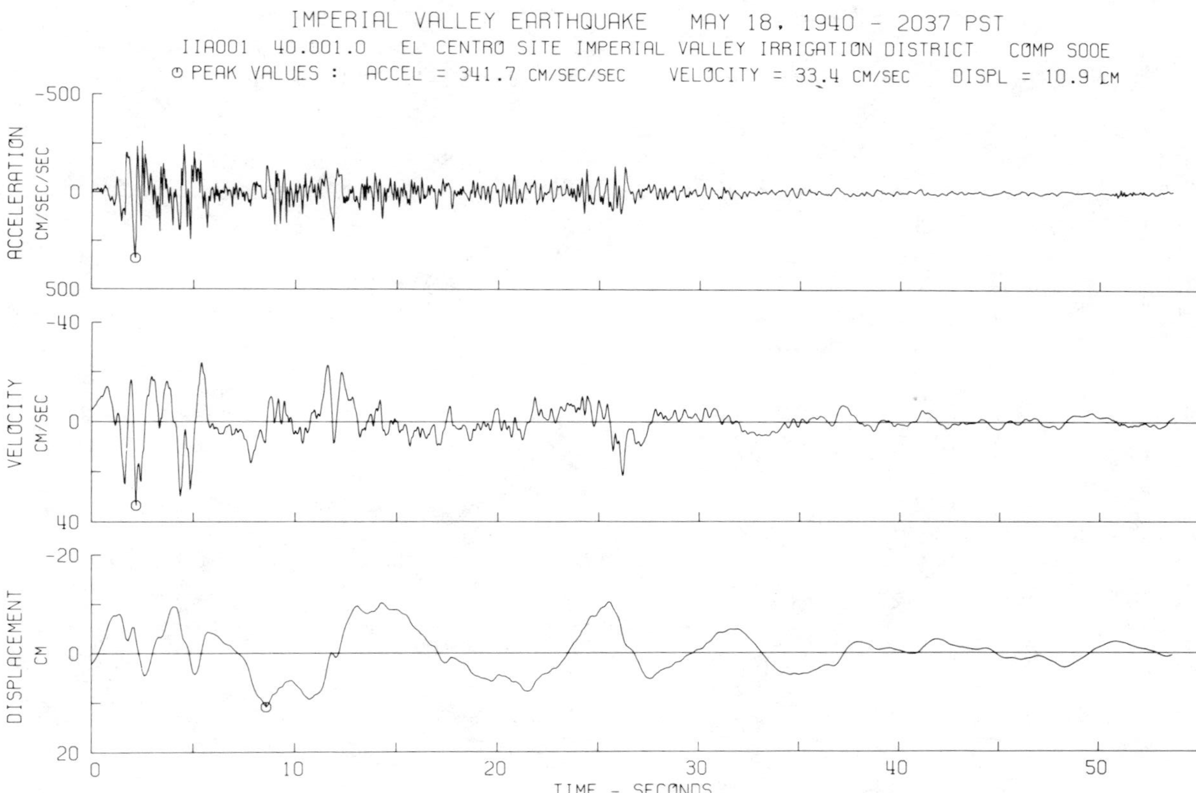

Figure 20.2. *Ground acceleration, velocity and displacement curves for the El Centro earthquake. (D. E. Hudson, et al.,* Strong Motion Earthquake Accelerograms—Vol. II—Corrected Accelerograms and Integrated Ground Velocity and Displacement Curves—Part A, *Earthquake Engineering Research Lab., California Institute of Technology, 1971.)*

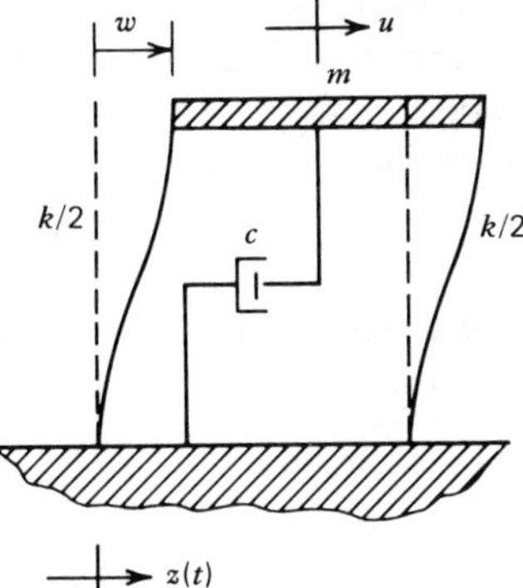

Figure 20.3. *SDOF system subjected to base motion.*

and

$$m\ddot{w} + c\dot{w} + kw = -m\ddot{z} \tag{20.2}$$

In Sec. 6.2 expressions for the maximum relative displacement w_{max} and maximum absolute acceleration (of an undamped system) were obtained. It is customary to ignore the minus sign in Eq. 20.2 [in effect reversing the sense of $z(t)$] and to neglect the small difference between ω_n and ω_d. Thus, the Duhamel integral solution of Eq. 20.2 becomes (see also Eq. 6.12)

$$w(t, \omega_n, \zeta) = \left(\frac{1}{\omega_n}\right) W(t) \tag{20.3}$$

where

$$W(t) = \int_0^t \ddot{z}(\tau) e^{-\zeta\omega_n(t-\tau)} \sin \omega_n(t - \tau)\, d\tau \tag{20.4}$$

The maximum value of relative displacement occurs at time t_m. This is customarily given the symbol S_d and is called the *spectral displacement.*

$$S_d(T, \zeta) = w_{max} = \left(\frac{1}{\omega_n}\right) W(t_m) \tag{20.5}$$

A plot of S_d versus the natural period of the system, $T = 2\pi/\omega_n$, is called the *displacement response spectrum.* The integral in Eq. 20.4 is seen to have the dimensions of velocity. The maximum value of this integral is called the *spectral pseudo-velocity,* S_v. A plot of S_v versus T is called the *pseudovelocity response spectrum.*

$$S_v(T, \zeta) = W(t_m) = \omega_n S_d \tag{20.6}$$

In Eq. 6.16 it was shown that for an undamped system the maximum absolute acceleration is given by $\ddot{u}_{max} = \omega_n^2 w_{max}$. Although this relationship does not hold for a damped system, the maximum absolute acceleration of a lightly damped system can be approximated by the *spectral pseudoacceleration,* S_a, where

$$S_a(T, \zeta) = \omega_n^2 S_d = \omega_n S_v \tag{20.7}$$

A plot of S_a versus T is called the *pseudoacceleration response spectrum.* The usefulness of S_a stems from the fact that the maximum spring force is given by kS_d, and thus

$$(f_s)_{max} = kS_d = \left(\frac{k}{\omega_n^2}\right) S_a = mS_a \tag{20.8}$$

that is, the maximum spring force is obtained by multiplying the spectral pseudoacceleration, S_a, by the mass. Figure 20.4 shows a pseudovelocity response spectrum plotted to linear scales. This gives the maximum pseudovelocity as a function of the natural period of the structure for several values of damping.

The sharp peaks and valleys in Figure 20.4 are the result of local resonances and antiresonances of the ground motion. For design purposes these irregularities can be smoothed out and a number of different response spectra

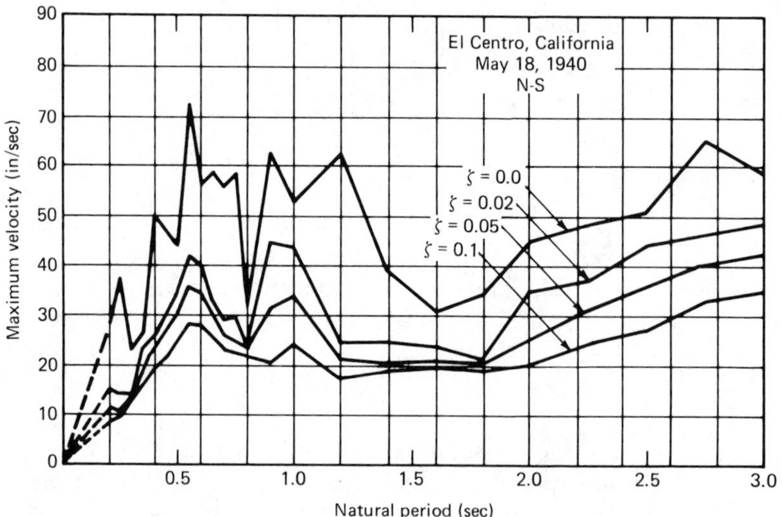

Figure 20.4. *Pseudovelocity response spectrum for the N-S component of the El Centro earthquake of May 18, 1940. (G. W. Housner, "Strong Ground Motion,"* Earthquake Engineering, *R. L. Wiegel, ed., Prentice-Hall, Englewood Cliffs, NJ, 1970.)*

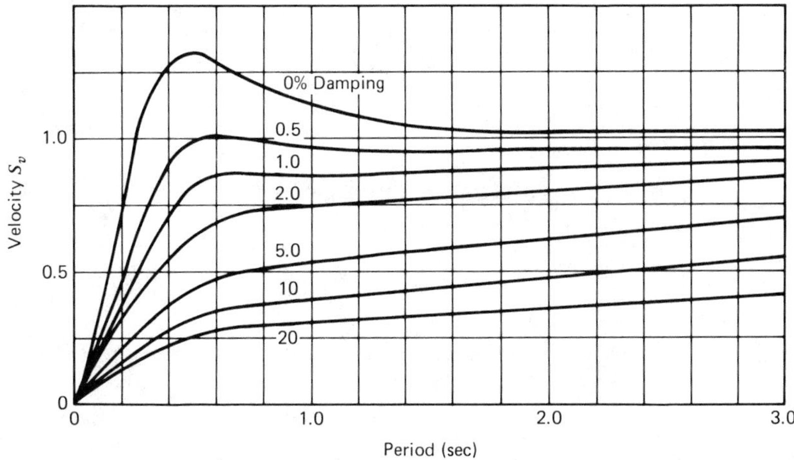

Figure 20.5. *Average velocity response spectrum, 1940 El Centro Intensity. (G. W. Housner, "Design Spectrum,"* Earthquake Engineering, *R. L. Wiegel, ed., Prentice-Hall, Englewood Cliffs, NJ, 1970.)*

averaged after normalizing them to a standard intensity. Figure 20.5 gives such an average velocity response spectrum.

One of the advantages of using pseudovelocity and pseudoacceleration results from the simple relationships given by Eq. 20.7. These relationships make it possible to plot all three response spectra simultaneously on a tripartite log-log graph, as seen in Fig. 20.6.

Earthquake response spectra such as those given in Figs. 20.4 through 20.6 make it possible to calculate the maximum response of any SDOF system to the earthquake from which they were derived. Example 20.1 illustrates such calculations for a lumped-mass SDOF system.

Example 20.1

A SDOF frame structure such as the one shown in Fig. 20.3 has a period $T = 1.0$ s and a damping factor of 5%. The weight is $W = mg = 1500$ lb. Using Fig. 20.6, determine:

a. The maximum relative displacement w_{max}.

b. The maximum base shear f_s.

Solution

For $T = 1.0$ s and $\zeta = 0.05$, Fig. 20.6 gives the following response values.

$S_d = 1.7$ in., $\quad S_v = 10.5$ in./sec, $\quad S_a = 0.17g = 66$ in./sec^2

From Eq. 20.5,

$$\boxed{w_{\max} = S_d = 1.7 \text{ in.}}$$

From Eq. 20.8,

$$(f_s)_{\max} = mS_a = \left(\frac{W}{g}\right) S_a = 1500(0.17) = 255 \text{ lb}$$

$$\boxed{(f_s)_{\max} = 255 \text{ lb}}$$

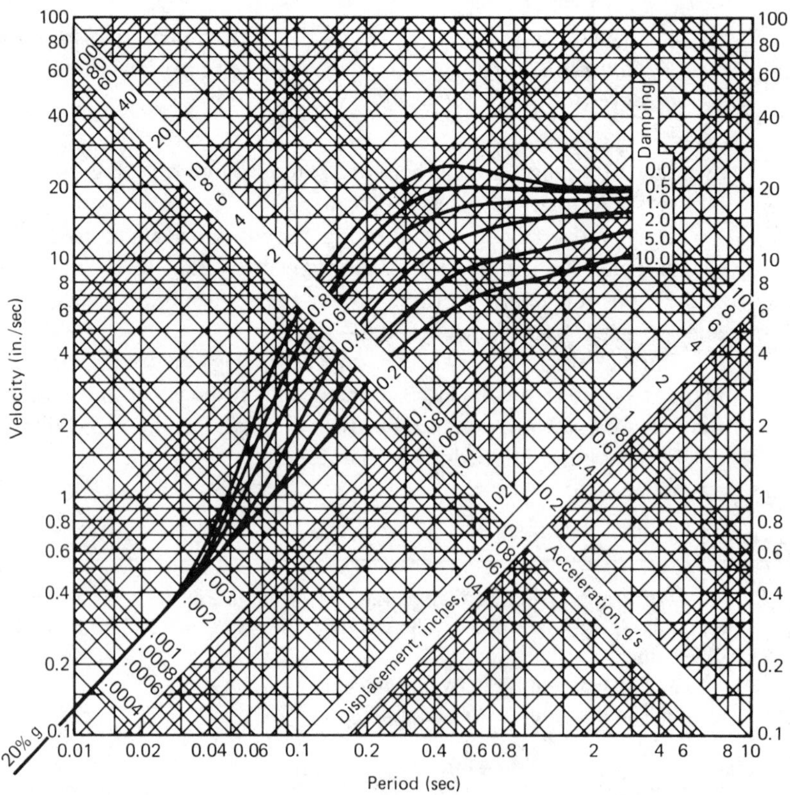

Figure 20.6. Tripartite plot of design spectrum scaled to 20%g at T = *0. (G. W. Housner, "Design Spectrum,"* Earthquake Engineering, *R. L. Wiegel, ed., Prentice-Hall, Englewood Cliffs, NJ, 1970.)*

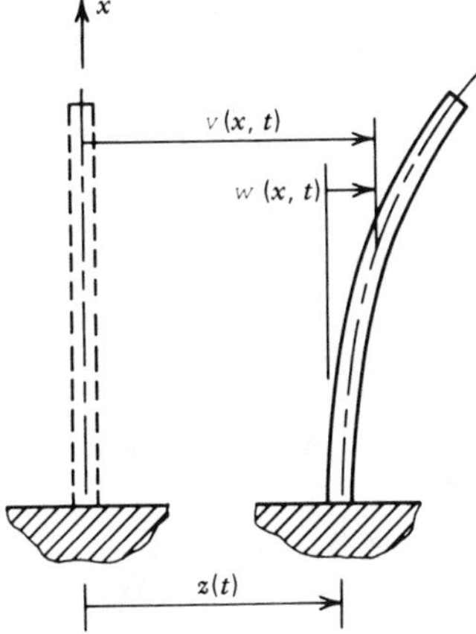

Figure 20.7. *Earthquake excitation of a SDOF generalized-coordinate cantilever column.*

In Example 2.9 a generalized-parameter model was obtained for transverse vibration of a Bernoulli-Euler beam. That model can easily be extended to beam-like structures that can be modeled by a SDOF generalized-parameter model. Figure 20.7 shows such a system. Following Example 2.9 let the relative displacement $w(x, t)$ be represented by the following assumed-modes form

$$w(x, t) = \psi(x)w(t) \tag{20.9}$$

If $\psi(x)$ is normalized so that $\psi(L) = 1$, $w(t)$ represents the relative motion, $w(L, t)$, at the top of the column. From Eq. 2.22 the virtual work is given by

$$\delta W' = \delta W_{nc} - \delta V + \delta W_{inertia} \tag{20.10}$$

If there are no distributed external loads, as is the case in Fig. 20.7, $\delta W_{nc} = 0$. The strain energy is a function of the relative displacement, so, from Eq. 2.24b,

$$\delta V = \int_0^L (EIw'') \, \delta w \, dx \tag{20.11}$$

The inertia force depends on the absolute acceleration. Thus,

$$\delta W_{inertia} = \int_0^L -\rho A(\ddot{w} + \ddot{z}) \, \delta w \, dx \tag{20.12}$$

Combining Eqs. 20.9 through 20.12 we obtain

$$(m\ddot{w} + kw + \ddot{z}\mu) \, \delta w = 0 \tag{20.13}$$

Since $\delta w \neq 0$ in general, then the equation of motion for the SDOF generalized-parameter system is

$$m\ddot{w} + kw = -\ddot{z}\mu \tag{20.14}$$

where, as before,

$$\begin{aligned} m &= \int_0^L \rho A \psi^2 \, dx \\ k &= \int_0^L EI(\psi'')^2 \, dx \end{aligned} \tag{20.15}$$

The base acceleration causes an effective force

$$p_{\text{eff}} = -\ddot{z}\mu \tag{20.16}$$

where μ is an *earthquake participation factor* given by

$$\mu = \int_0^L \rho A \psi \, dx \tag{20.17}$$

If damping proportional to relative velocity is included in the generalized-parameter model, Eq. 20.14 becomes

$$m\ddot{w} + c\dot{w} + kw = -\ddot{z}\mu \tag{20.18}$$

(Note that the right-hand side of Eq. 20.18 involves μ rather than m, which appears in Eq. 20.2.)

For a lumped-mass SDOF system the equation of relative motion is Eq. 20.2 and the maximum relative displacement is given by Eq. 20.5. By comparing Eq. 20.18 to Eq. 20.2 it can be seen that for a distributed mass system

$$w_{\max} = \left(\frac{\mu}{m}\right) S_d = \left(\frac{\mu}{m\omega_n}\right) S_v \tag{20.19}$$

where Eq. 20.6 has also been used.

It is of interest to determine the base shear and overturning moment due to earthquake-related base motion. The base shear will be determined here. From Eqs. 9.16 and 9.17 the shear could be expressed in terms of the third derivative of $w(x, t)$. However, since the assumed-mode solution in Eq. 20.9 is only approximate, the third derivative would not provide an acceptable approximation of the shear. In Eq. 20.8 it was shown that the maximum spring force can be obtained for a lumped-mass SDOF system by multiplying the mass by the spectral pseudoacceleration, which is ω_n^2 times the spectral displacement. For the generalized-parameter model of Fig. 20.7 we can define an effective acceleration $\ddot{w}_e$ by the equation

$$\ddot{w}_e = \omega_n^2 w(x, t) \tag{20.20}$$

Combining Eqs. 20.9 and 20.19, we get

$$\ddot{w}_e = \omega_n^2 \psi(x) w(t) \tag{20.21}$$

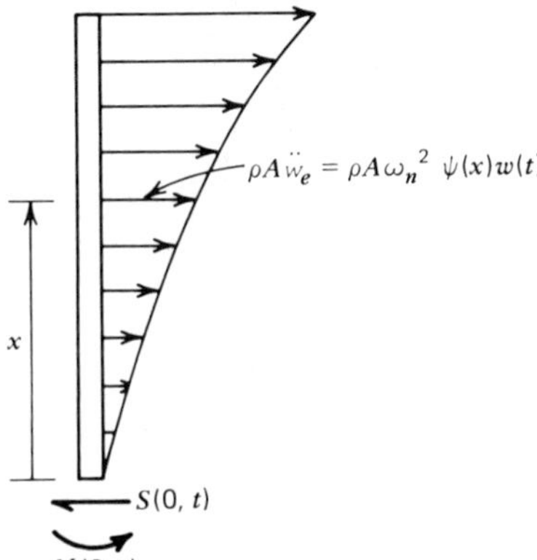

Figure 20.8. Effective inertia loads on a cantilever column.

From Fig. 20.8 the base shear $S(0, t)$ is given by

$$S(0, t) = \omega_n^2 w(t) \int_0^L \rho A \psi(x)\, dx = \mu \omega_n^2 w(t) \tag{20.22}$$

From Eqs. 20.19, 20.22, and 20.7 the maximum base shear is given by

$$S_{\max}(0, t) = \mu \omega_n^2 \left(\frac{\mu}{m}\right) S_d = \left(\frac{\mu^2}{m}\right) S_a \tag{20.23}$$

The base overturning moment can be determined from Fig. 20.8 in a similar fashion.

Example 20.2

A uniform cantilever column similar to the ones shown in Figs. 20.7 and 20.8 has a weight W, a period of 1.0 s, and a damping factor of 5%. Assume a deflection shape function

$$\psi(x) = \left(\frac{x}{L}\right)^2$$

and use the response spectrum of Fig. 20.6 to determine:

a. The maximum displacement at the top of the column.

b. The maximum base shear as a fraction of the total weight of the column.

Solution

a. Since the period and damping factor are the same as those of Example 20.2,

$$S_d = 1.7 \text{ in.}, \qquad S_v = 10.5 \text{ in./sec}, \qquad S_a = 0.17g$$

From Eqs. 20.9 and 20.19

$$w_{\max_t}(L, t) \equiv w_{\max} = \left(\frac{\mu}{m}\right) S_d \tag{1}$$

where, from Eqs. 20.15a and 20.17

$$m = \int_0^L \rho A \psi^2 \, dx \tag{2}$$

$$\mu = \int_0^L \rho A \psi \, dx \tag{3}$$

$$m = \frac{W}{gL} \int_0^L \psi^2 \, dx = \frac{W}{g}\left(\frac{1}{5}\right) \tag{4}$$

$$\mu = \frac{W}{gL} \int_0^L \psi \, dx = \frac{W}{g}\left(\frac{1}{3}\right) \tag{5}$$

Therefore,

$$w_{\max} = \frac{(1/3)}{(1/5)} (1.7) \tag{6}$$

or

$$\boxed{w_{\max} = 2.8 \text{ in.}} \tag{7}$$

b. From Eq. 20.23

$$S_{\max}(0, t) = \left(\frac{\mu^2}{m}\right) S_a = \frac{(1/3)^2 (W/g)^2}{(1/5)(W/g)} (0.17g) \tag{8}$$

so

$$S_{\max}(0, t) = \left(\frac{5}{9}\right) (0.17W) = 0.094W \tag{9}$$

or

$$\boxed{S_{\max}(0, t) = 9.4\% W} \tag{10}$$

20.3 Response of MDOF Systems to Earthquake Excitation

The methods described in Chapters 15 and 18 for determining the response of MDOF systems to dynamic excitation can, of course, be applied to earthquake problems. That is, for specified ground motion time histories the response of

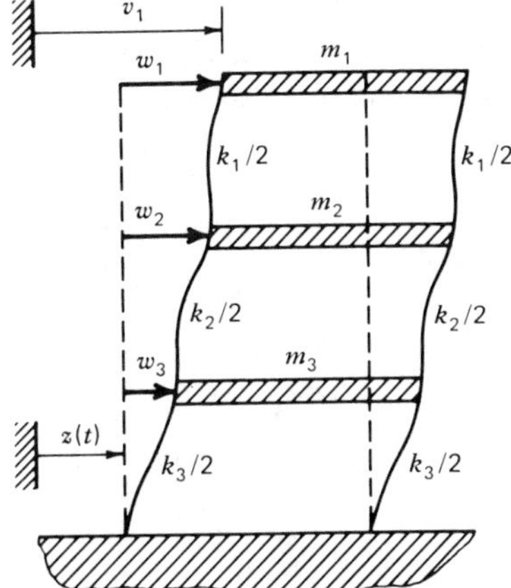

Figure 20.9. Multistory building subjected to earthquake excitation.

MDOF systems can be determined by using mode-superposition or direct integration of the equations of motion. However, some of the simplifications made possible for SDOF systems by the introduction of response spectra can be extended to MDOF systems. In this section relatively simple MDOF systems such as the lumped-mass multistory building model of Fig. 20.9 will be treated. In Sec. 20.4 some of the limitations of this simple model will be discussed briefly. Because response spectra are available for SDOF systems, the mode-superposition method of Chapter 15 will be applied.

For the building model of Fig. 20.9 it is assumed that there is no rotation of the base and that the base moves as a rigid body with displacement $z(t)$. The equations of motion can easily be shown to have the form

$$\mathbf{m}\ddot{\mathbf{v}} + \mathbf{c}\dot{\mathbf{w}} + \mathbf{k}\mathbf{w} = \mathbf{0} \tag{20.24}$$

if viscous damping proportional to the relative displacement is assumed. In vector form

$$\mathbf{v} = \mathbf{w} + \mathbf{1}z \tag{20.25}$$

where $\mathbf{1}$ is the column vector of 1's. Hence, Eq. 20.24 becomes

$$\mathbf{m}\ddot{\mathbf{w}} + \mathbf{c}\dot{\mathbf{w}} + \mathbf{k}\mathbf{w} = \mathbf{p}_{\text{eff}}(t) \tag{20.26}$$

where

$$\mathbf{p}_{\text{eff}}(t) = -\mathbf{m}\mathbf{1}\ddot{z}(t) \tag{20.27}$$

The use of mode-superposition to solve for the response of MDOF systems was treated extensively in Chapter 15. In particular, Eqs. 15.8 and 15.9 apply to MDOF systems with viscous damping, and Eqs. 15.31 through 15.33 restrict the damping to a form for which the equations of motion in principal coordinates are uncoupled. That is, a mode-superposition solution

$$\mathbf{w} = \mathbf{\Phi}\boldsymbol{\eta} = \sum_{r=1}^{N} \boldsymbol{\phi}_r \eta_r(t) \tag{20.28}$$

leads to the uncoupled equations

$$\ddot{\eta}_r + 2\zeta_r\omega_r\dot{\eta}_r + \omega_r^2\eta_r = \left(\frac{1}{M_r}\right) P_r(t) \tag{20.29}$$

where, from Eqs. 15.17 and 20.27

$$P_r(t) = \boldsymbol{\phi}_r^T \mathbf{m} \boldsymbol{1} \ddot{z}(t) \tag{20.30}$$

(As before, the negative sign in Eq. 20.27 is neglected in Eq. 20.30). By analogy with Eqs. 20.16 and 20.17 we can define a *modal earthquake participation factor* μ_r such that

$$\mu_r = \boldsymbol{\phi}_r^T \mathbf{m} \boldsymbol{1} = \boldsymbol{1}^T \mathbf{m} \boldsymbol{\phi}_r \tag{20.31}$$

and

$$P_r(t) = \mu_r \ddot{z}(t) \tag{20.32}$$

Finally, by analogy with Eqs. 20.2 through 20.4

$$\eta_r(t) = \left(\frac{\mu_r}{M_r\omega_r}\right) W_r(t) \tag{20.33}$$

where

$$W_r(t) = \int_0^t \ddot{z}(\tau) e^{-\zeta_r\omega_r(t-\tau)} \sin \omega_r(t - \tau)\, d\tau \tag{20.34}$$

Finally, from Eqs. 20.28 and 20.33,

$$\mathbf{w} = \sum_{r=1}^{N} \boldsymbol{\phi}_r \left(\frac{\mu_r W_r(t)}{M_r\omega_r}\right) \tag{20.35}$$

Although Eq. 20.35 gives a very straightforward means of determining the time history of the relative motion of a MDOF system subjected to earthquake excitation, it involves the integration of the various modal response equations of the form of Eq. 20.34. Even though the response of a many-DOF system can be acceptably represented by the response of the first few modes, Eq. 20.35 represents a tedious procedure which only produces the response to one particular base motion acceleration history, $\ddot{z}(t)$.

The maximum relative displacements due to the rth mode can be expressed in terms of response spectra by

$$\max_t |\mathbf{w}_r(t)| = |\phi_r| \left(\frac{|\mu_r|}{M_r \omega_r} \right) S_v(T_r, \zeta_r) \tag{20.36}$$

where the absolute value notation $|\mathbf{w}_r(t)|$ and $|\phi_r|$ implies the magnitude of each DOF. The difficulty arises from the fact that each mode in Eq. 20.35 may reach its maximum, given by Eq. 20.36, at a different time, and hence

$$\max_t |\mathbf{w}(t)| \neq \sum_{r=1}^{N} \max_t |\mathbf{w}_r(t)| \tag{20.37}$$

A very satisfactory estimate of the maximum of any response quantity can frequently be obtained by the root-mean-square (rms) method, that is

$$\max_t |Q(t)| \doteq \left[\sum_{r=1}^{N} (\max_t |Q_r(t)|)^2 \right]^{1/2} \tag{20.38}$$

where Q is any response quantity. Hence, the maximum relative displacement response can be approximated by the rms response

$$\max_t |\mathbf{w}(t)| \doteq \left\{ \sum_{r=1}^{N} \left[\left(\frac{|\phi_r||\mu_r|}{M_r \omega_r} \right) S_v(T_r, \zeta_r) \right]^2 \right\}^{1/2} \tag{20.39}$$

The rms value may overestimate or underestimate the true value of maximum response. This estimate is generally good for systems with well separated frequencies, but may be quite poor if the system has closely spaced frequencies. Reference 20.2 discusses methods for estimating the maximum response of such systems for which Eq. 20.38 does not hold.

In order to calculate the modal forces resulting from earthquake excitation of a MDOF system such as the building in Fig. 20.9 an effective modal acceleration can be defined which, by analogy with Eq. 20.20, is given by

$$(\ddot{\mathbf{w}}_r)_e = \omega_r^2 \mathbf{w}_r \tag{20.40}$$

The inertia force due to the effective acceleration is given by $\mathbf{m}(\ddot{\mathbf{w}}_r)_e$. For a structure such as the one in Fig. 20.9 the contribution of the rth mode to the base shear is, by analogy with Eq. 20.22,

$$\begin{aligned} S_r(t) &= \sum_{i=1}^{N} m_i(\ddot{w}_{ir})_e = \sum_{i=1}^{N} \omega_r^2 m_i w_{ir} \\ &= \mathit{1}^T \mathbf{m} \phi_r \omega_r^2 \eta_r(t) \\ &= \mathit{1}^T \mathbf{m} \phi_r \left(\frac{\mu_r \omega_r}{M_r} \right) W_r(t) \\ &= \left(\frac{\mu_r^2 \omega_r}{M_r} \right) W_r(t) \end{aligned} \tag{20.41}$$

The maximum base shear due to the rth mode can thus be expressed as

$$\max_t |S_r(t)| = \left(\frac{\mu_r^2}{M_r}\right) S_a(T_r, \zeta_r) \tag{20.42}$$

As with the displacement maximum, the base shear maximum can be approximated by the root-mean-square value

$$\max_t |S(t)| \doteq \left\{\sum_{r=1}^{N}\left[\left(\frac{\mu_r^2}{M_r}\right) S_a(T_r, \zeta_r)\right]^2\right\}^{1/2} \tag{20.43}$$

Example 20.3

For the four-story building of Example 15.1 determine the following:

a. S_d, S_v, and S_a for each mode, if each mode has 5% damping and if Fig. 20.6 gives the response spectra.

b. A root-mean-square estimate of the maximum displacement of the top mass.

c. A root-mean-square estimate of the maximum base shear.

Solution

a. From the ω_r values given in Example 15.1 the periods T_r are calculated. The corresponding points on the 5% damping curve of Fig. 20.6 are determined and the values of S_d, S_v, and S_a are read from the curves. These values are tabulated below.

r	T_r(sec)	S_d(in.)	S_v(in./sec)	S_a(g's)
1	0.47	0.60	8.0	0.28
2	0.21	0.14	4.0	0.30
3	0.15	0.07	2.7	0.28
4	0.11	0.03	1.6	0.24

b. From Eq. 20.39,

$$\max_t |w_1(t)| \doteq \left(\sum_{r=1}^{4}\left[\left(\frac{|\phi_{1r}||\mu_r|}{M_r\omega_r}\right) S_{vr}\right]^2\right)^{1/2} \tag{1}$$

From Eq. 20.31,

$$\mu_r = \boldsymbol{\phi}_r^T\mathbf{m}\mathbf{1} = \mathbf{1}^T\mathbf{m}\boldsymbol{\phi}_r = \sum_{i=1}^{N} m_i\phi_{ir} \tag{2}$$

Therefore, using values from Example 15.1 we get

$$\mu_1 = 4.2565, \qquad \mu_2 = -1.5919$$
$$\mu_3 = -1.3425, \qquad \mu_4 = -0.6526$$

Then,

$$\max_t |w_1(t)| \doteq \left[\left(\frac{1(4.2565)(8.0)}{2.873(13.294)}\right)^2 + \left(\frac{1(1.5919)(4.0)}{2.177(29.660)}\right)^2 + \left(\frac{0.9015(1.3425)(2.7)}{4.367(41.079)}\right)^2 + \left(\frac{0.1544(0.6526)(1.6)}{3.642(55.882)}\right)^2\right]^{1/2}$$

$$\max_t |w_1(t)| \doteq (0.7949 + 0.0097 + 0.0003 + 0)^{1/2}$$

$$\boxed{\max_t |w_1(t)| \doteq 0.897 \text{ in.}}$$

Note that the maximum response due to mode 1 alone is 0.8916. Hence, mode 1 contributes practically all of the response due to this earthquake excitation.

c. The maximum base shear can be approximated by using Eq. 20.43

$$\max_t |S(t)| \doteq \left(\sum_{r=1}^{N}\left[\left(\frac{\mu_r^2}{M_r}\right) S_{ar}\right]^2\right)^{1/2} \tag{3}$$

Thus,

$$\max_t |S(t)| \doteq \left[\left(\frac{(4.2565)^2(0.28)(386)}{2.873}\right)^2 + \left(\frac{(1.5919)^2(0.30)(386)}{2.177}\right)^2 + \left(\frac{(1.3425)^2(0.28)(386)}{4.367}\right)^2 + \left(\frac{(0.6526)^2(0.24)(386)}{3.642}\right)^2\right]^{1/2}$$

$$\boxed{\max_t |S(t)| \doteq 696.3 \text{ k}}$$

Again, most of the base shear is contributed by the first mode.

20.4 Further Considerations

The treatment of response of structures to earthquake excitation presented in this chapter is, of necessity, brief and limited in scope. As has been indicated, however, the tools necessary for linear analysis have been presented in detail and some of the computational tools necessary for nonlinear analysis have also been presented. The references at the end of the chapter provide resources for the study of more difficult problems in design of earthquake-resistant structures, for example, deterministic analysis of linear response, nondeterministic analysis of both linear and nonlinear response, multiple-support excitation, soil-structure interaction, base rotation, and so forth.

Whereas much attention has been given to the analysis and design of building structures to resist earthquakes, design of the equipment housed within the buildings to withstand dynamic excitation has received far less attention. Equipment such as pumps, compressors, piping systems, and power generators must be functional in the aftermath of a disaster such as a major earthquake. Reference 20.15 treats one class of problems associated with equipment design, the design of lightly damped relatively light-weight equipment.

Some of the aspects of design of earthquake-resistant structures are also present in other structural design problems. For example, the design of spacecraft and space payloads for transient environments shares some of the design aspects of earthquake problems, namely transient type base excitation, stochastic type excitation which is relatively poorly defined for early flights of a particular booster-payload system, and so forth. Reference 20.16, for example, presents a generalized modal shock spectra method for spacecraft loads analysis. Reference 20.17 presents a survey of the problem of design of space payloads for transient environments.

References

20.1 R. L. Wiegel, ed., *Earthquake Engineering,* Prentice-Hall, Englewood Cliffs, NJ (1970).

20.2 N. M. Newmark and E. Rosenbleuth, *Fundamentals of Earthquake Engineering,* Prentice-Hall, Englewood Cliffs, NJ (1971).

20.3 S. Okamoto, *Introduction to Earthquake Engineering,* Wiley, New York (1973).

20.4 R. W. Clough and J. Penzien, *Dynamics of Structures,* McGraw-Hill, New York (1975).

20.5 M. Barazangi and J. Dorman, "World Seismicity Maps Compiled from ESSA, Coast and Geodetic Survey, Epicenter Data, 1961–1967," Bull. Seism. Soc. Amer., v. 59, n. 1, 369–380 (1969).

20.6 D. E. Hudson et al., *Strong Motion Earthquake Accelerograms—Vol. II—Corrected Accelerograms and Integrated Ground Velocity and Displacement Curves—Part A,* Earthquake Engineering Research Laboratory, Report No. EERL 71-50, California Institute of Technology, Pasadena (1971).

20.7 J. M. Biggs, *Introduction to Structural Dynamics,* McGraw-Hill, New York (1964).

20.8 A. H-S. Ang, "Probability Concepts in Earthquake Engineering," *Applied*

Mechanics in Earthquake Engineering, AMD-v. 8, ASME, New York, 225–259 (1974).

20.9 M. A. Sozen, "Hysteresis in Structural Elements," *Applied Mechanics in Earthquake Engineering,* AMD-v. 8, ASME, New York, 63–98 (1974).

20.10 W. D. Iwan, "Application of Nonlinear Analysis Techniques," *Applied Mechanics in Earthquake Engineering,* AMD-v. 8, ASME, New York, 135–161 (1974).

20.11 M. A. Biot, "Analytical and Experimental Methods in Engineering Seismology," *Trans. ASCE,* v. 108, 365–408 (1943).

20.12 G. W. Housner, "Characteristics of Strong-Motion Earthquakes," *Bull. Seism. Soc. Amer.,* v. 37, 19–29 (1947).

20.13 D. E. Hudson, "Response Spectrum Techniques in Engineering Seismology," *Proc. World Conf. on Earthquake Eng.,* Berkeley, CA, 4-1, 4-12 (1956).

20.14 N. M. Newmark, "Current Trends in the Seismic Analysis and Design of High-Rise Structures," Ch. 16, *Earthquake Engineering* (R. L. Wiegel, ed.), Prentice-Hall, Englewood Cliffs, NJ (1970).

20.15 J. L. Sackman and J. M. Kelly, *Rational Design Methods for Light Equipment in Structures Subjected to Ground Motion,* Earthquake Engineering Research Center, Report No. DCB/EERC-78/19, University of California, Berkeley, CA (1978).

20.16 M. Trubert and M. Salama, *A Generalized Modal Shock Spectra Method for Spacecraft Loads Analysis,* Publ. 79-2, Jet Propulsion Laboratory, Pasadena, CA (1979).

20.17 B. K. Wada, "Design of Space Payloads for Transient Environments," *Survival of Mechanical Systems in Transient Environments,* ASME, AMD-v. 36 (1979).

20.18 G. W. Housner, "Strong Ground Motion," Ch. 4 in *Earthquake Engineering* (R. L. Wiegel, ed.), Prentice-Hall, Englewood Cliffs, NJ (1970).

20.19 G. W. Housner, "Design Spectrum," Ch. 5 in *Earthquake Engineering* (R. L. Wiegel, ed.), Prentice-Hall, Englewood Cliffs, NJ (1970).

Problem Set 20.2

20.1 Using the values from the $\zeta = 0\%$ curve of Fig. 20.5, sketch a curve of S_a versus T_n for $0.2 \leq T_n \leq 2.0$ Hz.

20.2 A SDOF frame structure such as the one shown in Fig. 20.3 has a period of 0.8 s and a damping factor of 2%. The mass is $m = 10$ lb-sec^2/in.

(a) Using Fig. 20.5 determine the value of the spectral velocity S_v which applies to this structure.
(b) Determine S_a and S_d.
(c) Determine the maximum displacement of the mass.
(d) Determine the maximum spring force.

20.3 Using Fig. 20.8 determine an expression for the maximum overturning moment $M(0, t)$.

20.4 Repeat Example 20.2 if the period of the column is 0.8 s, the damping factor is 2%, and $\psi(x)$ is chosen to have the form

$$\psi(x) = 1 - \cos\left(\frac{\pi x}{2L}\right)$$

Problem Set 20.3

20.5 Using the three-story building of Problem 14.2, whose natural frequencies and mode shapes are given in Problem 15.2, determine the following:

(a) The values of S_d, S_v, and S_a for each mode if each mode has 2% damping and if Fig. 20.6 gives the response spectra for the earthquake excitation under consideration.

(b) A root-mean-square estimate of the maximum displacement of the top mass.

(c) A root-mean-square estimate of the maximum base shear.

20.6 Reduce the stiffnesses of the building in Fig. P14.2 to 50% of their present values, increase the intensity of the response given in Fig. 20.6 by a factor of 2, and repeat Problem 20.5.

Appendix A UNITS

Problems in structural dynamics are based on Newton's second law, that is,

$$\text{Force} = (\text{Mass})(\text{Acceleration}) \tag{A1}$$

and these quantities, plus others directly related to them, must be expressed in a consistent system of units. At the present time engineering practice is in the process of a conversion from English engineering units (United States Customary System) to the International System of Units (SI)[A1]. In English engineering units dimensional homogeneity is obtained in Eq. A1 when the force is given in 1b_f, the mass in slugs, and the acceleration in ft/sec^2. This is the English ft-lb$_f$-sec system of units. The slug is a derived unit, and from Eq. A1 1 slug = 1 lb$_f$-sec^2/ft.

The units of force and mass frequently lead to confusion because of the use of the term *weight* as a quantity to mean either force or mass. On the one hand, when one speaks of an object's "weight," it is usually the mass, that is, quantity of matter, that is referred to. On the other hand, in scientific and technological usage the term "weight of a body" has usually meant the force which, if applied to the body, would give it an acceleration equal to the local acceleration of free fall. This is the "weight" that would be measured by a spring scale. If the "mass" of a body is given in pounds (lb$_m$) it must be divided by the acceleration of gravity $g \simeq 32.2$ ft/sec^2 to obtain the mass as referred to in Newton's second law, that is, a force of 1 lb$_f$ is exerted on a mass of 1 lb$_m$ by the gravitational pull of the earth.

In structural dynamics and vibrations the in.-lb$_f$-sec system is frequently used. In this system $g = 32.2(12) = 386$ is used.

The International System of Units (SI) is a modern version of the metric system[A1]. It is a *coherent** system with seven *base units* for which names, symbols, and precise definitions have been established. Many *derived units* are defined in terms of the base units. Symbols have been assigned to each, and, in some cases, they have been given names. In addition there are *supplementary* units.

*A *coherent* system of units is one in which there are no numerical factors that must enter into an equation employing numerical values, for example, Eq. (A1) is valid whether it is written in symbols or in numerical values—there is an implied multiplicative factor of 1 on the right-hand side.

The base units are regarded as dimensionally independent. The ones of interest in structural dynamics are the meter (m), kilogram* (kg), and second (s). One great advantage of SI is that there is one and only one unit for each physical quantity—the meter for length, the kilogram for mass, second for time, and so forth.

*Table A1. **Some Derived and Supplementary SI Units***

Quantity	*Unit (Name)*	*Symbol*	*Formula*
Derived Units			
Frequency	hertz	Hz	s^{-1}
Force	newton	N	$kg \cdot m/s^2$
Pressure, stress	pascal	Pa	N/m^2
Energy, work	joule	J	$N \cdot m$
Supplementary Unit			
Plane angle	radian	rad	

The unit of length, the meter, "is the length equal to 1 650 763.73 wavelengths in vacuum of the radiation corresponding to the transition between levels $2p_{10}$ and $5d_5$ of the krypton-86 atom. The kilogram is the unit of mass; it is equal to the mass of the international prototype of the kilogram."[A1] This prototype kilogram is preserved by the International Bureau of Weights and Measures near Paris, France. The second is the unit of time. It "is the duration of 9 192 631 770 periods of the radiation corresponding to the transition between the two hyperfine levels of the ground state of the cesium-133 atom."[A1]

Reference A1 gives extensive tables of conversion factors and a number of rules to be followed in writing numbers and their units in correct SI form. Table A2 gives the prefixes for the multiples and submultiples of SI units, while Table A3 provides conversion factors for conversion from English units to SI units.

A number of recommendations have been made in order to standardize the writing of numbers, symbols, and so forth in SI[A1]. The following recommendations apply to frequently encountered situations.

1. SI Prefixes

a. When expressing a quantity by a numerical value and a unit, prefixes should preferably be chosen so that the numerical value lies between 0.1 and 1000.

*In SI units the kilogram is a unit of mass. The kilogram-force (from which the suffix force is often omitted) should not be used.

Table A2. ***Prefixes for Multiples and Submultiples of SI Units***

Multiple	*Prefix*	*Symbol*	*Submultiple*	*Prefix*	*Symbol*
10^{12}	tera	T	10^{-1}	deci	d
10^{9}	giga	G	10^{-2}	centi	c
10^{6}	mega	M	10^{-3}	milli	m
10^{3}	kilo	k	10^{-6}	micro	μ
10^{2}	hecto	h	10^{-9}	nano	n
10	deca	dc	10^{-12}	pico	p
			10^{-15}	femto	f
			10^{-18}	atto	a

Table A3. ***Examples of Conversions[a] from English to SI Units***

To Convert from	*to*	*Multiply by*
foot (ft)	meter (m)	3.048 000 E−01
horsepower (550 ft-lb_f/sec)	watt (W)	7.456 999 E+02
inch (in.)	meter (m)	2.540 000 E−02
kip (1000 lb_f)	newton (N)	4.448 222 E+03
pound-force (lb_f)	newton (N)	4.448 222 E+00
pound-force-inch (lb_f-in.)	newton meter (N·m)	1.129 848 E−01
pound-force/inch (lb_f/in.)	newton per meter (N/m)	1.751 268 E+02
pound-force/in.^{2} (psi)	pascal (Pa)	6.894 757 E+03
pound-mass (lb_m)	kilogram (kg)	4.535 924 E−01
pound-mass/inch^3 (lb_m/in.^{3})	kilogram per meter^3 (kg/m^3)	2.767 990 E+04
slug	kilogram (kg)	1.459 390 E+01

[a]Conversions should be handled with careful regard to the implied correspondence between the accuracy of the data and the number of digits.

b. Normally the prefix should be attached to a unit in the numerator. One exception to this is when the kilogram is one of the units.

c. No space is used between the prefix and the unit symbol.

2. Unit Symbols

a. Unit symbols should be printed in roman (upright) type regardless of the type style used in the surrounding text.

b. Unit symbols are not followed by a period except when used at the end of a sentence.

c. In the complete expression for a quantity, a space should be left between the numerical value and the unit symbol.

3. Units Formed by Multiplication and Division

With unit names:

a. Product—use a space (newton meter) or a hyphen (newton-meter).

b. Quotient—use the word *per* and not the solidus (/).

c. Powers—use the modifier *squared* or *cubed* placed after the unit name (meter per second squared).

With unit symbols:

a. Product—use a raised dot ($N \cdot m$).

b. Quotient—use one of the following forms:

$$\mathrm{m/s} \quad \text{or} \quad \mathrm{m \cdot s^{-1}} \quad \text{or} \quad \frac{\mathrm{m}}{\mathrm{s}}$$

4. Numbers

Outside the United States the comma is sometimes used as a decimal marker. To avoid this potential source of confusion, recommended international practice calls for separating the digits into groups of three, counting from the decimal point toward the left and the right, and using a small space to separate the groups.

Reference

A1 An American National Standard ASTM/IEEE *Standard Metric Practice,* ASTM E 380-76, IEEE Std 268-1976, ANSI Z210.1-1976, Institute of Electrical and Electronics Engineers (1976).

AUTHOR INDEX

Amnar, A., 322, 323
Anderson, R. G., 413, 432
Ang, A. H-S., 498
Archer, J. S., 393, 432
Ashley, H., 351, 375

Bajan, R. L., 475
Bampton, M. C. C., 331, 414, 475, 478
Barazangi, M., 499
Bathe, K. J., 7, 147, 316, 322, 323, 452, 453, 455, 459, 461
Beckwith, T. G., 92
Belytschko, T., 452
Benfield, W. A., 469, 478
Bert, C. W., 16
Bickley, W. G., 217
Biggs, J. M., 128, 349, 498
Biot, M. A., 498
Bishop, R. E. D., 354
Bisplinghoff, R. L., 351, 375
Bodley, C. S., 478
Bouwkamp, J. G., 2
Brigham, E. O., 180
Brock, J. T., 95
Bucciarelli, L. L., 393, 432
Buck, N. L., 92

Chang, C-J., 470, 478, 487, 488, 491, 492
Chang, T-C., 429, 431, 432
Change, N. D., 95
Chopra, A. K., 151
Clough, D., 9
Clough, R. W., 9, 140, 448, 497, 498
Cooley, J. W., 180
Cornwell, R. E., 355
Craig, R. R., Jr., 331, 414, 429, 431, 432, 468, 470, 475, 478, 487, 488, 491
Crede, C. E., 88, 128
Cunningham, W. J., 151

Done, G. T. S., 337
Dorman, J., 499
Dowell, E. H., 492

Emero, D. H., 9

Farhoomand, I., 453
Felgar, R. P., Jr., 311, 315
Feng, C. C., 475
Flanagan, P. F., 322, 323
Fung, Y. C., 388

Gallagher, R. H., 402, 433
Gladwell, G. M. L., 354
Goldenberg, S., 469
Greenwood, D. T., 262
Guyan, R. J., 413, 428

Halfman, R. L., 351, 375
Harris, C. M., 88, 128
Hilber, H. M., 147, 459
Hinkle, R. T., 95
Hintz, R. M., 458, 474
Housner, G. W., 128, 498, 502, 503, 504
Hruda, R. F., 469
Hudson, D. E., 498, 500
Hughes, T. J. R., 147, 459
Hurty, W. C., 328, 388, 448, 475
Huzar, S., 492

Irons, B. M., 413, 432
Iwan, W. D., 498

Jennings, A., 337
Johnson, C. P., 413

Kana, D. D., 492
Kelly, J. M., 514
Kennedy, C. C., 362
Klein, L. R., 492
Klosterman, A. L., 2, 342

Langhaar, H. L., 199, 244
Leckie, F. A., 393, 432
Lee, B., 433

Leissa, A. W., 226
Lindberg, G. M., 393, 432

McGuire, W., 402
MacNeal, R. H., 479
Marcal, P. V., 452
Meirovitch, L., 199, 222, 252, 262, 327, 343
Mojtahedi, S., 448
Morosow, G., 478
Morse, I. E., 95

Newmark, N. M., 147, 322, 323, 497, 498, 511
Nickell, R. E., 455, 459, 461

Oden, J. T., 452
Okamoto, S., 497, 511
Osias, J. R., 452

Pancu, C. D. P., 362
Pendered, J. W., 362
Penzien, J., 140, 497, 498
Peterson, F. E., 7
Pian, T. H. H., 393, 432
Pipano, A., 322, 323
Potter, R., 95
Przemeniecki, J. S., 390

Raibstein, A., 322, 323
Ramm, E., 452
Ramsey, K. A., 95, 180
Rangacharyulu, M. A. V., 337
Richardson, M., 95, 180
Rivello, R. M., 388
Rosenbleuth, E., 497, 511
Rubin, S., 479, 487
Rubinstein, M. F., 328, 388, 448

Sackman, J. L., 514
Salama, M., 514
Shapiro, M., 469
Smith, S., 180
Soni, S. R., 448
Sozen, M. A., 151, 498
Stearns, S. D., 180
Stephen, R. M., 2
Stoker, J. J., 151
Stroud, W. J., 337

Taylor, R. L., 147, 459
Temple, G., 217
Thomson, W. T., 128, 351, 368
Timoshenko, S. P., 343
Tong, P., 393, 432
Trubert, M., 514
Tse, F. S., 95
Tukey, J. W., 180
Turner, M. J., 337

Wada, B. K., 514
Warburton, G. B., 448
Weaver, W., Jr., 343, 409
Wiegel, R. L., 497, 498
Williams, D., 351
Wilson, E. L., 7, 147, 316, 322, 323, 452, 453
Wilson, J. S., 92, 95

Young, D., 211, 215
Young, D. H., 343

Zienkiewicz, O. C., 413, 432

SUBJECT INDEX

Accelerometer, 94
Admissible function, 34
Algebraic eigensolvers, 321-337
  matrix transformation methods, 322, 323
  vector iteration methods, 322-331
Algebraic eigenvalue problem, 274, 297
Analysis, 2
  deterministic, 1
Argand plane, 53, 86
Assumed mode, *see* Shape function
Assumed-modes method, 33, 251-261
  finite element version, 381-415. *See also* Finite element method
  global version, 251-264, 381
Average acceleration method, 147-150, 457

Base motion, 20, 87, 89-95, 133, 240, 241
Beat phenomenon, 76
Bode plot, 81
Boundary conditions:
  for finite element models, 406-409
  generalized, 204
  geometric, 33, 199
  natural, 201

Characteristic equation, 50, 209, 274, 297
Circle-fit method, *see* Kennedy-Pancu method
Complementary solution, 50
Complex frequency response, 83-92, 172-175, 355
  relationship to unit impulse response, 179, 180
Complex frequency response plot, 86, 356
Complex mode shape, 274
Complex stiffness, 101
Component modes:
  attachment modes, 470, 478-482
  for constrained components, 469, 470
  constraint modes, 469, 470
  normal modes, 469
  rigid-body modes, 478
Component mode synthesis, 467-492
  Craig-Bampton method, 475-478
  Craig-Chang method, 488-492
  for free vibration of undamped systems, 470, 478
  Hintz method, 492
  Hurty method, 475
  MacNeal method, 492
  Rubin method, 479, 487
Computer graphics, 7
Consistent mass matrix, *see* Mass matrix, consistent
Constant-average acceleration method, *see* Average acceleration method
Constrained coordinates, 261-264
Constraint, 25
  applied to finite element models, 409-417
  equations of, 244
Continuous systems, 33, 187-234. *See also* Model, continuous
  mathematical models of, 189-205
Coordinate, displacement, 25
  generalized, 26, 244
Coupling:
  inertia, 243
  stiffness, 239

d'Alembert force, *see* Inertia force
Damping, Coulomb, 65
  equivalent viscous, 97-101
  generalized proportional, 451
  in MDOF models, 447-452
  numerical, 459
  proportional, 448
  Rayleigh, 449
  structural, 101-103
  viscous, 16
Damping factor, 96
  structural, 101

Damping level:
  critically-damped, 55, 58
  experimental determination of, 59-65
  overdamped, 55, 58
  underdamped, 55
Damping matrix, 261
Degrees-of-freedom (DOF), 4
Design, 2, 33
Design spectrum, for earthquake analysis, 503
Direct integration, 447-463
  for response of MDOF systems, 461-463
  for response of SDOF systems, 455-461
Direct stiffness method, 399-406
Discrete Fourier transform (DFT), 180
Displacement transformation:
  for plane frame element, 397
  for plane truss element, 395, 396
  for three-dimensional truss element, 397, 398
Displacement vector, 239
Distortion:
  amplitude, 95
  phase, 94
Duhamel integral, 123-127
Dynamic stresses by mode-superposition, 366-368
Dynamical investigation, 2
Dynamical matrix, 324

Earthquake excitation, 10, 87
Earthquake participation factor, 506
Earthquake response, 497-514
  of MDOF systems, 508-513
  of SDOF systems, 498-508
Eigenfunction, 210
Eigenvalue, 210, 274, 297
Eigenvalue equation:
  for continuous systems, 208
  for MDOF systems, 297
  for 2DOF systems, 274
Eigenvalue separation theorem, 316, 317
Eigenvector, 274, 297
  linearly-independent, 304
Element reference frame, 383
Euler's equation, 51
Expansion theorem:
  for continuous systems, 225
  for MDOF systems, 309

Explicit integration methods, 461

Fast Fourier transform (FFT), 180-182
Finite element method, 381-415
  for axial motion, 383-385
  for Bernoulli-Euler beam, 385-387
  displacement method, 393
  force method, 393
  mixed methods, 393
  for three-dimensional element, 390-393
  for torsion, 388, 389
Finite elements, tapered, 433
Forced response, 7
Force transmissibility, 87, 88
Force vector, 239
Force vector polygon, 78
Fourier integral, 175-179
Fourier series:
  complex form, 169-175
  real form, 163-169
Fourier transform, *see* Fourier integral
Free vibration, 7, 50
  axial motion, 207-210
  of Bernoulli-Euler beams, 210-217
  of continuous systems, 207-229
  of MDOF systems, 295-317
  of SDOF systems, 49-70
  of SDOF systems with Coulomb damping, 65, 66
  of thin, flat plates, 226-229
  of Timoshenko beams, 219-221
  of undamped SDOF systems, 51-54
  of undamped 2DOF systems, 273-289
  of viscous-damped SDOF systems, 54-65
Frequencies, closely-spaced, 362
Frequency:
  circular natural, 274
  damped circular natural, 55
  natural, 297
  undamped circular natural, 49, 52
  undamped natural, 95
Frequency domain analysis, 163-182
Frequency ratio, 73
Frequency response, 78
Frequency response function, 73

Generalized coordinates, *see* Generalized displacement coordinates

Generalized displacement coordinates, 26, 34, 244
Generalized external force, 40
Generalized geometric stiffness coefficient, 40
Generalized forces, 26, 244-246
Generalized mass, 38
Generalized mass matrix, 287, 308
Generalized stiffness coefficient, 39
Generalized stiffness matrix, 308
Generalized viscous damping force, 38
Geometric nonlinearity, 152
Geometric stiffness matrix, 260
Givens' method, 322
Gram-Schmidt orthogonalization, 327
Guyan reduction method, 413, 428-433

Half-amplitude method, 61
Half-power method, 96
Hamilton's principle, 198-204, 244
Harmonic excitation:
  of continuous systems, 208, 211
  mode-superposition solution for, 286-289
  response of SDOF systems to, 71-110
  response of undamped SDOF systems to, 72-76
  response of viscous-damped systems to, 76
  of 2DOF systems, 274
Householder's method, 322
Hysteresis loop, 99

Impulse response, 117-120
Inertia force, 2, 17, 20
Inertial reference frame, 17
Initial conditions, 275, 277, 278
Interpolation:
  piecewise constant, 140, 141
  piecewise-linear, 141-146
Inverse iteration, 323
  with spectrum shift, 330, 339
ISMIS (Interactive Structures and Matrix Interpretive System), 331-337, 426-441

Jacobi method, 322

Kennedy-Pancu method, 97, 110
Kinetic energy, 17
Kinetic friction, coefficient of, 65

Lagrange multipliers, 261-264
  in component mode synthesis, 471
Lagrange's equations, 243-264
  applied to continuous models, 251-261
  applied to lumped-parameter models, 247-251
Linear acceleration method, 161
Loading:
  dynamic, 1
  prescribed, 1
  random, 2
Locator matrix, 401
Locator vector, 401
Logarithmic decrement, 61
Lumped-mass model, 411-413

Magnification factor:
  steady-state, 73, 78
  total dynamic, 76
Mass matrix, 239
  consistent, 254, 432
  element, 383-393
  system, 401-404
Material nonlinearity, *see* Nonlinearity, material
Matrix deflation, 327
Maxwell's reciprocity relationship, 296
MDOF systems, 235-516
  mathematical models for, 237-272
Measurement, vibration, 92
Modal damping matrix, 343
Modal force vector, 287, 343
Modal initial conditions, 344
Modal mass matrix, 287, 298, 308, 343
Modal matrix, 287, 308
Modal static deflection, 349
Modal stiffness matrix, 287, 308, 343
Modal stress vector, 366
Mode-acceleration method:
  for response of undamped MDOF systems, 350-353
  for systems with rigid-body modes, 369
Mode-displacement method:
  for response of finite element models, 433-442
  for response of undamped MDOF systems, 344-350

for systems with rigid-body modes, 368
Mode shape, *see* Natural mode
Mode shapes:
for axial motion, 210
of MDOF systems, 299
Mode-superposition method, 309
for earthquake response analysis of MDOF systems, 509
for response of MDOF systems, 341-377
for response of viscous-damped systems, 353-366
Model:
analytical, 4
continuous, 4
discrete-parameter, 4
generalized-parameter, 15, 33
lumped-mass, 5
lumped-parameter, 15
mathematical, 2, 6
physical, 3, 8, 9
SDOF mathematical, 15-47
Multiple-degree-of-freedom systems, *see* MDOF systems

Natural frequencies:
finite element solution for, 423-433
properties of, 295-317
Natural frequency, *see* Frequency, natural
Natural modes, 274, 297
properties of, 221-226, 295-317
Newmark method, 147, 161, 461
Newton's laws, 15
applied to axial deformation, 189-192
applied to Bernoulli-Euler beams, 192-198
applied to lumped-parameter models, 237-243
Node, of vibration mode, 305
Node lines, 228
Node point, 210
Nonconservative force, 34
Nonlinear SDOF systems, 151-160
numerical solution for response of, 155-160
step-by-step solution for response of, 452-455
Nonlinearity:
geometrical, 152
material, 153
Normal mode method, *see* Mode-superposition method
Normal modes, *see* Natural modes
Normalization of modes, 222, 223, 298
Numerical integration, operator formulation, 455-459
amplitude accuracy, 459
period accuracy, 459
step-by-step methods, 146-160
Nyquist plot, *see* Complex frequency response plot

Optimization, 337
Orthogonality:
of modes of continuous systems, 223, 224
of modes of MDOF systems, 303
Orthonormal modal vectors, 308

Particular solution, 50
Period:
damped, 55
undamped natural, 52
Periodic excitation, response of SDOF systems to, 163-175
Phase angle, 77
Positive-definite matrix, 296
Positive-semidefinite matrix, 296
Principal coordinates, 287, 341-344
Principle of virtual displacements, 27
applied to continuous systems, *see* Assumed-modes method
applied to SDOF lumped-parameter models, 25-32
Pseudo-static response, 351

QR method, 322

Ramp loading, response of undamped SDOF systems to, 115-117
Rank, of a matrix, 299
Rayleigh quotient, 218, 325
for continuous systems, 225
for MDOF systems, 313
Rayleigh-Ritz bounds, 426-429
Rayleigh-Ritz method, convergence of, 317
for MDOF systems, 314-316
Rayleigh's method:
for continuous systems, 217-219

for MDOF systems, 314-316
Rectangular pulse, response of undamped SDOF system to, 113-115
Recursion relation, 456
Reduced stiffness matrix, *see* Stiffness matrix, reduced
Reference frame:
element, 383
global, 394
Repeated frequencies, 304
Residual flexibility, 483-492
Resonance, 76
Response:
forced, 50
natural, 50
Response spectra, 127-136
in earthquake response analysis, 498-508
Response spectrum:
displacement, 501
pseudoacceleration, 502
pseudovelocity, 501
Rigid-body modes, 301, 307
of beams, 231
of MDOF systems, 283-285
in mode-superposition solution, 368-375
transformation to reduce out, 413-415
Rise time, 115
Rotating unbalance, 106, 108, 266
Rotating vector, 53, 77

SAP, 7
SDOF systems, 13-185
Semi-definite eigenvalue problem, 285
Shape function, 34
for axial element, 384
for transverse motion of Bernoulli-Euler beams, 385, 386
SI units, 517-520
Simple harmonic motion, 52
Single-degree-of-freedom systems, *see* SDOF systems
Spectra, *see* Response spectra
Spectrum, *see* Frequency domain analysis
Spring, 15
Stability, of numerical integration operators, 458, 459
conditionally stable, 459
unconditionally stable, 459
unconditionally unstable, 459
Steady-state amplitude, 77
Steady-state response, 72
Step input, response of viscous-damped systems to, 111-113
Stiffness coupling, 239
Stiffness matrix:
element, 383-393
reduced, 411
system, 401-404
Strain energy, 16, 35
Subspace iteration method, 322
Substructure, 468
Substructure coupling, *see* Component mode synthesis
Sweeping matrix, 328
Symmetric modes, 294, 310
Symmetry, of mass and stiffness matrices, 296

Testing, 2
dynamical, 8
ground vibration, 8, 9
Time constant, 64
Time step, 147
Timoshenko beam, mathematical model of, 202-204
free vibration of, 219-221
Transducer, 92
seismic, 92
Transfer function, 356
Transmissibility, *see* Force transmissibility
Trapezoidal rule, 147
Tridiagonal reduction method, 323

Unit impulse response, *see* Impulse response

Vector response plot, *see* Complex frequency response plot
Vibration absorber, 266
Vibration isolation, 87-92
Virtual displacement, 25
of a continuous system, 33
Virtual work, 26
for finite element load vectors, 404-406
Viscous damping factor, 49

Wilson method, 453-455